# Concepts in Integrated Pest Management

**Robert F. Norris**
*University of California, Davis*

**Edward P. Caswell-Chen**
*University of California, Davis*

**Marcos Kogan**
*Oregon State University*

Prentice
Hall

Upper Saddle River, New Jersey 07458

**Library of Congress Cataloging-in-Publication Data**

Norris, Robert (Robert F.)
    Concepts in integrated pest management / Robert Norris, Edward Caswell-Chen,
Marcos Kogan.
       p. cm.
    Includes bibliographical references.
    ISBN 0-13-087016-1
       1. Pests—Integrated control. I. Norris, Robert (Robert F.) II. Caswell-Chen, Edward.
III. Kogan, M. (Marcos) IV. Title.

SB950 .N638 2003
632'.9—dc21
                                        2002070022

**Editor-in-Chief:** Stephen Helba
**Executive Editor:** Debbie Yarnell
**Associate Editor:** Kimberly Yehle
**Production Editor:** Lori Dalberg, Carlisle Publishers Services
**Production Liaison:** Janice Stangel
**Director of Manufacturing and Production:** Bruce Johnson
**Managing Editor:** Mary Carnis
**Marketing Manager:** Jimmy Stephens
**Manufacturing Buyer:** Cathleen Petersen
**Design Director:** Cheryl Asherman
**Senior Design Coordinator:** Miguel Ortiz
**Cover Designer:** Marianne Frasco
**Composition and Interior Design:** Carlisle Communications, Ltd.
**Printing and Binding:** Courier Westford

Pearson Education LTD.
Pearson Education Australia PTY, Limited
Pearson Education Singapore, Pte. Ltd.
Pearson Education North Asia Ltd.
Pearson Education Canada, Ltd.
Pearson Educación de Mexico, S.A. de C.V.
Pearson Education—Japan
Pearson Education Malaysia, Pte. Ltd.

10 9 8 7 6 5 4 3 2 1
ISBN 0-13-087016-1

*This book is dedicated to the farmers and pest control advisors involved in producing food and fiber for the world. Without the pest management knowledge, skills, and dedication of the few, many more persons would have to be directly involved with food and fiber production. Society owes those who manage pests a debt of gratitude, and we hereby acknowledge their efforts.*

# BRIEF CONTENTS

# DETAILED CONTENTS

# CHAPTER 3

## Historical Development of Pest Management   47

# CHAPTER 4

## Ecosystems and Pest Organisms    66

# CHAPTER 5

## Comparative Biology of Pests    90

## CHAPTER 6

### Ecology of Interactions Between Categories of Pests   128

## CHAPTER 7

### Ecosystem Biodiversity and IPM   155

# CHAPTER 8

# CHAPTER 9

## CHAPTER 10

### Pest Invasions and Legislative Prevention   214

## CHAPTER 11

### Pesticides   242

# CHAPTER 12

## Resistance, Resurgence, and Replacement    314

# CHAPTER 13

## Biological Control    337

# CHAPTER 17

## Host-Plant Resistance and Other Genetic Manipulations of Crops and Pests    443

# CHAPTER 18

## IPM Programs: Development and Implementation    471

# CHAPTER 19

## Societal and Environmental Limitations to IPM Tactics    501

## CHAPTER 20

### IPM in the Future   514

# PREFACE

Progress in pest control technologies has contributed to the improved yield and quality of food, fiber, and ornamental crops that have occurred during the twentieth century. However, the development and widespread adoption of some pest-control technologies did not occur without environmental impacts and societal concerns about food safety. Integrated pest management (IPM) arose in the second half of the century as the paradigm of choice for pest control, and stressed the need to incorporate basic ecological concepts in the design and implementation of pest control systems. Integrated pest management requires detailed understanding of pest biology and ecology, including interactions at the community and ecosystems levels.

This book is intended as a text for use in teaching the concepts of integrated pest management to upper-level undergraduates and graduate students that have successfully completed introductory biology. If the students have had more specialized courses in botany, entomology, invertebrate zoology, or vertebrate zoology, so much the better. This book explains the concepts upon which integrated pest management programs are based. We have gone beyond disciplinary boundaries, and consider IPM concepts relative to all pest categories, including: pathogens, weeds, nematodes, insects, mollusks, and vertebrate pests. Where possible, we consider interactions among pest categories.

The book emphasizes the complexity of managing pests in economically viable production systems while avoiding detrimental impacts on the environment and society. Integrated pest management is a work in progress, and we have attempted to create a book that will aid in teaching IPM from a broad perspective, toward the goal of true integrated pest management. The ultimate goal of IPM is to achieve complete integration of management tactics, breaking the traditional barriers often imposed by the pest disciplines. In reality, the application and realization of IPM varies among cropping systems, with some agroecosystems being managed in a more integrated manner than are others. The future of IPM is promising, with opportunities and challenges ahead. Improvements in IPM will depend on the continued efforts of plant pathologists, weed scientists, nematologists, entomologists, and applied vertebrate zoologists working together to achieve the highest possible level of integration.

This book is not a "how-to" manual for management of specific pests; we stress concepts and principles that, if understood and practiced, will help managers to design systems for the agroecological conditions prevailing within their regions. We have used international examples of IPM; however, details of how to manage a specific pest in a particular crop, in a given region, are found in specialized publications. We have provided reference lists that specify some such publications. The book shows how principles in pest management can be applied across ecosystems, how each strategy relates to the different categories of pests, and how these strategies impact ecosystems and human society. We have not written the book using a literature cited format as we would have if this were a review paper for the primary literature. Rather we acknowledge and direct students to resources and our source material in the recommended reading section that concludes each chapter. We hereby offer sincere thanks to all those authors whose work has provided the foundation for the development of IPM, and for our efforts herein.

Many people have contributed to the development of and collective thinking about integrated pest management. To list them would run the risk of inadvertently omissions. We opt, therefore, to offer a collective acknowledgment to all the pioneers of IPM and the many who have contributed the building blocks of what has become one of the great advances in agricultural sciences of the twentieth century.

The first three chapters of the book are introductory. Chapter 1 deals with pests and human society, defines the term pest, and the losses attributed to pests. Chapter 2 provides a more in-depth introduction to the different pest categories. Chapter 3 reviews the history of pest control. The next four chapters (4–7) cover the biology and ecology of pests. Chapter 8 describes monitoring and how information is used to make management decisions. Pest management tactics are introduced in Chapter 9, and Chapters 10 through 17 explore each of the tactics in greater detail. Chapter 18 discusses IPM programs, and presents case histories from lettuce, cotton, and pome fruit crops. Chapter 19 discusses societal influences on IPM, and a brief look to the future of IPM is presented in Chapter 20.

This book has a strong component that relates to sustainable agricultural production. It must be recognized that large-scale agriculture, as currently practiced in most industrialized countries, is not sustainable in the long-term. Given that the world's population continues to grow, this is a problem. Agriculture is human manipulation of the environment, and if human management and inputs are removed, the system will eventually revert to the native climax vegetation of the region. Perhaps the best-known example of the potential impact of insect pests on the sustainability of agriculture is offered by the history of cotton production in some of the coastal valleys of Peru, in particular, the Cañete Valley. In the 1920s, growers in the valley shifted from sugar cane to cotton production. Yields were low, about 300–400 lb. per acre, but stable. With the demand for cotton during the Second World War and the advent of DDT, cotton production was intensified and yields nearly doubled—but for a very short time. Soon, secondary pests began to appear and well-established pests became resistant to all known insecticides. The number of sprays increased but yields still dropped to levels that were no longer economical. Many growers were ruined and crops abandoned until new techniques of integrated control were introduced. The drama of the Cañete Valley is a reminder of the delicate balance that exists among components of an ecosystem. In this book we attempt to present pest management in the context of maintaining economically and environmentally sustainable agricultural systems.

---

The mention or description of particular management tactics or products for use in management is not an endorsement or recommendation of those tactics or products by the authors.

# ACKNOWLEDGMENTS

We thank Jim Carey, Tobi Jones, Ben Sacks, Regina Sarracino, Dale Shaner, Desley Whisson and two anonymous reviewers for reviewing chapter drafts. Mistakes that remain are ours.

Preparation of this book has been a time-consuming task that demanded long hours. Thanks to Lori Dalberg and Kim Yehle, of Carlisle Communications and Prentice Hall respectively, for their editorial assistance and their patience in bringing this book to completion. We also acknowledge those who have made it possible for us to complete this project and those who have inspired us in our professional careers.

I thank Roswita for her everlasting support of my career, and especially during the several years required to complete this book. I also must acknowledge John Harper for his insights into plant ecology, and Harry Lange and George Nyland for helping to shape my thinking about IPM—*Robert F. Norris.*

I thank my parents and family for their support and encouragement, and offer special thanks and recognition to Yvonne and Taylor for their patience. I extend my deep appreciation to I. J. Thomason and G. W. Bird for sparking and fostering my interest in nematology and in IPM—*Edward P. Caswell-Chen.*

For her patience, sacrifices, and support over 48 years of companionship, my gratitude to Jenny. At various stages of my career I owe a great deal, in Brazil, to A.M da Costa Lima and Cincinnato Gonçalves, and in the USA to Paul DeBach, Harry Shorey, Robert L. Metcalf, and William H. Luckmann—*Marcos Kogan.*

# PESTS, PEOPLE, AND INTEGRATED PEST MANAGEMENT

## CHAPTER OUTLINE

▶ Introduction
▶ Pest status
▶ Importance of pests
▶ Pest management

## INTRODUCTION

A pest has six legs and crawls!

This statement represents a common misconception held by the majority of people in western society. It does not, however, represent the diversity of organisms considered pests by those involved in pest management. The term *pest* is understood, in some form, by most people, yet is difficult to define, because pests originate in the context of human activities and objectives. Therefore, each observer defines what is, and what is not, a pest for a particular situation. A homeowner may consider spiders as pests, whereas a specialist in biological insect control sees them as beneficials.

A common thread that unites humanity is we all need food to survive; however, most people in developed countries are no longer involved in producing the food and fiber they require (Figure 1–1). They have little understanding of the limits that pests impose on agricultural production, or the costs that pest control places on human society. Most people in industrialized countries have never spent days hoeing weeds to obtain enough food to live, nor have they experienced the almost total loss of food caused by a disease epidemic, such as late blight of potatoes. Most people have not experienced the frustration of growing a crop only to have it decimated by an insect attack just before harvest. Crop losses to pests in developed countries average about 30% despite our best efforts to control them. In less developed countries, losses to pests are often substantially higher.

**The term *pest* is anthropocentric, and is defined differently by diverse segments of the human population. There are no pests in an ecological sense; in the absence of humans, all organisms are just part of an ecosystem.**

1

**FIGURE 1–1**

Trends for number of
persons involved in
farming in the United
States and worldwide.

Sources: (a) data for the
United States since 1790
from the USDA
(Anonymous, 2000),
(b) for the world during
the last 50 years by
country, and (c) by world
region from the Food
and Agriculture
Organization (FAO)
(1999).

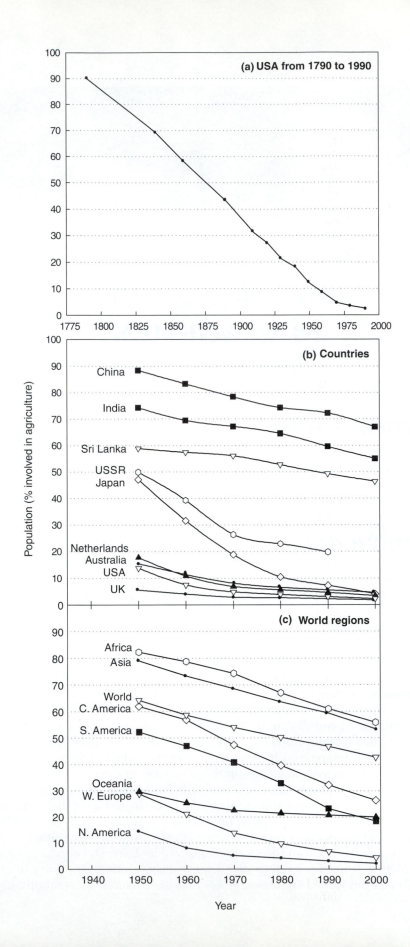

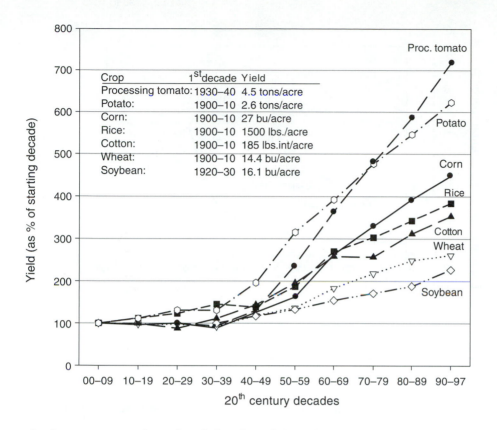

| Crop | 1st decade | Yield |
|---|---|---|
| Processing tomato: | 1930–40 | 4.5 tons/acre |
| Potato: | 1900–10 | 2.6 tons/acre |
| Corn: | 1900–10 | 27 bu/acre |
| Rice: | 1900–10 | 1500 lbs./acre |
| Cotton: | 1900–10 | 185 lbs.int/acre |
| Wheat: | 1900–10 | 14.4 bu/acre |
| Soybean: | 1920–30 | 16.1 bu/acre |

**FIGURE 1–2**

Percent yield increases for major crops grown in the United States during the last century.

Source: Warren, 1998.

In the 1800s, people realized that they did not have to suffer or die because of the vagaries of diseases and malnutrition. This realization resulted from rapid improvements in the understanding of key factors in human health and crop production. The latter led to major increases in crop yields (Figure 1–2) after centuries of only slight improvements. Exponential growth of human populations (Figure 1–3) in the past two centuries certainly was aided, if not directly promoted, by improved medical knowledge and food supply. Pest management has played a significant part in the latter. The growing population has in turn intensified demands on the planet's ecosystem resources to provide food. Beirne (1967) noted that humans cannot afford to feed pests in addition to themselves. Reducing the losses caused by pests is essential to continue to feed the ever-increasing human population (Figure 1–3).

This book explains the concepts and principles involved in the management of pests in agricultural crop-production systems. We will not consider organisms that directly attack humans, or domestic or farm animals, although they are mentioned to illustrate certain concepts. Although household arthropod pests (e.g., cockroaches, ants, termites) and farm animal pests (e.g., flies, mosquitoes, fleas) are targets of pest management programs, they often vector diseases and are therefore in the realm of human health or veterinary medicine. Such pests are outside the scope of this book, which is restricted to agroecosystem management.

**The ecological scope of this book is the agroecosystem, limited herein to crops and their environment, to the exclusion of domestic animals.**

## Definition of Pest

So, what is a pest?

Current dictionaries provide a confusing answer. The *Oxford English Dictionary* uses archaic definitions that state pests cause deadly epidemics or disease

**FIGURE 1–3**
Estimated world
human population
from 10,000 B.C. to
A.D. 2150.
Source: United Nations,
1999.

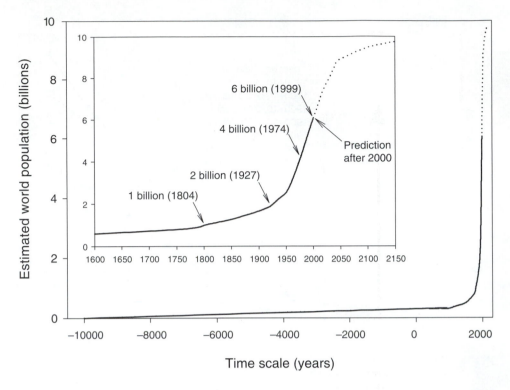

Time scale (years)

or plague, or are any thing or person that is noxious, destructive, or troublesome. *Longman's Dictionary of Contemporary English,* third edition, and the *Random House Webster's Unabridged Dictionary* perpetuate the belief that a pest is "an insect or small animal that harms or destroys garden plants, trees, etc." (*Random House*) or a "small animal or insect that destroys crops or food supplies" (*Longman's*). *Webster's Third New International Dictionary* (1981) and *Chamber's 21st Century Dictionary* (1996, updated in 1999) provide definitions that reflect current pest management thinking. *Webster's* states that pests are "[a] plant or animal detrimental to man or to his interests," whereas *Chamber's* states that a pest is "[a] living organism, such as an insect, fungus, or weed, that has damaging effect on animal livestock, crop plants, or stored products." The latter definition notes the broad spectrum of organisms that can be pests and reflects current usage in pest management.

In the United States, the Federal Insecticide Fungicide and Rodenticide Act (FIFRA) provides a useful definition (more about this act in Chapter 11). The definition shown in the margin note is probably as good as any currently in use, and is adopted throughout this book. Accordingly, as defined, pest includes all of the following categories of organisms:

**The definition of *pest* by FIFRA: "Any organism that interferes with the activities and desires of humans."**

Pathogens (including fungi, bacteria, mollicutes, and viruses)
Weeds (all classes of vascular plants)
Nematodes (roundworms)
Mollusks (slugs and snails)
Arthropods (including insects, mites, crustaceans, and other joint-legged
    invertebrates)
Vertebrates (including amphibians, reptiles, birds, mammals)

The characteristics, biology, and role of each of these different pest categories are explained in greater detail in Chapter 2. At the practical level, all

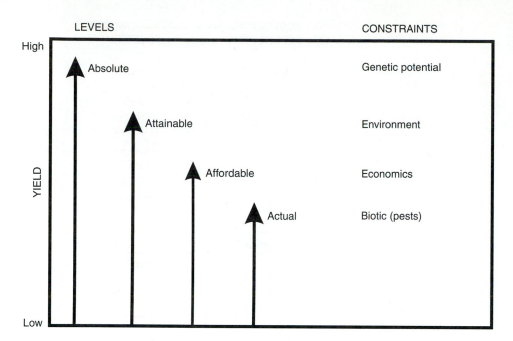

**FIGURE 1–4**
Relationship between crop yield and various constraints that limit crop yield.

pests have to be managed. A farmer cannot ignore one type of pest because it may destroy the crop. The absolute yield of a crop is determined by genetic yield potential of the crop variety (Figure 1–4). The absolute yield is normally never achieved due to various constraints. The local environmental conditions of a region and the economically feasible cultural practices place limits on the attainable yield. The *combined* effect of *all* organisms affecting the crop is also a serious constraint on crop production. From this a working definition of pest can be derived as *all organisms within the crop environment that cause injury to the crop and are capable of reducing yield or quality.* This book examines the principles underlying management of all the different categories of pests in agricultural systems, and emphasizes interrelated and interacting phenomena.

## Applied Ecology

**Integrated pest management (IPM)** is applied ecology. Ecology is the study of organisms, their relationships with each other and with their biotic and abiotic environments. Pest management occurs within agricultural ecosystems, or agroecosystems, which consist of the crop and its environment. Agroecosystems can be understood at different levels of resolution as shown in Figure 1–5. Research on pest management occurs at each level of resolution, thereby providing the requisite understanding of the crop, the pests, the environment, and the management activities of people who allow the development of truly integrated pest management.

## PEST STATUS

The status of an organism as a pest within an agroecosystem is not fixed. The same species can vary from being a pest and causing substantial losses, to being of no consequence. The factors that modify the status of a pest are shown in Figure 1–6,

**FIGURE 1–5**
Ecosystem organization from ecological, human, and integrated pest management perspectives.

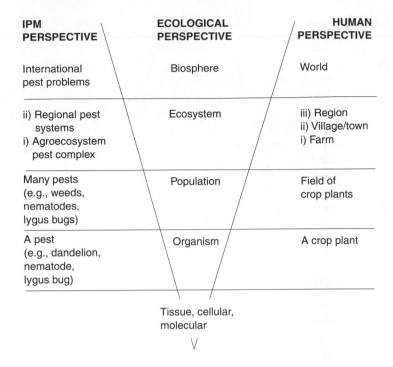

| IPM PERSPECTIVE | ECOLOGICAL PERSPECTIVE | HUMAN PERSPECTIVE |
|---|---|---|
| International pest problems | Biosphere | World |
| ii) Regional pest systems<br>i) Agroecosystem pest complex | Ecosystem | iii) Region<br>ii) Village/town<br>i) Farm |
| Many pests (e.g., weeds, nematodes, lygus bugs) | Population | Field of crop plants |
| A pest (e.g., dandelion, nematode, lygus bug) | Organism | A crop plant |

Tissue, cellular, molecular

\/

**FIGURE 1–6**  The pest tetrahedron concept for integrating the effects of pest organism, environment, host, and time, as determinants of the status of the organism as a pest.

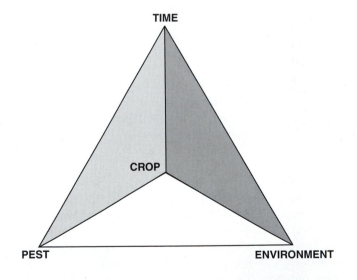

and include the organism involved, the crop, the environment in which the crop grows, and time. These interactions of host plant, potential pests, the environment, people, and time can be pictured as representing a pest tetrahedron. This concept has been widely used in plant pathology, but it applies to all types of pests. The potential pest organism only becomes an actual pest problem when all vertices of the tetrahedron are present in an appropriate state. A suitable host has to be available, the environment must be correct for both the pest and the host, the time must be adequate for the pest to interact with the host, and an objective of human activity with respect to the host must be compromised.

## Pest Organisms

There are several different broad categories of pest organisms. Nematodes, mollusks, arthropods, and vertebrates feed on plants and thereby injure them. If the amount of injury results in a loss of yield or quality, then the animal is said to cause damage to the crop and thus is considered a pest. Pathogenic organisms such as fungi, mollicutes, bacteria, and viruses invade the host tissue(s) and in so doing disrupt the normal physiological processes. This also results in damage to the host crop and loss of yield. Weeds must interfere with crop growth sufficiently to cause yield reduction, or impact human activities in other ways. Any organism must be able to increase to damaging populations before it will be classified as a pest. The ways in which populations are naturally regulated greatly influence whether an organism will reach the status of being considered a pest, and a limitation to population increase is why most organisms never become pests.

**Pest Categories**
**pathogens**
**weeds**
**nematodes**
**mollusks**
**arthropods**
**vertebrates**

## Crop (Host Plants)

The plants in the managed ecosystem determine the pest status of an organism in several ways. Lygus bugs, for instance, are not considered pests in alfalfa hay fields, but are very serious pests in alfalfa seed fields. The difference in status is due to the susceptibility of hay versus seed relative to the injury caused by the same density of lygus bugs. Many aphids are not a serious problem in low-value wheat crops, but when aphids infest a lettuce crop at harvest they can result in total crop loss, because most consumers in the United States will not buy a head of lettuce if it has a single aphid on its leaves. Aesthetics may be a major reason for weed management on a golf course, but plays no role in the status of the same weed species present in a pasture. Even crop plants can become pests. Volunteer crops can be extremely difficult weed problems in subsequent crops; examples include volunteer potatoes, wheat in sugar beets (Figure 1–7), or corn in soybean fields.

All categories of animal pests and plant pathogens can be further classified into two major types, depending on which plant part is injured. Indirect pests

**FIGURE 1–7** Sugar beet field. There are a few plants visible bottom left, almost completely smothered by volunteer wheat.
Source: photograph by Robert Norris.

are those organisms that feed on, or infect, the vegetative organs of the plant, such as foliage, stems, or roots. Damage is caused indirectly through disturbance of normal functions of the organs attacked. Direct pests, on the other hand, are those that attack the plant part that is harvested, such as fruits, seeds, or tubers. For example, the European corn borer, boring into corn stalks, is an indirect pest; the corn earworm, which feeds on corn kernels, is a direct pest.

## Environment

The environment can alter the status of an organism from pest to nonpest. Humidity and temperature regulate the ability of many pathogens to attack crops. In a dry climate, a pathogen such as *Phytophthora infestans* (late blight of potatoes) cannot attack the crop and is not usually a pest (e.g., in California), but the same fungus is a serious pest in moist temperate climates (e.g., in Ireland). Many pests of paddy rice, a crop that is kept flooded for most of the season, are of no consequence in dryland (or upland) rice, and water hyacinth is never a problem in nonaquatic habitats.

## Time

Time is significant in relation to pest population development. Many organisms can be present in low numbers and not result in sufficient impact on human activities to reach pest status. However, the same organism can, given time, increase sufficiently to be considered a pest. Time also may be significant in relation to duration of conditions required for attack or infection to occur. Many pathogens require several hours of specific environmental conditions in order to infect the host; insufficient time results in no infection.

## IMPORTANCE OF PESTS

All organisms within each category of pests are not equally important in relation to human activities. Within a pest category, organisms often are classified into the following designations. These designations do not apply equally to all pest categories.

1. Major or key pests. These pests occur routinely and will typically influence crop yield. The term *key pest* is frequently used for insect pests of crops. Key pests occur in damaging levels consistently across growing seasons; they often are direct pests for which there is low tolerance. The codling moth on apples and pears is a classic example of a key pest; the larvae bore into the developing fruit and a single larva in an apple either kills the fruit or makes it a cull (Figure 1–8). If not controlled they can cause total crop loss.
2. Minor pests. These pests occur routinely but usually only cause minor damage to crops, and hence insignificant yield losses.
3. Secondary pests. These pests have the potential to cause serious damage but they are usually under adequate control by natural enemies. The term is used most commonly for arthropod pests. If the natural enemies are disrupted by agricultural practices, the secondary pest can increase and produce economic losses. An example, also from fruit crops, is spider

**FIGURE 1–8**
Damage to a red
Delicious apple by a
codling moth larva.
Source: photograph by
D. Wilson, USDA/ARS.

mites that are normally under adequate control by predators. If, however, the crop is treated with a nonselective insecticide that kills the predators, the spider mites will quickly increase, or flare up, in damaging populations.

4. Occasional pests. These pest organisms are not always present, but can occur as a problem from time to time. This only applies to pests that are mobile or that die out at the end of the season. The bean leaf beetle on soybean in the Midwestern United States is a good example. Conditions that favor overwintering survival may produce a larger than normal colonizing population in the spring and the species may build up during the season into damaging populations. The ideal environmental conditions do not occur frequently, so the bean leaf beetle only occasionally becomes a serious pest of soybean in the Midwest.

5. Potential pests. These are similar to secondary pests in that they are typically not a problem, but if conditions change they can become a pest.

6. Migrant pests. These pests are highly mobile and can infest crops for short periods of time through movement. The locust is the classic example of this type of pest.

7. Nonpests. This designation includes organisms that never achieve status of being a problem to humans; many in fact are considered beneficial. This designation contains most of the organisms in the world.

## Overall Losses Caused by Pests

Is it possible to say which category of pest is more or less important? No, because the concept of pest is anthropocentric, which means that the relative importance depends on the specific situation. Uncontrolled, any pest from any category has the capability to limit, or even eliminate, useful crop production.

The necessity for pest management depends on the crop, pest category and density, and the environment. Occurrence of many arthropod and vertebrate pests is sporadic or variable. Severe outbreaks of locusts, for instance,

**Pests may cause enormous crop losses. Loss of food is the main impact, but there are other types of losses.**

**FIGURE 1–9**
Worldwide estimated percentage crop losses attributed to animals (this category includes nematodes and losses attributable to pathogens vectored by arthropods), pathogens, and weeds, (a) by eight major food and cash crops, and (b) by major regions in the world.
Source: Oerke, 1994; all data are based on estimated monetary losses.

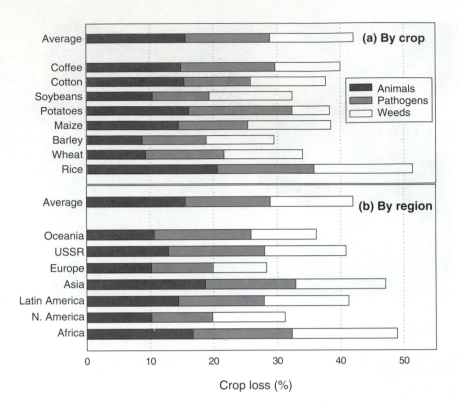

occur only once every few years. Even populations of relatively frequent pests, such as corn borer, are not the same from year to year. Weather conditions have to be appropriate before late blight of potatoes becomes epidemic. Many aerial pathogens vary in disease severity between years. In contrast, most soil-borne pests such as weeds, most nematodes, and some pathogens typically occur every year once land has been infested. This is due to the presence of a weed seedbank, or nematode and pathogen inoculum, that persist from year to year in the soil (more about this in Chapters 5 and 8).

Estimates of crop losses to pests are difficult to make (Oerke, 1994). Data on crop losses due to pests are presented here to provide the reader with an idea of the magnitude of the problem. The data in Figure 1–9 are based on monetary value of the crop rather than yield, due to the wide discrepancies in yield between regions. Another difficulty is that many crops are produced for direct consumption and thus the yield and market value often are not established. No data for vegetables are presented because most are not cash crops (on a world-wide basis) and because they tend to be more intensively managed. There is, however, no reason to assume that the losses in vegetables are vastly different from those presented for the staple crops.

## PEST MANAGEMENT

What is pest management? Is it killing unwanted insects? Although this is often how the expression is used, it obviously is not correct if the definition of *pest* presented earlier in this chapter is used. Is pest management beating nature into submission? Certainly not, although some approach control of pests as if they are something to be bludgeoned to the point of oblivion. It is unlikely that this approach will yield sustainable long-term solutions.

There are numerous definitions of pest management (Morse and Buhler, 1997). Sixty-seven definitions have been compiled and can be accessed at the Integrated Plant Protection Center's website (Bajwa and Kogan, 1996). It is important to agree upon a basic definition because not all pest control programs are IPM systems. The simple definition, "pest management is the manipulation of pest population(s) so that they no longer cause unacceptable losses to humans," is vague and does not capture the essence of the IPM concept. This definition also does not state whether one method of control is better than another, nor does it imply integration of practices.

## IPM Defined

What is *integrated* pest management? The expression "integrated control" was originally coined to describe the use of insecticides in a way that was compatible with the biological control of insects. The concept of IPM was first advanced in the 1960s as an expansion of the idea of integrated control, and has since gained widespread acceptance by much of society. We will use IPM throughout this text as defined in the accompanying margin note.

**IPM is a decision support system for the selection and use of pest control tactics singly or harmoniously coordinated into a management strategy, based on cost-benefit analyses that take into account the interests of and impacts on producers, society, and the environment.**

Other definitions often include specific components of IPM, such as population monitoring, utilization of thresholds, use of biological control agents, cultural practices, and judicious use of pesticides. Perhaps the most distinguishing aspect of IPM, however, is its direction toward ecological and economic consequences of crop-pest interactions. Based on those considerations, IPM provides a set of decision support rules that allow pest managers to optimize selection and timing of management tactics.

In 1996, the United States National Academy of Sciences pest management review recommended that the concept of IPM be replaced by Ecologically Based Pest Management or EBPM (Anonymous 1996). The EBPM model has, however, not seen much acceptance. It did not advance any ideas that had not already been incorporated into the IPM concept, and it has the limitation that it does not implicitly include integration of either control strategies or pest management disciplines.

In 1998, the United States Department of Agriculture defined IPM with a further acronym, PAMS. It proposed that IPM be a combination of *p*revention, *a*voidance, *m*onitoring, and *s*uppression of pests. This definition also deviates from the original proposal by Stern et al. (1959) in that it fails to implicitly integrate control strategies.

## Our Concept

The overall conceptual view of pest management used in this book is presented in Figure 1–10. Pest management does not exist unless there are pests, which are here presented at the base of the conceptual diagram. Understanding the ecology and population biology of these pests is the fundamental underpinning in developing an IPM program (see Chapters 4 through 8). IPM implies certain technologies that can be used for managing the pest populations (Chapters 9 through 17). The choice (decision making) of tactics used to control a pest organism is not fixed, but rather reflects variations in both

**FIGURE 1–10**
Conceptual framework
for IPM shows the
various components of
an IPM program and
their interrelationships.

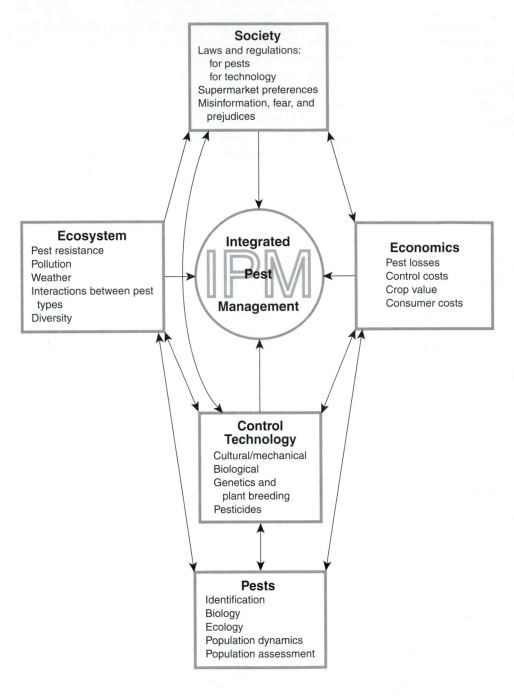

the economics of the situation and the overall environmental context of the
IPM system. These factors are considered here as they impact both the pests
and the control strategies. Specific components of IPM, such as pesticide resistance in pests (Chapter 12) and environmental contamination by pesticides (Chapter 19), are sufficiently important issues that they are addressed
as separate topics. We attempt to show how everything comes together to
form an IPM program in Chapter 18. IPM is not carried out in a vacuum and
human society dictates certain constraints for the practice of pest manage-

ment (Chapter 19). Societal views and constraints are represented at the top of the conceptual diagram, because decisions at the societal level can impact all aspects of an IPM program.

## Historical Perspective

American poet and philosopher George Santayana (1863–1952) said, "Those who cannot remember the past are condemned to repeat it." We discuss the historical development of pest control (Chapter 3), with the hope that society will not repeat pest management errors. People often seem to forget history, however, and those in pest management appear to be no exception. The continuing problems with pesticide resistance (Chapter 12) serves as an example. The phenomenon has been known for at least 50 years, but it would seem that the historical and the ecological perspectives are again being ignored. Crops that have been genetically engineered for resistance to certain pests are being widely deployed (more in Chapters 12 and 17) despite the potential for pests to develop resistance. The problem associated with pests being spread around the world is another example of forgetting history. Although the problem is recognized, accidental and intentional introductions of organisms still occur. The current pace of pest dissemination, if maintained, indicates worldwide distribution of many pest species in the near future, despite efforts to curtail their spread (see Chapter 10).

## Human Population Perspective

Pests of all types cause enormous losses to human society, whether people realize it or not. In developed countries, the burden of managing pests falls on the professional farmers and pest control advisors. In less developed countries, responsibility rests primarily with the families that work the land. If current pest management technology were to disappear, it is likely that more of the world population would suffer starvation. At the turn of the twentieth century, prior to the development of current farm management, the world supported a population of about 2 billion (see Figure 1–3). Without inputs of fertilizers, energy, and agricultural chemicals, the world could probably not support a population much greater than this, although demographers project that human population will grow to 8.9 billion by 2050 (FAO, 1999).

## Pesticide Perspective

The idea that agriculture should eliminate agricultural chemicals and revert to the pest management approaches of the past is not feasible given the world population. The complete exclusion of pesticides from IPM systems would be, in many instances, problematic for populations in the developed countries, and would offer people in less developed countries no reprieve from the drudgery and starvation in their current lives. What is critical, however, is that pesticides be used rationally within the context of IPM, and with a thorough understanding of ecological impacts.

This book explores the current concepts behind management of pests. It is our hope that it will contribute to an understanding of the complicated processes by which humans manipulate nature to maintain their food supply, and manage the pests to reduce the pest "share" of that food supply.

# SOURCES AND RECOMMENDED READING

Anonymous. 1996. *Ecologically based pest management: New solutions for a new century.* Washington, D.C.: National Research Council, National Academy Press, xiii, 144.

Anonymous. 2000. A history of American agriculture 1776–1990. Farmers and the land. http://www.usda.gov/history2/text3.htm

Bajwa, W. I., and M. Kogan. 1996. Compendium of IPM definitions (CID). http://www.ippc.orst.edu/IPMdefinitions/.

Beirne, B. P. 1967. *Pest management.* London: L. Hill, 123.

FAO. 1999. Agricultural database. http://apps.fao.org/default.htm

Morse, S., and W. Buhler. 1997. *Integrated pest management: Ideals and realities in developing countries.* Boulder, Colo.: Lynne Rienner Publishers, ix, 171.

Oerke, E. C. 1994. *Crop production and crop protection: Estimated losses in major food and cash crops.* Amsterdam; New York: Elsevier, xxii, 808.

Stern, V. M., R. F. Smith, R. van den Bosch, and K. S. Hagen. 1959. The integrated control concept. *Hilgardia* 29:81–99.

United Nations. 1999. The world at six billion. http://www.un.org/popin/wdtrends/6billion.

Warren, G. F. 1998. Spectacular increases in crop yields in the United States in the twentieth century. *Weed Technol.* 12:752–760.

# PESTS AND THEIR IMPACTS

## CHAPTER OUTLINE

▶ General impact of pests
▶ Plant pathogens
▶ Weeds
▶ Nematodes
▶ Mollusks
▶ Arthropods
▶ Vertebrates

At this point it is appropriate to examine the classification and morphology of the different categories of pests, and to discuss their impacts on managed ecosystems and the ways in which they cause losses in crop production, to the detriment of human society.

## GENERAL IMPACT OF PESTS

The effects of pests from several pest categories are quite similar, and are presented here in general terms and are not listed in detail for each pest category. Following are impacts that are common to more than one group of pests.

### Consumption of Plant Parts

The ingestion of plant cytoplasm or tissue represents an energy gain by the pest and an energy loss by the plant. For example, consumption of leaves (Figure 2–1a) and roots (Figure 2–1b) results in reduced photosynthetic area, which then results in yield loss. Pests may directly consume fruits and seeds, or damage the roots, which reduces top growth. A small amount of cutworm, slug, or snail feeding on seedlings often results in death of the seedling (Figure 2–1c). Nematodes, insects, mollusks, and vertebrates may consume plant sap, cytoplasm, plant parts, or entire plants, which can result in plant death (Figure 2–1d).

**Crop losses due to insects, pathogens, and nematodes may result from plants responding to chemical signals released by insects (e.g., rosy apple aphid), pathogens (crown gall), or nematodes (root knot). Plants may release chemicals that affect other plants, a phenomenon called allelopathy.**

# Chemical Toxins, Elicitors, and Signals

The different pest categories include many pests that produce chemical compounds which can damage host plants or induce host physiological responses such as defensive reactions. Some of these compounds are toxic to host tissue, and produce direct, local host responses such as stress or necrosis at the site where the pest attacks the plant. In such cases, the host response is generally thought to reduce the capacity of the pest to obtain nutrition from the host. Other pest secretions or metabolites may act as signals to the host, and such signal molecules may elicit localized or even systemic host defensive responses. Molecules that stimulate plant defensive mechanisms are called elicitors. The diverse chemicals produced by pests result in a tremendous range of host responses; however, as a general rule, when pest-produced chemicals disrupt host plant physiology and homeostasis, plant growth and yield are reduced.

Numerous pathogens produce chemical compounds that are released into the host plant and cause changes in growth. For example, many *Phytophthora* species produce elicitors that induce hypersensitive necrosis in tobacco leaves, and the elicitors also induce in tobacco a systemic acquired resistance to bacterial and fungal pathogens. *Taphrina deformans,* the fungus that causes peach leaf curl, produces auxins, which result in deformed leaf growth in the spring (Figure 2–1e). Foolish seedling disease of rice results from infection of the seedling by the fungus *Gibberella fujikuroi,* which releases gibberellins into the plant and thus causes the plant to grow spindly and weak.

Weeds may release chemical compounds called allelochemicals (see Chapter 14) which inhibit the growth of competing plants. Several nematode species inject chemical signals into host plants and cause physiological responses in the host, including the abnormal growth seen as galls produced on roots following infection by the root-knot nematode (see Figures 2–7f and g later in the chapter).

Many people are familiar with the "oak apples" that sometimes appear on oak tree leaves. The oak apples result from abnormal leaf growth caused by growth-regulating chemicals introduced into oak leaves by Cynipid gall wasp larvae. Many pests alter the physiology of their host plants through the action of chemical signals, such as toxins, that are released by the pest. Many arthropods with sucking mouthparts, such as aphids, cause abnormal host-plant growth due to the effect of plant growth–regulating chemicals in their saliva. This growth response may be much more serious than the effect of direct consumption of plant cell contents. The rosy apple aphid, which causes deformed growth of apples, is a good example of an insect causing this type of problem (Figure 2–1f).

# Physical Damage

The feeding of arthropods and nematodes makes wounds or holes that serve as entry points for pathogens. The resulting disease then damages the produce or even kills the plant. Physical damage can also cause loss of plant structural strength; corn, for example, can collapse in response to corn stem borer damage (Figure 2–1g).

# Loss of Harvest Quality

Physical damage leads to loss of harvest quality due to accelerated rotting during shipping and storage.

## Cosmetic Damage/Aesthetics

People in most developed countries do not like to eat fruits or vegetables that have galls or necrotic spots, leaves that have been chewed by insects or that contain frass (entomological term for *feces*), or—worse yet—actual insects (Figure 2–1h)! The implications of cosmetic standards are discussed in greater detail in Chapters 8 and 19. Nematode infection may also cause cosmetic damage that makes the commodity unmarketable, such as lesions in potatoes and forked roots and galls on carrots caused by root-knot nematodes (Figure 2–1i).

## Vectoring of Pathogens

Many pathogens, especially viruses, fastidious bacteria, and phytoplasmas, require an arthropod, nematode, or fungus to carry the inoculum from one plant to another and thus transmit the disease. Pathogens vectored by arthropods are an extremely important problem, and are often the most serious impediment to increasing crop yields in less developed countries. The concept is explored further in Chapters 5, 6, 7, and 16.

## Direct Contamination

Insects, insect parts, or mollusks can be present in the harvested product; most developed countries consider this as filth. Weed seeds can be present in harvested products, such as cereal grains. Ergot fungus sclerotia also are direct contaminants of rye and other cereals (see Figure 3–4).

## Costs Incurred to Implement Control Measures

Application of a control tactic has costs, such as purchase of diesel fuel for cultivation, purchase of beneficial insects to release, purchase of pesticides, or the cost of machinery and labor to apply a pesticide.

## Environmental and Social Costs

Measures to control pests may result in undesirable effects, such as erosion following tillage, or ground or surface water contamination by pesticides. These costs are often hard to quantify. Other costs occur in the form of worker safety in relation to control tactics, such as accidents relating to equipment use, or by exposure to pesticides.

## Embargo, Quarantines, and Shipment Costs

The presence of pests can result in imposition of an embargo such that a product cannot be shipped internationally. Embargoes can also occur between regions of a country, such as oranges between Florida and California. Shipment costs can be increased when treatment such as fumigation has to be done prior to shipping to kill pests present in the produce. All pest categories cause such losses.

    The remaining sections in this chapter are a review of each pest category. We have chosen to include pathogens first as they represent the "simpler" forms of life. Plants are considered second. We then discuss animal pests, starting with the nematodes. Mollusks are considered next, followed by arthropods, and finally vertebrates in increasing order of organism complexity.

**FIGURE 2–1**  Examples of different types of damage caused by pests. (a) Removal of leaf tissue; the example here is sugar beet leaf tissue eaten by fall army worm; (b) reduced growth of sugar beets due to root feeding damage by sugar beet cyst nematode (center two rows not protected from nematode attack); (c) cutworm damage to soybean seedlings; (d) plants killed by pest attack; here peach trees were killed by oakroot fungus; (e) production of growth regulator by peach leaf curl fungus causes peach leaf distortion; (f) apple leaves curled in response to toxins injected by rosy apple aphid infestation; (g) physical damage to corn by stem borers makes crop prone to collapse; (h) example of cosmetic damage; corn earworm in a sweet corn ear where it causes little yield loss but the infested sweet corn is not marketable; and (i) example of cosmetic damage to carrots by root-knot nematode; essentially no yield loss but most roots cannot be marketed.

Sources: photographs (a), (c), and (g) by Marcos Kogan; (b) by Edward Caswell-Chen; (d), (e), (f), and (h) by Robert Norris; and (i) by John Marcroft, with permission.

# PLANT PATHOGENS

## Description of Organisms

Pathogens are parasitic organisms that cause biotic disease or illness in other host organisms. They belong to several different phyla, and their morphology and cellular structure vary considerably.

***Fungi*** Fungi are eukaryotic organisms, are a single cell or multicellular, and have a cell wall. The body is called a mycelium, and it is composed of thread-like hyphae. The mycelium can be aggregated into several different macroscopic forms, such as mushrooms, brackets, and sclerotia. Fungi are not autotrophic, which means they do not survive without access to organic matter. They are either saprophytes, which live on dead or decaying organic matter, or parasites, which live in or on other living organisms. Many saprophytic fungi are useful as food (e.g., mushrooms) or producers of antibiotics (e.g., *Penicillium*). Parasitic fungi are the most common pathogens. Examples of a few common diseases caused by fungal pathogens include the rusts in cereals, smut in corn (Figure 2–2a), mildews on many plants, and oakroot fungus that attacks many fruit and nut trees. *Sclerotinia minor* causes a disease called lettuce drop, which can kill lettuce just before it is ready to harvest (Figure 2–2b).

***Phytoplasmas*** Phytoplasmas are prokaryotic organisms, and do not possess a cell wall. The cytoplasm is bounded only by a unit membrane. Discovered in the early 1970s, phytoplasmas typically affect the vascular system in plants, especially in tree crops. Most are fastidious. They were originally called mycoplasma-like organisms (MLOs), but this term is no longer used. Examples of diseases caused by phytoplasmas include corn stunt, pear decline (Figure 2–2c), and peach yellow leaf roll.

***Bacteria*** Bacteria are ubiquitous, prokaryotic, single-celled organisms. Many are beneficial, but some can attack living plants. Most pathogenic bacteria are not able to penetrate intact plant tissues; they must enter through wounds or be injected by a vector organism. Some gain entry into plants through natural openings including stomata or hydathodes. Most bacteria are free-living and can survive extended periods in the absence of a host, but some are fastidious and cannot survive without a living host. Bacteria are responsible for most "soft rots" in fruits and vegetables (Figure 2–2d). Examples of other bacterial diseases include speck on tomatoes, fire blight of apple and pear trees, Pierce's disease of grapes, and crown gall on trees (Figure 2–2e).

***Viruses*** Viruses consist of DNA or RNA strands contained inside a protein coat, but require a host to grow and reproduce. Viruses are spread in infected plant tissue or by means of another organism called a vector (see Chapter 5), which not only transmits the virus but also "injects" it into host tissue. When present in living cells, the viral replication within the host usually results in stunting, malformation of growth, and often almost complete elimination of useful yield. Squash, a vigorous, high-yielding vegetable, produces no useful fruit if it is infected with cucumber mosaic virus when the plants are young (Figure 2–2f). Other examples of virus diseases include tobacco mosaic, lettuce mosaic, sugar beet yellows, and grape fan leaf. Viroids are similar to viruses in that they are infectious RNA, but they do not possess a protein coat.

When the normal physiological functions of a plant are disrupted, the plant is said to be diseased. If a biological agent (pathogen) causes the disruption, then it is termed a biotic disease; if the disruption results from environmental stress, then it is an abiotic disease.

Fastidious microorganisms cannot be cultured on ordinary culture media.

***Higher Plants***   Some higher plants are parasitic on other plants, and thus fall under the general definition of biotic disease. Examples include mistletoes and dodder, which are stem parasites, and broomrapes and witchweed, which are root parasites. We present these parasites in the section on weeds, because they are higher plants.

***Abiotic Diseases***   Abiotic diseases are disorders not caused by a living organism, hence the term *abiotic*. Examples include such things as excess or insufficient nutrient or water, and environmental stresses such as smog, high ultraviolet (UV) radiation, frost (Figure 2–2g), high temperatures, and toxicity of elements such as boron, chlorine, or sodium.

Four component factors must occur for a disease to develop. There must be a pathogen present, and there must be suitable host for the pathogen to infect. A third requirement is that the environment must be suitable for the pathogen. These three factors have been referred to in the plant pathology literature as the

**FIGURE 2–2**   Examples of disease problems caused by different types of pathogen. (a) Swollen fungal spore masses in sweet corn due to infection by the smut fungus; (b) lettuce drop fungal infection causing collapse of cos lettuce just prior to harvest; (c) pear tree in foreground almost dead due to infection with the pear decline phytoplasma; (d) soft rot of onion (dark area) caused by the bacterium *Erwinia amylovora;* (e) crown gall on a poplar tree being grown for native species restoration project results in destruction of the tree; (f) cucumber mosaic on squash; note stunted plant on right; and (g) abiotic disease; leaf damage in center of photograph caused by frost damage on potato.
Sources: photographs (a), (b), (e), (f), and (g) by Robert Norris; (c) by Jack Clark, University of California statewide IPM program; and (d) by Mike Davis, with permission.

disease triangle. Most plant pathologists now recognize that time is a fourth factor, and the concept is diagrammed as a pyramid (see Figure 1–6). In the absence of any one of these factors there will be no disease development.

## Special Problems Caused by Plant Pathogens

The following impacts are in addition to those listed earlier in the chapter.

1. Loss of crop production. Pathogens absorb food and nutrients from the host plant. The diseases caused result in damage or death of leaves, roots, stems, flowers, and fruits. In some cases, the entire plant is killed. Losses range from partial to total; under epidemic conditions crop loss can be catastrophic, as happened when late blight attacked potatoes in Ireland in 1848 (caused famine and emigration), or coffee rust attacked coffee trees in Ceylon in the late nineteenth century (the crop could no longer be grown).
2. Abnormal growth. A few pathogens can inject part of their DNA into the host cells and cause abnormal growth; the crown gall bacterium (*Agrobacterium tumefaciens*) is an example of such a pathogen (Figure 2–2e). Viruses can cause abnormal growth of the host plant.
3. Rotting/spoilage of harvested products. Many pathogens infect ripe fruit and vegetables during storage and shipping, and cause them to rot (Figure 2–2d); total loss typically occurs. It has only been modern postharvest handling of produce that has permitted such fruits as peaches, strawberries, and cherries to be stored and shipped long distances without rotting.
4. Poisonous products. Some fungi produce chemicals that are toxic to animals that ingest them. They are called mycotoxins, and make the harvested product poisonous to people when eaten. Examples include aflatoxins in peanuts caused by *Aspergillus* spp., and alkaloids from ergot in cereals caused by *Claviceps purpurea* (see Figure 3–4).
5. Allergic reaction to spores. Many people are allergic to mold spores. Most people in North America are familiar with the spore (mold) counts provided by television in the spring and fall. Some of these spores can be from plant-pathogenic fungi.
6. Vectors for plant viruses. Fungi can carry some plant viruses; for example, the virus that causes rhizomania in sugar beets is vectored by a soil-borne fungus *Polymyxa betae*.

Following are examples of disease epidemics that have had severe impacts on plants or crops.

1. Dutch elm disease killed most of the elm trees in Europe and North America.
2. Coffee rust in Ceylon in the mid-nineteenth century caused a shift from coffee to tea production, changing the beverage drinking habits of the United Kingdom.
3. Fire blight of pears limits commercial production to regions with relatively dry climates.
4. In the late 1840s, late blight of potatoes resulted in famine in Ireland and mass migration of the population. The pathogen still causes enormous losses in potato production.
5. Chestnut blight has eliminated chestnut trees from most of North America.

# WEEDS

## Description of Organisms

Of the many definitions of the term *weed,* one of the simplest is any green plant growing where it is not wanted. This means that virtually any plant can be a weed, even crop plant species when they are found where they are not wanted (Figure 1–7). The following plant groups include species considered to be weeds.

1. Algae. These plants can be a serious problem in aquatic ecosystems, including paddy rice, in irrigation and drainage canals, and in recreational facilities such as lakes and marinas.
2. Mosses and liverworts. These gametophytic nonflowering lower plants are most often considered weedy when found growing in turf or in nursery conditions.
3. Ferns and horsetails. These nonflowering, spore-producing higher plants, of which there are several genera and species, are often considered as weeds, such as bracken fern in pastureland and horsetails in horticultural crops (Figure 2–3a).
4. Gymnosperms. These naked seed plants are mostly trees (often called conifers); most are not considered to be weeds, but can be a problem in some range and forest systems.
5. Angiosperms. These are flowering plants that produce seeds. The majority of weeds are in the flowering plants. Weeds may be annuals, biennials, or perennials. Table 2–1 provides a list of the world's 18 most important weed species. Many of the important weeds are in the monocotyledon (single seed leaf) families; fewer species of weeds are in the dicotyledon (two seed leaves) families.

**Most plants are never considered to be weeds.**

How many weed species exist in the world? It has been estimated that perhaps 1,000 species are typically designated as weeds. In a single region, the number is probably between 100 and 300 species.

What are typical weed densities? There are literally millions per acre in arable land if the seedbank (seeds in the soil) is included. It is physically difficult to have more than about 200 to 300 adult plants per square yard (plants/yard$^2$), unless they are very small plants. Weed biomass per area (biomass/A) can be enormous. For example, it has been estimated that fiddleneck can produce over 10 tons/A of dry matter, and similar quantities of wild oats and mustard species are assumed to be common.

## Special Problems Caused by Weeds

Many of the general impacts of pests discussed at the start of this chapter do not apply to weeds, because weeds do not "feed" on crop plants.

1. Economic impacts
   1.1. Yield loss is the single most important effect of weeds (Figure 2–3a through f). Interference of crop growth by weeds can result in yield losses up to 100% in tomatoes and sugar beets; or in more competitive crops, such as cotton and corn, yield losses of up to 50% can occur when weeds are not controlled. Yield losses of this

| TABLE 2–1 | The 18 Most Important Weeds in the World | | |
|---|---|---|---|
| Latin binomial | USA common name | | Ecosystem invaded |
| Cyperus rotundus | Purple nutsedge | P[1] | World's number one weed; many crops in warmer climatic regions |
| Cynodon dactylon | Bermudagrass | P | Many cropping systems, especially perennial, in warmer climates |
| Echinochloa crusgalli | Barnyardgrass | A | All cropping systems in warmer climates, especially rice |
| Echinochloa colona | Junglerice | A | Many cropping systems in warmer climates |
| Eleusine indica | Goosegrass | A | Many cropping systems in warmer climates |
| Sorghum halepense | Johnsongrass | P | Most nonaquatic cropping systems in warm climates |
| Imperata cylindrica | Cogongrass | P | Most crops in tropical climates |
| Chenopodium album | Lambsquarters | A | Most crops in temperate climates |
| Digitaria sanguinalis | Large crabgrass | A | Many crops in warmer climates, especially in turf |
| Convolvulus arvensis | Field bindweed | P | Many ecosystems in temperate climates, especially cereals |
| Portulaca oleracea | Common purslane | A | Many crops in warmer climates |
| Eichhornia crassipes | Water hyacinth | A | Aquatic ecosystems and lakes |
| Avena fatua | Wild oat | A | Many cropping systems in temperate climates, especially cereal crops |
| Amaranthus hybridus | Smooth pigweed | A | Many crops in warmer climates |
| Amaranthus spinosus | Spiny pigweed | A | Many crops in tropical climates |
| Cyperus esculentus | Yellow nutsedge | P | Many crops in warmer climates |
| Paspalum conjugatum | Sour paspalum | A/P | Many crops in humid tropics |
| Rottboelia exaltata | Itchgrass | A | Many crops in tropical countries |

*Adapted from Holm et al., 1977.
[1]Life form: A = annual, P = perennial.

magnitude are routine in most arable crops in the absence of weed control, and is the overriding reason why weeds must be controlled in most crops.

1.2. A rather specialized type of yield loss occurs when the weed is a parasitic flowering plant that derives some or all of its nutrients from the host (see also under plant pathogens). Severe infestations of parasitic weeds can stunt or even kill the host crop. Examples of parasitic weeds include the following:

1.2.1. Dodder (*Cuscuta* spp.) are stem and leaf parasites on such crops as alfalfa, tomatoes (Figure 2–3e), and sugar beets.

1.2.2. Mistletoes are also stem parasites, usually on trees. Dwarf mistletoes (*Arceuthobium* spp.) are some of the most destructive pests of conifer forests because they cause deformed growth of the main trunk of infected trees. In the western United States it is estimated that infection by dwarf mistletoes annually causes in excess of 3 billion board feet of lost lumber production.

**FIGURE 2–3**    Examples of weeds in various crops. (a) Horsetail, a nonflowering plant, as a weed in a celery crop in Denmark; (b) nutsedge has here eliminated cotton in San Joaquin Valley, California; (c) onions barely visible under a canopy of horse nettle and other weeds in Morocco; (d) young orchard in California dominated by weeds; (e) the parasitic weed dodder infesting a processing tomato crop can cause almost 100% loss; and (f) the yield of grain from an area infested with wild oat and an area from which wild oat was controlled with a herbicide.
Source: photographs by Robert Norris.

       1.2.3. *Striga* spp. are root parasites on grass crops such as corn, sorghum, and millets. They cause devastating losses in the semiarid tropics, especially in Africa.

       1.2.4. Broomrapes (*Orobanche* spp.) are root parasites on Apiaceae (carrot family) crops and Fabaceae (peas, beans, and pulse) crops. The two types of root parasitic weeds seriously limit crop production in many tropical and subtropical countries; crop losses often reach close to 100%.

    1.3. Reduction in operational efficiency results from clogged planters and harvesters, and difficulty in storage due to moist seed.

    1.4. Reduction in land value results especially where heavy infestation of perennial weeds such as johnsongrass or nutsedges occurs.

  2. Environmental impacts

    2.1. Weeds have serious ecological consequences when they replace desirable native vegetation. Examples of invasive species include Scotch broom in California, *Miconia* in Hawaii, and prickly pear in Australia (see Figure 3–8).

    2.2. Weeds can be used as a food source by other organisms, and thus serve as alternative hosts for all other types of pests. Weeds may also harbor beneficial insects (see Chapter 7).

    2.3. Weeds can clog aquatic ecosystems such as irrigation and drainage canals (Figure 2–4a), marinas, and reservoirs. Uncontrolled weed growth can reduce water flow in canals by greater than 90%. Whole irrigation systems have been rendered inoperable by weeds. Weeds can eliminate use of recreational aquatic systems. Water hyacinth has been called "million dollar weed" because of the costs required to control it.

   2.4. Control technologies used for weed management may also have environmental impacts; soil erosion and dust result from cultivation, and herbicides have been found in ground and surface water.

3. Aesthetic impacts

   3.1. Weeds are unsightly in turf and ornamentals.

   3.2. Scattered weeds in field crops are often considered unsightly by the farmer.

4. Health and safety impacts

   4.1. Weeds can be both poisonous and allergenic.

      4.1.1. Hay fever in humans from pollen; ragweed in the Midwest is a good example. It is estimated that 3.3 workdays are lost per year per person due to hay fever.

      4.1.2. Dermatitis may come from contact with weeds. Many people are sensitive to poison oak (Figure 2–4b) or poison ivy and other plants; contact with the plant causes severe inflammation and blistering of the skin. Nettles in a hand-harvested crop such as lettuce present a severe problem as the workers are stung by the plants, causing severe skin irritation.

      4.1.3. People die every year from ingesting poisonous plants. Poison-hemlock was used to kill Socrates, and it still causes deaths today.

      4.1.4. Acute and chronic animal poisoning is another factor. Fiddleneck and common groundsel cause nonreversible liver cirrhosis in animals that feed on contaminated hay. *Halogeton* spp. have killed whole herds of sheep in rangeland in the western United States, and locoweeds cause many animals to go crazy. The pasture and rangeland weed yellow starthistle (Figure 2–4c) contains a potent neurotoxin that is potentially lethal to horses. Birth defects can occur in sheep, pigs, and cattle when poisonous weeds are eaten during gestation, such as false hellebore in western rangeland in the United States.

      4.1.5. Light sensitization occurs in animals that ingest weeds that contain chemicals leading to increased skin sensitivity to sunburn, especially to eyelids and lips. Saint-Johns-Wort (Klamath weed), in rangeland is an example.

   4.2. Dead and dry weeds can be a serious fire hazard especially along roadsides and railroad beds, in equipment and lumberyards, and in oil storage facilities. Fire hazard is the main reason why weeds on roadside shoulders are mowed, or sprayed with herbicide.

   4.3. Weeds can break up paved roads (Figure 2–4d) creating a hazard for drivers, and can block visibility of signs and safety lighting such as runway lights at airports, and thus decrease safety.

There is also a human cost to weeds in most of the less developed countries that goes unrecognized in the western world. Hand weeding in Africa is estimated to require between 20% and 50% of the total time spent in crop production; perhaps as much as 25% of the acreage of subsistence crops in Africa is

**FIGURE 2–4**    Examples of noncrop impacts of weeds. (a) Aquatic weed water hyacinth completely blocking an irrigation/drainage canal; (b) rash on forearm in response to contact with poison oak; (c) rangeland in the California Coast completely dominated (white seed heads) by yellow star thistle; and (d) weeds breaking up pavement and blocking edge of paved road.

Sources: photograph (a) by Lars Anderson, USDA/ARS; and (b), (c), and (d) by Robert Norris.

lost each year because of inadequate weeding. On estimate, weeding requires more time than any other component of crop production in less developed countries (see Figure 2–5). It is perhaps hard for those living in Western Europe or North America to even imagine the toll that weed management exacts on society in less developed countries. Holm (1971) eloquently describes the situation:

> The setting for my first story is a coconut plantation in an unnamed country; it is early morning and the shadows are still long. I wish to tell you about a girl who is a Tamil. These bright, handsome people from the south of India are seen in many places in Africa and Asia where they seek homes and work. The girl, who is 16 or 17 years of age, is already soaked with perspiration. This slight, wisp of a girl holds a thin knife 3 feet or more in length which she must raise back over her shoulder, swing the blade as she stoops to bring it parallel to the ground for weed cutting, and then follow through until it is behind her body on the other

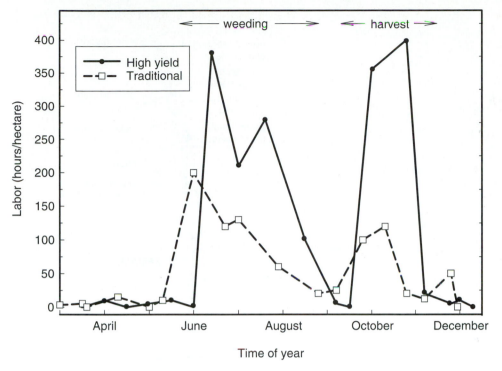

**FIGURE 2–5**
Graph of human time required for crop production in Africa under traditional and high-yield agricultural systems.
Source: data from Giampietro et al., 1992.

side. The weeds here are *Axonopus* and *Ischaemum* grass species and they may be knee deep when the cutters enter the plantation. She will swing the knife until early afternoon.

Many of the young people doing such work in Africa and Asia can never go to school. Tomorrow will be the same, and the next day, and the next day. Can you imagine such an occupation and future, for any young woman you know—daughter, wife, or sister?

# NEMATODES

## Description of Organisms

Nematodes are a group of nonsegmented, wormlike invertebrates. They are sometimes referred to as eelworms, or simply as worms. They are appendageless, bilaterally symmetric roundworms and they are essentially aquatic, living in water, moisture films, or the tissues of a host. Nematodes occur in many different habitats and environments, from mountaintops to deep ocean sediments, from deserts to rain forests. Nematodes are the most numerous multicellular organisms on earth, and different nematode species feed on many different food sources.

Nematodes are considered as pests because some species are parasites of plants, humans, and animals. It is important to consider that not all nematodes are detrimental to humans or human activities. For example, certain nematode parasites of insects are used as biological control agents to manage insect pests (see Chapter 13). Other nematodes feed on bacteria (bacterivores) and fungi (fungivores) in the soil, and so are significant contributors to nutrient cycling in soils.

Nematodes range in size from the smallest, a marine nematode that is only 80 μm long, to the longest, an 8 m nematode that lives in the placenta of whales.

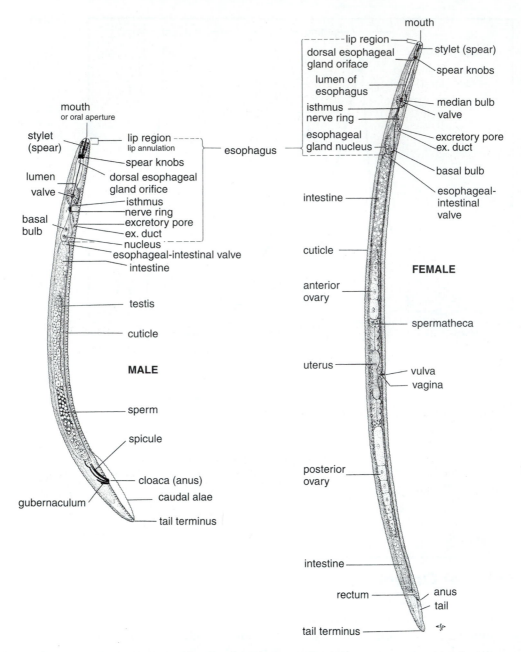

**FIGURE 2–6**    General anatomy of male and female plant-parasitic nematodes.
Source: drawings by Charlie Papp and Sadek Ayoub, Div. Plant Industry, California Dept. of Food and Agriculture.

Most plant-parasitic nematodes are less than 2 mm long. Nematodes are animals, and possess nervous, digestive, reproductive, muscular, and excretory systems but lack specialized circulatory and respiratory systems (Figure 2–6). The nematode body is covered by the cuticle, which is a noncellular, flexible, multilayered, protective structure that is normally shed by molting four times during the life cycle. All plant-parasitic nematodes possess a small, protrusible oral spear called a stylet, which is used to penetrate plant cells to obtain nutrition (Figure 2–7a). Most plant-parasitic nematodes are soil inhabitants and feed

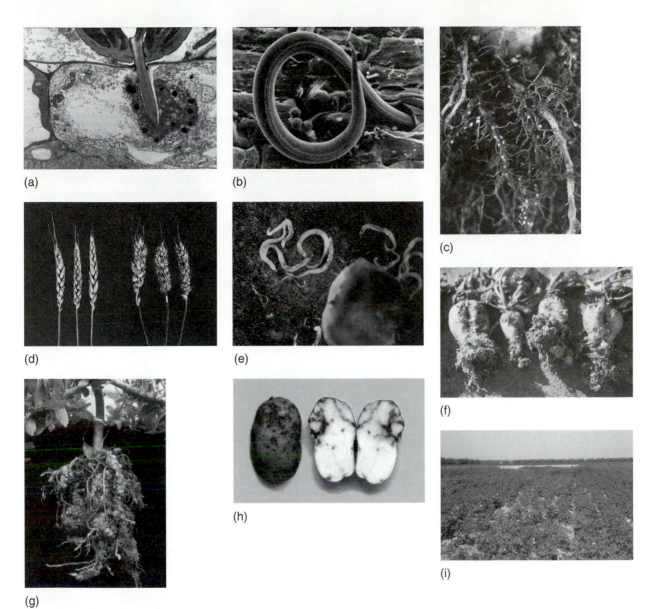

**FIGURE 2–7** Examples of nematodes and nematode damage. (a) Electron micrograph showing stylet of nematode penetrating a plant cell; (b) a scanning electron micrograph of a root-knot nematode juvenile penetrating a tomato plant root; (c) cyst nematode cysts on roots of sugar beet; (d) wheat heads, on left not infected and on right infected with *Anguina;* (e) *Anguina* nematodes extracted from wheat grain (bottom right); (f) sugar beets showing extensive galling damage by root-knot nematode; (g) extensive damage to root system of a *Protea* sp. by root-knot nematodes; (h) potato showing *tobacco rattle mosaic* symptoms transmitted by *Trichodorus* nematodes; and (i) cotton showing loss of resistance to *Fusarium* wilt in presence of root-knot nematode (more injury in center four rows).

Sources: photograph (a) by Michael McClure, with permission; (b) by William Wergin and Richard Sayre, USDA/ARS; (c) by Ivan Thomason, with permission; (d) and (e) by Jonathan D. Eisenback with permission; (f) by John Marcroft, with permission; (h) by G. Caubel with permission; and (g) and (i) by Edward Caswell-Chen.

| TABLE 2–2 | **The 10 Most Important Plant-Parasitic Nematodes in the World** | |
|---|---|---|
| **Genus** | **USA common name** | **Examples of important host crops** |
| *Meloidogyne* | root-knot nematode | potato, cotton, field corn (maize), soybean, dry beans, wheat, rice, sorghum, tomato, sugarcane, peanut, tobacco, alfalfa, apple, citrus, coffee, grape, cucurbits (melons), carrots, sweet potato, onion, beet, cereals |
| *Pratylenchus* | lesion nematode | field corn, soybean, dry beans, wheat, rice, potato, sorghum, sugarcane, peanut, alfalfa, apple, peach, strawberry, citrus, coffee, cherry |
| *Heterodera* | cyst nematode | soybean, dry beans, sugar beet, broccoli, Brussels sprout, cabbage, carrots, wheat, chickpea, pigeonpea |
| *Ditylenchus* | stem nematode | rice, potato, alfalfa, beet, oats, flower bulbs, mushrooms |
| *Globodera* | cyst nematode | potato, tobacco |
| *Tylenchulus* | citrus nematode | grape, citrus |
| *Xiphinema* | dagger nematode | sugarcane, grapes, apple, cherry, strawberry |
| *Radopholus* | burrowing nematode | banana, citrus, tea |
| *Rotylenchulus* | reniform nematode | cotton, soybean, papaya, sweet potato, pigeonpea |
| *Helicotylenchus* | spiral nematode | banana, pepper, eggplant, tea |

*Adapted from Koenning et al., 1999; Sasser and Freckman, 1987.

on roots (Figures 2–7b and c), but some species feed on stems and leaves, or even cause galls on flowers or fruits (Figures 2–7d and e). Some species produce signal molecules and inject them into the host plant to induce physiological changes in the host that serve to support the nematode, and may cause abnormal host growth, such as the galling produced by the root-knot nematode on sugar beet (Figure 2–7f) and *Protea* sp. (Figure 2–7g).

Plant-parasitic nematodes are obligate parasites, meaning that they must have an appropriate host plant to complete their life cycle and reproduce. The more important plant-parasitic nematodes are listed in Table 2-2. The common names of plant-parasitic nematodes are usually based on the symptoms they elicit in host plants, on the general morphology of the nematode, or on a combination of the characters; for example, as shown in Figures 2–7f and g, the root-knot nematode causes galls to form on roots. Host specificity is a factor among nematode species, meaning that for a particular nematode species, some plants are nonhosts. Some nematodes, such as the root-lesion nematode, have broad host ranges and can parasitize a large number of host-plant species from several plant families (Table 2–2). Other nematodes, such as the dagger nematodes, have narrower host ranges and can parasitize a more limited number of host species.

For the purpose of general discussion, the plant-parasitic nematodes are divided into the following two general types, based on feeding habits and the relationship between the nematode and its host.

1. **Endoparasites.** The nematode enters the plant and feeds on internal plant tissues.
2. **Ectoparasites.** The nematode body remains outside of the plant while it is feeding.

Depending on the species, as nematodes develop through their life cycle, they either remain vermiform (worm shaped) or they become swollen and sedentary. Vermiform nematodes are typically motile throughout their life, and are referred to as migratory. There is a range of life histories within the plant-parasitic nematodes, and the endo- and ectoparasitic types may be sedentary or migratory. Examples of important plant-parasitic nematodes include the root-knot and cyst nematodes. If fields are severely infested with these nematodes, annual yield losses of 10% to 50% may result.

Nematodes are animals with unique attributes, so they are considered as a separate phylum, the Nematoda. The phylum includes two classes, the Secernentea and the Adenophorea. Within the Secernentea, the orders Tylenchida and Aphelenchida include plant-parasitic species. Within the Adenophorea, only one order, the Dorylaimida, contains plant parasites, and the only nematodes known to vector plant viruses occur in the Dorylaimida. Nematode parasites of insects occur in the Secernentea and the Adenophorea.

There are approximately 20,000 named nematode species, and by some estimates, there may be as many as 1,000,000 total nematode species. The evolutionary relationships among nematodes are being investigated using comparisons of DNA sequence data. The taxonomic groupings of nematodes will undoubtedly change somewhat as further insights into the evolutionary relationships among nematodes are obtained.

## Special Problems Caused by Nematodes

The following are in addition to those problems listed under general impacts at the start of the chapter.

1. Nematodes may transmit, or vector, other pathogens. Several nematodes can carry and transmit plant viruses; for example, *Xiphenema* is the vector of *fanleaf virus* in grapes, and *Trichodorus* transmits *tobacco rattle virus* to potato (Figure 2–7h).
2. Plant-parasitic nematodes may interact with other pathogens to cause disease complexes that are worse than the disease caused by either pathogen alone. The root-lesion nematode interacts with the fungus *Verticillium* to cause potato early-dying disease, and root-knot nematodes may interact with the fungus *Fusarium* (see Chapter 6 and Figure 6–4).
3. Nematode infection may increase host-plant susceptibility to other environmental stresses such as high temperatures or moisture.
4. Nematode infection may nullify the host plant's resistance to other pathogens, for example, *Fusarium* resistance in cotton (Figure 2–7i).
5. Some nematode diseases may be moved from one host to another by insect pests. The pine-wilt nematode is carried from diseased to healthy pine trees by the longhorn beetle, and red-ring of coconut is carried by palm weevils.

# MOLLUSKS

## Description of Organisms

Mollusks are animals in the phylum Mollusca; slugs and snails belong to the Gastropoda. The body structure of slugs and snails is essentially identical except that snails produce a shell (Figures 2–8a and b). Mollusks possess a single "foot" on which they move. They create a slime layer between themselves and the substrate over which they are moving, resulting in the familiar silvery trail. They have a thin skin and can easily desiccate under dry conditions. Slugs and snails typically require a relatively cool, moist environment; snails are better able to tolerate hotter and drier conditions than slugs.

## Special Problems Caused by Mollusks

1. Feed on plants. Slugs and snails cause economic losses to ornamentals and other crops. They are a problem in greenhouses, home gardens, and certain field crops such as grass seed production. They are generally more important in seedling crops than in mature crops. A small amount of slug or snail feeding on seedlings often results in death of the seedling. Snails are especially serious in Citrus (where as many as 3,000 snails have been

**FIGURE 2–8**
Examples of pest mollusks. (a) Slug on a Primula plant; (b) garden snail sliding across soil; and (c) collection of snails on a branch of a Citrus tree in Morocco.
Source: photographs by Robert Norris.

(a)

(b)

(c)

recorded in a single tree) (Figure 2–8c), artichokes, and nurseries. They are much more serious in no-till farming than conventionally tilled systems due to the amount of cover and food that remains on the soil surface in no-till systems. A single slug, *Limax flavis,* weighing just under 0.5 oz ate up to 0.16 oz of pumpkin, cucumber, or potato in 24 hours. An equivalent biomass would translate into many dead seedlings.

2. Vector pathogens or nematodes. Slugs or snails have been shown to be capable of spreading *tobacco mosaic virus,* the fungus causing black root rot of cabbages, and foot rot of *Piper* spp. among others.

# ARTHROPODS

## Description of Organisms

In terms of species, arthropods are numerically more abundant than any other group of organisms. Over 1 million different arthropods have been named, which exceeds all other organisms combined. The number of insect species alone that are still unknown to science may be as high as 20 million. It has been estimated that there are typically over 40 million insects in an acre of land; this is equivalent to 200 million insects for each human. It has also been estimated that in the United States there are about 400 pounds of insect biomass per acre compared with about 14 pounds of humans per acre. This quantity of insect biomass is still low compared with plant biomass per acre; many weeds can exceed 5,000 pounds per acre in bad infestations.

Arthropods belong to the animal kingdom; the name of the phylum, Arthropoda, means "jointed foot." They all possess a more or less hardened external integument called an exoskeleton; consequently, they have to molt to increase in body size as they grow from one larval stage to the next until they molt into the adult stage. The following arthropod classes have members that are important as pests.

*Insecta*    Arthropods in the Insecta have one pair of antennae and a body in three major segments—head, thorax, and abdomen. For this reason the science that studies insects is known as entomology, from the Greek *tomos,* meaning "segments." Most insects also have three pairs of legs on the thorax, and many have wings (Figure 2–9; Table 2–3). Most insects reproduce by laying eggs, and thus are said to be oviparous; but some give birth to small nymphs or larvae, and are called viviparous. Nymphs or larvae emerge from eggs, depending on whether they resemble the adult, or imago; nymphs resemble the adult, whereas larvae do not. From egg through the adult stage, insects undergo several molts in a process called metamorphosis (discussed in Chapter 5).

The way in which an insect feeds depends on the structure of its mouthparts, and has considerable significance to pest management. Some insects have chewing mouthparts (Figure 2–10a), which means that the insect feeds by consuming pieces of the plant. Other insects have piercing-sucking mouthparts (Figure 2–10b) which they insert into their food and suck out cell contents or sap. A third group of insects has rasping-slurping mouthparts (Figure 2–10c) with which they scrape the surface of their food and then slurp up the contents of damaged cells.

**FIGURE 2–9**
External morphology
of a generalized
insect. The left pair of
wings are omitted for
clarity.

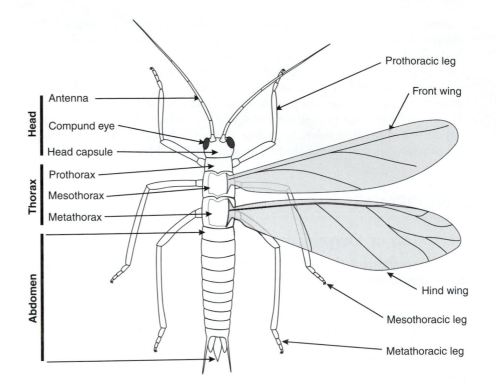

**FIGURE 2–10**   Simplified generic diagrams representing different insect mouthparts: A chewing, in which opposing mandibles cut and crush food (e.g., locust); B piercing/sucking, stylets inside the protective sheath are inserted into the host tissue permitting extraction of cell contents (e.g., stinkbug); C rasping/slurping (e.g., flies), and D siphoning (e.g., butterflies and moths), used to suck up free liquids without actually penetrating the host.

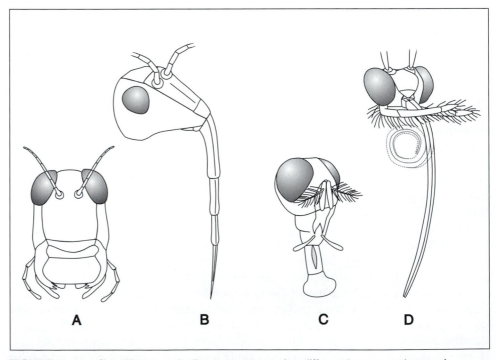

| Order | Common names | Example | Mouthparts/ metamorphosis | Other diagnostics | Pest information |
|---|---|---|---|---|---|
| Ephemeroptera Odonata | Mayflies Dragonflies and damselflies | | | | Not considered pests |
| Plecoptera Embioptera Blattodea | Stoneflies Web spinners Cockroaches | | Chewing. Partial metamorphosis | Long threadlike antennae | Not considered pests Not considered pests Not considered pests Human nuisance pests pests, not usually a problem in agriculture |
| Mantodea | Mantids | | Chewing. Partial metamorphosis | Large grasping front legs, praying stance | Beneficials |
| Isoptera | Termites | | Chewing. Partial metamorphosis | Relatively large head to accommodate jaw muscles; no eyes | Structural pests; large mounds cause serious problem in agriculture |
| Orthoptera | Grasshoppers Locusts Crickets Katydids | | Chewing. Partial metamorphosis | Strong back legs | Can be serious pests; locusts are in this order |
| Dermaptera | Earwigs | | Chewing. Partial metamorphosis | Large pincer appendages at tail | Serious damage to intensive flower crops and seedlings in small-scale farming |

*continued*

35

**TABLE 2–3**    **Major Orders of Pest and Beneficial Insects, Plus Examples of Other Arthropods** (*continued*)

| Order | Common names | Example | Mouthparts/ metamorphosis | Other diagnostics | Pest information |
|---|---|---|---|---|---|
| Thysanoptera | Thrips | | Rasping/slurping. Partial metamorphosis | Small insects with fringed wings | Small group of pest and beneficial species |
| Neuroptera | Lacewings | | Chewing as larvae, slurping as adult. Full metamorphosis | Large transparent wings with complicated vein pattern. | Larvae are beneficials |
| Coleoptera | Beetles | | Chewing. Complete metamorphosis | Front wings are hard covers called elytra | Many pests and many beneficials |
| | Snout beetles, or weevils | | Chewing. Complete metamorphosis | Front of head extends forward to form a snout bearing the antennae and mouthparts | Many serious pests; e.g., cotton boll weevil |
| Diptera | Flies; examples<br>Fruit<br>Flower<br>House<br>Mosquitoes<br>Midges<br>Leaf miners | | Slurping, sucking. Complete metamorphosis | Only one pair of wings | Many pests and beneficials |
| Homoptera | Aphids<br>Scale insects<br>Whiteflies<br>Leafhoppers<br>Cicadas | | Piercing and sucking. Partial metamorphosis | Many aphids exhibit parthenogenic reproduction | Many serious pests; many transmit viruses |

| Order | Common names | Example | Mouthparts/ metamorphosis | Other diagnostics | Pest information |
|---|---|---|---|---|---|
| Homoptera, cont. | | | | Aphids can be apterous (without wings) or alate (with wings) | |
| Lepidoptera | Butterflies and moths | | Larvae chewing; adults slurping. Complete metamorphosis | Relatively large wings covered with scales, often have characteristic markings. Larvae are a caterpillar (often called worm) | Many serious pests |
| Hemiptera | True bugs | | Piercing and sucking. Partial metamorphosis | Front wings form hardened covers | Many pests and beneficials |
| Hymenoptera | Bees Wasps Ants | | Chewing. Complete metamorphosis | Typically have a constricted waist between thorax and abdomen. Many have sting | Numerous pests and many beneficials |
| Collembola | Springtails | | Chewing. No metamorphosis | Lack wings. Abdomen has prong (furcula) folded back under body with which insect can spring. Colophore on first abdominal segment. Very small | A few are soil-borne pests |
| Arachnida | Spiders Spider mites Ticks Scorpions | | Rasping and slurping. No metamorphosis | | Many pest species and many beneficials |
| Symphyla | Garden centipedes | | Chewing. No metamorphosis | Multiple legs; do not possess wings | A few are soil-borne pests |

Within the class Insecta are approximately 30 orders, which vary in importance in relation to pest management. Orders that include either pest or beneficial species are shown in Table 2–3, and are listed below.

**Collembola**   Collembolans, or springtails, are mostly soil inhabiting, small insects that are important as prey for soil predators. A few species in the family Sminthuridae are pests of forage crops in Europe.

**Orthoptera**   Grasshoppers, crickets, locusts (see Figure 3–1), and mole crickets are orthopterans. They are mostly plant feeding and have chewing mouthparts.

**Mantodea**   The familiar praying mantis (Figure 2–12a) is in the order Mantodea, although in some textbooks it is placed under the Orthoptera, a close relative. Mantodea are general predators with chewing mouthparts.

**Blattodea**   Cockroaches are the most important insects in the Blattodea. The many species of plant-feeding cockroaches have chewing mouthparts. They are mostly known by the species that are household pests.

**Isoptera**   The Isoptera are mostly wood feeders with chewing mouthparts, and includes termites. Some tropical species cause damage to crops and the presence of termite mounds remove crop and pasture soils from economic use.

**Dermaptera**   The small order of Dermaptera contains the earwig. Some are plant feeding but most are predators with chewing mouthparts.

**Thysanoptera**   The Thysanoptera are a small order of insects called thrips (Figure 2–11a). They are mostly plant feeding, but a few species are predators, with rasping mouthparts. An unusual characteristic of this order is the presence of a fringe around the wings.

**Neuroptera**   The Neuroptera is a small order that contains the lacewings. The larvae are very good generalist predators (Figure 2–12b) with chewing pincerlike jaws.

**Hemiptera**   The Hemiptera contains the true bugs (which is actually an entomological term). Some textbooks include the true bugs in suborder Heteroptera, with the Homoptera (see below) forming a second major suborder. Many families are strictly plant feeding (Figure 2–11b); others have both plant feeding and predacious species (Figure 2–12c). All have piercing/sucking mouthparts.

**Homoptera**   The Homoptera is a large order of insects that are strictly plant feeding with piercing/sucking mouthparts. It includes aphids (Figure 2–11c), scale insects, whiteflies (Figures 2–11d and e), planthoppers, and leafhoppers. The taxonomy of this group is currently under review; based on new evidence, some are suggesting that all organisms in the order Homoptera should be included in the Hemiptera.

**Coleoptera**   The order Coleoptera includes the most diverse group of living organisms. This large order contains the beetles and weevils. Many are plant-feeding (Figures 2–11f and g) and predacious species (Figure 2–12d), and all have chewing mouthparts. The larvae are often called grubs.

**Lepidoptera**   The Lepidoptera contains the butterflies and moths. They are mostly plant feeding; the larvae have chewing mouthparts and the adults have a coiled structure adapted for sucking or slurping (Figure 2–10d). The larvae are frequently called caterpillars or more commonly worms (Figure 2–11h).

**FIGURE 2–11** Examples of pest arthropods from different orders. (a) Western flower thrips; (b) lygus bug nymph feeding on broccoli floret; (c) large colony of aphids on a rose stem; (d) whiteflies on a squash leaf; (e) closeup of a whitefly adult; (f) spotted cucumber beetle; (g) cotton boll weevil; (h) cotton boll worm larva; and (i) colony of two-spotted mites. Sources: photographs (a), (f), and (i) by Jack Clark, University of California statewide IPM program; (b) by Allen Cohen, USDA/ARS; (c) and (d) by Robert Norris; (e) and (h) by Scott Bauer, USDA/ARS; and (g) by USDA/ARS, photographer not identified.

**Diptera**   Flies and mosquitoes are dipterans, which are characterized by having only a single pair of wings, hence the name Diptera (see Figure 10–3a). The second pair of wings is reduced to small clubbed structures called halteres. There are many plant-feeding species as well as predacious or parasitic species. The larvae have rasping mouthparts; adults have rasping, slurping, or sucking mouthparts. The larvae are often referred to as maggots (see Figure 10–3b).

**Hymenoptera**    The Hymenoptera contains the wasps, ants, and bees. A few families are plant feeding, but most are predacious or parasitic (Figure 2–12e). All have chewing or chewing-lapping mouthparts. A characteristic of this order is the presence of a narrow, constricted waist between the thorax and the abdomen in some of the most common families.

**FIGURE 2–12**   Examples of selected beneficial arthropods. (a) Praying mantis, here laying an egg mass (ootheca) on a wooden beam; (b) lacewing larva; (c) bigeyed bug; (d) ladybird beetle devouring an aphid; (e) parasitic wasp laying an egg in a gypsy moth caterpillar; (f) orb spider; and (g) predatory mite feeding on two-spotted mite egg. Sources: photograph (a) by Robert Norris; (b) by Jack Dykinga, USDA/ARS; (c), (d), and (g) by Jack Clark, University of California statewide IPM program; (e) and (f) by Scott Bauer, USDA/ARS.

***Arachnida***   The class Arachnida includes arthropods with no antennae or gills, typically with four pairs of legs, and a fused cephalothorax. This class includes about eight orders, of which the following contain the species which are of greatest economic importance. The Araneae contains the spiders, which are often generalist predators (Figure 2–12f). The Acarina contains the mites, which are both phytophagous pests (Figure 2–11i) or predaceous beneficials (Figure 2–12g), and the ticks. The Phalangida or opilionids are the daddy longlegs, and the Scorpionida contains the scorpions.

***Crustacea***   Arthropods in the Crustacea class have two pairs of antennae, a fused head and thorax called a cephalothorax, and most are aquatic and breathe with gills. The class includes shrimps, lobsters, and crabs. Examples of pest species include sowbugs (Figure 2–13a) and crayfish (Figure 2–13b).

***Symphyla***   The Symphyla is a class that includes wormlike or centipedelike species that have simple antennae, and from 10 to 20 pairs of legs. They are typically soil-inhabiting organisms.

Additional classes include Diploda (millipedes) and Chilopoda (centipedes). They are not usually considered to be pests.

The term *pest* is frequently used to mean insect pest, but "insect" is often omitted. This use conflicts with the definition of pest used in this book, and often results in confusion. The pest definition adopted here, one that includes pathogens, weeds, nematodes, mollusks, vertebrates, and insects, is the most compatible with the IPM concept.

Most people not involved in pest management have a low tolerance for arthropods of any kind. The aversion to insects is termed entomophobia, and some even refer to the phenomenon as irrational entomophobia. Entomophobia leads to the perception of all insects as pests. As discussed, most pathogens, plants, nematodes, and vertebrates are typically not pests; likewise, most insects are in fact *not* pests. Most arthropods are simply part of the ecosystem, are useful to human endeavors, and are classed as beneficials (see Chapter 13) or

**Most arthropods are beneficial. They eat other arthropods, assist in decomposition/scavenging, pollinate many plants, and produce honey and silk. They are key components of ecosystem function and should be preserved.**

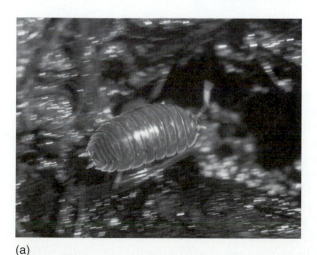

(a)

(b)

**FIGURE 2–13**   Examples of crustaceans. (a) Sowbug or pillbug; and (b) crayfish emerging from a rice paddy. Source: photographs by Robert Norris.

contribute to ecosystem processes (e.g., pollinators, decomposers). In reality, only a relatively few arthropod species harm crops or otherwise interfere with human activities.

## Special Problems Caused by Arthropods In Relation to Plants

The following problems are in addition to those listed under general impacts at the start of the chapter. Arthropod impacts on crops fall into four major categories.

1. Economic impacts in addition to those mentioned earlier in the chapter, include
   1.1. Mining within the layers of cells of the mesophyll of leaves (leaf miners).
   1.2. Chewing or boring into grain, tubers, or other plant products in storage. It is estimated that about 20% of stored products are lost annually to arthropod feeding worldwide, 10% in North America and Europe and up to 30% in developing countries.
   1.3. Protecting other insects that feed on the plant from their predators (ants that protect aphid colonies).
   1.4. Excreting liquid frass (honey dew) that accumulates on leaves and serves as a medium for growth of black, sooty mold. Heavy black mold blocks light for photosynthesis (scale insects, aphids).
2. Environmental impacts. Environmental or ecological impacts of insects result mainly from human interference in normal ecosystem level processes, or through actions taken to control pest insects.
   2.1. The uncontrollable invasive species have been particularly noticeable in forest ecosystems (e.g., gypsy moth).
   2.2. Invasive vectors of new diseases include the example of the smaller European elm bark beetle that invaded the United States in the early 1900s. The fungus *Ophiostoma ulmi* is the causal agent of the Dutch elm disease and is vectored by the bark beetles. Most stands of American elms in the eastern United States were destroyed by the disease.
   2.3. Contamination by insecticides used to control pests include DDT and other chlorinated hydrocarbon insecticides which were responsible for the decline of many birds species populations in the 1940s and 1950s.
   2.4. Unforeseen effects of introduced, exotic natural enemies occur on nontarget native species (see Chapter 13).
   2.5. Nesting structures include ones built by certain tropical termite species. They build hardened earthen nests that reach 4 to 5 feet above the ground. These nests may be so abundant that entire regions are rendered worthless for agricultural production
3. Health and safety impacts. Of the many health- and safety-related impacts of arthropods, most are irrelevant to agriculture and are not considered here.
   3.1. Arthropods can annoy people and domestic animals. They can cause severe discomfort to people working in crops. Grape leafhopper infestations can result in workers refusing to enter the vineyards, as the insects are inhaled when present in large numbers. High fly populations in a dairy operation may cause reduced milk production.

# VERTEBRATES

## Description of Organisms

Vertebrates are animals with a backbone (Figure 2–14). Some are herbivores, some are carnivores, and many are omnivores. Herbivorous vertebrates are often extremely serious pests in some systems. The lack of development of arable agriculture in the African continent prior to colonial times has been attributed to elephants (Parker and Graham, 1989; Barnes, 1996; Hoare, 1999). Humans could not stop elephants from destroying crops or storage facilities. Loss of entire villages as a result of activity of these "pests" was probably a common occurrence. The following groups of vertebrates include those deemed to be pests.

***Reptiles*** Reptilian pests include some lizards and snakes, crocodiles, and frogs. These pest species are mostly tropical. Most reptiles are not herbivores, and usually do not cause agricultural problems.

***Birds*** The ability to fly, the formation of flocks (Figure 2–14a), and their relatively long life make certain birds serious pest problems. Information about losses is "annoyingly vague" or nonexistent in many cases (Anonymous, 1970). Blackbirds consume enormous quantities of corn in the Midwest, and starlings cause major crop losses in the United Kingdom. The horned lark can eliminate a stand of newly emerged crops such as lettuce and sugar beets, necessitating the replanting of fields. Birds are a particularly serious problem in home gardens or small truck-farming operations.

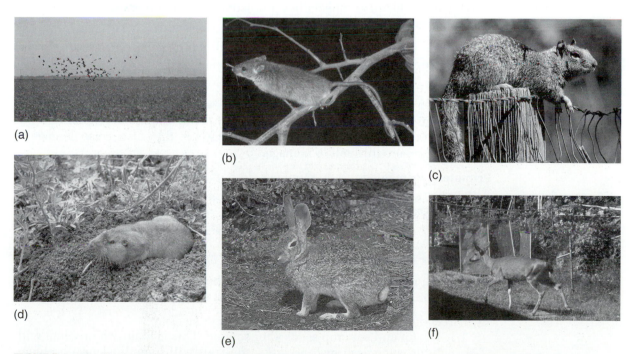

(a)

(b)

(c)

(d)

(e)

(f)

**FIGURE 2–14** Photographic examples of vertebrate pests. (a) Flock of birds over a wheat field, (b) roof rat on a tree branch, (c) California ground squirrel, (d) pocket gopher, (e) cottontail rabbit, and (f) deer in a garden.
Sources: photographs (a), (b), (c), (d), and (e) by Jack Clark, University of California statewide IPM program; (f) by Robert Norris.

***Mammals***   Mammals are a large, diverse group of vertebrate animals characterized by live birth of the young. Many important pest species are in several different orders.

**Marsupialia**   A group of animals that are born alive, but not fully developed, are marsupials. Development of the young is completed within a marsupium, or pouch, on the female. Pests in this order include opossums and kangaroos, which are insectivores and herbivores, respectively.

**Rodentia (Rodents)**   Certain mammals (rodents) are characterized by their gnawing front teeth, such as rats (Figure 2–14b), squirrels (Figure 2–14c), chipmunks, marmots, prairie dogs, mice, voles, gophers (Figure 2–14d), and beavers.

**Chiroptera**   The order Chiroptera includes some of the few truly flying mammals—the bats. Many species are predacious and important biological control agents. Others are fruit feeders and may cause damage to fruit crops.

**Lagomorpha**   Lagomorphs or Duplicidentata are essentially rodents with a second pair of small incisor teeth, with the main pair growing just behind them. Examples include cottontail rabbits (Figure 2–14e) and jackrabbits (hares).

**Perissodactyla**   The relatively large-size mammals of Perissodactyla have a characteristic tooth structure for chewing forages, and have an odd number of toes encased in hooves (horses, tapirs, rhinoceroses).

**Arctiodactyla**   The Arctiodactyla are large mammals, usually with an even number of toes. The order includes pigs, deer, cattle, goats, sheep, hippopotamuses, and camels. Deer (Figure 2–14f) can cause considerable crop loss through direct feeding, especially a problem for small farmers and backyard gardeners in rural areas.

**Carnivora**   Carnivores have well-developed canine teeth. The brain is also well developed. Most are predators but may also use food of plant origin (dogs, cats, bears, raccoons, weasels, mongooses, hyenas, seals, walruses).

**Proboscidea**   The large animals of Proboscidea are characterized mainly by the extension of the nose into a long proboscis or trunk. Elephants, as herbivorous animals, can be serious pests. Due to their large size they also cause damage by trampling and are difficult to manage.

**Primates**   Primates include monkeys, lemurs, gorillas, chimps, and humans. Although monkeys can be a nuisance pest in certain areas, humans are at times the worst pest through vandalism, malicious or gratuitous damage, and theft.

## Special Problems Caused by Vertebrates

Due to their relatively large size and intelligence, vertebrate pests create several problems unlike those attributed to other pest categories.

1. Direct damage occurs to crop in addition to feeding.
   1.1. Dig up and eat crop seeds after sowing. This is mainly a problem with birds, such as crows and larks; squirrels will sometimes be a problem. The result is the need to replant fields, which is both a monetary cost and a cost in reduced length of crop season.

1.2. Debarking trees (see Figure 6–7). This severe problem leads to death of the trees. Examples include damage done by deer, ground squirrels, rabbits, and mice.

1.3. Eat fruits and seeds. Birds and fruit bats can devastate ripe crops such as plums and cherries. Blackbirds in the Midwest United States have been estimated to cause an average loss of 4 bushels of corn per acre. Mice can forage in grain fields and carry entire ears away to their food cache.

2. Physical damage occurs from the activities of rather large vertebrate pests.

2.1. Creating soil mounds. Their burrowing activity causes problems for equipment, especially harvesters.

2.2. Trampling, pulling plants up. This is a serious problem with large animals such as elephants and hippopotamuses.

2.3. Digging to obtain food, such as roots or grubs. Skunks on golf courses looking for white grubs in the turf are a good example of this type of problem.

3. Transportation and dissemination of other pests occurs when rodents and birds cache weed seeds. Birds can eat weed seeds that are later deposited in feces at a different location.

4. Burrowing animals such as gophers and rabbits construct underground tunnels, burrows, and warrens. This activity can make surface irrigation difficult and can lead to failure of flood-control levees. The importance to flood control cannot be overstated, as levee failures pose a high cost to society.

5. Vandalism can be a serious problem in parks and golf courses. This type of damage is difficult to manage; excluding the culprits can be one solution. This is the only instance when the "pest" can be taken to court and tried.

Many of these problems have been increasing in recent years due to conservation programs that result in increased habitat suitable for vertebrate pests.

## SOURCES AND RECOMMENDED READING

All general textbooks for each pest discipline provide much greater detail than we have presented here. Titles of such texts are provided in the general reference list at the end of this book.

Anonymous. 1970. *Vertebrate pests: Problems and control.* Washington, National Academy of Sciences, National Research Council (U.S.), Committee on Plant and Animal Pests, Subcommittee on Vertebrate pests, p. 153.

Barnes, R. F. W. 1996. The conflict between humans and elephants in the Central African forests. *Mammal Review* 26:67–80.

Giampietro, M., G. Cerretelli, and D. Pimentel. 1992. Energy analysis of agricultural ecosystem management—Human return and sustainability. *Agric. Ecosyst. Environ.* 38:219–224.

Hoare, R. E. 1999. Determinants of human-elephant conflict in a land-use mosaic. *J. Appl. Ecol.* 36:689–700.

Holm, L. G. 1971. The role of weeds in human affairs. *Weed Sci.* 19:485–490.

Holm, L. G., D. L. Plucknett, J. V. Pancho, and J. P. Herberger. 1977. *The world's worst weeds: Distribution and biology.* Honolulu: University Press of Hawaii, xii, 609.

Koenning, S. R., C. Overstreet, J. W. Noling, P. A. Donald, J. D. Becker, and B. A. Fortnum. 1999. Survey of crop losses in response to phytoparasitic nematodes in the United States for 1994. *J. Nematol.* 31:587–618.

Parker, I. S. C., and A. D. Graham. 1989. Elephant decline (part I). Downward trends in African elephant distribution and numbers. *Int. J. Environ. Studies* 34:287–305.

Sasser, J. N., and D. W. Freckman. 1987. A world perspective on nematology: The role of the society. In J. A. Veech and D. W. Dickson, eds., *Vistas on nematology*. Hyattsville, Md.: Society of Nematologists, 7–14.

# 3

# HISTORICAL DEVELOPMENT OF PEST MANAGEMENT

## CHAPTER OUTLINE

## INTRODUCTION

Since the dawn of agriculture at about 10,000 B.C.E., severe food shortages and famines have been a constant threat to human populations around the world. Food shortages with the associated malnutrition increase susceptibility to disease, which has been largely responsible for the very slow growth of human population during most of history. The potential for such catastrophic loss still exists. The greatest crop losses were caused by inclement weather and diseases. There were, and still are, periodic insect outbreaks, such as locusts, armyworms, stem borers, and planthoppers, that cause problems from time to time (Figure 3–1). The early literature of China and Japan contains records of insect pest outbreaks and methods for their control. Weeds have always been an important problem in agricultural production, but probably never were a cause of famines, because weeds can be controlled by manual labor. Vertebrate pests were an ever-present source of food loss.

Major improvements in the knowledge about pests occurred over the past 200 years. Much of the technology used in current pest management is the result of advances over the last 60 years or so, and was not available prior to 1940. Current developments in molecular biology and precision agriculture will provide technological advances that promise to further revolutionize pest control

**FIGURE 3–1**
Locust swarm (about
3 to 6 km² ) milling
over a harvested millet
field in Bombay.
Source: photograph by
M. de Montaigne, FAO.

tactics. These are discussed throughout and revisited at the end of this book. This chapter provides a brief review of historical developments in biology and agriculture that have influenced pest management.

## ANCIENT TIMES (10,000 B.C.E. TO 0)

### Summary of Events

During this period, people in different parts of the world domesticated the major crop species, such as wheat, barley, oats, rice, maize, soybean, beans, potato, sweet potato, yam, manioc, date palm, coconut, sorghum, grapes, and cotton. Crops were grown in fields, in soil that was tilled, mainly with large hoes. Tilling must have been a major chore, and may explain the biblical account of Adam being cursed by God: "And unto Adam he said, 'Because thou hast hearkened unto the voice of thy wife, and hast eaten of the tree, of which I commanded thee, saying, Thou shalt not eat of it: cursed is the ground for thy sake; in sorrow shalt thou eat of it all the days of thy life'" (Genesis 3:17). The chore must have been made a bit lighter when simple plows were introduced about 4000 B.C.E. Harvested produce, mainly cereal grains, was stored for various lengths of time. Both spoilage and destruction by rats, mice, and insect pests were constant problems.

Lacking understanding of the biology of crop production and associated pests, humans were subject to the vagaries of nature for the whole of this period. Lack of food, due to the depredations by pests, was a major constraint to human population growth. There are numerous historical accounts of pest problems during this time, and pest outbreaks were often considered as retribution visited on humans by angry gods.

Insects have been a problem for humans since the beginning of crop domestication. Plagues of locusts and other insects occurred, but there was little one could do about them. Unable to control these pests, people prayed for relief or resorted to folk remedies, which were often useless. Insect problems were usually cyclic, however. If people could store food, then they could survive a bad year; if not, then their family, tribe, or village starved to death.

**FIGURE 3–2**
Hoeing weeds in corn in Central America; the man with hoe is symbolic of agriculture and the need for weed control.
Source: photograph by Marcos Kogan.

Pathogens also damaged and killed crops, and, like insects, most could not be controlled. Diseases dictated what crops people could grow, and because plant disease is a function of environment, plant diseases could even influence where people lived.

The consumption of stored grain products by vertebrates, especially rats and mice, was especially important. Rodents living in proximity to humans created other problems, because they acted as hosts for vectors of human diseases. The bubonic plague was a devastating disease transmitted by a flea from house rats.

Weeds were different from other pests. They could be considered unique, because throughout history weeds have been one type of pest that could actually be controlled, even in ancient times. In fact, manual weed control was absolutely necessary because weed seed banks in the soil resulted in weeds being present each year. If weeds were not controlled, production of staple crops would decrease substantially. For millennia, the image of the farmer was one of endless weeding. Man, woman, and child holding a crude hoe symbolized human farm labor, and still does today in some regions of the world (Figure 3–2). Weed management, to a great extent, dictated how many people were needed to successfully produce sufficient food and fiber from a given area of land. This is still true in many less developed countries (see Figure 2–5). In effect, farmers could farm only as much land as they had family members to weed it.

## Important Dates During Ancient Times

| | |
|---|---|
| 4700 B.C.E. | The Chinese domesticate the silkworm, an indication of understanding basic knowledge of how insects grow and reproduce. |
| 3000 B.C.E. | Rats are an ever-present problem. The following is a quote from Indian scripture: |

> O Ishwini, kill all the burrowing rodents which devastate our food grains, cut their heads, break their necks, plug their mouths so that they can never destroy our food. Rid mankind of them.

Mousetraps were already in use in the Turanian civilization.

| 2500 B.C.E. | Insecticidal properties of sulfur are discovered, and the element is used to control insects and mites by the Sumerians. |
| 1000 B.C.E. | Egyptians and Chinese learn of "botanical insecticides" to protect stored grains. These plant-derived chemicals have insecticidal properties, such as nicotine and pyrethrin (see Chapter 11). |
| 370 to 286 B.C.E. | Theophrastus, the father of botany, observes differences in disease epidemiology between crops in low areas and those on a hillside. This observation leads to the concept of growing susceptible crops on the slopes rather than in the valley. We can now attribute this differential susceptibility to changes in drainage and humidity. Theophrastus also recognized the existence of weeds, but discussed no control methods other than hoeing or pulling. Except for minor additions there were essentially no further advances in understanding pests and their management for the next 2,000 years. |
| 250 B.C.E. | Idea of the above-ground granary with "mushroom" posts is developed to stop rats and mice from climbing up into the stored grain (Figure 3–3). These are still in use in many regions of the world. |
| 0 | The Old Testament describes many pest problems, but includes no discussions of control: |

Tares are weeds.

Blasting and mildew are diseases.

Plagues probably refer to locusts. Other arthropods specifically mentioned in the Bible are ants, bees, fleas, flies, grubs, lice, maggots, mole crickets, moths, scorpions, spiders, and worms (in general).

Vermin are rats and mice.

The fiery serpent was probably a human-parasitic nematode.

**FIGURE 3–3**
Granary in Switzerland built on "mushroom stones" to deter rodent invasion.
Source: photograph by Robert Norris.

# C.E. 1 TO MIDDLE AGES

## Summary of Events

Not much is known about agricultural developments during this period, because little was recorded. What is known, however, is that plagues of rats, cockroaches, locusts, and disease epidemics leading to famines were frequent occurrences.

Although people recognized the importance of pests, the lack of understanding, even of basic pest biology, limited their ability to develop control tactics. Without the tools to see microscopic organisms, and lacking an understanding of genetics, the progress was minimal. Some of the early attempts to control nonweed pests were based on superstition. The Romans had a god of rust, Robigus, who was appeased with a feast each spring in the hope that this would reduce the amount of rust disease in cereals. The Romans also had a god of weeding, Runcina, who was supposed to make controlling weeds easier. Some societies used sacrifice, including human, as a means to keep pests away.

It is now assumed that such conditions as ergotism, also known as St. Anthony's fire, were constant problems. Ergots, hard black structures that replace the grain in cereals, especially rye, result when the fungal pathogen *Claviceps purpurea* causes the developing grain to form a sclerotium, which is actually an overwintering structure (Figure 3–4). When as little as 0.3% of rye grain is ergotted, the flour made from such grain can cause illness or even death in humans when eaten. The fungus produces compounds related to lysergic acid, which cause hallucinations, gangrene, and other afflictions. In medieval times this pathogen was common on cereal crops.

In 1691 to 1692, ergot may well have been a contributor to why people were thought of as witches in Salem, Massachusetts. The people suffering symptoms of bewitchment could have literally been high on ergotted grain. An outbreak of ergotism occurred as recently as the 1920s in England and even more recently in France. The fungus is still thought to cause illness in cattle and may account for unexplained illness in humans. (How would a person in an industrialized country today know if they were eating flour made with ergotted grain?)

**FIGURE 3–4**
Cereal rye head showing large black ergotted grain due to infection by the fungus *Claviceps purpurea*. The ergot grain is replaced by the fungus and contains toxins related to lysergic acid.
Source: photograph by Jack Clark, University of California statewide IPM program.

**FIGURE 3–5**
"Bridges" between Citrus trees encourage predacious ants. This biological control technique was developed by the Chinese about A.D. 800.
Source: modified from Olkowski and Zhang, 1998.

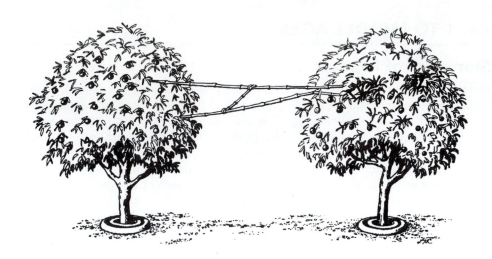

## Important Dates

| | |
|---|---|
| 400 | In China, silkworm droppings are used for control of rice pests. |
| Circa 800 | The connection is made between crop phenology (developmental stage) and insect attack, which leads to the concept of crop planting date as a means to avoid an insect problem. Another important development in China during this time is the use of ants for biological control in Citrus. The Chinese even constructed "bridges" of bamboo between trees, so the ants could readily move from tree to tree (Figure 3–5). Ant nests were sold near Guangzhou (Canton) for biological control purposes. |
| 1100 | In China, soaps are used to control some insects. |
| circa 1200 | Albertus Magnus observes that mistletoes are parasites, and finds that the "sick" host recovers when the parasite is removed. His discovery was not recognized and, as noted by Horsfall and Cowling, humans were doomed to another 650 years of suffering due to plant parasites. |
| 1476 | In Berne, Switzerland, cutworms are taken to court, pronounced guilty, excommunicated by the archbishop, and then banished. This procedure demonstrates that pests were recognized, but people were desperate without a method to control them. |

## SEVENTEENTH CENTURY

### Summary of Events

The seventeenth century saw the beginnings of the scientific method and the invention of scientific equipment, such as the microscope, which permitted major advances to occur. That which was discovered, however, was often misinterpreted. Control of most pests remained at the folklore level. An important advance was the concept that insects arose from eggs rather than spontaneously from decaying matter. The seventeenth century also saw an increased need for agriculture due to the expansion of urbanization. The improvement of agriculture permitted increased relocation of populations to more urban settings, creating the need for further advancements in agricultural production systems. A shift toward greater urbanization first started to occur in Europe.

## Important Dates In the Seventeenth Century

1640s        van Helmont demonstrates that plants do not consume soil to grow; he found that a 169 lb willow tree grown for 5 years in a pot had "used" only 2 oz of soil. This was a major conceptual step in the understanding of how plants grow, although at the time he thought that the tree must have "sprung from the water." It was not until 100 years later that Priestly and other workers realized that plants took up $CO_2$ and gave off oxygen, eventually leading to the concept of photosynthesis.

Mid 1600s    The invention of the compound microscope leads to the discovery of causal agents of many diseases. Having the ability to see the organisms, however, did not necessarily explain their roles (see below). Some very small organisms, such as viruses, were not observed directly until the invention of the electron microscope (1930s).

            About this time, farmers in France realize the connection between the barberry bush and incidence of rust disease in wheat, leading to legislation requiring removal of barberry bushes for disease management. This was probably the first example of what is now called areawide IPM.

1665         Robert Hooke observes the teliospores of a rust fungus for the first time. Hooke mistook what he saw to be the result of the disease, not the cause. This error persisted for nearly 200 years.

1666         Encyclopedias of natural history are published in Japan with abundant information on insect biology and descriptions of species.

1683         van Leeuwenhoek discovers bacteria, although it was almost 200 years later when their role was correctly elucidated.

1690         Infusions of tobacco are discovered to be insecticidal. Arsenic is used as an insecticide. Although toxicity problems were encountered, people continued to use arsenicals well into the twentieth century as fear of hunger is more powerful than fear of poisons.

# EIGHTEENTH CENTURY

## Summary of Events

Carl Linnaeus (1708–1778) introduced the Latin binomial system of nomenclature for all living organisms, and thus greatly facilitated progress in pest control. Without a standardized name for a pest, it was not possible to store and retrieve the available information on its biology and methods for its control.

    Two additional discoveries occurred that further advanced the foundations for the scientific control of arthropod pests. The connection was made between heat summation, or "degree days," and insect development, which is a key to understanding rates of individual growth, development, and reproduction. These are key factors in regulating population dynamics (see Chapter 5). The existence of plant natural defenses against insects was also recognized, which led to increased development and use of botanical insecticides such as pyrethrin, rotenone, and nicotine. These botanical insecticides are still in use. The realization that plants could be resistant to insects coincided with world exploration, resulting in collection of plant material that had insecticidal properties.

The first real advance in weed control occurred when Jethro Tull invented the rotary seed drill. The drill allowed crops to be planted in straight lines, which permitted improved cultivation between the rows with horse-drawn equipment.

Plant-parasitic nematodes were first observed during this time, although knowledge of the causal organisms of plant diseases was minimal in this century. It was not until the nineteenth century that major advances in understanding these organisms started to occur.

## Important Dates in the Eighteenth Century

1729   Micheli notes that dust from a diseased melon causes the disease to develop on previously disease-free melons. He thought the dust was "seeds" of the fungus.

1731   *The New Horse Houghing Husbandry* is published by Jethro Tull, and provides a major step in the evolution of weed management. His invention was not widely adopted for almost another century, however. Once adopted, Tull's invention accelerated the change from human energy to animal energy in agriculture, particularly for planting and weed control. The reduction in human labor needed for the control of weeds became a factor associated with the industrial age. People freed from the weeding chores in the fields were able to move to the cities.

1743   Needham first observes plant-parasitic nematodes in galled wheat kernels (see Figures 2-7d and e).

1755   Tillet notes that smut disease is worse if he inoculates wheat with smut "dust," and that treating seeds with copper sulfate before planting reduces the disease. The fungus causing the disease was later named after him (*Tilletia* causes covered smut, or bunt, of cereals). This was the first unequivocal demonstration of a contagious plant disease, although Tillet thought that a poisonous principle caused the disease, not the microorganisms. Tillet's work was not immediately accepted, however, and another 100 years would pass before the theory of spontaneous generation was finally rejected.

## NINETEENTH CENTURY

### Summary of Events

During the nineteenth century, societal assumption of control over the environment accelerated, as did the experimental approach to solving agricultural problems, with the establishment of agricultural experimental stations in Europe and North America. The Rothamsted Experimental Station in England was started with private funds in the 1840s to learn more about plant nutrition and the use of fertilizers. In the United States, land grant universities were started with public funding in the 1860s. Following the Hatch Act in 1887, each state established an agricultural experiment station within the land grant university system. Similar agricultural research systems were started in most of the developed countries early in the twentieth century. Scientific knowledge about agriculture, agricultural pests, and pest control began to advance more rapidly.

In 1845, M. J. Berkeley argued that an Oomycete, considered to be a fungus at the time, was the cause of potato late blight disease. This was the first convincing argument that a microorganism caused disease. The realization that fungi caused plant disease was a major advance in understanding diseases and their management. Discovery of bacterial pathogens was also made. In the 1860s, Louis Pasteur provided experimental evidence that convincingly demonstrated living organisms did not arise from nonliving matter, but rather, living organisms arise from other living organisms. Pasteur used heat to kill the microbes in his experiments, thus disproving the theory of spontaneous generation. Killing microbes with heat resulted in sterilization procedures that provided effective microbe control.

In 1876, Robert Koch discovered that he could take blood from a sick animal and transmit disease to a healthy animal, and that this process could be continued from one animal to another. In this way he proved that bacteria could cause disease. He formulated a set of criteria, now referred to as Koch's postulates (see margin note), which are still followed to demonstrate that a particular organism causes a particular disease. It is difficult to apply Koch's postulates if the pathogen cannot be cultured. Viruses cannot be cultured, which is one reason why viruses were still not considered to be pathogens even in the early twentieth century. Development of Koch's postulates in the nineteenth century should have put to rest the idea that diseases were spontaneous; however, the debate continued into the 1920s. Koch's postulates represented a major advance and provided a sound scientific foundation for establishing the causes of disease, and are still used today.

The latter part of the nineteenth century saw the initial examples of chemical therapy to control pests, and set in motion the whole concept of the "silver bullet" approach to pest control. The serendipitous discovery that a mixture of copper sulfate and lime controlled grape downy mildew was an important breakthrough for disease control. The first pesticide application equipment was developed in France (Figure 3–6) to apply this fungicide. These early findings contributed to the belief that a simple cure could be found for any major pest problem. This misconception persisted well into the IPM era and into today.

The invasion of pests, especially insects, into new regions started to become a serious problem because of increasing human travel and exploration of foreign regions. The introduction of the San Jose scale into the United States from China is an early example. The spread of pest organisms through human activity remains a most serious problem today, particularly with the ease and growing volume of global trade and tourism (see Chapter 10).

The first books devoted to control of insects and diseases were written during the nineteenth century. Weed management progressed very little during this period except for incremental development of cultivation equipment.

**Koch's postulates**
Step 1: Obtain the pathogenic organism from a diseased plant.
Step 2: Grow the organism in culture
Step 3: Inoculate the cultured organism back to a healthy plant.
Step 4: Establish that the inoculated plant becomes diseased and shows similar symptoms.
Step 5: Reisolate the same pathogen from the now-diseased inoculated plant.

**silver bullet**
Term applied to the thinking that an almost magical single cure can be found for a pest problem.

## Important Dates In the Nineteenth Century

1807            Prevost repeats Tillet's earlier work, and concludes that the spores actually produce the smut disease, and that environment alters the severity and rapidity of development of the disease. He was not believed and the French scientific

**FIGURE 3–6**
Application of
Bordeaux mixture was
(a) initially by hand
using a brush,
(b) sprayers had
already been
developed in France
by the late nineteenth
century.
Source: drawings copied
from Lodeman, 1896.

(a)

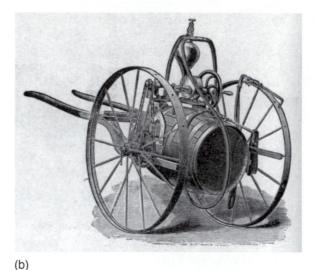

(b)

academy ruled that his conclusions were not acceptable; diseases were still considered "self-generating" and fungus was a result of the disease rather than its cause. Prevost actually observed with a microscope that dilute copper sulfate stopped the fungus spores from germinating, and concluded that copper controlled the disease. His observation and conclusion went unnoticed until Millardet rediscovered copper as a fungicide 70 years later.

1831    First record of a plant resistant to an insect pest. The apple variety 'Winter Majetin' is resistant to the wooly apple aphid, in England.

1845–1846    Devastating epidemics of late blight on potatoes in Ireland (and other parts of Europe) result in famine in Ireland and mass emigration to the United States. Berkeley argued convincingly that late blight was caused by a microbial pathogen (he considered it to be a fungus), but this concept was still not accepted.

1850–1860s    Berkeley observes root-knot nematode galls on the roots of greenhouse-grown cucumbers, bulb nematodes on teasel, and cyst nematodes on sugar beet roots.

1853    DeBary proves that fungi causes the rust diseases in cereals rather than results from disease. The scientific proof of a pathogen/disease association was an important conceptual discovery to support progress in the control of plant pathogens. Science was slowly winning over spontaneous generation.

1858    Charles Darwin and Alfred R. Wallace jointly present the concept of evolution by natural selection to the Royal Society in London.

1860s    Louis Pasteur helps demonstrate that spontaneous generation did not occur, and that heating could kill microbes.

Paris green (copper acetoarsenite) is introduced and widely used in the United States for control of Colorado potato beetle and other insects.

|  | The grape Phylloxera, a North American species of homopteran insect, is transported to Europe and becomes a serious threat to the French wine industry. This invasion leads to the first organized attempt to use legislative measures to prevent future pest invasions in Europe. |
|---|---|
| 1861 | DeBary proves that the pathogen causing the late blight disease of potatoes was *Phytophthora infestans*. |
| 1862 | The United States Department of Agriculture (USDA) is established. |
| 1866 | Gregor Mendel publishes his classic paper on genetics of population segregation in peas. Although not recognized at the time, he lays the foundation of modern genetics and all plant breeding for crop improvement, including pest resistance. |
| 1875 and later | Techniques are developed to permit microorganisms to be cultured. This technology leads to development of Koch's postulates, which are still used today. The postulates are a set of criteria that must be met in order to accept that a particular pathogen is the causal agent for a particular disease (see margin note). |
| 1878 | Downy mildew of grapes reaches France from America. This biological invasion results in a devastating disease epidemic on grapes and almost destroys the French wine industry. |
| 1878 | First demonstration that bacteria cause diseases in plants; fire blight disease of pear, discovered by Burrill in the United States, is caused by a bacterium; anthrax was found in cattle 2 years earlier by Pasteur and Koch. |
| 1882 | Millardet, an observant grape researcher in France, notices that grapes treated with a mixture of lime and copper sulfate to deter pilferers have less downy mildew than adjacent rows that have not been treated. |
| 1885 | Millardet perfects a mixture of copper sulfate and hydrated lime to control the mildew. The mixture has been called Bordeaux mixture since that time, because it was first used in the Bordeaux region of France. This discovery gave great impetus to research on plant diseases, because of the possibility that they could be controlled. |
| 1882 | Ward, working on coffee rust, makes the first observation that a monoculture of a crop increases disease incidence. |
| 1886 | For the first time, Mayer transmits a disease (*tobacco mosaic virus*) by taking juice of one plant and injecting it into another. |
| 1888 | The Vedalia beetle is introduced into California from Australia for biological control of the cottony cushion scale that had been accidentally introduced into Citrus growing areas in 1868 (Figure 3–7). By 1880, the scale was threatening to eliminate the citrus industry. The Vedalia beetle importation and release was the first example of effective classical, biological control. This biocontrol system is still working today, although during a period of heavy pesticide use in the 1960s, the beetle was nearly eliminated from California groves. |
| 1890s | Several pathogens are shown to be vectored by insects. It was conceptually an important idea to know that insects could |

**FIGURE 3–7** The Vedalia beetle was introduced into California about 1887 to provide biological control of the Citrus cottony cushion scale. This was the first example of effective biological control with an imported insect.

Source: photograph by Jack Clark, University of California statewide IPM program.

| | |
|---|---|
| | transmit the disease-causing agents. The cotton boll weevil invaded southern Texas from Mexico. The weevil spread to the rest of the cotton belt in the following 30 years almost destroying the cotton industry in the South. |
| 1892 | Ivanowski observes that a disease-causing agent passes through a filter that stops bacteria. He erroneously concludes that the "agent" was a toxin. |
| 1898 | Beijerink shows that the disease agent with which Mayer and Ivanowski had worked could be killed by heat and thus must be a "contagious living fluid." He coined the term *virus* for it. |
| 1895 to early twentieth century | Arguments continue to rage about whether bacteria caused or were the result of plant disease. Record of this controversy is one of the best-documented scientific disagreements in the literature. |

## EARLY TWENTIETH CENTURY (1900 TO 1950)

### Summary of Events

By the turn of the century, five major pest control methods were in use, namely legislative, cultural, biological, genetic, and chemical. All have been refined over the years and only a few others were added, mainly for insect control. These technologies are described in more detail in Chapters 10 through 17.

Expansion of knowledge of crop and pest biology occurred with the increased numbers of scientists working in agriculture. Until the mid-nineteenth century, interested amateur scientists conducted most agricultural research. Only late in the nineteenth century had professional scientists been hired in the disciplines of entomology and plant pathology. By the 1920s and 1930s, yields of all crops in the developed countries started to increase dramatically and rapidly (see Figure 1–2 for the United States). This rise in yields is attributed to plant breeding for increased yield, increased use of fertilizers, and improving pest control.

The realization that disease resistance was inherited occurred early in the century, and rapidly led to advances in plant breeding for increased resistance to pathogen and arthropod pests.

A profound change occurred early in the century when people switched from human and animal energy to fossil fuel energy for mechanical weed control.

Toward the middle of the period the rapid development of chemically based pest control occurred following the discovery of synthetic organic chemical pesticides. The two decades from the mid-1940s to the mid-1960s were the epitome of chemical silver bullet thinking. The period was dominated by the belief that pesticides were *the* solution to pest management, and that a new chemical could be developed to solve any pest problem. This thinking persisted until the mid-1960s (see below), and has yet to entirely disappear.

## Important Dates In the Early Twentieth Century

| | |
|---|---|
| 1900–1920 | Development of the internal combustion engine occurs in the 1890s and is employed in farm equipment (tractors), effectively substituting petroleum energy for animal energy. Tractors were more powerful and quicker than horses (or other draft animals), and did not require the farmer to raise feed or provide care as he did for the animals. Using the new machinery, one person could farm much larger areas, meaning that fewer people were required for a farm to be productive. In the western developed world, mechanization accelerated the shift from agrarian to the urban society. |
| Early 1900s | Inorganic chemicals, such as copper sulfate, sodium chlorate, and iron sulfate, are used to kill plants. The problem was the treatments killed most plants, and the application rates were high. These chemicals were never widely used except in situations like railroad ballast. Fires along railroad tracks are a problem because the wooden ties and dead, dry weeds form excellent fuel and are ignited by sparks from brakes. Selective chemicals for weed control in crops such as wheat and barley also found limited use. |
| | Insecticide dusts (many based on lead arsenate) are widely used in crops such as cotton for boll weevil control, and apples for codling moth control. Problems included low efficacy, poor safety to workers and crops, and soil contamination. In some situations so much arsenic accumulated in the soil of treated orchards that trees replanted in the orchards died. |
| 1905 | The first discovery that resistance to diseases is inherited (e.g., rust on wheat) lays the foundation for breeding pest-resistant crops. |
| 1911 | First demonstration that there is host-range variability among isolates of a pathogen. Such variability is now the basis for defining races, pathovars, biotypes, or strains. |
| 1912 | Concept of control by legislative action is established in the United States with development of quarantines to prevent introduction of foreign pests. |

(a)

(b)

**FIGURE 3–8**   Photographs of prickly pear cactus in Australia (a) prior to release of the biological control moth *Cactoblastis cactorum,* and (b) the same area two years after release of biocontrol agent. This was the first example of successful weed control with an introduced insect.
Source: photographs reproduced with permission of the Department of Natural Resources and Mines, Queensland.

| | |
|---|---|
| 1913 | Riehm uses organo-mercury compounds as seed treatments to prevent diseases. Such treatments were widely used until the 1960s when the toxicity of mercury was recognized, and all uses were phased out. |
| | First comprehensive taxonomic studies of nematodes begin. |
| 1914 | First report is given of resistance in insects to an insecticide: the San Jose scale to lime-sulfur. |
| 1919 | The term *biological control* is introduced, based on the idea that natural predators and parasites could control pest organisms. |
| 1921 | The first aircraft application of a pesticide occurs (against Catalpa sphinx moth in Ohio). |
| 1926 | The first successful wide-scale biological control of a weed occurs through the introduction of *Cactoblastis* moth into Australia to control prickly pear cactus (Figure 3–8). |
| 1929 | First record is given of a nematode infecting an insect larva. |
| 1933 | Tanaka shows for the first time that pathogens release chemical compounds that result in disease symptoms on plants (filtrates of *Alternaria* caused spots on pears similar to those caused by the pathogen). It was not until 1952 that an actual toxin was isolated (bacterial wildfire toxin of tobacco). |
| 1934 | Tisdale discovers the dithiocarbamate fungicide Thiram. This discovery led to rapid development of many synthetic contact fungicides. It was not until over 30 years later in 1966 that a systemic fungicide was discovered. |
| 1930 to 1940s | First discovery occurs of organic chemicals that kill plants. The early ones (e.g., dinitro-ortho-cresol) were not selective and thus found limited use. Discovery of 2,4-D during World War II changed everything by allowing farmers to kill weeds in the crop without killing the crop (Figure 3–9). Selectivity of the herbicides is fundamental to modern crop production (see Chapters 9 and 11). The concept of selectivity was later |

**FIGURE 3–9**
Selective weed control with a phenoxy herbicide. Field was sprayed with 2,4-D, but the weedy area was not sprayed.
Source: photograph by Robert Norris.

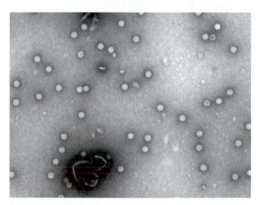

(a)

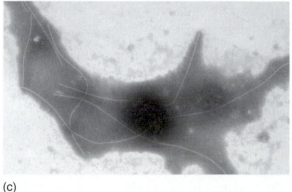

(c)

(b)

**FIGURE 3–10**
Modern electron micrographs of virus particles. (a) Icosahedral virions of *tomato bushy stunt virus* (TBSV); (b) rigid rods of *tobacco mosaic virus* (TMV) virions; and (c) long flexuous rod-shaped virions of *lettuce mosaic virus.*
Sources: photograph (a) by R. Harris, B. Falk, and R. Gilbertson, with permission; (b) and (c) by R. Harris and R. Gilbertson, with permission.

extended to insecticides with the introduction of the principles of integrated control.

1939    Virus particles are observed for the first time using an electron microscope. It was not until the 1950s that the protein coat and the infective nucleic acid inside were actually seen (Figure 3–10).

| | |
|---|---|
| 1939 | Swiss chemist Paul Mueller discovers insecticidal properties of DDT (although the compound had been synthesized during the nineteenth century). Mueller was awarded the Nobel Prize in 1948. The discovery of DDT, and of benzene hexachloride, marked the onset of the insecticide era. The absolute dominance of insecticides for insect control lasted until the 1960s when disillusionment set in because of problems documented by entomologists who pioneered the integrated control concept, as compiled by Rachel Carson in her book, *Silent Spring.* Killing insects had become a preoccupation rather than producing crops or maintaining environments. |
| 1943 | Walter Carter discovers the first effective fumigant nematicide. Nematicides allowed clear documentation of the crop loss caused by plant-parasitic nematodes and launched the era of chemically based nematode control. |
| 1946 | First report is given of insect resistance to the new synthetic organic chemical insecticides; houseflies resistant to DDT were observed in Sweden and Denmark. |

## LATE TWENTIETH CENTURY (1951 TO 2000)

## Summary of Events

A major change in thinking occurred rather early during this period following the realization that pesticides had serious limitations, and in some cases produced numerous undesirable side effects. The search for silver bullet chemical solutions to pest problems began to be replaced with a more balanced ecological view of pest control. This change continues to this day.

Resistance of all types of pests to pesticides was observed during this period, rendering many pesticides ineffective for controlling the intended target pest. The problems with pesticide resistance were initially observed, and were more severe in insect pests, but later resistance proved to be a problem for all types of pests, with the possible exception of nematodes. Pesticide resistance became so severe that all areas of pest management had to devote considerable effort to develop a new approach to the use of pesticides, called resistance management. Pesticide resistance is an ongoing problem and the topic is discussed in detail in Chapter 12. The phenomenon of pest resurgence and release of minor pests due to killing of beneficial insects was documented.

At about the same time that pesticide resistance became a problem, ecologists also observed that pesticides might have effects on nontarget species, such as eggshell thinning in several bird species and earthworm kills. The problem of biomagnification of organochlorine insecticides in the food web was recognized. So far there is no evidence of herbicide or nematicide biomagnification in food webs, but their movement through the soil profile into groundwater became an important pollution problem. The multitude of problems associated with careless pesticide use forced a change in the pest control paradigm from silver bullet to an understanding of the biology and ecology of crop-pest interactions. The biological and ecological information is used to design integrated management strategies that we now call integrated pest management (IPM).

The concept of integrated control was developed in the late 1950s and it consisted mainly of the use of insecticides in a manner that was compatible with biological control of insect pests. The concept was later expanded to encompass all tactics to control pests of all categories. Again, the expanded concept is what is now universally known as IPM. Implementation and acceptance of the IPM idea has been relatively slow, and numerous proposals have been made to increase adoption of the IPM paradigm. Emphasis on understanding pest biology and ecology has resulted in major improvements in the management of most pests. Attempts to restructure pest management on an ecological basis continue to be made around the world.

In 1953, Watson and Crick described the structure of DNA. Toward the end of the twentieth century, molecular biology presented new options for manipulating resistance to pests and pesticides and opened new avenues for pest management. Acceptance of the new technology, however, varies from little initial concern in the United States, to rejection in Europe and other parts of the world where a more cautious approach has been taken.

## Important Dates in the Late Twentieth Century

| | |
|---|---|
| 1950s | The use of antibiotics for bacterial disease control become widespread. Resistance to new classes of pesticides start to appear in insect populations (see Chapter 12). |
| 1950s to present | Chemical industry synthesizes many new pesticides. |
| 1953 | Watson and Crick describe the structure of DNA, which sets the stage for modern molecular biology with implications for plant breeding and genetic engineering. |
| 1955 | E. Knipling, of the United States Department of Agriculture, first describes a method for suppressing insect reproduction in the field by sterile insect release. |
| 1958 | Howitt, Raski, and Goheen first document virus transmission by a nematode, showing that *Xiphenema* is a vector of *grape fanleaf virus*. |
| 1959 | Karlson and Butenandt identify the first insect sex pheromone, for silkworm moth. Sex pheromones are key elements of the reproductive behavior of many pest insects and have been successfully used in the tactic called behavioral control (see Chapter 14). By the mid-1980s, over 1,000 sex pheromones had been identified. |
| 1962 | Rachel Carson writes *Silent Spring,* a book that brought the problems caused by pesticides to the attention of the public and forever changed the way pesticides are viewed by the general public and dealt with by scientists. |
| Late 1960s | Resistance to a herbicide is first documented in a weed population, and the number of weeds showing resistance increases rapidly in the next two decades, to the extent that resistance has now started to change the herbicide paradigm. First documented field resistance to fungicides also occurs. The problem became more widespread as new fungicides with site-specific activity were introduced over the next 30 years. |

| | |
|---|---|
| 1967 | Phytoplasma organisms are first observed in relation to a disease. Originally called mycoplasma-like organisms or MLOs, this term is now discontinued. |
| 1968 | Dale Newsom publishes a paper in which he is the first to use the expression "Integrated Pest Management systems." |
| 1969 | United States National Academy of Sciences formalizes the term and defines the principles of pest management |
| 1972 | The full expression—integrated pest management (IPM)—formally appears in a message from President Richard Nixon to Congress in which he transmits a program for environmental protection which leads to the creation of the United States Environmental Protection Agency (EPA). The discovery is made that some pathogenic bacteria could live in xylem or phloem. They are referred to as fastidious because they cannot survive outside a living organism. The first observation of spiroplasmas is also made at this time. |
| 1980 | Cause of crown gall of trees was determined to be due to bacterial DNA inserted into the plant genome by the crown gall bacterium *Agrobacterium tumefaciens,* causing overexpression of growth regulators. The ability of this bacterium to insert DNA into a higher plant genome formed the initial basis for genetic transformation of crop plants. |
| 1980s to present | Rapid development occurs of molecular biology and application of transgenic crops (also known as genetically modified organisms, or GMOs) to pest management. |
| 1994 | First transgenic squash varieties are released with resistance to viruses. First glyphosate-resistant transgenic varieties of corn and soybean crops are released. |
| 1995 | Release of insect-resistant cotton, corn, and potatoes using genes derived from *Bacillus thuringiensis* that encode endotoxin production. |
| mid 1990s to present | Societal concerns about engineered transgenic crops (GMOs) increase in several regions of the world, especially Europe. Public pressure dramatically slows adoption of the technology in IPM systems. |

The discerning reader may note the absence of historical information about the understanding or control of mollusks. This is due to the lack of information on these pests in any of the treatises dealing with the history of farming or pest management.

## SOURCES AND RECOMMENDED READING

The book by Evans (1998) is a highly readable account of the history of agriculture, much of which involved development of pest management, set in the context of world population development. Other books dealing with crop domestication and the history of the various pest disciplines are listed here.

Ainsworth, G. C. 1981. *Introduction to the history of plant pathology.* Cambridge; New York: Cambridge University Press, xii, 315.

Campbell, C. L., P. D. Peterson, and C. S. Griffith. 1999. *The formative years of plant pathology in the United States.* St. Paul, Minn.: APS Press, xvii, 427.

Carson, R. 1962. *Silent spring.* New York: Fawcett Crest, 304.

Chapman, R. F. 2000. Entomology in the twentieth century. *Annu. Rev. Entomol.* 45:261–285.

Dethier, V. G. 1976. *Man's plague? Insects and agriculture.* Princeton, N.J.: Darwin Press, 237.

Evans, L. T. 1998. *Feeding the ten billion: Plants and population growth.* Cambridge, UK: Cambridge University Press, xiv, 247.

Harlan, J. R. 1992. *Crops & man.* Madison, Wis.: American Society of Agronomy: Crop Science Society of America, xiii, 284.

Harlan, J. R. 1998. *The living fields: Our agricultural heritage.* Cambridge; New York: Cambridge University Press, xi, 271.

Kritsky, G. 1997. The insects and other arthropods of the Bible, the New Revised Version. *Am. Entomol.* 43:183–188.

Lodeman, E. G. 1896. *The spraying of plants; a succinct account of the history, principles and practice of the application of liquids and powders to plants, for the purpose of destroying insects and fungi.* New York; London: Macmillan and Co., xvii, 399.

Matossian, M. A. K. 1989. *Poisons of the past: Molds, epidemics, and history.* New Haven, Conn.: Yale University Press, xiv, 190.

Olkowski, W., and A. Zhang. 1998. Habitat management for biological control, examples from China. In C. H. Pickett and R. L. Bugg, eds., *Enhancing biological control: Habitat management to promote natural enemies of agricultural pests.* Berkeley, Calif.: University of California Press, 255–270.

Smith, R. F., T. E. Mittler, and C. N. Smith, eds. 1973. *History of entomology.* Palo Alto, Calif.: Annual Reviews Inc., vii, 517.

Timmons, F. L. 1970. A history of weed control in the United States and Canada. *Weed Sci.* 18:294–307.

# 4

# ECOSYSTEMS AND PEST ORGANISMS

## CHAPTER OUTLINE

▶ Introduction
▶ Ecosystem organization and succession
▶ Definitions and terminology
▶ Trophic dynamics
▶ Limiting resources and competition

*So naturalists observe, a flea*
*Hath smaller fleas that on him prey;*
*And these have smaller still to bite on 'em;*
*And so proceed* ad infinitum.

*(Jonathan Swift, 1667–1745)*

## INTRODUCTION

Robust ecological principles are essentially applicable to all living organisms, including pests and beneficials. For instance, organisms that require the same resources and coexist in an area will compete. An important biological or ecological principle may be relevant to one pest category in ways that are much different than for another. Broad, general statements about the applicability or value of an ecological principle relative to pest management are difficult to derive, because the specifics of each management situation are unique with respect to the pest categories present, the ecosystem, the human goals, economics, and societal implications.

The fundamental underpinning to all pest management is an understanding of the biology and ecology of the organisms that are to be managed. The biological features of organisms to consider relative to pest management include:

1. Taxonomy: the classification and naming of organisms;
2. Physiology: the metabolic processes of organisms;
3. Morphology and anatomy: physical structure of organisms;
4. Genetics: characteristics that are passed from generation to generation;
5. Reproduction and population dynamics: factors that determine population size; and
6. Ecology: how organisms interact with their biotic and abiotic environment.

In this chapter we concentrate on general principles. Information specific to individual pests can be found in textbooks for each pest category (Sources and Recommended Reading), and in locally developed management guidelines.

We cannot overemphasize that correct pest identification is an essential component of any pest management program. Identification is sufficiently complex that entire texts are devoted to the taxonomy of each pest category. The scientific Latin name of an organism is the key to the existing information on its biology. Botanical and zoological classifications are hierarchical and imply evolutionary relationships. The evolutionary relationships among organisms, as reflected in their taxonomic relationships, allow information on a known pest to be generally applicable to a new, closely related pest.

Physiological processes and anatomical features of different pests are likewise not covered in this text. Such information can be found in reference books that cover general entomology, botany, mycology and plant pathology, and vertebrate zoology. Genetics of pests and crops plays a major role in pest management; pest genetics is covered in Chapter 5, and Chapter 17 covers crop genetics.

Pest reproduction and population dynamics are key components to developing sound pest management and are presented in the next chapter. Understanding the ways pests participate in ecosystem processes is important to developing effective pest management and is discussed in this chapter.

**Knowledge of the pest biology is fundamental to developing a sound pest management program.**

# ECOSYSTEM ORGANIZATION AND SUCCESSION

Understanding the levels of ecosystem organization is fundamental to pest management. We do not consider levels of organization that are lower than the individual. The reader is assumed to have a general understanding of cell structure. The levels of organization that are important to pest management are shown in Figure 1–5:

*Individual:* a single organism such as a bacterium, a weed, a nematode, or a lygus bug

*Population:* all individuals of a species within a defined area, usually isolated to some degree from other similar groups, such as weeds or boll weevils on an acre of cotton field.

*Community:* the assemblage of all species occurring together within a limited geographic area. Within an agricultural field, the community would include the crop plants, weeds, pathogens, bacteria, fungi, nematodes,

**Organisms in a community interact with each other. The direction and intensity of the interactions are governed by both abiotic and biotic environmental factors.**

insects, and vertebrates. Although typically restricted to a specific geographic area, such as a crop field or a ditchbank, it is possible to refer to a community over a larger area, such as the alfalfa community in San Joaquin Valley, California. Communities are composed of taxonomic assemblages, the groups of all species of a particular taxonomic designation that occur together within a particular geographic area. For example, one might refer to all the insects that occur in a particular field as the insect assemblage, or the insect community, for that site.

*Ecosystem:* a community of organisms together with the abiotic environment in a specified region. The ecosystem is dimensionally undefined, in that ecosystem boundaries are specified relative to the objective of consideration and time scale. In the pest management context, ecosystems can be thought of as the crop fields and the surrounding ditches, hedges, woodlands, and so forth, within a defined area such as a valley or a watershed. Agricultural ecosystems are termed agroecosystems.

*Ecological regions or ecoregions:* The concept of ecoregions denotes areas of general similarity in ecosystems and in the type, quality, and quantity of environmental resources. Ecoregions are designated to serve as a spatial framework for the research, assessment, management, and monitoring of ecosystems and ecosystem components. Ecoregions are delineated by geographic characteristics such as climate, topography, soil, vegetation, and land use. A distinguishing factor in the concept of ecoregions is the inclusion of human settlement patterns and resource use in the delineation of ecoregions. Although the term has been in use since the mid-1970s, there has been a recent increase in interest in the concept and a tendency to adopt the ecoregion classification as the basis for resource management, assessment of ecological impacts, and even IPM. The vagueness of the spatial scales inherent in the concepts of ecosystem and watershed has been one of the reasons for the adoption of the concept of ecoregions (Omernik, 1995).

*Biosphere:* inclusive term often used to describe the global ecosystem, and represents the broad-scale assemblage of interrelated ecosystems that make up the biological world.

The processes that occur in agroecosystems greatly influence the development and management of pests. Ecosystems have dynamic inputs of energy and nutrients, and outputs represented by the production of living organisms including crop plants. Inputs and outputs may be described in terms of water, energy, and nutrient cycles. Examples of these cycles can be found in most general ecology texts (see Recommended Reading at the end of this chapter).

## Concept of Succession

Succession is the chronological sequence of successive species in a biotic community, from an open system with small permanent biomass, to a stable climax community. After a forest fire much of the established community of plants and animals die or are driven away. Slowly new life springs up; first invasive, quick developing plants are established. They are followed by shrubs and later by new trees. Animals return to the area as the plant assemblage is restored. This scenario illustrates the concept of succession within an ecosystem. After a fire, the return to the original tree cover may take many years, but at that point the community is said to have returned to its climax condition. The climax represents a relatively stable community in which the rate of extinction of species is in equilibrium with the rate of colonization by new species, given the existing environmental conditions.

The actual climax community that is realized in an area depends on factors such as seasonal average temperatures and rainfall, but in areas with low rainfall climax communities are typically grassland, and in areas with adequate or high rainfall climax communities are some type of forest. Deserts and tundra represent the climax when water and temperature limit organism growth and survival. The problem created by succession, from a human perspective, is that the net productivity of the system is zero when the climax vegetation is realized (Figure 4–1). Stated another way, when the climax condition exists, the gain in biomass from photosynthesis is balanced by the loss to respiration of all the organisms in the community.

Agriculture represents human attempts to arrest succession prior to a climax state, so that the net increase in biomass (energy) can be harvested for human benefit. Arresting succession requires human inputs to the system, and the greater the input, the earlier in the successional stage the system can be arrested, and the greater the return that humans can harvest (Figure 4–1).

The long-term continuous wheat experiment on the field named Broadbalk, at the Rothamsted Experimental Station in England, provides a dramatic example of succession (Figure 4–2a). The experiment was initiated in 1843 and is still underway. Wheat has been grown continuously since the inception (Figure 4–2b), and has required the typical human intervention of plowing, fertilizer application, and pest management, especially weed control. The typical annual harvest of a staple food grain has been sustained, even in the absence of additional fertilizer. In 1882, a section at the end of Broadbalk was abandoned. Lawes (1884), one of the researchers who created the Broadbalk experiment, told the 1882 wheat crop prior to harvest:

> I am going to withdraw all protection from you, and you must for the future make your own seed bed and defend yourself in the best way you can against the natives, who will do everything in their power to exterminate you.

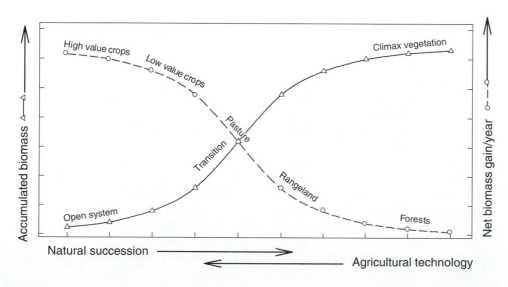

**FIGURE 4–1**
Diagram showing relationship between ecological succession and harvestable productivity per season.
Source: redrawn from Shaw, 1961; Flint and Van den Bosch, 1981.

**FIGURE 4–2**  The Broadbalk long-term experiment at Rothamsted in England. (a) Overview of experimental area with wilderness area abandoned in 1882 in background; (b) wheat on Broadbalk after 150 years of continuous cropping using nominal agricultural practices; (c) wilderness area from which trees were removed and that has been grazed since 1957; (d) wilderness area with no inputs since 1882 showing reestablished woodland.

Source: photographs by Robert Norris, August 2000.

Without human input, wheat ceased to grow after four seasons! Tree seedlings were grubbed from one section, and it has been grazed annually by sheep since 1957. It is now a pasture (Figure 4–2c) that supports grazing animals which provide meat and wool. Another section of the abandoned area has never been managed nor has it received any human input, and it is now mature woodland (Figure 4–2d). Apart from aesthetics and ecosystem processes, it has not provided any direct commodities for human consumption. A more complete description of the Broadbalk Wilderness (as it is called) is provided by Kerr et al. (2000).

Virtually all of the world's current crop plants would suffer the same fate as the wheat on Broadbalk if management of agroecosystems ceased. Even a reduction in the level of input has the potential to decrease the yield and economic return of high value, intensive cropping systems.

The intent of the human activity called agriculture is to manipulate nature to achieve the goals of human society. Pest management is a technology that humans use to achieve the goals by disrupting succession. Reestablishment of various plant communities, as noted, is the first step in succession because they are the bases of the food webs (see below). Most of the plants in the early successional stages of arable land are what people designate as weeds. Accordingly, weed management is a major component of the technology used to maintain ecosystems at the level of harvestable output required by society.

To keep nature from reaching equilibrium and thus to allow agricultural production, normal nutrient and energy cycles are altered by heavy subsidies of high-energy inputs. The levels of inputs required to achieve high yields from agroecosystems have raised the question of whether agricultural production as it is now practiced in industrialized countries is sustainable. A vigorous movement started in the late 1970s, known as sustainable agriculture, refers to agricultural practices that foster long-term production without deleterious consequences to agroecosystems or the biosphere. The development of sustainable agriculture requires that the biology and ecology of agroecosystems be considered and that production practices are planned accordingly. IPM is a key component of sustainable agriculture.

# DEFINITIONS AND TERMINOLOGY

To discuss pest management it is necessary to use terminology appropriate to the different pest organisms. The following terms are often used.

## Herbivore and Herbivory

A herbivore is an organism that destructively consumes plants or plant parts. Herbivory is the act of feeding on plants. A synonym of herbivory is *phytophagy* (from the Greek *phyton*, meaning "plant," and *phagos*, meaning "eats"). Plant-eating insects are referred to as phytophagous insects. Special modalities of phytophagy are phyllophagy (*phyllon* means "leaf" or "leaf eating") and xylophagy (*xylon* means "wood" or "wood eating"). Herbivores are usually mobile organisms, and an individual will typically feed on more than one plant during its life. Herbivores are considered to be pests when they cause damage to desirable crop plants. Generalist herbivores feed on many different plant species, and specialists feed on a single plant species or group of related species. Pathogens feed on plants, but because of the intimate relationship that they usually establish with a single host plant, they are considered parasites, not herbivores.

## Carnivore

A carnivore is an organism that eats other animals (herbivores or other carnivores). Like herbivores, they can be generalists or specialists.

## Omnivore

Organisms that feed on plants and animals are both herbivores and carnivores, and are therefore called omnivores. Most humans are in this category, so are many birds and insects, especially beneficial organisms. Omnivores feed at multiple trophic levels (see later in this chapter).

## Monophagous

The term monophagous applies to organisms that eat only one or a few species in a genus or in related genera. Monophagy is a desirable trait in most biological control agents. This is discussed further in Chapter 13.

## Oligophagous

Organisms that eat a limited range of usually related organisms are referred to as oligophagous.

## Polyphagous

Polyphagous organisms eat a range of taxonomically unrelated species, or even species in different trophic levels.

## Host

The term host describes organisms that serve as both a food source and habitat for parasitic organisms.

## Prey

Organisms that serve as a food source for predators are known as prey.

## Predators

Predators eat all or part of their prey, resulting in its death. Predators usually kill many prey individuals during their life cycle.

## Parasites and Parasitoids

Organisms that live in or on a host, from which they obtain nourishment, are called parasites or parasitoids. Parasites generally cause harm to their hosts, although their activities usually do not kill the host. In the process of developing in a host, parasitoids kill the host. Parasites and parasitoids are usually associated with a single host individual during the majority of their life cycle.

## Hyperparasites and Hyperparasitoids

Parasites of parasites are known as hyperparasites. They are sometimes called secondary parasites. From an arthropod management perspective, these organisms are a problem because they may destroy parasites that are acting as helpful biological control agents.

## TROPHIC DYNAMICS

## General Concepts

Trophic relationships describe the energy and food resources of organisms, and the way in which particular organisms serve as food for other organisms.

Trophic relationships may be considered from the perspective of organismal activities, or from the perspective of flows of energy through organisms. Trophic dynamics describe the temporal flux or organism numbers over time, or the flow of energy and resources through the ecosystem.

A food chain is a simplified depiction of a linear sequence of organisms in a community that feed on one another in successively higher trophic levels, and so the food chain describes the transfer of energy through those trophic levels. The idea of a simple food chain can be misleading, because it depicts exclusive feeding relationships among organisms. The trophic relationships and interdependence among organisms are better described as a food web, or a characterization of the multitude of possible "who eats who" relations among the different organisms in a community.

Food webs may be depicted as diagrams that specify the interconnections between the various organisms in the community at different trophic levels. The concept is a particularly useful way to analyze the position of different types of pests in the ecosystem, and to understand the relationships and connections between different categories of pests. Food webs and trophic dynamics show how management of different classes of pests is interrelated at the community/ecosystem level.

Environmental resources provide the inputs for all food chains (Figure 4–3). The major resources are water, mineral nutrients, energy derived from sunlight, and carbon dioxide. Green plants are the only organisms, with the exception of a few chemosynthetic bacteria, that can directly utilize mineral nutrients, carbon dioxide, and the sun's energy. Through the process of photosynthesis, green plants capture the energy in sunlight and use it to convert carbon dioxide and water into simple carbohydrates that form the building blocks for all other biochemicals needed for life. In the trophic dynamics terminology, plants are called producers, and are described as autotrophic (which literally means "self feeder").

All other organisms in the ecosystem, from bacteria to elephants, are consumers. They cannot make their food from the ecosystem resources, but rather they must eat something else in order to acquire the biochemicals and energy

**Plants are the base of food webs. Therefore weeds, as producers, are in a unique position in comparison with other pests, which are all consumers.**

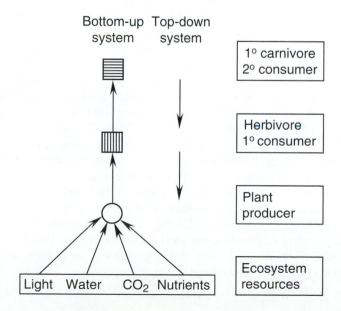

**FIGURE 4–3**
Diagram of simple theoretical food chain; energy and resources flow in direction of arrows.

that they need to live. They are called heterotrophs ("other feeders"). Consumers can be segregated into several different types depending on their food source. If a consumer feeds on plants it is called a herbivore (which literally means it eats herbs) and is considered to be a primary consumer (Figure 4–3). In the simple food chain depicted in the figure, a further trophic level is shown as a secondary consumer. Such organisms feed directly on herbivores. Because they are eating an animal, they are referred to as carnivores. Although the simple food chain depicted does not extend beyond the first level of carnivore, all organisms above this level in the food chain are also carnivores, although some, such as humans, are omnivores.

## Bottom-Up versus Top-Down Processes

**Populations of organisms may be regulated by interactions between trophic levels, and the outcome of interactions can be driven by producers (bottom-up) or by consumers (top-down) depending on the particular system.**

In Figure 4–3, the arrows between the boxes indicate a resource flow from producers to primary consumers to higher consumers. If the producers in the lowest trophic level (the primary producers) limit the population growth of the next higher level (the primary consumers), and so on through the food web, the populations of constituent organisms are said to show "bottom-up" dynamics. This means that plant resources regulate the herbivore populations, and that the herbivore resources regulate the carnivores and so on. Ultimately all ecosystems must function from the bottom up, at least part of the time, because plants form the base of the food web. Without plants there could be no consumers.

A top-down system is one in which the organisms in the highest trophic level (the top consumers) limit the population growth of the next lower level (their prey), and so on down the food chain. If higher-order carnivores, such as predators and parasites, are removed from a system, then their prey will increase, indicating that the herbivore or carnivore populations eat enough to determine the population dynamics of organisms lower in the food chain (Figure 4–3).

Whether systems function primarily in a bottom-up fashion or in a top-down manner is debated by ecologists, and much current research attempts to define how different types of systems function. Both top-down and bottom-up forces are probably acting together in many systems. Most ecological discussions focus on the interactions and dynamics that occur in natural ecosystems, and the top-down and bottom-up ideas have not been considered as thoroughly for disturbed agroecosystems. However, the concept of population regulation has considerable relevance to pest management. When weeds determine the outcome of a system it is a clear example of bottom-up dynamics. Two examples of top-down-driven systems in IPM are presented here. Many crops cannot be successfully grown in all regions of the world. This is typically attributed to problems related to climate. However, pests may be involved because ecological conditions may allow a pest organism to flourish and become so devastating that the crop can no longer provide an economic return. Pear trees are susceptible to the fireblight bacterium where the climate is wet and warm during the bloom period. The damage to the pear trees is so severe (Figure 4–4) that the crop cannot be grown economically in regions of the world with such a climate. The dynamics of the system, in this case pear productivity, is being driven top-down by the pathogenic bacterium. Biological control serves as a second example. When biocontrol is functioning well it can be argued that the system is in a top-down-driven mode, because the higher-level carnivore is determining how the system performs (see Chapter 13).

**FIGURE 4–4**
Severe damage to a pear tree in response to infection by the fireblight bacterium. The tree has been pruned to remove infected wood, which is the only control available after infection has occurred, resulting in almost complete destruction of the scaffold.
Source: photograph by Wilbur Reil, University of California IPM program.

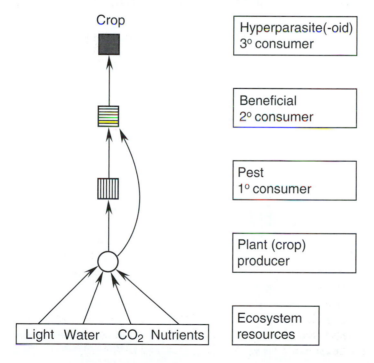

Crop

| Hyperparasite(-oid) 3° consumer |
| Beneficial 2° consumer |
| Pest 1° consumer |
| Plant (crop) producer |
| Ecosystem resources |

Light   Water   CO₂   Nutrients

**FIGURE 4–5**
Diagram of simple food chain with the terminology changed to reflect that used in pest management. Energy/resource flow is in the direction of the arrows.

# Basic Food Chain (Arthropods, Nematodes, Vertebrates)

Figure 4–3 represents the basic components of a simple food chain. Viewed from a pest management perspective, the terminology for the trophic levels is different from that used in a strictly ecological sense (Figure 4–5). The producer is the crop in agroecosystems. Herbivores are thus feeding on the crop, and are designated as pests because their feeding damage results in reduced crop vigor and yield. Reduced crop growth, vigor, and yield are referred to as crop loss. Carnivores that feed on pests are useful from the pest management perspective because they decrease the numbers of herbivores. They are predators or parasites that act as biological control agents of pest organisms, and such predators and parasites are referred to as antagonists or beneficials.

From a pest management viewpoint, a tertiary consumer can be very important. These are carnivores that prey on the second-level consumers, which are the beneficials. By such predation, tertiary consumers can limit the effectiveness of biocontrol agents. The role of both secondary and tertiary consumers is discussed in detail in Chapter 13.

The scheme in Figure 4–5 provides a reasonably accurate model for most arthropod pests, mollusks, nematodes, herbivorous vertebrate pests, and a few pathogens. It has little relevance to many pathogens, and does not apply to weeds. In the latter two cases a different model must be used to describe trophic relationships.

## Pathogen Food Chain

The food chain involving pathogens in plant-based cropping systems is different from those presented for the other types of pests (Figure 4–6). In most cases, and especially for aerial pathogens, there are few higher-order consumers that actively seek out and utilize the pathogen as a food source. The food chain thus terminates at the primary consumer level in such cases (secondary and tertiary consumers depicted in gray). There are some notable exceptions; for example, there are fungi that parasitize soil-borne plant-pathogenic fungi, and such parasitic fungi have potential in pest management. The horizontal arrows between two pathogens at the primary consumer level are intended to demonstrate the concept of competition between pathogens at the same trophic level. This principle is utilized for biological control of pathogens (see Chapter 13).

**FIGURE 4–6**
Diagram of simple food chain showing resource flow when a pathogen is the primary consumer.

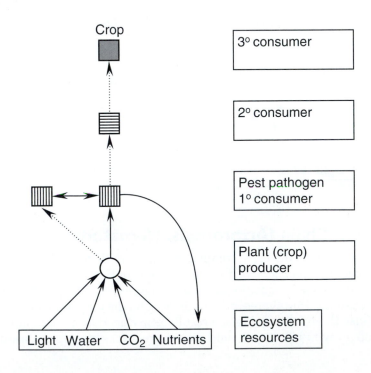

Crop

3° consumer

2° consumer

Pest pathogen
1° consumer

Plant (crop)
producer

Light   Water   $CO_2$   Nutrients

Ecosystem
resources

## Weed Food Chain

Unlike all other pests, weeds do not exert a direct physical or mechanical influence on the crop, because weeds do not eat crop leaves, flowers, or fruits. The influence of weeds is indirect in that they reduce the availability of ecosystem resources to the crop because they occupy the same trophic level as the crop (Figure 4–7). When a weed competes for and uses a resource that is available in finite amounts, such as nitrogen, the amount of that resource available for crop plant use is decreased.

The weed in the system depicted in Figure 4–7 has no direct effect on the food chain and consumer organisms supported by the crop. All interactions are driven by changes in crop growth due to altered resource availability. As a producer, the weed can also serve as a food source and support its own food chain. Organisms in the weed food chain that eat only the weed and do not feed on the crop, form the ecological basis of biological control of weeds. A biological control agent for a weed is a herbivore rather than a primary carnivore. When such herbivores are used in the biological control of weeds, it is imperative to ensure that they do not consume desirable plants. Thus, the intentional introduction of herbivores for the biological control of weeds is subject to legislative and regulatory restrictions (more about this in Chapter 13).

## Food Webs

The diagrams in the preceding sections depicted simple food chains for clarity of presentation. In reality, organisms usually do not feed on only one food source, nor does only one other organism feed on them. Rather, there are numerous possible feeding connections between organisms in different trophic levels. The interconnections may be described and graphically depicted as food

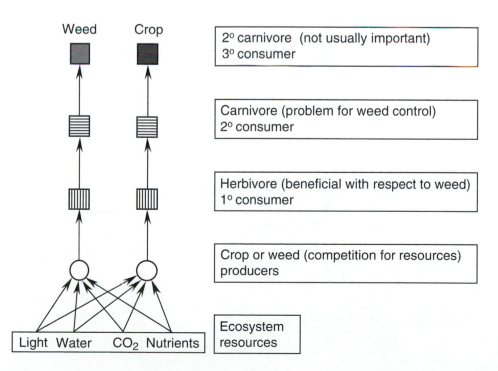

Weed    Crop

| 2° carnivore (not usually important) |
| 3° consumer |

| Carnivore (problem for weed control) |
| 2° consumer |

| Herbivore (beneficial with respect to weed) |
| 1° consumer |

| Crop or weed (competition for resources) |
| producers |

| Ecosystem |
| resources |

Light  Water    $CO_2$  Nutrients

**FIGURE 4–7**
Diagram of simple food chain reflecting competition for resources between crops and weeds; note the necessity to include a food chain for each type of plant.

webs that show the interdependence and interactions of feeding relationships among organisms. Figure 4–8 provides examples of the increased complexity of food webs with multiple pest species and beneficials feeding on various plant sources, with varying overlap in prey choice, and with cannibalism. Tritrophic interactions involve organisms in three different trophic levels, and are depicted by an arrow within the herbivore pest (D). Some of the important conceptual interactions are explained below using Figure 4–8:

1. Weeds can serve as alternative hosts for herbivore pests (A and D). Note that pest D is supported on a plant outside of the managed crop ecosystem; this type of trophic connection is important for areawide pest management.
2. Weeds can serve as alternative hosts for a beneficial insect B through provision of prey living on the weed that is growing within the crop agroecosystem. Similarly, herbivore prey that is feeding on a plant outside the managed ecosystem supports beneficial D. Plants growing outside of the agroecosystem may be important for areawide pest management (discussed in Chapter 18).
3. Weeds can serve as direct food sources for beneficial organisms; for example, weeds within the crop are food for beneficial A, and plants outside the managed ecosystem are food for beneficial B.
4. The crop has multiple herbivore pests, and the likelihood that any of the pests will be a significant problem depends on what beneficial organisms are present. Pest organisms A and D are maintained on weeds or noncrop plants. Pest A might be root-knot nematodes that are able to parasitize weed hosts such as pigweed and lamb's-quarters. Pest D might be leafhoppers, or a virus such as *lettuce mosaic,* living in or on weeds in areas that surround managed fields.

**FIGURE 4–8**
Diagram of simple food web showing resource flow between crops and weeds, plants outside the agroecosystem, and their herbivore pests and carnivore natural enemies.

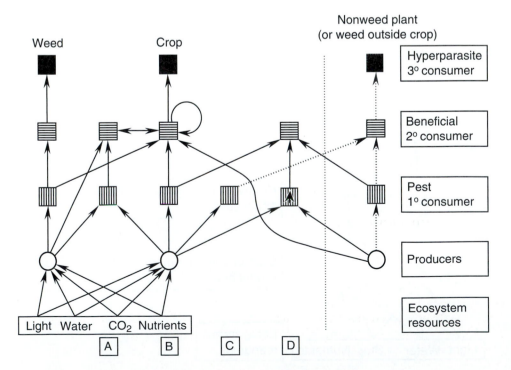

5. Beneficial organism B is probably not very effective, because any of the following could alter its population dynamics sufficiently to stop the population from ever increasing to the point where it provides effective control of pest organism B:

    5.1. Beneficial B requires plant food, such as nectar or pollen, and the appropriate food-source plants may or may not be present.

    5.2. Beneficial B can be eaten by another beneficial organism (A). Many generalist predators feed at the same trophic level (e.g., spiders).

    5.3. Beneficial B is cannibalistic and will feed on its own species (this can be a serious limitation for many generalist predators) and is depicted by the curved arrow back onto itself (e.g., lacewing and lady beetle larvae).

    5.4. Beneficial B is a food source for a hyperparasitic tertiary consumer (B).

6. Hyperparasites from outside the managed ecosystem would probably limit the effectiveness of beneficial organisms.

7. The activity of beneficial D may be modified by what entomologists refer to as a tritrophic interaction, which by definition involves the passage of plant-derived chemical compounds across trophic levels. The vertical arrow within pest organism D represents the passage of chemical compounds derived from the crop plant to the next higher trophic level. However, as beneficial D is also supported by in-crop weeds and by plants external to the system, it could function normally.

In reality, most food webs involve many interactions and so are much more complex than those described here. Note that parts of the system may include bottom-up processes, whereas other parts may be simultaneously responding in a top-down manner.

## Animal-Based Food Web

When an animal-based agricultural system is considered, the importance of different trophic levels, and the positive and negative connotations of the terminology used to describe them, changes considerably from that used to describe plant-based systems (Figure 4–9). Animal-based systems differ in the following ways.

1. Herbivorous primary consumers may be the desired organism in the system. Examples include cows, sheep, or chickens.
2. The primary herbivorous crop pest(s) are now at the same trophic level as the desired herbivorous animal, which can influence the pest management tactics used. For example, foraging cattle may consume pesticides applied to a pasture.
3. The secondary consumer carnivore organism is a pest (e.g., horn fly), and the tertiary-level consumer becomes the beneficial organism.
4. Because the weed is a producer, and the desired organism is a herbivore, it is now possible that weeds become a food source for the desired animal. This trophic relationship is exploited in many peasant cultures where weeds are fodder for domestic animals. In most industrialized countries, the use of weeds as a food source for animals is unimportant; however, weeds that are eaten by livestock are important if the weeds are poisonous, such as common groundsel.

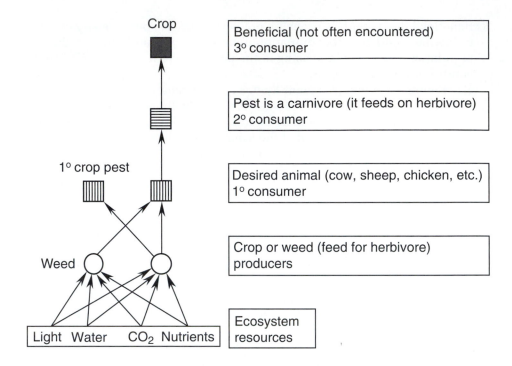

**FIGURE 4–9**
Diagram of simple
food web showing
energy/resource flow
in an animal-based
agricultural system.

## Summary of Trophic Relationships

Figure 4-10 shows how different food web possibilities form a complex network when considered at the agroecosystem level. Dotted lines depict organisms feeding at more than one trophic level, such as humans and beneficial insects. It is important to realize that pest management involves three different types of primary producers:

- the crop or desired plants;
- weeds that may be in direct competition with the crop plants; and
- weeds and other plants external to the managed system.

The appropriate management for each of these three types of producers varies, and the tactics used can have far-reaching impacts on the management of other pest categories (see Chapters 6, 7, and 16).

## LIMITING RESOURCES AND COMPETITION

The preceding sections covered the trophic connections between different pest organisms in the ecosystem. The next section introduces the concept of ecosystem carrying capacity, r- and K-selected organisms, competition between organisms for their vital resources, and density-dependent phenomena.

## Carrying Capacity and Logistic Growth

The concept of environmental carrying capacity for organisms is fundamental to understanding major constraints to population growth. Density is the number of individuals per unit area. The number of organisms in a population changes over time as a result of the relationship between births, deaths,

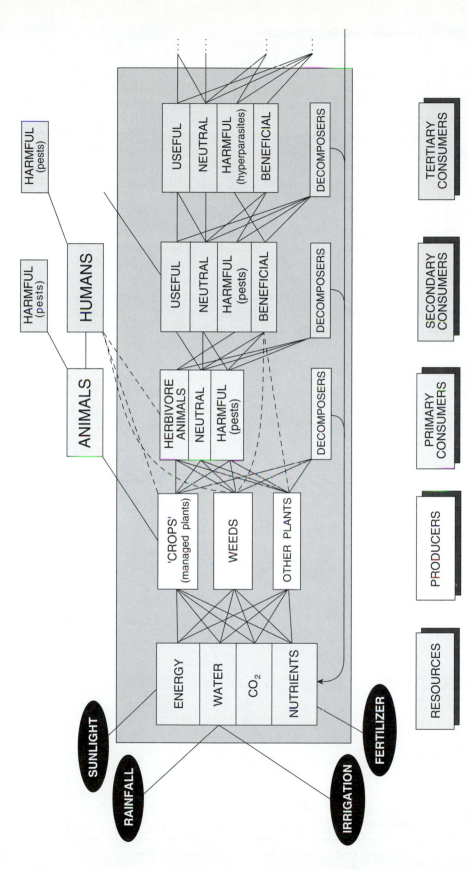

**FIGURE 4–10** Summary of food web connections for a hypothetical agroecosystem. Note that all resource and energy flow is from left to right, except for nutrient recycling.

immigration, and emigration. Without any limits on reproduction and survival, any reproducing population will eventually increase exponentially (Figure 4–11). Unlimited exponential population growth is not possible for long periods, because the population density eventually becomes sufficient to use all available resources, resulting in decreased reproduction and increased mortality. The maximum population size that can be supported in a particular environment is termed the carrying capacity of the system. Food web interactions, including competition, predation, and parasitism, combined with resource limitations determine the carrying capacity.

Population growth that is initially fast but is followed by a gradual slowing until the carrying capacity is reached can be depicted by an S curve, and is termed logistic growth (Figure 4–11). The logistic growth pattern is an idealized depiction of how populations grow when they are not constrained by external biotic factors, and logistic growth should be considered a useful heuristic model. Populations should not be expected to closely follow smooth logistic dynamics for many reasons. For example, the environment that organisms experience is not necessarily constant, but rather is continually changing and, as a consequence, carrying capacity and the population growth rate may change. Suffice to say that the logistic growth model is based on assumptions that may be inappropriate for many systems; however, the general phenomenon of limited population growth described by the logistic model is extremely useful. For the purpose of this book we assume that pest populations, generally speaking, increase according to the logistic growth pattern. Logistic growth is a sigmoid

**FIGURE 4–11**

Comparison of exponential versus logistic theoretical population growth.

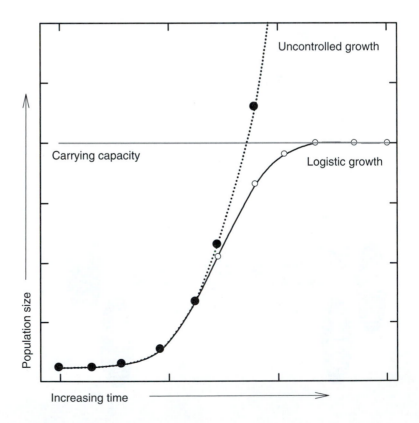

or s-shaped time-density curve, and may be described mathematically by the logistic equation (Figure 4–11). The equation, discovered by Verhulst in 1838, and independently derived by Pearl and Reed in 1920, can be expressed in different ways; the form that relates the number of individuals to time is:

$$N_t = \frac{K}{1 + \left(\dfrac{K}{N_o} - 1\right)e^{-rt}}$$

where: $N_t$ = the number of individuals at time t
   $N_o$ = number of individuals at arbitrary time zero
   K = the carrying capacity of the system
   r = the intrinsic or maximum unrestricted population growth rate
   t = time
   e = base of natural logarithms

According to the logistic model, when the population is small it shows an accelerating rate of increase. As the population increases, intraspecific competition acts as a constraint to slow the rate of increase, until the population eventually reaches the maximum density that the environmental resources can support (the carrying capacity). The population cannot increase above this level without experiencing increased death rate and/or decreased birth rate.

If the availability of environmental resources changes, the carrying capacity of the system changes and results in a different equilibrium population density. The horizontal lines at A, B, and C in Figure 4–12 represent changes in the

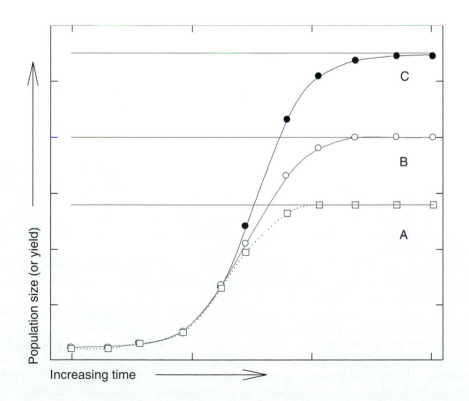

**FIGURE 4–12**
Comparison of theoretical logistic population growth at different levels of limiting resources: A = low, B = medium, C = high.

available resources, essentially the removal of a constraint to population growth. The resources referred to might be the number of prey for a carnivore, the number of suitable host plants for a herbivore, or the level of nutrients for a plant. When the carrying-capacity concept is applied to crops, it is referred to as the law of constant yield, which is how it is understood in relation to weed growth.

The logistic equation illustrates a major problem encountered in attempting to manage pests. Tactics aimed at lowering the population of a pest by direct action against the organism lowers the population to numbers where population growth is less constrained. Following the management action the population is thus more likely to increase, unless the control strategy also reduces the resources available for population growth. This is one of the major drawbacks to the use of pesticides, as they do not resolve the underlying reason why the population had increased sufficiently to attain pest status.

## r-Selected versus K-Selected Organisms

The life history of an organism includes attributes such as age at maturity, size at maturity, age-specific fecundity, survival rate, number of offspring, and size of offspring. The parameters of the logistic equation (shown earlier in the chapter) may be used as shorthand descriptors of life history traits of organisms, including their reproductive strategies and competitive abilities. With reference to the parameters of the logistic equation, different pest organisms may be considered as exhibiting features that place them somewhere on a continuum from small, rapidly reproducing organisms that have been "r-selected," to larger-bodied organisms that reproduce more slowly and compete effectively that have been "K-selected."

***r-Selected Organisms***    The "r" in this term refers to the parameter *r*, which is the reproductive rate, from the logistic equation. Organisms described as being r-selected are considered to produce large numbers of offspring as quickly as possible, are short-lived, and have poor competitive ability. Such attributes are thought to allow rapid use of resources. Organisms of this type are usually well adapted to fluctuating environments and can rapidly colonize the disturbed habitats typical of many agroecosystems. Many insects, most annual weeds, many plant pathogens, and most nematodes are considered to be r-selected.

***K-Selected Organisms***    The "K" in this term refers to the parameter *K* from the logistic equation, the carrying capacity. K-selected organisms are good competitors, are long-lived, and utilize resources to produce only a few offspring. These attributes are considered to arise from selection for competitive ability in relatively stable environments. Examples include most mammals, many perennial plants, some insects, and some pathogens.

A group of organisms considered to be stress tolerators are considered as another life history type. These organisms have characteristics that allow them to survive in harsh environments.

The differences between these types of population dynamics are important to modeling efforts—and in the design of pest management strategies, particularly for arthropod pests. For instance, the codling moth is an example of a K-selected insect pest of apples and pears. It has discrete generations over the growing season, which permit planning of monitoring and control activities following well-defined degree-day-driven phenological models (see Chapter 5).

# Competition

As the density of organisms in an area increases, the result is food shortage for consumers, or ecosystem resources for producers. Competition is the simultaneous demand by two or more organisms for a common resource. Examples of resources include nutrition, space, or mates, and organisms may compete through their relative efficiencies at procuring a limited resource, termed exploitative competition, or through direct detrimental interactions, termed interference competition. Competition occurs when there are enough individuals present that their combined demand for the resources exceeds the available supply. Competition cannot occur unless resources are limiting. Competition is typically divided into two types:

**Competition can occur when a resource supply falls below the total demand of all the individuals using that resource.**

1. Intraspecific competition occurs among individuals within a single species. An example is the competition among individual plants in the crop for nutrients.
2. Interspecific competition occurs between individuals of different species. Competition between weeds and a crop is a good example of interspecific competition, and competition for aphid prey between lacewing larvae and lady beetle larvae provides an arthropod example.

**Competition between individuals of the same species = intraspecific. Competition between individuals of different species = interspecific.**

The concept of competition applies to all pest categories, but differs considerably for producers versus consumers.

1. Producers. Due to their trophic position weeds compete with crops for ecosystem resources, including light energy, water, and mineral nutrients. Weeds also compete among themselves in many situations. Some authors include competition for space, but typically resource limitation occurs before competition for physical space. When water and nutrients are utilized they are not available to other plants in the system. Light is an unusual resource in that it is not "used up"; there will be as much tomorrow as there is today. Competition for light therefore occurs by interference competition (blocking access), rather than through exploitative competition.
2. Herbivores. Due to their trophic position, herbivores compete for adequate sources of host plants. Except under extreme conditions their food source is generally not considered limiting, and thus pest herbivores typically only compete when they have caused severe damage to the crop. It is a very important consideration that the food source of herbivores, the plant, is not mobile. Plants therefore cannot hide or move away from herbivores, and must have defense mechanisms in order to avoid damage when herbivore numbers are great. Most plants are not attacked by most herbivores due to the presence of protective chemical or physical defenses (more about these in Chapter 17).
3. Carnivores. These organisms feed on another animal. It is quite feasible that the prey population may be inadequate to support the carnivores, and individuals compete for prey. The food source for carnivores is typically mobile, so searching and catching become part of the competition equation. This has resulted in two definitions of competition for carnivores:
   3.1. Real. This competition implies that the prey source is actually limited and that true competition for food is occurring.
   3.2. Apparent. This competition occurs when two organisms appear to compete, as each has an indirect negative influence on the other

because of its positive effect on the abundance of a shared predator or parasite. In pest management terminology, apparent competition also implies that the prey source may be adequate, but that the carnivore cannot locate it and is hence prey limited.

Competition is important relative to the population dynamics of many pests, but is particularly important with respect to weeds growing in crops. The consequences of competition to management of nonweed pests are not obvious, but some think that changes in competition among parasites or predators is of little relevance to pest management because the most competitive species will be the one that drives the system. Alternatively, evidence suggests the general competitive ability of biological control agents influences their ability to become established in soils; and therefore, competition is involved in determining the outcome of attempts at biological control in soil. Competition among beneficial arthropods can be a problem if, under a situation of prey limitation, they feed on each other.

Competitive exclusion, or displacement, has been at the center of arguments in biological control of insect pests. Competitive exclusion postulates that no ecological homologues can coexist in the same location, even if resources are not obviously limited. According to this theory, one species ends up displacing the other and dominating that niche in the community. Several species of ground beetles of European origin have displaced native American species in crop fields, and the ten-spotted lady beetle is thought to have displaced several native lady beetles.

**The principle of competitive exclusion states that two species with identical ecological requirements cannot indefinitely coexist in the same location.**

## Density-Dependent Phenomena

Density-related phenomena must be considered from the aspect of producers (weeds) versus herbivore pests, and carnivores, as the response differs between pest categories.

*Weeds*    As population density increases, individuals eventually experience competition, as reflected in the logistic growth of populations. Competition for light, nutrients, and water is central to weed science. Competition of weeds with crop plants results in decreased crop growth and development that is observed as yield loss. Figure 4–13 shows typical crop yield response in relation to increasing competition caused by increasing weed density. The rectangular hyperbolic function shown in Figure 4–13 is considered a good model to represent most cases of yield loss in relation to weed density. It is important to recognize that substantial yield loss can occur at very low weed densities. The actual loss varies, depending on the following:

1. Weed species. Some species such as mustards, barnyardgrass, and nutsedges are very competitive, and others such as purslane and chickweed are much less competitive.
2. Crop species. Low-growing crops, such as sugar beets and lettuce, are much less able to tolerate competition from weeds than are tall crops such as maize and rubber.
3. Abiotic environment. Surroundings can modify or alter the outcome of competitive interactions. Examples are temperature (cool-season weeds do not compete well in warm-season crops), moisture level (some crops

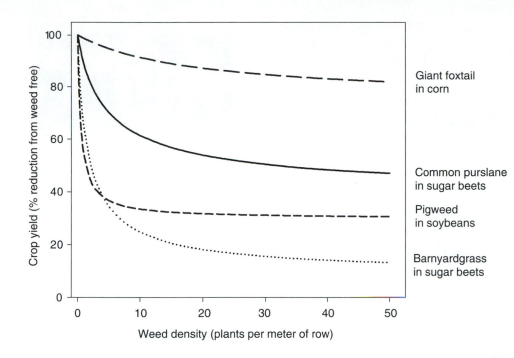

**FIGURE 4–13**
Examples of crop yields, expressed as a percentage of weed-free controls, in response to increasing weed densities. All curves are generated from actual published research data.

can tolerate drought better than weeds), and nutrients (nitrogen placement can alter competitive interactions in several crops).

4. Duration. Although not depicted in Figure 4–13, the length of time that the crop is exposed to competition is also important. Generally, weeds that germinate later than the crop do not compete as intensely as those that germinate at the same time as the crop. Conversely, the longer that weeds remain in the crop, the greater the crop losses due to competition.

Self-thinning is an important concept for plant population dynamics. Self-thinning results from competition-induced mortality, and may result in logistic growth. The phenomenon applies to nonmobile organisms that must obtain their resources from the limited area directly around them. As the density of individuals increases, the demand for resources in a defined area also increases. When resources are no longer adequate to sustain the population, the weaker individuals die, permitting the others to continue to grow. As individual size increases with age, the weaker individuals continue to die, which frees up further resources for the survivors. The process is referred to as self-thinning in plants, and may also occur in other nonmobile pests such as pathogens and nematodes.

***Herbivore Pests*** As pest density increases, the impact on the crop also increases. For example, sugar beet size and yield (Figure 4–14) decrease in relation to increasing at-planting densities of sugar beet cyst nematodes. Similarly, defoliation of alfalfa increased in relation to increasing numbers of cutworms (see Figure 6–8). Damage intensity related to increasing herbivore density is one of the main concepts underlying the use of economic thresholds (see Chapter 8).

When arthropod populations become large, such as occurs under outbreak situations, r-selected arthropod pests may overwhelm a food resource to the

**FIGURE 4–14**

Tap root weights of individual sugar beets grown in soil with increasing at-planting egg densities of the sugar beet cyst nematode. Graph shows average responses and the inset photograph shows representative beets, from left to right, to 0, 0.5, 1.0, 2.0, 3.0, 4.0, 5.0, and 10.0 eggs per gram of soil.

Source: data and photograph by Edward Caswell-Chen.

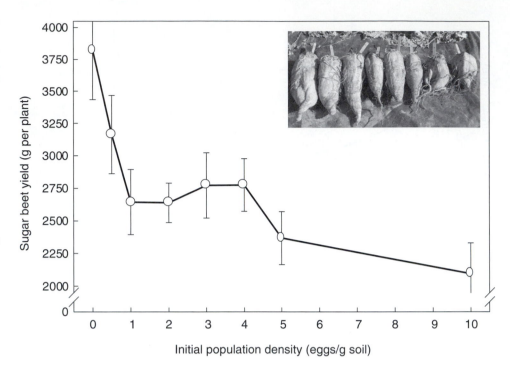

point that the pest population crashes, leaving only few survivors. In agroecosystems, such outbreaks are a consequence of the resource concentration produced by agricultural practices. Outbreaks may occur in nature, but do so much less frequently than is seen in agroecosystems. Mobile animals have evolved life histories that result in their avoiding or responding to intraspecific competition. One behavioral mechanism in vertebrates is related to the concept of territoriality. Many vertebrates clearly mark and actively defend their territory to reduce competition, although the act of maintaining a territory can be considered a form of interference competition. Some insect species deposit chemical cues or odors on plants when they oviposit, and the presence of the marker on the plant deters other females from ovipositing on the same plant, thus avoiding competition among the offspring.

## SOURCES AND RECOMMENDED READING

Ecology texts provide more thorough coverage of many of the points made in this chapter, such as books by Begon, Harper, and Townsend (1996), Krebs, (2001), Morin (1999), Pianka (2000), Ricklefs and Miller (2000), and Stiling (1999).

Begon, M., J. L. Harper, and C. R. Townsend. 1996. *Ecology: Individuals, populations, and communities.* Oxford, UK; Cambridge, Mass.: Blackwell Science, xii, 1068.

Flint, M. L., and R. Van den Bosch. 1981. *Introduction to integrated pest management.* New York: Plenum Press, xv, 240.

Kerr, G., R. Hermer, and S. R. Moss. 2000. A century of vegetation change at Broadbalk wilderness. *English Nature* 34:41–47.

Krebs, C. J. 2001. *Ecology.* 5th Edition, San Francisco: Benjamin Cummings, XX, 695.

Lawes, J. G. 1884. In the sweat of thy face shalt thou eat bread. *The Agricultural Gazette* 23:427–428.

MacArthur, R. H., and E. O. Wilson. 1967. *The theory of island biogeography.* Princeton, N.J.: Princeton University Press, xi, 203.

Morin, P. J. 1999. *Community ecology.* Malden, Mass.: Blackwell Science, viii, 424.

Omernik, J. M. 1995. Ecoregions: A spatial framework for environmental management. In W. S. Davis and T. P. Simon, eds., *Biological assessment and criteria: Tools for water resource planning and decision making.* Boca Raton, Fla.: Lewis Publishers, 49–66.

Pianka, E. R. 2000. *Evolutionary ecology.* San Francisco: Benjamin Cummings, xv, 512.

Ricklefs, R. E., and G. L. Miller. 2000. *Ecology.* 4th Edition, New York, W.H. Freeman & Co., xxxvii, 822.

Shaw, W. C. 1961. Weed science—Revolution in agricultural technology. *Weeds* 12:153–162.

Stiling, P. D. 1999. *Ecology: Theories and applications.* Upper Saddle River, N.J.: Prentice Hall, xviii, 638.

# 5

# COMPARATIVE BIOLOGY OF PESTS

## CHAPTER OUTLINE

▶ Introduction
▶ Concepts of pest population regulation
▶ Dissemination, invasion, and colonization processes
▶ Pest genetics

## INTRODUCTION

**The growth of any population is regulated by biotic and abiotic factors, otherwise populations would continue to increase indefinitely.**

Populations of organisms have tremendous potential for growth, and if provided with ideal environmental conditions, unlimited resources, and protection so that all progeny survived, it would not take many generations until any population of organisms would occupy the earth's entire surface. Not too long after that, the growth would expand to occupy the entire known universe. Exactly how long it might take to reach such an impressive end would depend on two things: how many progeny the organism produced per reproductive bout, and how frequently reproduction occurred.

Unlimited survival and reproduction leads to dramatic exponential population growth (Figure 4–11). The dramatic nature of exponential population growth is embedded in entomological folklore, in the estimate that under ideal conditions, a single pair of houseflies could produce enough offspring in seven months to cover the earth with a 40-foot-deep coating of flies! Obviously, in reality many factors contribute to regulate populations so that unlimited population growth does not occur. The ecological theory of population growth as described by the logistic model holds that populations will increase only to a density that the environment can support, and the upper limit of population density is termed the carrying capacity of the environment for that organism. An important concept in IPM is that organisms become pests only when their population densities exceed certain threshold levels. This chapter will cover some of the important ecological factors involved in the regulation of pest populations.

# CONCEPTS INVOLVED IN PEST POPULATION REGULATION

## Reproduction

The life cycles and reproductive patterns of organisms are termed *life history schedules* by ecologists. How often and when reproduction occurs during the life of an organism is important to survival in particular environments, and the mode of reproduction has a profound effect on the pest potential of an organism. The two fundamentally different reproductive modes in organisms are sexual and asexual reproduction.

***Sexual reproduction***    Sexual reproduction (or amphimixis) involves cross-fertilization, in which meiotic (reduction) division of the nucleus generates haploid gametes (sperm and eggs), and the gametes fuse to produce a diploid zygote, which later develops into the mature organism. Sexual reproduction is important to pest management, because it facilitates recombination of genes. The significance of recombination is that it generates individuals with new, unique combinations of genes. Relative to pest management, such novel gene combinations may lead to the development of biotypes, races, or resistance to pest control tactics.

In the different pest categories, different structures result from sexual reproduction. Examples of the structures are shown in Figure 5–1.

1. Various types of spores (e.g., zygospores, ascospores, basidiospores) are produced by fungal and bacterial pathogens.
2. Seeds are produced by weeds, and are shed singly or remain encased in the parental ovary in the form of a fruit.
3. Eggs are laid by arthropods, mollusks, nematodes, and birds. Eggs can be laid singly, in clusters, or enclosed into hardened capsules called oothecae (arthropods) or cysts (nematodes). For example, cockroaches, mantids, and certain grasshoppers lay clusters of eggs protected by oothecae, while the dead bodies of female cyst nematodes serve to protect eggs contained within.
4. Live young may be produced by mammals, a few insects, and some nematodes (see the following section).

**Vivipary**    Vivipary is a form of reproduction in which the developing progeny are nourished internally by the parent, and young are born in an active condition. Vivipary is usually referred to as "live" birth, although seeds and eggs are alive. Most mammals, some insects, and a few plants and nematodes show this form of reproduction. True vivipary is rare among insects, but a remarkable example is the tsetse fly, the vector of the causal organism of the African sleeping sickness. Larvipary or ovovivipary occur among some insects and nematodes, in which eggs develop and eclose within the female body, and nymphs, first-instar maggots, or juveniles emerge from the female. Many insect-parasitic flies present this kind of reproduction.

***Asexual reproduction***    Asexual reproduction may occur in a number of ways, but generally is the result of reproduction by a single individual organism, without a requirement for the involvement of two sexes. Asexual reproduction results in offspring that have the same genotype as the parent and, generally speaking, there is little or no genetic recombination. The resulting progeny are referred to as clones. Asexual reproduction may be advantageous

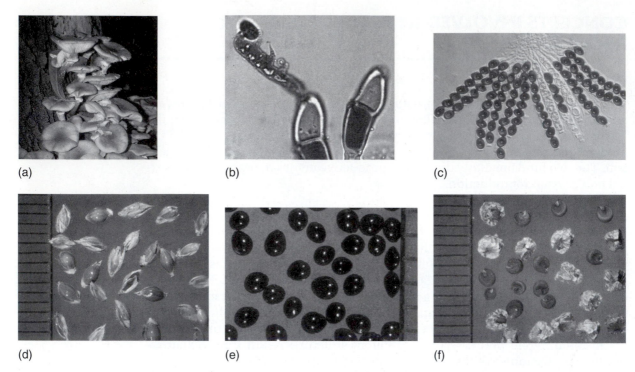

**FIGURE 5–1** Examples of reproductive structures of different pest categories derived from sexual reproduction. (a) Mushrooms (basidiocarps) of oakroot fungus, *Armillaria mellea*. (b) teliospore and germinating teliospore of *Puccinia* sp. A basidiospore is still on the sterigma of the metabasidium. (c) asci containing ascospores of *Sordari fumicola*. (d) barnyardgrass seeds (they are actually a fruit called a caryopsis), (e) redroot pigweed seeds (note size), (f) mallow, or cheeseweed, seeds (note coiled embryo).

Sources: photographs (a), (c) by John Menge with permission; (b) by James E. Adaskaveg with permission; (d), (e), (f), (g), (j), and (l) by Robert Norris; (h) by Ed Caswell-Chen; (i) by Marcos Kogan; and (k) by Jack Clark, University of California statewide IPM program.

because it perpetuates exact copies of successful genotypes. A genotype that is not adapted to prevailing conditions will be at a disadvantage and may be eliminated. Asexual reproduction occurs in perennial plants, pathogens, nematodes, and insects. Several different forms of asexual reproduction are important to pests.

**Apomixis**    The form of seed production without fusion of gametes is apomixis. It occurs in some weedy plants (e.g., dandelion). The significance to pest management is minor.

**Asexual Spore Production**    Many pathogens can produce enormous numbers of vegetatively produced spores (e.g., conidiospores) without going through the process of gamete production. Many fungal species can reproduce both sexually and asexually. In some fungi, sexual reproduction has not been observed, and their only known method of reproduction is through asexual spore production; such fungi are grouped in the Fungi Imperfecti. The implication is that they are not perfect, because a sexual reproductive stage has never been observed. Examples include *Alternaria* spp. (leaf spots), *Botrytis cinerea* (grey mold), and *Rhizoctonia solani* (stem and root rots). Conidiophores and sporangia are two forms of asexual spore-producing structures (Figure 5–2).

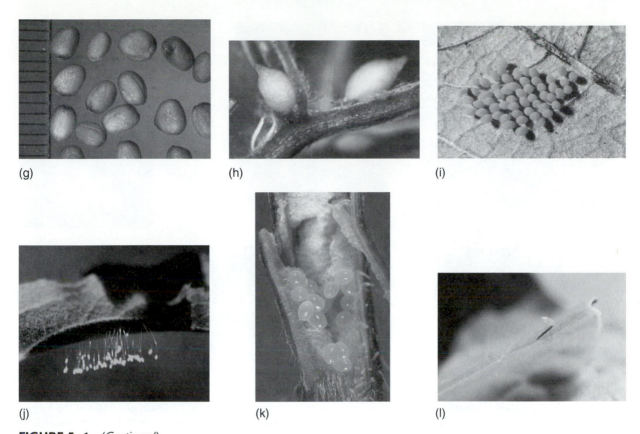

(g)  (h)  (i)

(j)  (k)  (l)

**FIGURE 5–1** (*Continued*)
(g) wild radish seeds, (h) cysts of cyst nematode which contain eggs, (i) eggs of Mexican bean beetle laid in a large mass, (j) lacewing eggs on stalks in a group, (k) alfalfa weevil eggs laid *inside* the hollow stem of the alfalfa plant, and (l) example of egg laid singly, here for imported cabbage butterfly. Scale for weed seeds is in millimeters.

**Vegetative Reproduction (Cloning)**   The ability to grow a new individual from part of a mature individual is called vegetative reproduction (cloning). Bacteria may reproduce asexually by budding, and fungi through fragmentation of hyphae, although these forms of reproduction are not usually referred to as vegetative reproduction.

There are two forms of vegetative reproduction in plants. Many perennial plants, including weeds, produce specialized vegetative structures such as tubers or stolons that persist through unfavorable environmental conditions. For example, nutsedges produce tubers (Figure 5–2d), johnsongrass produces rhizomes (Figure 5–2e), and field bindweed has creeping rootstocks.

Vegetative reproduction also occurs when some perennial plants are broken into fragments, with each fragment capable of regenerating an entire new plant. No specialized reproductive structures are involved. This form of vegetative reproduction is important to IPM, because human activities such as tillage cause fragmentation of weeds. Fragmentation is common in many aquatic weeds, and floating fragments disseminate the weeds.

Vegetative reproduction has important implications for weed control, because vegetative structures are typically larger than seeds. They establish new plants more rapidly because they have carbohydrate reserves that enhance rapid growth of the young plant.

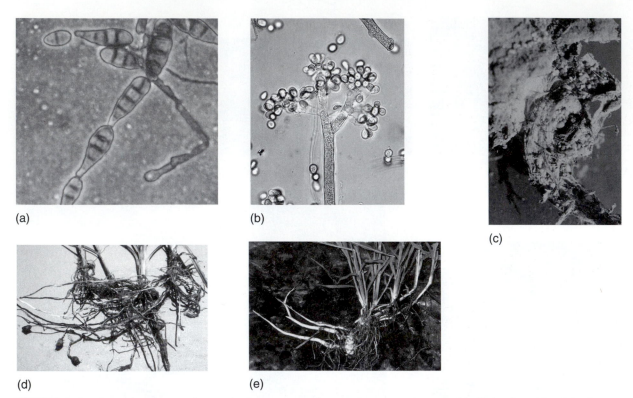

**FIGURE 5–2**   Examples of asexual reproductive structures from pathogens and weeds. (a) Chain of conidia of the fungus *Alternaria alternata,* (b) conidiophore and conidia of *Botrytis cinerea,* (c) sclerotia (small round structures) of *Sclerotium rolfsii* in a rotting sugar beet, (d) tubers at end of rhizomes in yellow nutsedge, and (e) johnsongrass rhizomes.
Sources: photograph (a) by John Menge with permission, (b) by James E. Adaskaveg with permission, (c), (d), and (e) by Robert Norris.

**Parthenogenesis**   The production of progeny without the fusion of compatible gamete nuclei to produce a zygote is parthenogenesis. The process is important for aphid population growth, and aphids show cyclic parthenogenesis (Figure 5–3). Obligate parthenogenesis occurs among some hymenopterous and coleopterous insect species in which the males are absent. Among social *Hymenoptera,* ants, wasps, and bees mating occurs but the reproductive female can control whether the eggs are fertilized. Fertilized eggs produce female workers or new queens. Unfertilized eggs produce males. This kind of reproduction is termed *haplodiploidy.*

Many important pest nematodes also reproduce by parthenogenesis. In some parthenogenetic nematodes, fertile eggs are produced through mitosis, while in others the eggs with the somatic chromosome number are produced by chromosome duplication during meiosis or by the fusion of the meiotic products (e.g., the egg with a polar body). Pseudogamy is another form of reproduction seen in some nematodes, and it occurs through the fusion of sperm and egg (plasmogamy) without subsequent fusion of sperm and egg nuclei (karyogamy); in pseudogamy, the development of the egg to produce a zygote is apparently stimulated by plasmogamy.

**FIGURE 5–3**  Adult female aphid giving birth to live young.

Source: photograph by Jack Clark, University of California statewide IPM program.

## Fecundity and Fertility

Fecundity is the potential reproductive capacity of an organism, and is measured by the number of gametes produced. The actual number of viable offspring produced per adult individual is referred to as fertility, and is the realized fecundity. Among the various organisms, differences in fertility lead to different rates of population increase per generation. A range of fecundity exists among different organisms. Some organisms produce a few, typically large, progeny at each reproductive bout, whereas others produce many, typically small, progeny. Using the parameters of the logistic growth equation, the former are referred to as K-selected organisms and the latter are referred to as r-selected organisms. The reproductive capacity of different pests may influence the tactics used for their management. Among pests, a range of levels of fecundity is observed:

1. One to 10 offspring per adult for most of the mammals
2. Hundreds to a few thousand progeny per adult for most arthropods (honey bees and termites are exceptions), nematodes, and mollusks. The fecundity for a boll weevil female may be up to 200 eggs, a cyst nematode up to 600 eggs, and corn earworm moths up to 3,000 eggs per female.
3. Many thousands to millions of propagules for most weeds and pathogens

Organisms with low fecundity require multiple generations to accumulate large numbers, whereas those with high fecundity can achieve high populations in just one or two generations (Figure 5–4). This characteristic is important when considering management strategies. Organisms with high fecundity per individual (many weeds and pathogens) mean that the concept of thresholds (see Chapter 8) may need to be applied with care, because a low population that is allowed to reproduce can maintain the overall pest population at damaging levels.

Most animals (arthropods, vertebrates, nematodes, and mollusks) produce a relatively constant number of offspring per individual, although nutrition and environmental stress may influence fertility. Litter size in mice typically falls

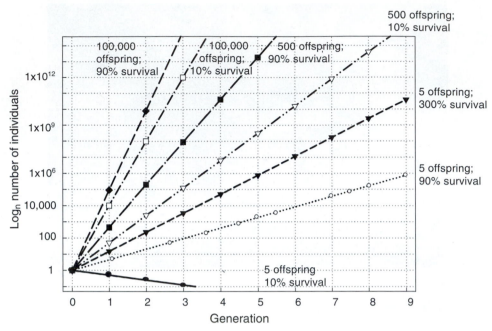

**FIGURE 5–4** Theoretical population growth in relation to fertility. Each line represents a different fecundity (5, 500, or 100,000 offspring per individual) with either 10% or 90% surviving to reproduce. An additional plot line shows five offspring per individual but with 300% survival, which represents theoretical population development for an animal that has three litters per adult female. Note that a fecundity of five with only 10% survival results in a population crash.

within the range 6 to 12, while the fertility of the sugar beet cyst nematode ranges from 50 to 600 eggs. As a general rule, most animals have a fertility that varies by less than one order of magnitude among individuals of the same species. This contrasts with range of fertility exceeding three or four orders of magnitude for many pathogens and weeds. A barnyardgrass plant growing under severe competition from corn may produce 1,000 seeds, but when growing in less competitive crops such as tomatoes, 100,000 seeds is typical, and without competition may produce nearly 1,000,000 seeds. Redroot pigweed shows a similar range in fertility (Figure 5–5). Plant pathogens can range from a few thousand spores to many millions, depending on environmental conditions; brown rot of peaches can produce from 2,200,000 to 7,500,000 ascospores per apothecium (reproductive structure).

Most organisms experience density-dependent alteration of fertility. The alteration of fertility is large for weeds and pathogens in which the body size of the adult is plastic. The phenomenon is observed in insects and nematodes; the number of eggs laid per female *Arytaina spartii* (a psyllid on broom) decreases as the density of insects per stem increases (Figure 5–6). This may not represent the change in realized fecundity due to competition for resources, but may reflect intraspecific competition for suitable oviposition space. Indeed, both inter- and intraspecific competition can reduce the realized fecundity per individual.

Variation in the fertility per individual has important implications for pest management decision making, because such variation decreases the ability of pest managers to anticipate or predict pest population growth. Estimates of number of individuals present will only permit accurate prediction of the size

(a)

(b)

**FIGURE 5–5**
Mature flowering redroot pigweed plants demonstrating extreme plasticity in size of individuals and fecundity. (a) About 1.5 m tall in the summer with high seed production, and (b) about 5 cm tall in the fall producing only a few seeds.
Source: photographs by Robert Norris.

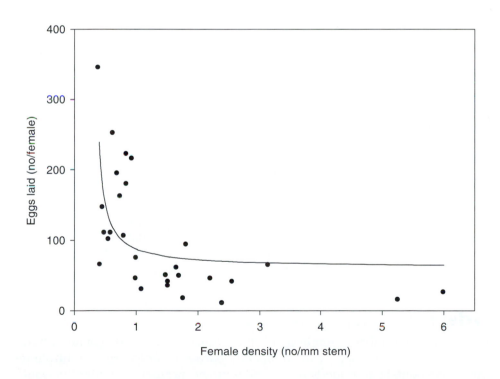

**FIGURE 5–6**
Relationship between density of the psyllid insect *Arytaina spartii* and the number of eggs laid per female on broom stems.
Source: data from Dempster, 1975, p. 26.

of the next generation if the fertility per individual is relatively constant, or the influence of density on fecundity is known or negligible.

Differences between fecundity or reproductive potential, and fertility or the realized reproduction, are the factors that determine the need for field monitoring of pest populations in IPM decisions (see Chapter 8). Factors that influence the progeny survival are dealt with later in this chapter.

## Population Generation Time

The average time required for a population to pass from birth to active reproductive status is termed the generation time. Populations of organisms with short generation times can increase rapidly, whereas populations with long generation times will increase more slowly. The length of time between generations varies greatly among the different pest categories, and among species within the same pest category. For example, insect species usually have short generation times with many generations occurring in a year, but one brood of the periodical cicada has a generation time of 17 years. Following are examples of typical generation times for different pest organisms under ideal conditions.

1. Hours: bacteria. This population generation time results in multiple generations per day.
2. Days: spider mites, aphids, some nematodes and pathogens. A generation time in days results in multiple generations in months.
3. Weeks: many insects, many nematodes, pathogenic fungi, small vertebrates (e.g., mice). A generation time in weeks results in multiple generations per season.
4. Months: some insects, some pathogens, some nematodes, larger vertebrates, annual weeds. This generation time may result in a single generation per season.
5. Years: large vertebrates, perennial weeds, a few arthropods. A generation time in years results in one or fewer generations per year. Long-lived parents or dormancy (see below) skews population decline, which results in considerable overlap of generations in time.

Generation times can be extended by orders of magnitude when suboptimal environmental conditions occur, particularly in poikilothermic organisms. Organisms such as bacteria can develop high populations within hours, requiring that control actions be taken frequently (typically at daily to every few day intervals under ideal conditions). Mite and aphid populations can achieve levels requiring action within a few weeks due to their short generation time. Short generation times make timing of control actions critical. Even organisms with a generation time of a few weeks can build up to very high populations within a season. For weeds and most pathogens, it is often better to manage the pest inoculum level than to attempt pest control after the population has increased. For organisms with longer generation times, control strategies should probably be designed with the goal of long-term population regulation.

## Cycles Per Season

In ecology, the number of reproductive bouts during the lifetime of an individual organism is described by the terms *iteroparous* or *polycarpic* (multiple reproductive bouts) and *semelparous* or *monocarpic* (a single reproductive bout). In pest management, the number of reproductive cycles, or generations, per year is vital, because it influences the rapidity of population increase and has serious implications for the management strategy employed. In several cases, the number of generations is not adequately considered. In the different pest disciplines, special terminology may be used to describe the number of generations per year, as follows:

| | |
|---|---|
| Pathogens | Single cycle, or monocyclic pathogen—one reproductive bout, a single pathogen generation, and one disease cycle per year (examples include some rusts, smuts, *Verticillium* wilt) |
| | Multicycle, or polycyclic pathogen—two or more pathogen generations per year (examples include some rusts, mildews, apple scab). In plant pathology, the spatial spread of pathogens within a season combined with multiple generations determines the loss caused by an epidemic. |
| Weeds | More than one cycle per year—no equivalent. Annuals: live only one season and reproduce once (monocarpic) (examples are barnyardgrass, redroot pigweed, and lambsquarters). Biennials: live for two seasons but only reproduce once in second season (monocarpic) (examples include prickly lettuce and milk thistle) Perennials: live for many years and typically reproduce once per year (polycarpic) (examples include nutsedges, johnsongrass, field bindweed). |
| Nematodes, mollusks, insects, and vertebrates | Univoltine—one reproductive cycle per year (examples include potato-cyst nematode, alfalfa weevil, ground squirrels). Multivoltine—two to several reproductive cycles per year (examples include sugar beet cyst nematode, spider mites, corn earworm, voles). |

Figure 5–7a shows typical single-season dynamics for a multivoltine insect (or polycyclic pathogen), where the progeny produced in successive generations begins to overlap more closely during the season. Each generation is often referred to as a cohort; the term is applied to the group of individuals born at the same time from a single event such as the spring hatch of insects, a litter of mice, or a germination event of a weed. For univoltine organisms or monocyclic pathogens with little carryover between cohorts, there is little population overlap and the first curve (Figure 5–7a, solid line) represents theoretical population size. Weeds reproduce only once per year or less, so the spacing of the peaks is uniform with time, but the skew to the right is increased because some individuals survive a long time and thus the overlap between generations increases (Figure 5–7b).

## Longevity and Mortality

Individual organisms age, senesce, and die. There are major differences in lifespan among pest organisms (Figure 5–8). Life span is important not only because of its influence on demographics, but also because it determines how long the pest is active and exerting a negative influence on the crop. The life cycle of some pest organisms includes stages capable of surviving in a quiescent state for extended time periods. If the adult organism lives longer than the average generation time, then overlapping generations result (Figure 5–7b), which is

**FIGURE 5–7**
Theoretical population size and age structure in relation to time. (a) A single season of a multivoltine or polycyclic pest showing increasing generation overlap with each successive generation; and (b) multiple years of a single-cycle pest which possesses long-lived propagules that persist from generation to generation (e.g., weed seedbank).

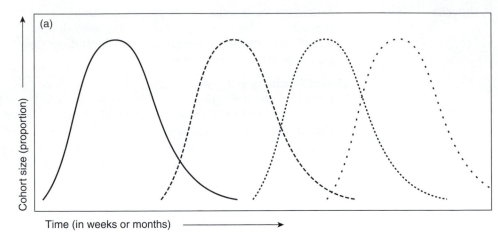

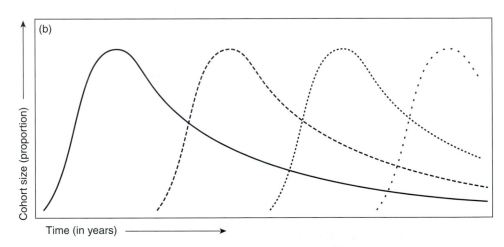

typical of most weeds and vertebrates. Many multivoltine insect pests have overlapping generations, particularly in warmer climates. Following are examples of the adult pest life spans under normal conditions (see Figure 5–8).

1. Pathogenic fungi have the following life span characteristics.
   1.1. Single-cycle pathogens live a year to a few years, and some are essentially perennial, living for many years (e.g., oakroot fungus).
   1.2. Multicycle pathogens live a week or two to several months.
   1.3. Fungi produce resting stages that can survive for years under appropriate conditions.
2. Bacteria typically live for hours to a few days, although bacteria may produce survival stages, such as spores, that are long-lived.
3. Life spans of growing weeds and the dormant seedbank must be considered separately, because the seedbank serves as a survival stage in the soil.
   3.1. Annual plants live from a few months to about one year. Short-lived perennials live for a few years, but many perennials live for decades or even centuries.

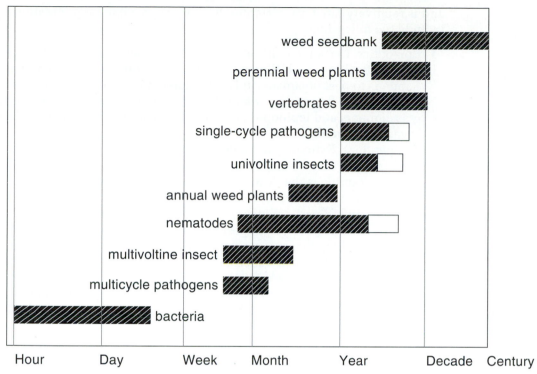

**FIGURE 5–8**   Typical longevity of a single cohort of different categories of pests; note the time scale is logarithmic.

3.2. Survival stage for a few weed species such as prickly lettuce and barnyardgrass, which have short-lived seeds, is typically five years or less. Many other species have seeds that can survive for decades (e.g., lambsquarters, knotweeds, and velvetleaf) or centuries (e.g., lupins).

4. Nematodes have life spans of months to a year to two. Nematodes have stages that can persist for years in the soil and, in effect, serve the same reproductive function as the weed seedbank. The eggs contained within cysts of the cyst nematodes are an example.

5. Arthropods have different life stages, but the stages do not necessarily have the same life spans, particularly when a life stage serves as a survival stage.

5.1. The life span of univoltine arthropods is typically a little longer than a year; some species survive for more than a year (e.g., cicadas, tadpole shrimps) and some termite queens live for more than 50 years.

5.2. The life span of multivoltine arthropods is typically a few weeks to perhaps a year.

6. Vertebrates typically live a year or two (e.g., mice, voles, and rats), to many years (e.g., rabbits, deer, elephants).

Usually, the maximum life span is not achieved. Arthropod pest populations sometimes experience catastrophic reduction (crash) over a short time period. Such crashes do not normally happen for most pathogens or weeds, and is relatively rare for vertebrates. Factors responsible for population crashes include the following:

1. Epizootics. Certain conditions permit a pathogen to build up to high levels, attacking and killing most of the pest population. When this happens to pest organisms, it is considered a form of natural control. Examples are fungi (e.g., *Entomphthora* spp.) that attack and kill aphids, grasshoppers and lepidopterous caterpillars, and viruses that kill larvae of several pest moths.
2. Temperature. Extremes in temperature can kill most individuals in a pest population if they are not in an appropriate quiescent or dormant phase. A good example is the death of green peach aphids and pea aphids that occurs in hot weather (above about 85°F). In temperate climates, early freezes may kill pest organisms, and the winter months reduce pest populations; however, most pests that adapt to temperate climates have stages that allow them to survive the winter.
3. Flooding. Soil organisms such as nematodes may be killed during flooding, because of a reduction in soil oxygen concentrations.
4. Beneficials. Under certain conditions beneficial organisms can rapidly build up to the point at which they cause a dramatic reduction in the target pest populations.

The survival of a cohort over time is described by a survivorship curve (or survivorship schedule). The three basic types of idealized survivorship curves are shown in Figure 5–9. The simplest form arises from a constant death rate and results in a steady decline in the numbers of living individuals in a cohort (typical of many plants and birds). A second form occurs when there is very little mortality until old age, which results in an almost constant population until the end of the life span when there is a rapid decrease in the number of living individuals (humans, univoltine insects). The third possibility occurs when

**FIGURE 5–9**

Theoretical population survivorship curves for three different individual mortality rates. (a) Steady mortality throughout cohort life span; (b) low mortality early in the cohort life with high mortality of mature adults; and (c) high initial juvenile mortality with long life of surviving adults.

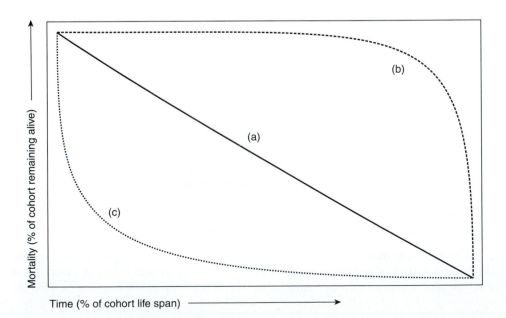

the death rate is high for juveniles and low for adults (typical of many pathogens, most insects, high densities of annual weeds, forest tree species).

Knowledge of the type of population decline is important to pest management, because it determines the best time to institute control action. For population decline type A, action needs to be taken once the population size and its rate of decline have been determined. For type B, it is necessary to act quickly as the population will not decrease much by natural means. For type C, it is better to delay action until the natural early mortality has occurred and the more stable population phase is occurring.

## Quiescence and Dormancy

Organisms typically have a resting or survival phase in their life cycle that provides a mechanism to withstand adverse environmental conditions (Figure 5–10). It is necessary to differentiate between those organisms that are quiescent (not active) due to adverse environmental conditions and those that are in diapause or are dormant. If the organism is quiescent, it resumes active growth when environmental conditions again become favorable. An organism that is in diapause or is dormant will not resume normal activity until specific physiological changes have occurred, often triggered by a specific set of environmental

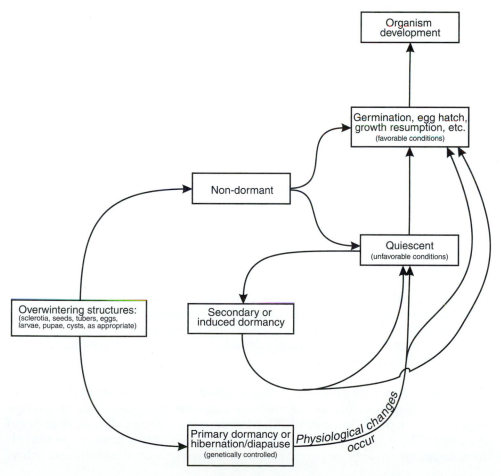

**FIGURE 5–10**
Relationships between dormancy/hibernation/diapause, secondary dormancy, quiescence, and germination on resumption of growth.

conditions (e.g., three months at a temperature below 42°F). When dormant, the organism does not necessarily resume normal activity if conditions become favorable prior to the requisite passage of time. The concept differs between pest categories.

***Quiescence***    Reduced growth or activity due to inappropriate environmental conditions is quiescence. Most organisms resume growth or activity once environmental conditions return to physiologically acceptable levels (Figure 5–10). The phenomenon takes two forms, hibernation and aestivation. Hibernation is the overwintering resting state when conditions are too cold to support growth; aestivation is the oversummering resting state under hot conditions. Many arthropods, some mammals, and some snails exhibit one of these phenomena. It often involves the adult organisms but in insects any of the basic developmental stages may undergo a quiescent phase, depending on the species.

Nematodes have different types of quiescence that are responses to environmental stresses, and if the stresses become sufficiently severe, then the nematodes enter a cryptobiotic ("hidden life") state of suspended animation in which there is no detectable metabolism. As soon as the stressful conditions abate, the nematode emerges from the cryptobiotic state and resumes normal activity. There is no fixed time required for quiescent periods as there is in diapause. The different types of cryptobiosis can be responses to cold (cryobiosis), oxygen deficiency (anoxybiosis), dehydration (anhydrobiosis), and osmotic shock (osmobiosis). For example, wheat-gall nematode may survive in an anhydrobiotic state for 32 years in wheat kernel galls (see Figures 2–7d and e), and anhydrobiotic stem nematode may survive for 23 years in plant tissue. Cryptobiosis is important to IPM because some management tactics may actually induce a quiescent state that allows the nematode to survive for an extended time.

Examples of quiescent resting structures include the following:

1. Adults. These include all vertebrate pests, some insects (e.g., alfalfa weevil), and snails.
2. Immature stages (larvae, prepupae, and pupae). These include some insects (e.g., peach twig borer, navel orangeworm) and nematodes.
3. Eggs. These include many insects and nematodes.
4. Juveniles and adults. Under conditions of anoxia, cold, or moisture stress, particular stages of some nematodes can enter a state of suspended metabolic activity called cryptobiosis that allows them to survive until environmental conditions again become favorable.
5. Dauer stages. Some nematodes have dauer stages that function in survival and dispersal, and have special adaptations, such as particularly thick cuticle, that allow them to survive adverse environmental conditions.
6. Cleistothecia, sclerotia. Survival structures of pathogens allow survival in adverse conditions and growth when conditions are suitable.
7. Seeds. Many weed seeds simply do not grow because environmental conditions are not correct.

***Diapause***    A period of arrested development in some insects and some nematodes is diapause. Insects may undergo diapause at any developmental stage (egg, larva or nymph, pupa or adult), whereas nematode eggs typically undergo diapause. The term *diapause* refers to an obligate adaptive mechanism that allows organisms to survive periods of adverse environmental conditions such as extremes in temperature or the absence of adequate food. Many species in temperate zones undergo an obligate winter diapause. In anticipation of the onset of cold temperatures the last generation of the year in a multivoltine species will respond to the decreasing day length in the fall (known as a photoperiodic response). This stage accumulates nutrient reserves (if a larva or adult), will search for a protected area, and remain quiescent until the following spring when temperatures rise and days are longer. Obligate diapause is hormonally regulated. The Colorado potato beetle is a good example of a species that undergoes diapause in the adult stage. They overwinter in the soil and become active in the spring. A similar response may occur when insects are faced with extremely hot temperatures in the summer; they undergo aestivation. Often under a facultative summer diapause, the insect remains active if the temperature does not exceed upper threshold extremes.

Some insects enter diapause when the host plant is no longer available, and the insects remain inactive until the appropriate host becomes available again. The Western corn rootworm used to diapause as eggs in the soil and emerge when corn germinated in the following season. A good pest management approach for rootworms in the Midwest USA was the rotation of corn with soybean. The overwintering population did not survive in the absence of the preferred host. Lately, however, some rootworm biotypes evolved a two-year diapause in the egg stage and thus survived to infest the corn planted the year following the soybean rotation. The pest management advantage of the corn–soybean rotation was significantly reduced by this development, and serves as a good example of resistance in an insect to a cultural management practice.

The potato cyst nematodes have a diapause that requires exposure to cold winter months prior to egg hatch. Eggs of the cyst nematode can remain viable in soil for many years.

***Dormancy***   An organism is said to be dormant if it does not resume active growth under appropriate environmental conditions. The dormant state can last from a few weeks to decades and even centuries. It occurs in weed seeds and vegetative structures (plus a few pathogens and nematodes). Innate dormancy means that seeds will not germinate (grow) even when environmental conditions are acceptable, which differs from dormancy enforced by conditions that do not support growth (Figure 5–10). Loss of innate dormancy in weed seeds is a slow process that occurs over a period of years; thus, viable weed seeds can remain in the soil for many years, resulting in the development of what is called the soil weed seed reserve, or seedbank.

**An old British adage states: "one year's seeding equals 7 years weeding."**

***Soil Weed Seedbanks, Nematode Eggbanks, and Fungal Sclerotiabanks***   Weeds, nematodes, and fungi have survival stages that can remain viable in soil for many years, and the survival of organisms in an inactive state may be considered as a storage effect. Weed seedbanks are the total accumulated viable seeds in the soil that have not germinated due to innate dormancy. The length of time that seeds in the seedbank remain viable (defined as the length of time that a single seeding event will lead to viable seeds) in the soil varies greatly with weed species involved; similarly, nematode and fungal species have long-term survival structures. Seedbank longevity ranges from about 4 to over 100 years, while eggbanks and sclerotiabanks typically survive from 2 to 20 years.

Due to the high seed output and great longevity of many weeds, the seed-bank size in arable soils can become very high. In well-maintained fields, values in the range of a few hundred to about 4,000 seeds/m$^2$ seem common. In poorly maintained fields, seedbanks in excess of 75,000 seeds/m$^2$ have been recorded in several examples. If weeds are not controlled, seedbanks in excess of 500,000/m$^2$ can be anticipated.

The budbank is analogous to the seedbank. A budbank is the total accumulated meristems (buds) on vegetative reproductive structures of weeds such as nutsedges, johnsongrass, and field bindweed. Each bud is capable of generating an entire new plant, and is thus numerically equivalent to a seed. The budbank tends to be shorter lived than the seedbank; 2 to 10 years is typical.

Seedbanks, budbanks, eggbanks, and sclerotiabanks have important implications for pest management. They necessitate that long-term impacts of control strategies be considered. The presence of the seedbank also means that the genetic makeup of the population is maintained over many years. This phenomenon has been referred to as genetic memory, because the genotypes of 20-year-old dormant seed reflect genotypes present when the seed was shed, not that of the current growing population. Genetic memory has implications to management of herbicide resistance in weeds (see Chapter 12).

## Heat Summation and Degree-Days

In cold-blooded organisms, termed poikilotherms, rates of metabolism and physiological processes are regulated primarily by environmental temperature. For normal physiological processes to proceed, and the organism to have normal development or activity, the ambient temperature must be above a critical lower limit called the lower threshold temperature (Figure 5–11). If the temperature is below the lower threshold, then normal physiological and metabolic processes slow and result in inactivity or cessation of development. An upper threshold temperature also exists, above which development stops. If temperatures are too low or too high, death of the organism may result through freezing or heat damage to proteins, cell membranes, and other cell components.

Poikilothermic organisms are interesting, because when temperatures are below their lower threshold the organisms are not active and thus do not develop; in effect they do not age. Accordingly, the chronological or calendar age of a pest organism may not correspond to the expected developmental stage of the pest. For example, in humans, a three-year-old can always be expected to display certain developmental characteristics. The development of a fly larva depends on the passage of time and the temperature during that time. The rate of development of poikilothermic organisms can be estimated by summing the time that has elapsed under specific temperature conditions (i.e., between the lower and upper threshold temperatures). By accumulating the hours per day (or the days) between the lower and upper thresholds, it is possible to predict the developmental rate of the organism (Figure 5–12). This provides a means of predicting generation time relative to environmental conditions. The method used to predict temperature-dependent developmental rates is based on the degree-day concept, which is often abbreviated as °D. A single degree-day is defined as a temperature 1° above the threshold maintained for 24 hours. Using degree-day accumulation has become central to management decision making for several major insect pests (see Chapter 8).

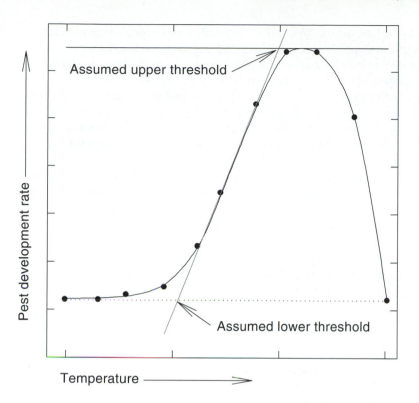

**FIGURE 5–11**
Generalized influence of temperature on the development rate of poikilothermic (cold-blooded) organisms.

Assumed upper threshold

Pest development rate

Assumed lower threshold

Temperature

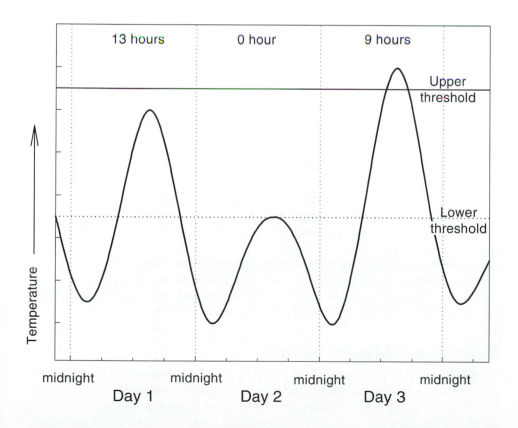

**FIGURE 5–12**
Graphical representation of how degree-day accumulation is calculated (see text for explanation).

13 hours          0 hour          9 hours

Upper threshold

Lower threshold

Temperature

midnight          midnight          midnight          midnight
Day 1          Day 2          Day 3

**Phenology is the sequence of growth and development of an organism over time.**

1. The degree-day accumulation is used to predict phenological events for several major arthropod pests, such as codling moth, several species of apple leaf rollers, predacious mites, San Jose scale, and cotton bollworms. Once the timing of major events in the life cycle can be predicted, the initiation of management activities can be directed at the most susceptible stage in the pest's life cycle. It is necessary to be aware that poikilothermic organisms may have different threshold temperatures for different activities in the life cycle. For example, the lower threshold temperatures for nematode activity may be different than the threshold temperatures for reproduction and development. Thus, it may be possible to plant when soil temperatures are low enough that nematodes are unable to infect seedlings.

2. Degree-day-based models for certain plant diseases also are available; examples are apple scab and fire blight of apple and pear. Although temperature is an important factor in the development of plant disease, moisture is also critical for pathogen infection and is, depending on the pathogen, a driving variable considered in such models.

3. The degree-day concept is only partially useful for predicting plant (crop or weed) growth, because soil water status can often become more important than temperature. In irrigated agriculture, however, degree-day accumulation can provide reasonably accurate prediction of phenological development. Light intensity may also play a critical role in determining plant development in some situations. Predictive growth models for crops are useful in IPM because they provide a means to match crop phenology with pest phenology, a critical piece of information in the application of the economic injury level concept (see Chapter 8).

4. The degree-day concept has no relevance to management of vertebrate pests, as the metabolic rate of warm-blooded animals does not depend on environmental temperature in the way that is used in the degree-day concept. Warm-blooded animals do, of course, show variation in activity in relation to temperature and seasons.

***Calculation of Degree-Days***   The computation of degree-days is based on maximum and minimum daily temperatures. A simple average degree-day uses the daily maximum (MAX) and minimum (MIN) temperature, a lower threshold ([TLOW]), such as 50°F, and the formula: degree-day = ([MAX + MIN]/2) − TLOW. So for a MAX and MIN of 100°F and 40°F, and TLOW of 50°F, the calculation is ([100 + 40]/2) − 50, or 70 − 50 = 20 degree-days for the 24-hour cycle. Degree-days are accumulated by adding the per day degree-days over consecutive days. More complex methods to compute degree-days have been developed, but we do not deal with those here. By using the World Wide Web it is now possible to download weather data directly from monitoring stations and obtain the degree-day accumulation for many pest species at specific locations. To use the degree-day models on the WWW, the only information required is the lower threshold temperature for the organism of interest.

## Molting and Metamorphosis

All organisms increase in size as they mature. Plants are referred to as modular organisms, and can increase in size at the periphery or by adding new modules (e.g., leaf, branch). Plant pathogens are able to increase in size in a manner similar to plants. Mammals have an internal skeleton that increases in size with growth.

Arthropods and nematodes, however, have an external skeleton, which cannot readily expand as the organism grows. Arthropods and nematodes need to molt in order to increase in size, and molts occur several times during their life cycles. Different stages in the life cycle are given different names, which are often the basis of understanding population development relative to the timing of management actions. Figures 5–13 and 5–14 provide examples of this important terminology. Many arthropods and nematodes also exhibit a change in gross morphology between juvenile and adult life stages, although nematodes do not undergo metamorphosis, but many insects do. There are three different types of arthropod metamorphosis:

1. **No metamorphosis (Ametabolous).** In the simplest form of arthropod development, juveniles resemble small versions of the adult (Figure 5–13). There is no change in gross morphology. Symphylans and silverfish are examples.

2. **Incomplete or partial metamorphosis (Paurometabolous).** A large group of arthropods exhibit only a partial change in morphology between juvenile and adult growth stages (Figures 5–13 and 5–14). The difference between juveniles and adults is most frequently manifested in absence of wings in the juvenile forms. Lack of wings in the juveniles has important pest management implications as it means the juvenile forms are much less

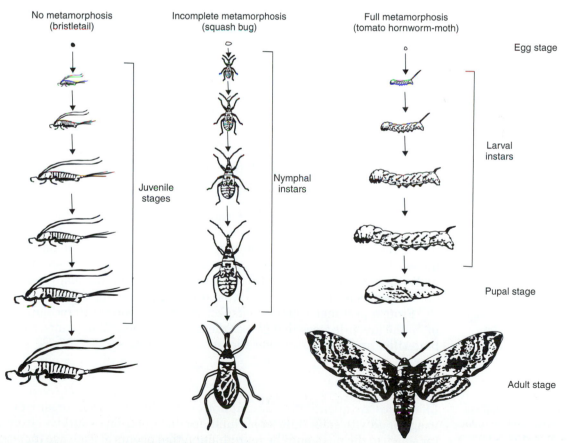

**FIGURE 5–13**  Examples of the three different types of insect development.
Source: redrawn from Evans and Brewer, 1984, with labels added.

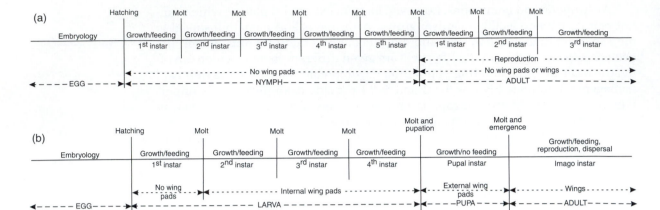

**FIGURE 5–14**   Comparison of the different life stages in a) ametabolous and b) holometabolous insect development, including comparative terminology.
Source: redrawn after Elzinga, 1997.

mobile than the adults. Examples are aphids, psyllids, lygus bugs, and grasshoppers.

3. Complete or full metamorphosis (Holometabolous). In this developmental pattern, the juvenile growth stages do not resemble the adult. There is typically a nonactive growth stage, called a pupa, during which profound morphological changes take place (Figures 5–13 and 5–14). The classic example of complete metamorphosis is the caterpillar larval stage that changes into a chrysalis, often within a cocoon, from which then emerges an adult butterfly or moth. Note the different terminology that is used with this type of development. All moths and butterflies (e.g., codling moth and cabbageworms), flies (e.g., Mediterranean fruit fly, apple maggot), and beetles (e.g., boll weevil, flea beetles) are examples.

These different forms of development can be depicted in a time line format, which shows how different stages compare for the different types of metamorphosis (Figure 5–14).

Some nematode species show variable morphology from the juvenile to the adult stages. In the cyst and root-knot nematodes, for example, the juveniles that infect the host are vermiform and mobile whereas the later juvenile stages and the adult females are swollen and immobile. In most of such species the adult males are vermiform. Such difference in form between the male and the female of the species is an example of sexual dimorphism (Figures 5–15 and 5–16).

From a pest management perspective, it is important to recognize that it is often the juvenile stage that causes the damage to desirable plants, including all butterflies and moths, some beetles, and most flies.

## Life Tables

**Life tables document survival and reproduction relative to time or the stages in the life cycle of an organism.**

Life tables are quantitative descriptions of the life history of organisms relative to age or growth stage. Life tables include consideration of survival from one age (or stage) to the next, and the reproduction that occurs at each age (or stage). The life stages of various pests are compared in Table 5–1. To construct a life table, the mortality during and between the different ages (stages) is determined,

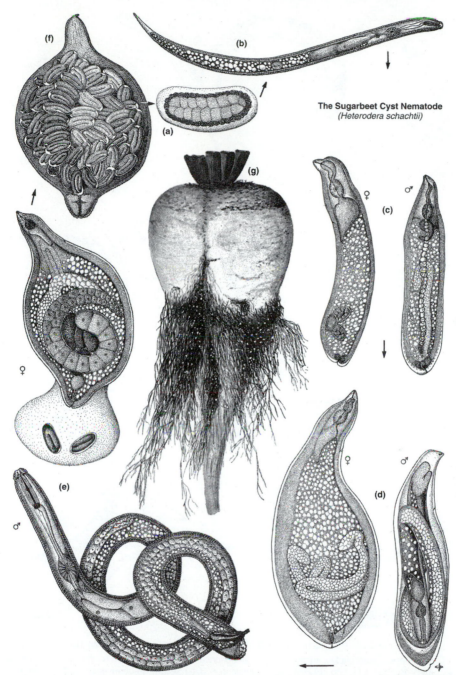

The Sugarbeet Cyst Nematode
(*Heterodera schachtii*)

**FIGURE 5–15**   Life cycle of a sugar beet cyst nematode (*Heterodera schachtii*), including sexual dimorphism. Second-stage juveniles (J2) are stimulated to hatch by host-root exudates. The J2 penetrate the host root, typically near the root tip, they enter the root, move to a position with their head near the vascular cylinder, and cease movement. The nematode feeds, and causes dissolution of the plant cell walls resulting in a multinucleate syncytium in the pericycle tissue of the vascular cylinder. The developing nematode obtains nutrition from the syncytium, sometimes called a nurse cell, and continues development through the juvenile stages to become an adult. Adult males are vermiform and are attracted to females by pheromones. As the female matures and swells, her posterior body emerges from the root surface while her head remains within the root. Females produce from 200 to 600 eggs, and the life cycle, from egg to egg, takes about 22 days at a temperature of 25° C.

Source: drawing by Charles Papp, California Department of Food and Agriculture.

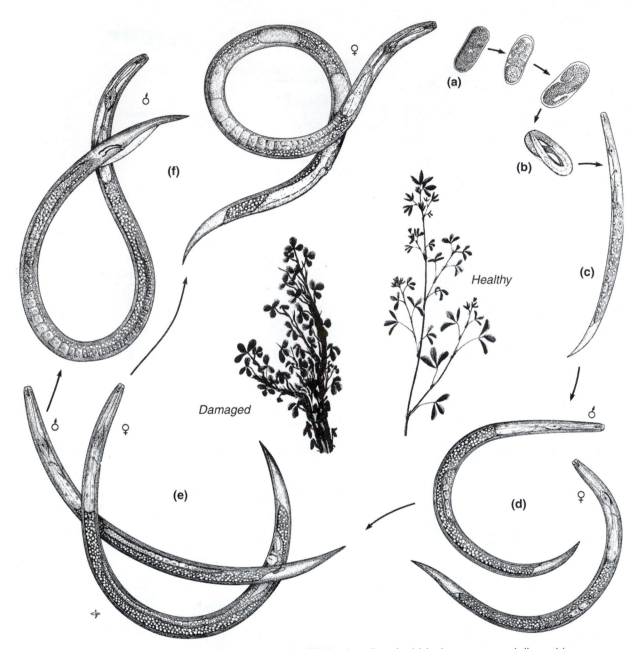

**FIGURE 5–16**   Life cycle of a bulb and stem nematode (*Ditylenchus dipsaci*) which shows no sexual dimorphism.
Source: drawing by Charles Papp, California Department of Food and Agriculture. The host plant is alfalfa, and the nematode infects leaves and stems, as well as tubers and bulbs. (a) Egg, (b) First-stage juvenile in egg, (c) Second-stage juvenile hatches from egg, (d) Third-stage juveniles, (e) Fourth-stage juveniles, (f) sexually mature adult ♂ and ♀.

which then permits derivation of survivorship curves. This information can be used to predict possible rates of population change, and can aid in identification of mortality factors operating at different ages (or stages) in the life cycle.

As an organism ages it goes through its various stages of development, and the sequence of organismal growth and development over time is referred to as its phenology. Some pest management strategies are aimed at a specific phenological stage of the pest.

| TABLE 5–1 | Life Table of Phenological Stages Compared for Different Categories of Pests | | | |
|---|---|---|---|---|
| **Pest Organism** | **Juvenile Stage** | **Mature Stage** | **Reproductive Unit** | **Overwintering** |
| Pathogens: fungi | Mycelium | Mushroom, sporangium, etc. | Spores, zygospores, conidia | Spores, zygospores; special structures (e.g., cleistothecia, sclerotia) |
| bacteria | Cell | Cell | Cells | Bacterial cells in host tissue |
| viruses | Virus particle | Virus particle | Virus particle | Virus particle in host tissue |
| Weeds: annual | Seedling | Flowering | Seed | Seed |
| perennial | Vegetative | Flowering | Seed or buds (stem pieces) | Seed or vegetative structure (rhizome, tuber, etc.) |
| Nematodes | Juvenile | Adult | Eggs | Eggs, juveniles |
| Mollusks | Juvenile | Adult | Eggs | All stages |
| Arthropods: incomplete metamorphosis | Nymph (instars) | Adult | Eggs (some live young) | Eggs, adults, some nymphs |
| full metamorphosis | Larva to pupa (instars) | Adult | Eggs (some live young) | Eggs, adults, some larvae, pupae |
| Vertebrates | Baby, young | Adult | (not present, N/A) | Adults |

# Basic Life Cycle Models

Most pest organism life cycles can be presented in a generic format, although as with most generalizations, important details are lost. Figure 5–17 represents four generic pest life cycles.

**Understanding the life cycle of a pest is helpful in designing an effective management strategy.**

1. Summer/fall reproduction (Figure 5–17a). Warm-season organisms are quiescent (hibernate) or are dormant in the winter, and include many insects, some pathogens, some nematodes, and many mammals. Summer annual weeds fit this cycle with recognition that the seedbank persists from year to year.
2. Winter/spring reproduction (Figure 5–17b). This cycle is the mirror image of Figure 5–17a. It represents cool-season organisms that are quiescent (aestivate) during the warm season. Examples include some insects (Egyptian alfalfa weevil), some mammals (ground squirrels), and most winter annual weeds (seedbank now persists through the summer). Cycles a and b represent univoltine or single-cycle pests. The degree to which these two cycles differ depends on climate zone. In the tropics there is essentially no difference, but in Mediterranean climates the two are very distinct, and toward the cooler regions there is only one cycle. In the latter, the b cycle organisms shift to the a cycle, and the warm-season organisms are typically not able to survive.
3. Summer/fall reproduction with additional reproduction within the growing season (Figure 5–17c). This type of life cycle represents those organisms that have short generation times, and can reproduce more than once within the growing season. Nematodes and multivoltine arthropods including aphids and spider mites, and the multicycle diseases such as

**FIGURE 5–17**   Four possible generic life cycles. (a) Univoltine or single-cycle organism that has a resting phase in the winter; (b) univoltine or single-cycle organism that has a resting phase in the summer; (c) multivoltine or multicycle organism with multiple generations during the growing season; and (d) univoltine or single-cycle organisms that have a resting stage that is persistent in the soil.

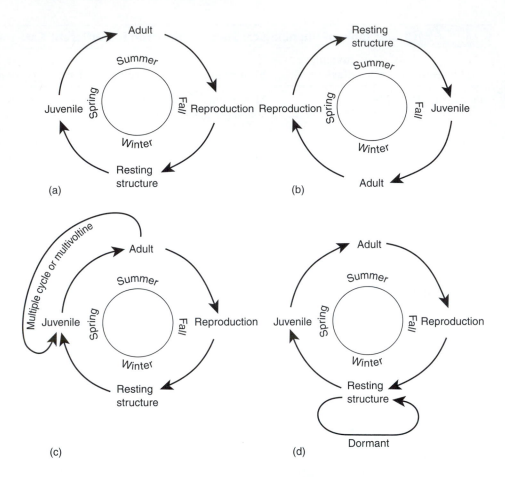

many rusts and mildews. Mice, other small mammals, and mollusks also have this type of life cycle. There are no weeds with this type of life cycle.

4. Dormant reproductive stages (Figure 5–17d). Dormant or diapausing resting plant structures or animal stages survive longer than the next generation cycle. Examples include all weed seedbanks, some pathogens (e.g., sclerotia of lettuce drop and microsclerotia of rice stem rot), and nematode eggs within cysts.

The specific life cycles of many pathogens are fairly complicated. Figures 5–18 through 5–21 demonstrate this complexity. The many variations need to be understood to implement effective disease management programs. Most viruses and fastidious bacteria require vectors for transmission (Figure 5–21). Other pathogen life cycles are described in plant pathology textbooks (see Agrios, 1997).

The generic life cycles shown in Figure 5–17 do not properly represent the life cycle for biennial weeds. These plants have a two-year reproductive life cycle in which the plant grows vegetatively in the first season and produces seeds in the second season.

***Alternate and Alternative Hosts***   Several pathogenic organisms, and some insects, require an obligate alternate host plant for completion of the life cycle.

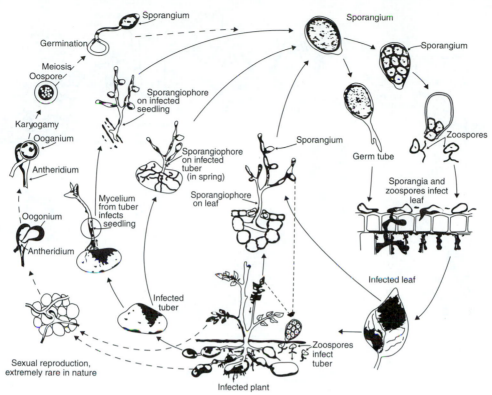

**FIGURE 5–18**   Life cycle of a multicycle fungus that does not require an obligate alternate host to complete its life cycle. The example here is for late blight of potatoes.
Source: redrawn from Agrios, 1997.

For example, wheat rust requires the barberry bush, and the lettuce root aphid requires Lombardy poplars.

Many pathogenic organisms cannot survive outside a living host (some bacteria, phytoplasmas, and all viruses), and when a primary host is not present in an agroecosystem, the pathogens may survive in other plants, described as alternative or reservoir host. Pathogens, nematodes, and arthropods that are not host specific can utilize many plant hosts. Such hosts may not be the primary host of the pest and are usually referred to as alternative or reservoir hosts. Although the entomological literature refers to plants able to serve as secondary hosts as alternate hosts, to avoid confusion with obligate alternate hosts, we designate secondary hosts as alternative hosts.

# DISSEMINATION, INVASION, AND COLONIZATION PROCESSES

The ability of an organism to spread, or be spread, is critical to its ability to become a pest. If an organism cannot spread, either through its own motility or through the activities of humans, the likelihood of its achieving pest status is relatively low, even if other attributes of population dynamics are suitable. In many cases, human activities effect the dissemination of organisms that otherwise would spread very little, so people create the pest (see Chapter 10).

**It is unlikely that an organism will be a serious pest unless it is disseminated to, and colonizes, a crop.**

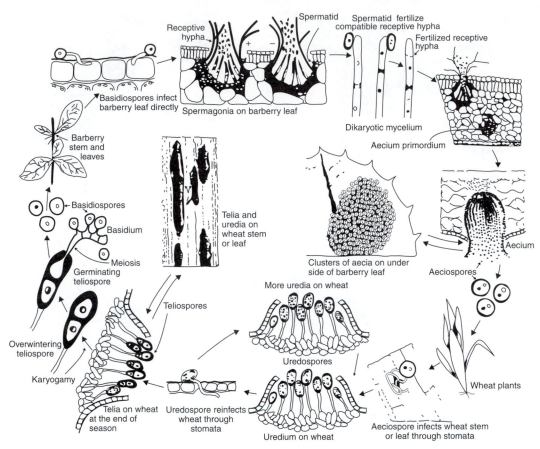

**FIGURE 5–19** Life cycle of a single-cycle disease that requires an obligate alternate host (not the crop) to complete its life cycle. The example here is for wheat rust.
Source: redrawn from Agrios, 1997.

## Mechanisms of Dissemination

**Soil-borne pest organisms (most weeds, many pathogens, nematodes) are frequently disseminated by human activity.**

The mechanism by which pests are disseminated varies widely in relation to pest categories. The mechanism by which one pest spreads may not apply to another pest. Following are three basic means by which organisms can spread.

1. Passive. The pest is disseminated in the air by wind, carried in water, or on an artifact (usually transportation or farming equipment) that is moved by people.
2. Active. The pest is capable of moving itself (by walking, flying, or swimming) from one location to another.
3. Vector and phoresis. The pest is taken up, usually internally, by a vector organism from an infected host and moved to an unifected host; in phoresis, the pest is moved by being physically attached to another mobile organism.

Each of these means act, to varying degrees, to disseminate the following pest categories.

***Pathogens*** Different classes of pathogens have different means of spread which greatly impact the way they can be managed.

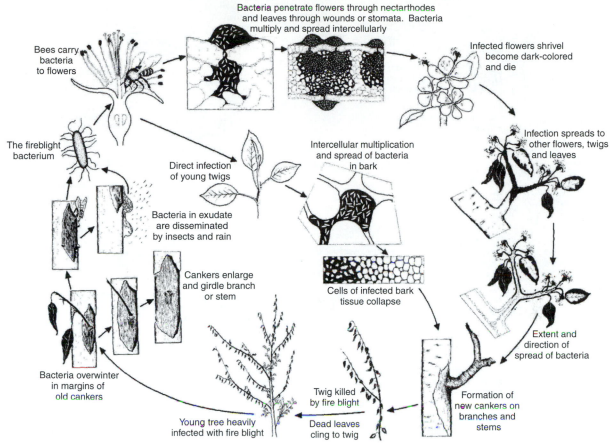

Bacteria penetrate flowers through nectarthodes
and leaves through wounds or stomata. Bacteria
multiply and spread intercellularly

Bees carry
bacteria
to flowers

Infected flowers shrivel
become dark-colored
and die

The fireblight
bacterium

Direct infection
of young twigs

Intercellular multiplication
and spread of bacteria
in bark

Infection spreads to
other flowers, twigs
and leaves

Bacteria in exudate
are disseminated
by insects and rain

Cankers enlarge
and girdle branch
or stem

Cells of infected bark
tissue collapse

Extent and
direction of
spread of bacteria

Bacteria overwinter
in margins of
old cankers

Twig killed
by fire blight

Formation of
new cankers on
branches and
stems

Young tree heavily
infected with fire blight

Dead leaves
cling to twig

**FIGURE 5–20**  Life cycle of a bacterial disease; the example is fire blight of pears.
Source: redrawn from Agrios, 1997.

**Spores**  Spores are the reproductive stage produced by many pathogens. They are often airborne or waterborne, and some are lighter than air and can drift for long distances before landing. Spores have been recorded at altitudes of up to 20,000 ft! Spore-producing pathogens can therefore spread independently and perhaps in spite of human activity. From a human management point of view there is almost no means of combating spread of these organisms, except by reducing the source of inoculum. Other spores in soil may be carried by machinery, spread by splash during rain events or sprinkler irrigation, or carried by insects (e.g., Dutch elm disease by elm bark beetle). Zoospores are motile spores that have a flagellum or cilia and can swim short distances, providing active dissemination.

**Mycelium**  Mycelium is the vegetative body of a fungus, and it can spread locally by invading adjacent plants (e.g., root grafting). Humans spread mycelium in or on plants.

**Bacteria**  These pathogens have no specific mechanisms of spread. They are spread by humans or other animals (phoresy). Some are locally spread by rain or sprinkler splash for short distances (e.g., fire blight in pears and bacterial speck of tomato), or may be spread by foraging insects.

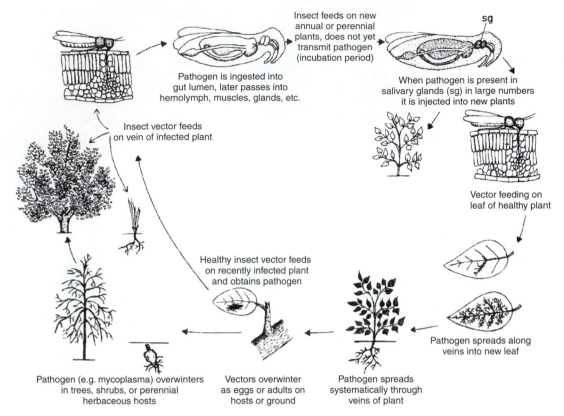

**FIGURE 5–21**    Generalized life cycle for a fastidious bacterium or virus; it requires the presence of a vector organism (here leafhopper) to transmit the pathogen from host to host. The figure shows this for a perennial tree crop where the host crop maintains the inoculum; a variant of this life cycle occurs for annual crops, where the inoculum is maintained in a noncrop alternate host plant (often a weed) when the crop is not present.
Source: redrawn from Agrios, 1997.

**Controlling the vector can be very effective in limiting the spread of virus diseases.**

**Viruses, Fastidious Bacteria, and Phytoplasmas**    Fastidious pathogens cannot exist outside a host and must be associated with living cells. There is no specific resting stage. They can only be spread with the aid of their host or a vector organism. Virus vectors include insects, nematodes, and fungi that take up the virus from one plant and move it to another. This is known as transmission by vector. Some are spread by physical contact, such as *tobacco mosaic virus,* in which case human activity may be involved. They can also be spread when the infected host is transported and vegetatively reproduced, such as grapevines. Some other viruses may be disseminated in infected seeds. The various ways in which viruses are transmitted are depicted in Figure 5–22, and the mechanism of transmission to a great extent determines how the virus must be managed.

**Serious weeds that have no dispersal mechanism include nutsedges, barnyardgrass, goosegrass, wild oats, pigweeds, lambsquarters, wild mustards, nightshades, and velvetleaf.**

*Weeds*    A widely held misconception is that most weeds have reproductive structures to facilitate long-distance dispersal. Most of the world's worst weeds have no special adaptations to aid dissemination, and have been widely spread only through human activity. A relatively few weed species possess special structures that aid spread over distances greater than a few meters.

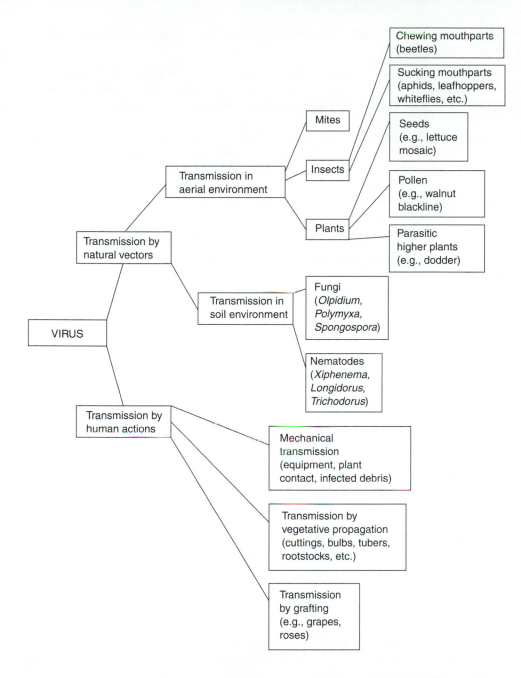

**FIGURE 5–22**
Diagrammatic representation of the different organisms involved in the transmission (vectoring) of viruses, and other fastidious organisms, from one host to another.

**Seeds**　Seeds or fruits possess special structures to facilitate dispersion by wind. Examples include a pappus (feathery structure) or wings. Some seeds are sufficiently small and light to be wind blown (e.g., broomrape species). In a few instances, the whole mother plant is dispersed, spreading seeds as it travels (e.g., tumbleweeds such as *Salsola* spp.). Most seeds float and can be dispersed by water, which creates a serious management problem when irrigation water is supplied using surface canals or after flooding events. Hooked or spiny fruits and seeds can become attached to, and spread by, animals, while fleshy fruits may be eaten by birds and other animals and have seeds that can survive passage through the gut.

**Vegetative Fragments of Plants**   Plant fragments such as pieces of rhizomes, tubers, or stolons may be broken and moved by tillage equipment. Pieces of aquatic plants break off, float downstream, and take root in a new area.

### Nematodes
Most nematodes have no direct capacity for active dispersal over long distances. They are spread readily in windblown soil and in water, including irrigation water. Human activities have spread nematodes over wide areas, either through the movement of infected plant material and soil, or on contaminated implements and machinery. Mobile animals such as birds or cattle may accidentally ingest nematodes, and because the nematodes may survive passage through the digestive system, such ingestion aids nematode dissemination.

**Eggs and Eggs in Cysts**   Obviously, cysts and eggs in cysts do not disperse by active movement, but they are light and can be carried for considerable distances in wind-blown soil. Under appropriate conditions, cysts float and can be carried in irrigation or floodwaters. Because cysts and the eggs they contain are persistent, they are spread by human and animal activity. Sugar beet cyst nematode is a good example of a nematode that was widely disseminated as a contaminant on sugar beet seed. Historically, beet seed was frequently gathered from the soil surface, thus inadvertently including nematode cysts containing eggs. Viable cysts of the sugar beet cyst nematode have been recovered from cattle and bird feces, and animals foraging on the ground may ingest cysts.

**Juveniles and Adults**   Juvenile and adult stages can move short distances through the soil, but do not move over distances greater than several meters in any given year. They can be carried in wind or in water. Infected planting material is an important means of nematode dissemination. Some plant-parasitic nematodes, such as the nematode that causes red-ring disease in coconut and the nematode that causes pine-wilt disease, are spread tree to tree by insects.

### Arthropods
The adults of many species are capable of active movement over relatively large distances, either by walking or flying. Immature stages do not possess wings and are thus much less mobile, but insects and many small arthropods are dispersed by air currents over long distances. The abundance of arthropods in the air at certain times of the year led to the concept of an aerial plankton.

**Eggs**   Although eggs are immobile they can, and have been, moved on field equipment, in soil, on or in produce, on prunings, in feed, in air, in water, by phoresy, and by human activity.

**Larvae and Nymphs**   Nymphs and larvae that have legs can move short distances, but this life stage is usually not involved in long-distance spread unless aided by human, or other animal, activity. The first instars of certain parasitic insects have special structures that facilitate phoresy (e.g., blister beetles).

**Pupae**   Arthropod pupae possess no ability to spread, but human activity can move them.

**Adults**   Adult insects that have wings often can fly over great distances; thus, their ability to spread is essentially independent of human activity. The ability of a species to disperse is called vagility. Species with low vagility tend to interbreed and cannot colonize new areas over long distances. A few organisms create special structures that increase their ability to spread, such as the silk strands used as "parachutes" by spider mites. Wind can carry insects long distances, and the aerial spread of insects is a major factor in the dissemination of certain species.

***Vertebrates (including human activity)***   Vertebrates are mobile organisms, and the adults can walk or fly long distances. Dissemination of pest vertebrates is not dependent on human activity in most cases, although humans have been responsible for the movement of many important vertebrate pests. Pests with high mobility are very difficult to exclude from agroecosystems.

Agriculture is a basic human activity, and humans have transported agricultural plant material as they migrated from region to region. Of course, many pests have hitchhiked along on such plant material. Industrialized agriculture and global trade and travel have accelerated both the temporal and spatial scales of pest dissemination. The difficulties caused by dissemination and invasion of pests is considered in detail in Chapter 10.

## Seasonal Migration and Movements

Some pests cannot survive the winter in the region where they are a problem during the growing season. They have to reinvade the area every year. In North America such organisms typically overwinter in Mexico or the southern states and migrate back into the more temperate northern states and Canada in the spring. This phenomenon cannot occur for weeds, soil-borne pathogens, or nematodes, as they do not possess the mechanism for long-distance movement. In other cases, pests do not migrate in the strict sense, but rather move over shorter distances as the seasons and environment change.

***Pathogens***   Several pathogens also reinvade North America from the south, moving northward each year as the growing season progresses; southern corn leaf blight and blue mold of tobacco are good examples.

***Arthropods***   The monarch butterfly is a classic example of insect migration between overwintering sites in South America and summer sites throughout North America. The velvetbean caterpillar and fall armyworm are good examples of seasonal pest migration in the eastern United States. Both pests are of tropical origin and cannot survive the winter temperatures. Another example is the migration of the brown planthopper from the tropical to the temperate zones in rice production areas in Asia. Knowledge of the patterns of seasonal reinvasion permits prediction of pest return, which in turn can improve the timing of control measures so that they coincide with the appearance of the pest. Such prediction is used successfully with the black cutworm in soybeans and corn in the Midwest United States.

Understanding insect vagility permits interpreting factors that have an impact on IPM. For example, soybean looper populations in Louisiana started to show signs of resistance to many insecticides. Because insecticide use, and hence selection pressure, on soybean was low, there was no reason for the rapid growth of a resistant endemic population. The explanation was that soybean loopers migrated into Louisiana soybean fields from breeding areas in southern Florida. Those populations in Florida developed on commercial ornamental plants where they received multiple insecticide treatments, thus being under considerable insecticide selection pressure.

**Many pests must overwinter in vegetation external to the managed system due to lack of suitable host plants.**

Due to the absence of host crops at certain times of the year, many mobile pests must overwinter on vegetation in noncropped areas and then move back to the crop in the spring. This type of movement is over relatively short distances, typically less than a few miles. Examples of such movement include cotton bollworms, various leafhoppers, leaf beetles, and lygus bugs. The concept of area-wide pest management is largely derived from knowledge of this short-range movement. In California, leafhoppers are controlled in overwintering sites in the foothills surrounding the Central Valley to reduce disease transmission to crops (e.g., beet leafhoppers and *curly top virus*). In the Southern USA, proposals have been made to control *Heliothini* on wild vegetation as a strategy to reduce the buildup of these insects in adjacent cotton crops. These tactics only apply to a pest that spreads back to the crop each year.

Attempts have been made to stop movement of pests, but most have not been very effective. Foremost among such attempts have been the various campaigns to control the migratory grasshopper in Africa.

***Vertebrates***   Annual long-distance migration of birds is a well-known phenomenon. Pest species such as blackbirds and starlings migrate substantial distances in relation to seasons.

# PEST GENETICS

The growth and development of all organisms results from the genes they possess, and the interaction of those genes with the environment. There are different forms of genes that control a particular function, and such variants are termed alleles. In some cases, more than one gene is involved with determining a particular trait, and such traits are termed polygenic to reflect this. The allelic variants of genes do not necessarily function at exactly the same level in different individuals, leading to variation in how the function is expressed. Good examples relative to pest management are alleles that result in altered enzyme function. Due to the genetic recombination that occurs during sexual reproduction, changes in the level of expression of an allele can also lead to subtle variations among individuals within a population. The complexities of how genes interact to form the many complex facets of individual organisms are still poorly understood, and such understanding is the goal of a new field in molecular genetics, functional genomics. Understanding the nature of intrapopulation genetic variability within pests is fundamental to many aspects of IPM.

**Genetic variability allows pests to evolve relative to the selection pressures of the environment, including management tactics.**

## Genetic Variability

A large amount of variability exists within the genotypes, or genetic makeup, of most pest species. This variability provides them with the ability to be successful in variable environments and under different management conditions.

The existence of genetic variability within pests is probably the most important aspect of pest biology in relation to long-term sustainability of IPM tactics. When a population of organisms with variable genotypes is exposed to an external stressor, such as high temperature or an unsuitable food source, the frequency of individuals in the population that tolerate the stress increases, while the frequency of intolerant types decreases. Eventually the genetic composition of the population shifts to genotypes that best tolerate the external stressor. Other genotypes will not necessarily be eliminated from the population, but rather their frequency is reduced to a low level.

This fact has important implications for pest management. The process by which genotypes increase in frequency in the population is termed selection (sometimes colloquially referred to as "survival of the fittest"), and is the mechanism for evolution by natural selection. The external stressor to which the organisms respond is termed *selection pressure.* The population-level responses of pests to the selection imposed by management tactics is an important manifestation of evolution.

Any pest management tactic exerts a selection pressure on the pest population, and will select for the genotype(s) in the pest population best able to tolerate that management tactic. This ecological process will occur regardless of the specific management tactic, but historically has been the most important in relation to the use of pesticides (see Chapter 12), and in breeding for host-plant resistance (see Chapter 17). The evolution of pests to overcome many of the tactics that have been developed to control them has been, and will continue to be, the most serious impediment to long-term sustainability of IPM.

The following brief discussion covers the terminology used to describe the genetic variability that occurs within pests relative to host specificity and mating compatibilities. The scientists conducting research on the different pest categories have established terminology used to indicate genetic variation within pest species, and comparisons of how some terms are used among the pest categories can be confusing. The terminology appropriate to each category of pest is explained in the following discussions.

## Pathogens

The inherent variability in pathogens provides them with the ability to attack different species of host plant and different host varieties, and to withstand environmental variation. In agricultural systems, genetic variability has also allowed pathogens to develop resistance to pesticides and other control tactics (see Chapters 12 and 17).

*Fungi*    Fungal pathogens usually have considerable host specificity. *Erisyphe polygonae,* for example, is a pathogen that causes a disease called powdery mildew, but only the host-specific *Erisyphe polygonae f. sp. betae* attacks sugar beets. The abbreviation *f. sp.* stands for *forma specialis,* which indicates that it is the sugar beet, hence *betae,* host-specific form of the fungus. Many fungi have such host-specific special forms. *Puccinia graminis f. sp. tritici* is the stem rust that attacks wheat (the genus for wheat is *Triticum,* hence the name *tritici*). Sometimes the *f. sp.* is omitted and the name is presented as *Puccinia graminis tritici.* For example, *Puccinia graminis hordei,* attacks barley (genus *Hordeum*).

Specific genotypes, called races, occur within some fungal species. These are host-specific forms of the fungus at the level of crop variety. For instance, over 200 races of wheat stem rust are adapted to different varieties of wheat. They are simply referred to as race 1, race 2, race 3, and so forth. The term *biotype* is applied to new variants within a race, so that the designations 1A and

1B are used for the biotypes. The same fungus species thus can have considerable genetic diversity which permits it to attack different hosts, and even different varieties within a host species.

**Bacteria**   Many bacteria are host specific, and genotypes that parasitize certain host species are called pathovars (pv.). *Pseudomonas syringae* causes leafspot diseases of many plants. *P. syringae* pv. *tomato* causes bacterial speck of tomatoes, *P. syringae* pv. *tabaci* causes wildfire of tobacco, and *P. syringae* pv. *phaseolicola* causes halo blight of beans. *Xanthomonas campestris* causes spots and blights on many crops, whereas *X. campestris* pv. *phaseoli* causes common blight of beans, *X. campestris* pv. *oryzici* causes leaf blight of rice, and *X. campestris* pv. *malvacearum* causes angular leafspot of cotton.

Many fungi and bacteria have developed resistance to fungicides and bactericides (see resistance discussion in Chapter 12).

**Viruses**   Viruses can also develop host-specific types, which are referred to as strains. When two strains are introduced into the same host, a type of genetic recombination may occur producing a new strain.

## Weeds

**Crop cultural practices select for weed biotypes that are "crop mimics."**

Most weeds are morphologically, phenologically, and physiologically variable. Leaf shape and color in field bindweed is, for example, extremely variable. Barnyardgrass and yellow foxtail can range from almost prostrate to upright, and lambsquarter produces dark- and light-colored seed that vary in dormancy. Weeds can also differ in their ability to tolerate herbivores and pathogens. Different genotypes of field bindweed exhibit almost complete tolerance, or susceptibility to powdery mildew and two-spotted spider mites. Differing genotypes of weeds that breed true are known as biotypes. It is ecologically unlikely that a plant with little genetic variability would ever reach the status of being designated as a weed by humans.

Inherent genetic variation in weeds, as in other pests, permits them to respond to external selection pressure. Weeds that are crop mimics are some of the oldest known examples of this type of selection. Selection pressure exerted by the cultural practices used for the crop selects for weeds that are also adapted to those cultural practices. False flax (*Camellina*) was selected so that it matured at the same time as the flax crop, an example of selection on phenology. Late watergrass in rice was selected by harvesting machinery to have seed that is essentially the same size as the rice kernel (phenological and morphological selection). Prostrate forms of yellow foxtail have been selected in alfalfa in California because they are adapted to the monthly mowing used to harvest the crop (morphological selection) (Figure 5–23a). Upright biotypes occur in regions where the weed grows in tall crops such as corn and cereals (Figure 5–23b).

## Nematodes

The considerable physiological variation that exists within nematode species and populations has been documented primarily through differences detected in host specificity, host preference, and host resistance–allele specificity. Defining the genes involved in physiological variation is an active research area in nematology. The interaction between host-plant resistance genes and nematode

(a)                                                                    (b)

**FIGURE 5–23** Biotypes of yellow foxtail. (a) Prostrate form that occurs in alfalfa in California, and (b) upright biotypes that occur in corn and cereal crops in the northern and eastern United States.
Source: photographs by Robert Norris.

virulence genes, for at least some nematodes, appears to be a gene-for-gene type of relationship.

Many nematologists, particularly in the United States, have used the term *race* to designate the infraspecific groups that can be distinguished by their ability to reproduce on particular genotypes of different host-plant species, or on different genotypes of particular plant species. For example, within the root-knot nematodes, numeric host-race designations have been assigned on the basis of a simple host-range differential test, which involves a bioassay to assess the ability of a nematode isolate to reproduce on designated cultivars of peanut, pepper, watermelon, tobacco, and tomato. For example, race 1, race 2, race 3, and race 4 of a particular nematode species might be defined using such a bioassay.

In Europe, the term *pathotype* has been used to describe and characterize the ability of potato cyst nematodes to reproduce on a series of *Solanum* clones that carry different nematode-resistance genes. The clones were developed to aid in characterizing variant populations of the potato cyst nematode. Some controversy exists concerning the appropriateness of the pathotype schema.

The soybean cyst nematode is a yield-limiting pathogen of soybean in many areas of the world, and is a good example of the difficulties and importance of defining genetic variability in nematodes. A successful management tactic for soybean cyst nematode has been the development of soybean varieties that have resistance to the nematode. Unfortunately, as the resistant soybean varieties were grown in fields, it was discovered that it did not take very long, sometimes only several growing seasons, before cyst nematode races were detected that were able to survive and reproduce on the resistant varieties. The initial response to the problem was to find other, different, resistance genes to put into new soybean varieties, but as this was accomplished, new resistance-breaking soybean cyst nematode races were detected. The detection of new races was an unanticipated consequence of the selection pressure produced by the deployment of resistance as a management tactic. Now, the soybean cyst nematode has many different races that are distinguished by their ability to reproduce on different resistance alleles in soybean.

A flaw in the race concept is that the number of resistance alleles or host species used in the defining bioassay determines the number of races. Using a bioassay host differential involving $n$ hosts or alleles, the potential number of nematode races that might be designated is, in principle, $2^n$. As new resistance genes are discovered and tested against nematodes, new races must be defined, and accordingly, the race composition within a field redefined. An important concept is that different nematode races may occur within a field, and the particular resistance alleles or host species deployed act as selection pressures that will change the frequency of the races within the field over time. This has ramifications for how resistance genes are deployed in fields over time.

The variability among and within nematode populations needs to be recognized and characterized in order to anticipate population responses to, and hence the efficacy of, nematode management tactics. Efforts are now underway to standardize and codify approaches for dealing with the genetic variability in plant-parasitic nematodes, including the root-knot nematodes and the soybean cyst nematode.

## Mollusks and Vertebrates

Undoubtedly biotypes exist, but their importance to pest management has not been documented.

## Arthropods

Understanding interactions between plants and their herbivorous arthropods has been of interest to ecologists for decades, and many books have been written about insect and host-plant interactions. Arthropods respond to plant characteristics such as succulence, toughness, hairiness, sticky glands, and chemical content that can act as selection pressures on arthropod populations. Arthropods thus exhibit many biotypes associated with particular crops, or even varieties within a crop. Hessian fly on cereals was the first insect in which variety-specific biotypes were observed as far back as 1779. There are 16 recognizable biotypes of the Hessian fly in cereals based on studies of interactions between the insect and different host plants. Studies on the Hessian fly provided the basis for the theory of the gene-for-gene resistance in insects (more about this in Chapter 17). According to this theory, for each gene, for virulence in the fly, there is a gene for resistance in the wheat, or vice versa, so no wheat variety is resistant to all biotypes of the fly and no biotype is capable of attacking all wheat varieties. This condition requires that breeders be constantly screening new germ plasm for sources of resistance.

There are at least 15 pea aphid biotypes that attack different crops and weeds. *Phylloxera* is a root parasite of grapes and is endemic to North America and causes little damage to the resistant native grape species. It was transported to France in the mid-1800s, where there was no resistance to the insect in the commercial wine grapes, resulting in tremendous losses to the wine industry. Grafting grapes onto resistant rootstocks eventually resolved the problem. Resistant rootstocks have been the standard tool to manage this insect in grapes for over 100 years. Starting in the late 1980s, a new strain of *Phylloxera* developed that could overcome the rootstock resistance used for grape production in California. It destroyed many vineyards, and has resulted in the need to replant many existing vineyards to a new resistant rootstock.

Intensive selection for brown planthopper resistance in rice led to the discovery that an insect biotype virulent to the resistant variety 'Mudgo' evolved within 10 generations. Three brown planthopper biotypes were characterized and named at the International Rice Research Institute. The evolution of biotypes of the brown planthopper forces the breeder to constantly search for new sources of resistance. Because of their enormous genetic variability, insects offer many examples of the rapid evolution of biotypes.

## Pest Genetics Summary

The overriding importance of pest genetic variability to sustainable pest management is the recognition that the pest will evolve in response to an applied selection pressure. The selection pressure results from factors such as change in environment or change in management practice. Examples of the latter include use of new resistant varieties, alteration in sowing date or harvest time, cutting height, pesticide use, or virtually any management activity.

## SOURCES AND RECOMMENDED READING

Agrios, G. N. 1997. *Plant pathology.* San Diego: Academic Press, xvi, 635.

Dempster, J. P. 1975. *Animal population ecology.* London; New York: Academic Press, x, 155.

Elzinga, R. J. 1997. *Fundamentals of entomology.* Upper Saddle River, N.J.: Prentice Hall, xiv, 475.

Evans, H. E., and J. W. Brewer. 1984. *Insect biology: A textbook of entomology.* Reading, Mass.: Addison-Wesley Publishing Co., x, 436.

# *6*

# ECOLOGY OF INTERACTIONS BETWEEN CATEGORIES OF PESTS

**(or Putting the "I" in *Integrated* Pest Management)**

## CHAPTER OUTLINE

▶ Introduction
▶ Energy/resource flow (trophic dynamics) interactions
▶ Habitat modification
▶ Summaries and importance of interactions driven by food source or habitat modification
▶ Interactions due to physical phenomena
▶ IPM implications and economic analysis of pest interactions

## INTRODUCTION

**Pest management should be viewed at an ecosystem level to account for interactions.**

Plant pathogens, weeds, nematodes, mollusks, insects, and vertebrate pests do not exist in isolation from each other. In most crops many pests of different categories will be present simultaneously. This is an important consideration for integrated pest management, because the use of tactics to manage one category of pest can potentially influence those in all other categories present in the ecosystem. Interactions can be diagrammatically represented by the pest hexagon shown in Figure 6–1. Pest management should be approached within this conceptual representation. The interactions among pest categories merit consideration, if management of one pest category is not to complicate or negate that for another. This is especially true for weed management, because weeds, due to their trophic position in food webs, can serve as alternative hosts to many pests and beneficials. Arthropods and nematodes, and plant pathogens, may show a range of interactions from subtle to dramatic.

Many complex kinds of interactions can occur in an agroecosystem. As mentioned, weeds can serve as alternative hosts for other pests and beneficials. Arthropods can carry spores, create wounds, and transmit (vector) nematodes

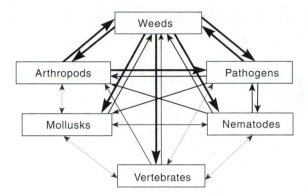

**FIGURE 6–1** The pest hexagon is a diagram representing potential interactions between different categories of pests. Thickness of arrows depicts approximate importance of the interaction.

and viruses, thus facilitating the spread of disease-causing agents. Plants that are dying from a pathogen attack may support fewer other organisms and are less competitive against weeds. Vertebrates interact with other pests by carrying fleas and ticks that vector diseases of domestic animals and humans, and they can disseminate weed seeds and pathogens. Nematodes vector viruses and bacteria, and mollusks serve as vectors of fastidious pathogens.

Interactions occur within a single class of pests, but just as often interactions occur between more than two pest categories. Multiple-level interactions between three categories of pests are fairly common; weeds can host viruses that are vectored by insect pests. There are examples of even more complex interactions.

## Example Interaction:

Following introduction of myxomatosis into Australia to control rabbits, the incidence of *lettuce necrotic yellows virus* increased. Why? Because previously rabbits had eaten sow thistles, an alternative reservoir host of the virus.

Interactions between pest categories are based on four different mechanisms (not including interactions classed as biological control). The four major mechanisms for interactions are as follows:

1. Interactions resulting from trophic relationships
2. Interactions due to environmental modification
3. Interactions due to mechanical phenomena
4. Interactions due to control tactics used, which can be subdivided into:
    4.1. Interactions in response to nonpesticidal tactics
    4.2. Interactions arising from the use of pesticides

The first three categories of interactions are driven by the pests themselves or the ecosystem, and are discussed in depth in this chapter. The interactions that occur because of the tactics used to manage a pest are discussed under the appropriate chapter for each tactic. Pests also have the potential to interact at the community level, thus contributing to ecosystem diversity. This topic is discussed in Chapter 7.

# ENERGY/RESOURCE FLOW (TROPHIC DYNAMICS) INTERACTIONS

The flow of energy and resources through the food web explains why several interactions between pest organisms should be anticipated (Figure 4–8). Following are possible reasons for such interactions.

1. Interactions can result when two different pest organisms at the same trophic level exploit the same host plant. Interactions of this kind are called multispecies attacks, or impacts of pest complexes. The pests could be from the same class of organisms (lygus bugs and aphids) or may be from different categories (leafhoppers and root rots, or nematodes and root rots).
2. Interactions can also occur between pest organisms at different trophic levels and involve primarily weeds and other pest organisms (Figure 6–2). In addition, important intertrophic-level interactions also may occur between animals and plant pathogens.
    2.1. Direct interactions occur when a pest organism utilizes a host at the next lower trophic level; weeds (producers) can serve as a host for many phytophagous pest organisms (Figure 6–2, weed and herbivores). Interactions between different levels of higher consumer organisms also occur and are considered in Chapter 13, as the significance to IPM is related to biological control.
    2.2. Indirect interactions occur when there is an intermediary organism between the target pest or beneficial and the organism causing the interaction. Beneficial insects, for example, may interact with weeds two trophic levels lower if the weeds provide the prey food source for the beneficials (Figure 6–2, beneficials and weed). Alternatively, beneficials influence crop plant growth by reducing the damage caused by herbivore prey.

**FIGURE 6–2**
Simple food chains demonstrating direct (black arrows) and indirect (gray arrows) interactions between pests.

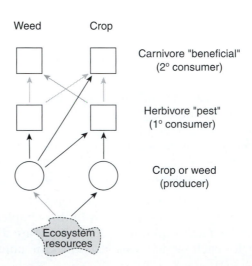

2.3. Additional interactions involve organisms that feed at different trophic levels during different parts of their life cycle. Such feeding behavior is common for many beneficial insects.

## Multiple Primary Consumer Attack

When two or more pest species attack the same host plant, they are sharing the same resource. This common interaction occurs in every crop; however, scientific research to assess the outcome of such multispecies attacks is difficult and usually limited in scope. Leafhoppers and *Fusarium* root rot attacking alfalfa illustrate the importance of assessing how multispecies attacks influence host plants. Leafhoppers, in the absence of the pathogen, cause only slight, nonsignificant stand loss (Figure 6–3). Root rot, in the absence of the leafhopper attack, likewise results in no significant stand loss. When the two pests attack the crop simultaneously, a significant 50% reduction in stand occurs.

Interactions, termed disease complexes, between nematodes and pathogenic fungi are common. Root-knot nematodes and *Fusarium* both parasitize melons, but the damage resulting from concomitant infection may be much greater than that resulting from infection with either pathogen alone (Figure 6–4). Similarly, the presence of root-knot nematode can overcome the resistance of cotton to *Fusarium* (see Figure 2–7i). In the absence of the nematode (outer rows), the cotton grows normally; but in the presence of the nematode (center rows), the cotton is damaged to a greater degree by the fungal pathogen.

Multispecies interactions, such as those just presented, demonstrate a difficulty associated with the use of damage or economic thresholds (see Chapter 8) as defined under conditions of single pest species attack. A synergistic interaction between pests results in damage or economic thresholds which are lower than those predicted for the single species. Such interactions can result in IPM decisions that are less than ideal.

Alternatively, pest attacks may be simultaneous, but the influences on the host can be independent. An example is the interaction of the soybean looper

**Multispecies pest attacks are the norm, but their combined impacts are poorly understood.**

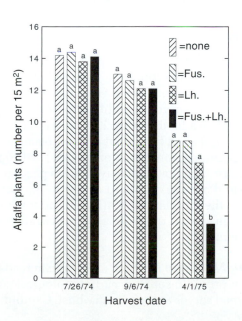

**FIGURE 6–3**
Interaction between *Fusarium* (Fus.) root rot and *Empoasca* leafhoppers (Lh.) on stand loss in alfalfa. Columns within a sampling date associated with the same letter do not differ at the $P = 0.05$ level of significance.
Source: redrawn from Leath and Byers, 1977.

**FIGURE 6–4**

Watermelon response to *Fusarium* in the absence and presence of root-knot nematodes.
(a) Noninfected control; (b) nematodes only; (c) *Fusarium* only; and
(d) nematodes and *Fusarium*.

Source: photograph by Ivan J. Thomason, with permission.

larvae, a foliage feeder, and the weed cocklebur, where insect defoliation of the crop and the effect of weed competition on the crop were independent, so that the damage threshold for defoliators did not require consideration of weed infestation.

Agricultural scientists conducting research on IPM recognize that multispecies interactions are significant; however, the need for more research in this area is great. In the meantime, most pest managers are forced to rely on experience. The limited documentation and understanding of multispecies interactions is a major stumbling block to adoption of integrated pest management with consideration of higher-level interactions. Unfortunately, most current IPM programs have been developed using information for a single pest species.

## Direct Interactions

Direct interactions occur when an animal pest or a pathogen has both crop plant and wild plant hosts. The noncrop plant is called an alternative, or reservoir, host. Most of these interactions are between weeds and other pests, because weed plants serve as a food source in addition to the crop (primary food). When an animal uses a weed as a food source, it is necessary to distinguish between animals that are crop pests and those that are not, because the pest status alters the importance of the interaction. Some pathogens, arthropods, and nematodes display host specificity such that they may attack a particular weed, but they do not attack the crop and are not classified as pests—in fact, they are beneficial. Problems arise when the animals and pathogens attack and survive on weeds, but they also attack the crop. If the crop is the preferred host, then the situation is particularly problematic.

***Direct Interaction: Not Crop Pest***   If an organism feeds on weeds, but is not a crop pest, it is then providing at least partial biological control of the weed. In this situation, the weed serves as the primary host. Carabid beetles that eat weed seeds are a good example of this interaction. Biological control of weeds is discussed in more detail in Chapter 13.

***Direct Interaction: Crop Pest***   If the organism that feeds on weeds is also a crop pest (Figure 6–2b), then the result of the interaction requires careful analysis. There are three possible outcomes:

1. The pest may provide biological control of the weed (see Chapter 13). The amount of weed control is typically low, and the deleterious impact of the organism as a crop pest usually outweighs any benefits resulting from biological weed control.
2. The weed may be an alternative, or reservoir, host for the pest when the primary crop host is not present, and thus supports pest survival between crop cycles. This interaction is important, especially in relation to pathogens, nematodes, and many species of arthropod pests. The pest population increases on the weed, and the pest then moves into the crop in larger numbers.
   2.1. Obligate alternate hosts are a special interaction, in which the weed is a necessary component of the pest life cycle. This interaction, as noted in Chapter 5, is particularly important for many plant pathogens.
3. The pest living on the weed can also serve as a host or prey for beneficial insects. This phenomenon is considered later in this chapter under the topic of indirect interactions.

**Alternative hosts serve as a general food source in the absence of the preferred crop host; (obligate) alternate host is a second host that is required for completion of the life cycle.**

When a pest organism uses a weed species as an alternative host, the damage to the weed is often minimal; in many cases, weeds infected with a pathogen do not show any symptoms. Nutsedge, for example, shows no symptoms when the tubers are infected with root-knot nematodes, likely because weeds have evolved with nematodes, resulting in constant selection for a capacity to tolerate nematode infection. The result is the weed survives and reproduces despite feeding by higher trophic-level organisms. Thus, the ability of weeds to tolerate attack by higher trophic-level organisms should be expected. If a plant could not tolerate such attack and thrive, it probably would never attain the status of being a weed.

Because many organisms designated as pests of crops do not harm weeds, their presence on the weeds is often undetected. If the pest numbers increase on the weed, when the pest eventually obtains access to the host crop, then their numbers can be high enough to cause crop loss. The lack of impact of pests on weeds makes the ability of weeds to host pests rather insidious. From the management perspective, it is important to determine the significance of weeds growing external to the cropping system versus those that are growing within the system.

**Weeds Growing External to the Cropping System**   Outside the crop system, pests can utilize weeds as a food source and then move back into the crop. This type of interaction only applies to mobile pests, or those that have propagules readily disseminated by wind (e.g., fungus spores, aphids, spider mites). The damage resulting from stinkbugs moving from Russian thistles into adjacent wheat fields in Canada provides a good example (Figure 6–5). The use of weed hosts by rust diseases of cereals serves as an example for pathogens. Many fastidious pathogens can utilize weeds as alternative, or

**Removal of weeds in the vicinity of lettuce crops is one of the most important aspects of *lettuce mosaic virus* management.**

**FIGURE 6–5**

Relationship between damage by stinkbugs in wheat and distance from alternative host-plant Russian thistle.

Source: drawn from data in Jacobsen, 1945.

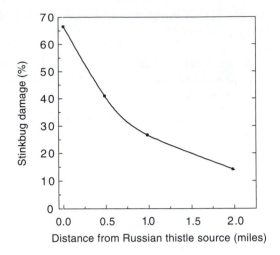

reservoir, hosts; in such cases there must be a vector to transfer the pathogen from the weed to the host crop. Pierce's disease of grapes, caused by a fastidious xylem-resident bacterium, is an example. Weedy vegetation in riparian areas serves as the reservoir hosts for the bacteria, and leafhoppers are the vector. Greater disease severity occurs in vineyards adjacent to riparian areas (Figure 6–6). Weeds surrounding fields can also serve as a food source for vertebrate pests, which then reinvade the fields when the crop is present. A standard recommendation for control of voles is to remove the weedy vegetation surrounding fields.

Common examples of interactions between pests and vegetation outside the crop field include lygus bugs, leafhoppers, and aphids that use weeds as alternative hosts, and many important viruses survive in weeds (Table 6–1). Weeds serving as reservoirs of viruses are one of the more important problems in virus management. Removal of alternative host plants (usually weeds) is a major strategy for managing diseases. Allowing weeds that can host viruses to grow around crops can result in the collapse of virus management programs.

**Weeds Growing within the Crop System**    Within the crop field, weeds can serve as alternative hosts for other pests. During certain periods in the growth of most crops the field may be fallow, or the crop has little or no canopy, such as during seedling stages of an annual crop or the dormant period in perennial crops. At such times, weeds may be present in large numbers, and are the main source of actively growing tissue that can serve as food for primary consumer organisms. In that way, weeds can serve as a bridge for pest organisms by providing nourishment when the crop is not present, and as a means for pest population increase during the early phases of crop establishment. Some examples include:

**The efficacy of rotation or fallow as a control tactic to reduce soil-borne pathogen and nematode numbers can be negated by the presence of weeds.**

1. Pathogens. Many pathogens, such as *Verticillium* wilt species, *Botrytis* grey molds, and *Sclerotium rolfsii,* are capable of infecting and reproducing on many weed species. *Rhizoctonia solani* (which causes damping-off in several crops) can transfer to many weed species when no crop is present; for example, the transfer from potatoes to nightshade and back to potatoes has been well documented.

**FIGURE 6–6** Aerial photograph of vineyard to show increased intensity of Pierce's disease, seen as numbers of missing vines, in relation to distance from adjacent riparian area to left of photograph.
Source: photograph by Jack Clark, University of California statewide IPM program.

| TABLE 6–1 | Weedy Reservoir Alternative Hosts of Some Important Viruses in Crops | |
|---|---|---|
| **Crop[1]** | **Virus** | **Examples of weed hosts (incomplete lists)** |
| Corn (maize) (also sorghum and millets) | *Maize dwarf mosaic virus* (MDMV) | Barnyardgrass, crabgrass, sprangletop, dallisgrass, foxtail species, johnsongrass, and other grasses |
| Barley (all cereals) | *Barley yellow dwarf virus* (BYDV) | Bromes, barnyardgrass, crabgrass, ryegrass species, dallisgrass, bermudagrass, and many other species |
| Rice | *Rice tungro virus* (RTSV) | Wild rice species, cutgrass species, goosegrass, and barnyardgrass and jungle rice (both symptomless) |
| Alfalfa (beans, clover, lettuce, peas, tobacco) | *Alfalfa mosaic virus* (AMV) | Lambsquarters, henbit, common purslane, common sowthistle, common chickweed (all important inoculum sources) |
| Tobacco (tomato, pepper, potato) | *Tobacco mosaic virus* (TMV) | Nettleleaf goosefoot, narrowleaf plantain, horseweed (all natural weedy hosts) |
| Cucumber (many other cucurbits) | *Cucumber mosaic virus* (CMV) | Almost unlimited host range; weeds include redroot pigweed, shepherdspurse, field bindweed, lady's thumb, and wild mustard |
| Lettuce (spinach, pea) | *Lettuce mosaic virus* (LMV) | Prickly lettuce, sowthistles, common groundsel, common chickweed, lambsquarters, and many other species |
| Sugar beet | *Beet curly top virus* (BCTV) | Many species in Chenopodiaceae, Asteraceae, Brassicaceae, and others |

[1]Additional crops that are also susceptible are listed in parentheses.
All data obtained from Sutic et al., 1999.

**2.** Nematodes. Root-knot, cyst, lesion, and ring nematodes are serious pests in numerous crops. They have wide host ranges, which include many weed species (Table 6–2). These alternative hosts present a serious challenge to the pest manager, as rotation to fallow or nonsusceptible crops is one of the standard recommendations for control of these nematodes. For such rotations to effectively reduce nematode numbers, there must be no alternative host weeds growing in the nonsusceptible crop. Effective weed control during rotation or fallow thus becomes an important consideration in any nematode management program. For nematodes, such as the sugar beet cyst nematode, which are managed by rotation to nonhost crops, weed management is critical.

3. Mollusks. Slugs and snails can also use weeds as an alternative food source; annual bluegrass and shepherd's purse are examples of weed species that serve as fodder for mollusks. Controlling weeds can result in slugs and snails moving into the crop to find food. Slugs living in wildflower strips used to increase infield diversity can create a hazard to the adjacent crop, as discussed in Chapter 7.

4. Arthropods. Many arthropod pests utilize weeds growing in crop fields as alternative hosts. Cutworms in asparagus are an example of this interaction (Table 6–3). Field experiments have shown that the number of cutworms was low in the absence of weeds. When either field bindweed or Canada thistle were present, the number of cutworms was much higher. Overall, weedy rows had between 250- to over 300-fold more cutworm larvae than did weed-free rows. The increase in cutworms was attributed to the weeds serving as a food source before the asparagus started to grow. Weedy mustard species grown as a cover crop can be the genesis of false chinch bugs as a pest problem; the insect builds up on the weed and then migrates to grapes. Enormous numbers of green peach aphids can build up on weeds (Table 6–4). Additional examples of weeds as alternative hosts include two-spotted spider mites on field bindweed, omnivorous leaf roller on pigweed and lambsquarters, and flea beetles on groundcherry and nightshade.

| TABLE 6–2 | **Examples of Reservoir Alternative Hosts[1] of Several Important Plant Pathogenic Nematodes** | | |
|---|---|---|---|
| **Common name** | **Latin name** | **Examples of weedy hosts** | |
| Root-lesion nematode | *Pratylenchus* spp. | *P. penetrans* has been recorded from over 55 weed species, including cocklebur, crabgrass, sowthistles, and bermudagrass. | |
| Stem and bulb nematodes | *Ditylenchus* spp. | Weed hosts include plantains, dandelions, cat's ear, chickweed, and wild oats. Several of these nematodes can be transported in wind-borne seeds. | |
| Sugar beet cyst nematode | *Heterodera schactii* | Weed hosts include lambsquarters, mustards, and pigweed. | |
| Soybean cyst nematode | *H. glycines* | Hosts are recorded from more than 1,100 plant species, including many weeds, such as common chickweed, henbit, common lespedeza, hemp sesbania. | |
| Root-knot nematode | *Meliodogyne* spp. | *M. hapla* has been recorded from over 70 species of weeds, including crabgrass, pigweed, shepherdspurse, green foxtail, and lambsquarters. | |
| Rice root nematode | *Hirschmaniella spinicaudata* | Weed hosts include several *Cyperus* spp., itchgrass, jungle rice, wild rice, bermudagrass, spreading dayflower, *Oxalis* spp., and many others. | |

[1]Weed lists in this table derived from Bendixen et al., 1979; Babatola, 1980; Manuel et al., 1980; Manuel et al., 1981, 1982; Bendixen, 1986.

| TABLE 6–3 | Impact of Field Bindweed and Canada Thistle on the Presence of Redback Cutworm in Asparagus | |
|---|---|---|
| Weed status | Distance assessed (ft) | Cutworms (quantity) |
| No weeds | 5,000 | 1 |
| Field bindweed | 360 | 81 |
| Canada thistle | 160 | 11 |

From Tamaki et al., 1975.

| TABLE 6–4 | Numbers of Green Peach Aphids Living on Weeds in the Spring in a Peach Orchard in Washington | |
|---|---|---|
| | Aphids (millions/acre) | |
| | Wingless | Alate (winged) |
| Clasping pepperweed | 0 | 0 |
| Common mallow | 5.5 | 0.69 |
| Hoary cress | 5.9 | 0.76 |
| Redroot pigweed | 8.6 | 0.35 |
| Lambsquarters | 41 | 5.8 |
| Field bindweed | 41 | 14 |
| Flixweed | 360 | 48 |

From Tamaki and Olsen, 1979.

Insect buildup on weeds within the crop causes a serious hazard in relation to weed control because of the major likelihood that the insect will be driven to the crop following any weed control practice. In field experiments, a cutworm population that built up on pigweed destroyed the bean crop following mechanical cultivation to control the weed.

**Insect pests living on weeds can be driven to the crop by weed control practices.**

5. Vertebrates. Weeds provide cover and feed for rodents such as voles and mice, or lagomorphs such as rabbits and hares. The animals are often attracted to weedy orchards, where they feed on the succulent bark of young trees, girdling and killing them (Figure 6–7). Removing the vegetation is considered a much better management practice than trying to poison the animals.

## Indirect Interactions

Indirect interactions occur when there is an intermediate organism between the one driving the interaction and the recipient, as in Figure 6–2 where organism A influences organism C through its influence on organism B. Such interactions are sometimes referred to as multitrophic interactions. The two possible interactions significant to IPM are:

1. Herbivorous pests and pathogens reduce crop growth.
2. Weeds support phytophagous pests that in turn support beneficials. There is no direct feeding by the beneficial on the plant; the interaction is therefore indirect.

**FIGURE 6–7** Bark damage to young tree trunk due to rodent gnawing in a weedy orchard; such damage rarely occurs in weed-free orchards.
Source: photograph by Robert Norris.

***Herbivore or Pathogen Damage to Crop***   Crop damage by a pest reduces the crop's ability to grow and compete, which indirectly affects any other pest associated with the crop. Changes in resources have the potential to impact all organisms higher in the food chain, and other plants in the system are impacted through altered utilization, and hence availability, of ecosystem resources.

An old adage states, "The best form of weed control is a healthy vigorous crop." The corollary is, a pest-damaged or weakened crop is less able to compete with weeds. In a trophic dynamics context, the injured crop uses less ecosystem resources than a healthy crop, which leaves more of the resources available for weeds.

Some pests are capable of killing the crop, resulting in stand loss. This is a fairly obvious impact of pest attack, that is, crop presence or crop absence. The result is gaps in the crop in which no resources are used by the crop. Examples of stand loss due to pests include feeding by gophers in alfalfa, cutworms that kill crop seedlings, and root pathogenic fungi that kill crops such as alfalfa, cotton, and established orchard trees. In the areas where the crop plants have died, weed growth is much more vigorous due to the absence of competition with the crop. Although this effect is well accepted, its actual extent and importance are not well documented.

Defoliation of the crop by any organism, such as foliar pathogen attack, insect herbivory, or rabbit grazing, results in decreased crop canopy. Decreased canopy allows more light to reach understory weed plants, resulting in more weed growth. Damage to the root system caused by nematodes, insects, or pathogens results in decreased root function and plant growth. Damage to the canopy or root system results in less water and nutrients being used by crop; thus, more water and nutrients are available for the weeds, again resulting in more weed growth. This is a good example of ecosystem feedback, and represents the expression of an interaction wherein higher level consumers determine how well producers are growing (i.e., a top-down trophic interaction) (see Chapter 4).

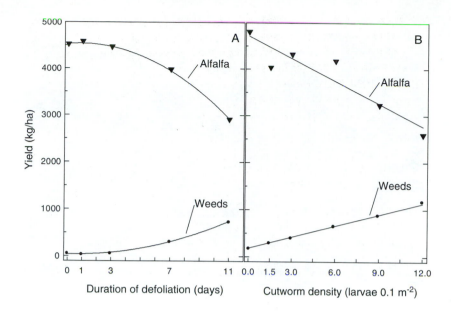

**FIGURE 6–8** Effect of variegated cutworm feeding on alfalfa yield and on the growth of competing weeds. (a) Effect of increasing duration of cutworm feeding, and (b) effect of increasing numbers of cutworm larvae.

Source: redrawn from Buntin and Pedigo, 1986.

There are several examples that document this type of interaction. The variegated cutworm defoliates alfalfa in the Midwestern United States. Feeding damage by this insect leads to increased weed growth (Figure 6–8), and the increased weed growth is closely correlated with increased insect damage (duration of feeding or number of insects). Defoliation of alfalfa by the Egyptian alfalfa weevil in California likewise resulted in increased yellow foxtail compared with areas where the insect was controlled. A similar phenomenon has been reported for wheat bulbfly injury to wheat in England, where damaged wheat had greater incidence of mayweed and blackgrass.

A variant of the defoliation interaction has been observed in California. Alfalfa is usually machine cut with a swath width of about 15 ft. Low populations of the Egyptian alfalfa weevil larvae present in the hay are thus concentrated about 5-fold in the 3-ft-wide swaths created by the harvester. The legless larvae cannot migrate, and they consume the alfalfa regrowth, resulting in severe damage under the swath. By late summer, lines of yellow foxtail occur only in those areas where the severe damage occurred beneath the swath (Figure 6–9).

The importance of these resource feedback interactions to IPM is that all efforts to maintain crop vigor through improved management of phytophagous pests and pathogens can lead to reduction in the effort required for weed control. The impact of crop insect management on weed growth, however, is not usually incorporated into economic thresholds used for insect management decision making (see Chapter 8). Although these interactions are recognized, they are usually not considered when developing economic thresholds.

### Beneficial Insects Utilize Prey Living on Weeds

An arthropod living on weeds can serve as prey for beneficial insects (Figure 6–2), which is the ecological basis for the use of biodiversity to stabilize insect populations (see Chapter 7).

It is important to differentiate between a prey that is also a crop pest and one that is not. In the latter case, the potential exists for the weed to serve as an

**FIGURE 6–9**

Yellow foxtail growing in strips in alfalfa fields in relation to weevil larval feeding under the windrow at the first cutting.

Source: photograph by Robert Norris.

indirect resource for arthropod biological control programs. Aphids on sow thistles are a good example, as the aphids are not crop pests; however, the aphids serve as prey and support beneficials such as syrphid-fly larvae, lady beetles, and parasitic wasps, which is clearly a useful interaction. If the phytophagous prey insect is providing control of the weed, it may be undesirable to have that insect attacked by the beneficial. In such a case, assessing the utility of the insect for biological weed control relative to serving as a food source for beneficials is difficult. As with pest arthropods, it is useful to consider interactions with beneficials from both outside and within-field perspectives.

**Outside Crop**　　The beneficial insect is maintained or increased on vegetation outside the field over the period when the crop cannot support the target prey. This interaction is part of the rationale for on-farm and regional diversity discussed in Chapter 7. The role of the blackberry leafhopper in supporting *Anagrus epos* during the winter is an example of such interaction (Figure 6–10). *A. epos* is a parasite of the grape leafhopper, but it cannot survive in vineyards in the winter because grapes are deciduous and thus there are no leafhoppers in the winter. The parasite can attack a leafhopper (*Dikrella* sp.) on blackberries, which retain leaves even in the winter, and can thus survive the winter. Parasitism of grape leafhoppers by the beneficial wasp is higher in vineyards adjacent to riparian areas where blackberries grow. Although the interaction has been well documented, it has proved difficult to implement the knowledge as a means to manage the pest.

**Within Crop**　　Weeds present in the crop may serve as hosts for arthropods that can be attacked by beneficials. There are many examples of such interactions. Carabid ground beetles, for instance, are almost always present at higher density in weedy crops than in weed-free crops (e.g., Figure 6–11). Beneficial insects in cole crops are almost always higher when weeds are present than when weeds are absent. The implications for IPM are discussed under the topic of in-field diversity in Chapter 7.

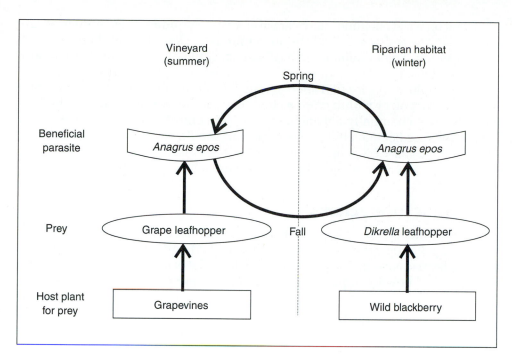

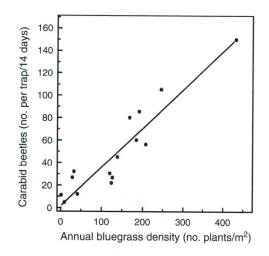

**FIGURE 6–10**
Diagram depicting the relationship between the overwintering host leafhopper *Dikrella* on blackberries in riparian areas, and the grape leafhopper parasitoid wasp *Anagrus epos*.

**FIGURE 6–11**
Relationship between annual bluegrass density and the number of carabid ground beetles caught in pitfall traps.
Source: redrawn from Speight and Lawton, 1976.

# Polytrophic Interactions (Direct and Indirect)

Some organisms feed at more than one trophic level. Human beings, for example, eat meat (carnivore at least at consumer level 2), and cereals, vegetables, and fruits (herbivore).

Many beneficial insects feed at different trophic levels at different stages in their life cycle. The larvae are usually carnivorous and feed on other insects, providing biological control of pests. The adults of many beneficial arthropods are not carnivorous but rather feed on nectar or pollen, and are thus herbivores. Adult female insects that obtain pollen or nectar lay more eggs and provide higher levels of biological control. Many weeds can serve as this nectar or pollen source, and thus interact directly with the beneficial. Examples of beneficials that feed on weeds include many parasitic wasps, syrphid and tachinid flies, and lacewings. Several generalist hemipteran predators supplement their diet with plant feeding. The minute pirate bug is an example of a predator that will also feed on plants.

The implications for IPM are that it may be desirable to leave certain weeds in the vicinity of, or within crops so that they can serve as food sources for beneficials, thus potentially enhancing biological control of pest insects. Leaving weeds for such beneficial insect management should be carefully assessed from the pest hexagon viewpoint (see Figure 6–1). This issue is addressed in greater detail under the topic of biodiversity in Chapter 7.

## Tritrophic Interactions

Tritrophic interactions happen when a plant-derived chemical compound is passed up through the food web so that a plant influences a carnivore two trophic levels above it. The importance to IPM relates mainly to how beneficials perform, and how hyperparasitoids can impact their prey.

# HABITAT MODIFICATION

The activities of pests can result in habitat changes that are not related to trophic dynamics. The following areas of habitat modification can result in pest interactions.

1. Altered resource concentration
2. Altered apparency
3. Altered microenvironment

## Altered Resource Concentration

If nonhosts are interspersed with hosts so that the host density is effectively reduced, then hosts are less likely to be located by the pest. This is the main rationale for diversity within crops to prevent epidemics of diseases and insect infestations. The hypothesis predicts that the rate of pest population increase and attack of managed plants will be higher in monoculture than in mixed culture. Altered resource concentration implies the following:

1. Pest immigration into a more concentrated area of host is likely to occur.
2. Pest emigration from a concentrated area of host is less likely to occur.
3. Pest population increase will be more rapid when the host is more concentrated.

The altered resource concentration hypothesis has led to the concept that growing a genetically uniform crop over large geographic areas (monocultures) is a mistake for the management of mobile arthropods and vertebrates, because they have sensory systems that allow them to readily detect and move to monocultures. The altered resource concentration hypothesis is only partially applicable to pathogens, and has no relevance to weeds per se. Weeds, in fact, can serve as the nonhosts interspersed with hosts due to their unique trophic position among pests. An additional problem associated with monoculture is that such cropping systems may exert a selection pressure on pest populations, leading to increased frequency of pests capable of finding and doing well on the host crop.

## Altered Apparency

The apparency hypothesis suggests that if the preferred host is nonapparent (less visible), then it may be less likely to be found by an animal and attacked. Host finding in most animals is usually mediated by sensory input, including chemical signals and visual cues. The chemical signals emitted by the host plants are called kairomones (see Chapter 14). Mixtures of different crops, or crop and weeds, could decrease the apparency of a preferred crop host, resulting in a reduced level of invasion and attack. The concept of altered apparency only applies to pests that actively search to locate a host, such as insects and nematodes, so it is not relevant for pathogens or weeds. Just as with the resource concentration hypothesis, weeds are involved because their presence may partially hide the preferred host plants. Altered apparency is central to the rationale that leaving weeds in fields can reduce arthropod attack on the crop. The apparency hypothesis supports the concept of using a polyculture as a means to combat pest movement to crop plants. The implications of this management approach are discussed in Chapter 7 on biodiversity in pest management.

## Microenvironment Alteration

The presence of crop or weed plants, or damage by other pests, causes changes in the habitat. The prime change is due to the presence of, or damage to, a canopy. Alteration of the canopy produces several changes in the microclimate. Typical effects are shown in Figure 6–12. Changes in root habitats also occur but are less understood or documented.

The significance to IPM involves the role of weed canopy in modifying microclimate, and the impact of crop defoliation on light, temperature, moisture/humidity, air movement/wind, and mineral nutrients.

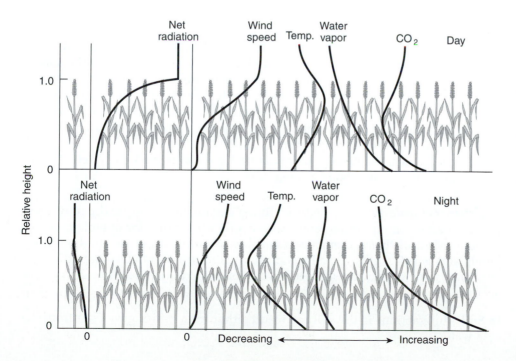

**FIGURE 6–12**
Diagrammatic presentation of the effects of a plant canopy on radiation, temperature, wind speed, relative humidity, and $CO_2$ distribution in the canopy during night or day.
Source: redrawn from Norris and Kogan, 2000.

***Light***    Damage to the crop canopy means that the crop intercepts less solar radiation, resulting in more radiation being available to other plants in the system (i.e., weeds; see indirect trophic interactions earlier in the chapter). The interception of light by the weed canopy is a major basis for competition between weeds and crops.

***Temperature***    The plant canopy intercepts radiation, and the growing canopy results in alteration of both air and soil temperatures. As weeds can be a substantial component of the canopy at certain times and in certain crops, their presence can lead to changes in microhabitat temperatures. Defoliation of crops by pests can also lead to changes in temperatures.

**In Air**    The presence of a plant canopy typically reduces air temperature within the canopy in the daytime, and increases the temperature during the night. The canopy surface temperature is usually higher than the ambient temperature in the daytime, and lower at night (Figure 6–10).

**In Soil**    A plant canopy, whether from crop plants or weeds, results in cooler soil in spring and summer, and warmer soil in fall and winter. These differences are typically ±5° to 10°F, but can be as great as 15°F. Diurnal temperature fluctuations are damped or offset a few hours by the presence of a canopy.

The major IPM implication of temperature on interactions between pests is environmental temperature regulates the metabolic activity of poikilothermic (cold-blooded) organisms (nematodes, mollusks, arthropods). Because weeds provide a plant canopy, in some situations they may have the potential to alter rates of pest development. Hypothetically, the presence of weeds could lead to less accurate prediction of pest population development by models using regional air temperature data. It is possible that the presence of weeds could render output of such models inaccurate in some situations. Although most models operate at an accuracy level that would probably not be significantly affected by relatively minor microclimatic effects, this kind of interaction between pests should be considered in predictive IPM.

Alteration of temperature can influence weed development, especially germination. Defoliation of a crop such as alfalfa in the spring allows the soil to warm more quickly, thus speeding weed germination. The presence of winter weeds in alfalfa can slow the increase in soil temperature in the spring, delaying emergence of summer weeds. Cover crops can also reduce soil temperatures, leading to delayed emergence of warm-season weeds and slowing development of all other soil-borne pests.

Microclimate alterations in temperature have less impact on vertebrates because they regulate their body temperature.

***Moisture/Humidity***    The body of living organisms is approximately 75% water. Changes in environmental water availability can alter organism development. Most of these interactions are due to either water use by both crop plants and weeds, or by the effects of their canopies on relative humidity. Interactions that occur in the soil need to be considered separately from those that occur in air.

**In Soil**    Interactions in the soil can result either from limiting water or from excess water.

1. Limiting water is mainly important for producers, because it intensifies competition for this basic ecosystem resource and thus is most important for weeds. Nematodes are aquatic organisms, and low soil moisture decreases their ability to move and survive. Low soil moisture resulting from use by weeds therefore has the potential to alter the degree of attack by primary consumers.
2. Excess water may occur during wet weather, and the presence of a canopy can retard drying of the soil upper surface layers. Many soil-borne pathogens, such as *Phytophthora* require wet conditions for infection to occur and thus the presence of a weed canopy could alter this process.

**In Air**   The presence of a canopy increases the within-canopy relative humidity (Figure 6–12) and many foliar pathogens require high humidity to infect host plant leaves. The presence of a weed canopy has the potential to increase humidity and thus alter pathogen infection. Increased incidence of downy mildew in grapes may be attributed to effects of johnsongrass canopy. Reducing the grape canopy to increase air exchange is a method used to manage bunch rots (*Botrytis*) in grapes (see Figure 15–3), and the presence of a weed canopy could negate this management tactic. A canopy also benefits slugs and snails as it protects them from desiccation and, in many situations, weeds could provide such protection.

*Air Movement/Wind*   Changes in air movement and wind due to the presence of a canopy may move pathogen propagules, can dislodge or dehydrate some insects, and can alter temperature and moisture.

*Nutrients*   Changes in the nutrients available to hosts can alter the capacity of the host to tolerate pest attack, and thus alter the importance of particular pests (see Chapter 16). Limited nutrient availability is one of the main causes of competition between weeds and crops, and changes in nutrients may drive interactions between pests. Alteration of nutrients has the potential to affect the food value of weeds for other pest organisms. Nitrogen metabolism may be altered by water stress, changing the growth of spider mite populations on crops and weeds.

*Shelter*   An organism's shelter encompasses many of the factors already discussed. Shelter provides protection from environmental extremes, cover, and reduced visibility. Many insects, vertebrates, and mollusks cannot survive in the open all day; they require protection from overheating (direct radiation) and desiccation. Cover allows pests to hide from predators. As noted, in many situations, weeds provide much of the plant canopy (cover), and so weeds and weed control alter the availability of shelter.

Examples of weeds providing cover for other pests include meadow mice (voles) in fence lines and around the base of trees and artichoke plants, gophers in noncropped areas, and slugs and snails in and around fields. Castor bean plants growing along cotton and soybean fields in Brazil provide shelter for large numbers of stinkbugs during periods when the crops are not in the fields. Even after the weeds die they can provide shelter for insects. The biomass of foliage and stems that covers the soil provides shelter for aggregations of insects that overwinter in the adult stage. Several species of coccinellid and chrysomelid beetles have this behavioral adaptation.

**FIGURE 6–13**

Interaction between quantity of the weed henbit and alfalfa weevil damage to alfalfa.

Source: redrawn from Waldrep et al., 1969.

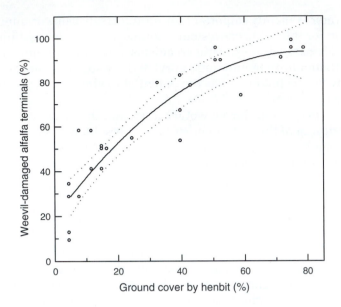

Weeds can provide a unique kind of shelter for some arthropods in the form of oviposition sites, even when the plant itself is not a food source. The alfalfa weevil normally lays its eggs inside hollow alfalfa stems (see Figure 5–1k). Henbit, a weed in the Lamiaceae, also has hollow stems that can serve as oviposition sites and provide shelter for alfalfa weevil eggs. The presence of the weed leads to elevated levels of weevil damage to alfalfa (Figure 6–13) through provision of shelter for the eggs.

## SUMMARIES AND IMPORTANCE OF INTERACTIONS DRIVEN BY FOOD SOURCE OR HABITAT MODIFICATION

### Pathogens

Most interactions between pathogens and other pests are negative in an IPM sense; the presence of the other pest increases the severity of the pathogen-induced damage. Controlling the other pest is thus likely to lead to improved disease control. This applies to organisms that are vectors of pathogens (see Figure 5–22), and weeds that are alternative hosts of pathogens. Weeds are often symptomless carriers of plant pathogens. Examples of weeds as hosts of pathogens include the following:

1. *Curly top virus.* The vectors are leafhoppers; hosts include Russian thistle and other weeds.
2. *Cucumber mosaic virus.* The vectors are aphids and whiteflies; hosts include common chickweed, stinging nettle, and other species. The virus can be transmitted through seed; 1% to 4% if chickweed seeds in the soil seedbank have been found to be infected with the virus. It has been reported to persist for up to 20 months in such seeds.
3. *Lettuce mosaic virus* is seedborne in common groundsel.

4. *Beet western yellows virus.* The vectors are aphids; hosts include numerous weed species.
5. Pierce's disease of grapes. The insect vectors are sharpshooters; hosts include johnsongrass and many other species.
6. Ergot in cereals. Alternative hosts include many weedy grass species, especially annual bluegrass and blackgrass.
7. *Rhizoctonia solani* in potatoes. There are many weed alternative hosts, such as nightshade species.
8. *Verticillium* wilt pathogens in cotton and other crops. There are about 300 alternative hosts, including numerous weed species.

Additional alternative reservoir hosts for several viruses are presented in Table 6–1.

# Nematodes

Similar to pathogens, almost all interactions between nematodes and other pests are negative relative to IPM. Nematode management in agricultural systems, where weeds are not well controlled, is considered to be particularly difficult because nematodes use weeds as alternative hosts, resulting in high numbers in the soil. Exceptions include the possible use of weeds as "trap crop" species and when some weeds may release compounds that are directly nematicidal; the latter concept is discussed under cultural controls in Chapter 16. Examples of weeds that are alternative hosts for nematodes are shown in Table 6–2.

# Mollusks

Interactions between mollusks and other pests are essentially all negative from a pest management sense. Weeds almost always increase mollusk problems. In some instances weeds may be preferred as a food source over the crop, and therefore presence of weeds may reduce damage to crop by slugs. Transportation of nematodes carrying viruses by mollusks is a limited, but serious negative interaction between pest categories.

# Arthropods

Interactions between weeds and arthropods can be beneficial or detrimental, depending on the ecosystem and the species involved. Following are examples of beneficial interactions where the weeds provide resources for beneficial arthropods.

1. Orchards. Phytophagous mites that are feeding on orchard floor weeds act as a food source for beneficial mites.
2. Grapes. Phytophagous mites on johnsongrass are prey for predatory mites.
3. Cole crops. The prey living on weeds supports several wasps and other predators.
4. Beans. Weeds may support leafhopper predators.
5. Grapes. The grape leafhopper parasite *Anagrus epos* overwinters in leafhoppers on blackberries.
6. Cereals. Aphid predators may attack prey on annual bluegrass.
7. Various crops. Some parasitic wasps feed on the energy-rich nectar/pollen they gather from weeds in field boundaries.

Given the previous points, the conclusion that reduced weed control would result in reduced numbers of insect pests appears valid; however, most of the examples noted do not show that the presence of the beneficial arthropod actually alters the numbers of the pest in the crop. Thus, the capacity of weeds to sometimes increase the numbers of beneficials has been documented, but a direct connection to concomitant crop pest reduction and change in yield loss due to the pest has not been recorded. An additional problem is competition and yield loss due to the presence of weeds in the crop, which may exceed any gain from the increased beneficial insects. Thus, the use of weeds to enhance biological control of arthropod pests must be carefully assessed (see Chapter 7).

Weeds may provide resources that support increases in pest arthropod numbers. In general, the IPM perspective reveals that weeds create as many arthropod problems as they might alleviate. Following is a list of examples of detrimental interactions with pest arthropods.

1. Cutworms are supported on pigweeds in beans and on Canada thistle and bindweed in asparagus.
2. Lygus bugs are capable of using many weed species, including pigweed and lambsquarters.
3. Beet leafhopper feeds on Russian thistle, redroot pigweed, and many other weeds.
4. Green peach aphid numbers increase on orchard floor weeds such as mustard and shepherd's purse; as many as 530 million aphids/acre have been recorded living on weeds.
5. False chinchbug numbers increase on mustard species and may then move into crops such as grapes.
6. Heliothis/Helicoverpa increase on numerous weeds around crops in early spring and then move to crop fields.

The conclusions drawn from these observations are opposed to the positive interactions noted previously, and argue that increased weed control often leads to improvements in insect management.

## Vertebrates

Most multipest interactions with vertebrates involve weeds. All such interactions are negative relative to IPM, because weeds serve as both cover and food source, or the vertebrate carries weed seeds, pathogens, or nematodes from one location to another.

# INTERACTIONS DUE TO PHYSICAL PHENOMENA

Some interactions between organisms in different pest categories occur strictly in response to the physical effects of one pest on another, or on the host crop. These interactions are based on physical damage to the target host plant by one pest that increases the host's susceptibility to another pest, or to pests serving as vectors.

## Physical Damage to Host

Many pathogens, especially bacteria, cannot penetrate an intact plant cuticle. Chewing damage by animal pests removes the cuticle barrier and provides physical openings for such pathogens, even if the pathogen is not actually carried by the organism doing the chewing. Following are examples of this type of interaction.

1. Insect larvae feeding on immature grapes can increase the disease incidence of grey mold.
2. Armyworm feeding on sugar beets can enhance disease caused by the *Erwinia* soft rot bacterium.
3. Root-knot nematode infecting cotton roots breaks the resistance to *Fusarium* of *Fusarium*-resistant cotton (see Figure 2–7i).

## Physical External Transport

The reproductive structures (bacteria, spores, seeds) of several pest categories can be transported externally and spread by a different category of pest. Probably one of the best documented examples of this interaction was the spread of Dutch elm disease spores from infected wood to healthy trees by the elm bark beetle. The beetle damages the bark creating entry sites for the fungus. Controlling the beetle was one tactic that was used to stop the spread of the disease, but it was not successful. Many insects, including honey bees and flower-visiting flies, carry the fire blight bacterium from blossom to blossom in pears.

**The vectoring of pathogens between hosts by other pests is probably the single most important interaction between pests in IPM systems.**

Many mammals can carry weed seeds in their fur, or nematodes in soil adhering to their paws. There is a rather unique interaction between birds and the familiar large-leaved mistletoes, which are actually serious parasites on many trees. The seeds of the mistletoes are fleshy and sticky. If the bird tries to eat them they become stuck to its beak. The bird later scrapes the seed off onto the tree bark, but not necessarily in the same tree. Birds may also eat seeds, and viable seeds pass through the digestive system and are then deposited in feces.

Human beings are especially important in relation to external transport of pests, either accidentally or intentionally. The significance to IPM is sufficiently great that the topic is discussed in detail in Chapter 10.

## Physical Internal Transport

Insects can transport nematodes from host to host. The palm weevil is a devastating insect pest of many palm crops. Its importance is increased by the fact that the weevil also transports the red-ring nematode over long distances. The nematode causes red-ring disease of coconuts and oil palms and kills trees within a few months of infection (Figure 6–14); its geographic distribution is still increasing. A similar situation exists with pine-wilt disease, which is caused by the pinewood nematode. The nematode is transported from tree to tree by a Cerambycid longhorn beetle in the genus *Monochamus*.

Numerous pathogens cannot survive outside living cells; they are said to be fastidious organisms (see Chapter 2). These categories of pathogens include all viruses and viroids, many bacteria, and most phytoplasmas. For these categories of pathogens to spread, a living organism must carry them from one host to another. This process is referred to as vectoring, and the agent carrying the pathogen is called a vector. The vector acquires the pathogen by feeding on an infected host plant, moves to a noninfected plant, and by feeding on the new plant transmits the pathogen into it. Vectoring of pathogens is a combination of

**FIGURE 6–14**
Palm tree stump infected with red-ring disease, with red-ring symptom seen as the dark ring inside the trunk. The disease is caused by a nematode transported from palm to palm by the palm weevil.
Source: photograph by E. Caswell-Chen.

physical transport plus provision of suitable 'environment' in which the pathogen can survive and, in some cases, even reproduce.

Vectoring is actually a complicated three-way interaction involving several different phenomena. The specific mechanics of how this transfer occurs vary depending on the pathogen, the vector, and the host plants. Species within all pest categories can serve as vectors (see Figure 5–22). Weeds, as noted earlier, serve as alternative hosts from which the vector can acquire the pathogen. Active vectoring by a weed is rare. One example of such vectoring is the parasitic plant dodder which has been used experimentally to transfer viruses from an infected plant to a healthy one. Examples of different pest categories serving as vectors include:

1. Pathogens. Several viruses are vectored by other pathogens. Rhizomania in sugar beets, for example, is caused by *beet necrotic yellow vein virus,* which is vectored by the soil-borne fungus *Polymyxa betae.*

2. Nematodes. Numerous viruses are vectored by nematodes (Table 6–5). The dagger nematode (*Xiphenema index*), for example, carries and transmits *grape fanleaf virus;* in the absence of the nematode there is no virus problem. Other nematode vectors are in the genera *Longidorus* and *Trichodorus* (see Taylor and Brown, 1997, for many other examples).

3. Slugs and snails. These provide an example of one of the more complex three-way interactions. Slugs carry a nematode that in turn carries a virus of alfalfa; the nematode is the actual vector but the slug provides the transport from plant to plant.

4. Insects. These vector a large number of viruses, phytoplasmas, and fastidious bacteria. Table 6–6 provides just a few examples of such insect/pathogen associations; the reader should be aware that there are many more cases than those listed here.

5. Vertebrates. These do not vector other pests involved in crop production, but are involved in vectoring of pathogens of other mammals, including humans.

| TABLE 6–5 | Examples of Nematode Genera and the Viruses That They Can Vector[1] |
|---|---|
| **Nematode Genus** | **Virus** |
| *Longidorus* | *Peach rosette mosaic* |
| | *Cherry rosette* |
| | *Raspberry ringspot* |
| | *Tomato black ring* |
| | *Mulberry ringspot* |
| *Paralongidorus* | *Raspberry ringspot* |
| *Trichodorus* | *Tobacco rattle* |
| | *Pea early-browning* |
| *Paratrichodorus* | *Tobacco rattle* |
| | *Pea early-browning* |
| | *Pepper ringspot* |
| *Xiphinema* | *Cherry rasp leaf* |
| | *Peach rosette mosaic* |
| | *Tobacco ringspot* |
| | *Tomato ringspot* |
| | *Arabis mosaic* |
| | *Strawberry latent ringspot* |
| | *Grapevine fanleaf* |

[1]Although there is specificity between nematode species and virus-vectoring capacity, that level of detail is not presented here.
Source: List from Taylor and Brown, 1997; and Weischer and Brown, 2000.

The internal transport of weed seeds that have been eaten by birds and mammals is much less important than vectoring of pathogens. This interaction does, however, result in long-distance spread of weeds in some situations. The ability of weed seeds to survive the animal digestive system is particularly problematic when animal manure is used to improve soil fertility. Animals that browse close to the ground may inadvertently ingest soil and the nematodes therein. Many viable nematodes move through the digestive track of the animal, and are deposited in feces at a new location.

# IPM IMPLICATIONS AND ECONOMIC ANALYSIS OF PEST INTERACTIONS

It is important to recognize that all the pest interactions discussed in this chapter are independent of the tactic(s) used for pest management. There is no difference, from an ecological perspective, whether a pest is controlled biologically, physically, culturally, or chemically, because it is the change in pest numbers that influence the interactions. Therefore, the interactions discussed earlier in this chapter may occur regardless of the pest management tactics in use. The significance of the outcome of these interactions should be judged in the IPM context:

**Interactions based on pest ecology, habitat modification, or physical processes are independent of the tactics used for control.**

1. Positive—interactions that decrease overall pest problems
2. Neutral—interactions that do not alter extant pest problems
3. Detrimental—interactions that make overall pest problems worse

| TABLE 6–6 | Examples of Insects and the Pathogens That They Can Vector |
|---|---|
| **Insect[1]** | **Pathogen** |
| Aphids | *Alfalfa mosaic virus* |
| | *Barley yellow dwarf virus* |
| | *Beet mosaic virus* |
| | *Celery mosaic virus* |
| | *Citrus tristeza virus* |
| | *Cucumber mosaic virus* |
| | *Lettuce mosaic virus* |
| | *Maize dwarf mosaic virus* |
| | *Onion yellow dwarf virus* |
| | *Plum pox virus* (a.k.a. *Sharka virus*) |
| | *Potato leafroll virus* |
| | *Soybean mosaic virus* |
| | *Sugarcane mosaic virus* |
| Leafhoppers | Aster yellows |
| | *Beet curly top virus* |
| | Celery aster yellows |
| | Corn (maize) stunt |
| | *Rice tungro virus* |
| | Western X disease of peach |
| Psyllids | Pear decline phytoplasma |
| Sharpshooters | Pierce's disease of grapes, almond leaf scorch |
| | Yellow leaf roll of peaches |
| Whiteflies | *Cucumber yellow vein virus* |
| | Leaf curl virus of several crops |
| | Yellow mosaic of many tropical crops |
| Beetles | *Bean and mottle virus* |
| | *Cowpea chlorotic mottle virus* |
| | *Cowpea mosaic virus* |
| | *Potato spindle tuber virus* |
| | *Squash mosaic virus* |
| Thrips | *Tomato spotted wilt virus* |
| | *Peanut bud necrosis virus* |
| Additional arthropod groups that transmit pathogens include mealybugs, piesmids, flies, and some mites. | |

[1]Insects are listed by generic common name; there are differences in the ability of individual species within each category in relation to efficiency as vectors.
List compiled from Gibbs, 1973; Harris and Maramorosch, 1980; Maramorosch and Harris, 1979, 1982; and Sutic et al., 1999.

It is therefore important to know if there are side effects from the interaction that should be considered. It is necessary to know how these relate to overall crop IPM.

Current information about the importance of ecological interactions between pests does not allow us to draw broad conclusions about the positive or negative aspects of interactions between pest categories. Each interaction is ecosystem and species specific and must be judged in relation to the particular cropping system and the goals of the manager. In addition, the economic ramifications of interactions between the crop and its pests should be considered, but data are limited on the economics of multiple pest interactions. It is therefore difficult to judge the importance of the interactions discussed in this chapter relative to applied IPM. Currently we must be satisfied to know that interactions are occurring, and to make management decisions that at least incorporate this knowledge to the extent possible.

The final caveat from this section is that for every management action there is potential for multiple effects through the food web and the whole agroecosystem (Figure 6–1). From an IPM perspective, the relative importance of all pests must be considered, and the interactions that can occur must be evaluated. Lack of consideration for the impacts of one pest category, or the interactions that occur, potentially can lead to total crop loss.

## SOURCES AND RECOMMENDED READING

The book *Pests, Pathogens, and Vegetation,* edited by Thresh (1981), is the only additional source of information that is not keyed to a specific pest category. There are numerous reviews of plant insect interactions; Norris and Kogan (2000) specifically address weed and arthropod interactions. Cultural changes in crops to enhance activity of natural enemies through habitat management are discussed in detail in a book edited by Pickett and Bugg (1998). For additional information on the role of weeds as alternative hosts for viruses, we suggest the reviews by Duffus (1971) and Bos (1981); and for specific viruses the handbook of viruses by Sutic, Ford, and Tosic (1999) is very useful. Richardson and Noble (1979) provide a list of pathogens that can be transmitted in seeds, of which many are weeds. The book edited by Khan (1993) is a useful source of information about nematode/pathogen interactions. Kranz et al. (1978) provide many examples of interactions between arthropods and pathogens of tropical crops.

Babatola, J. O. 1980. Studies on the weed hosts of the rice root nematode, *Hirschmanniella spinicaudata* Sch. Stek. 1944. *Weed Res.* 20:59–61.

Barnes, M. M. 1970. Genesis of a pest: *Nysius raphanus* and *Sisymbrium irio* in vineyards. *J. Econ. Entomol.* 63:1462–1463.

Bendixen, L. E. 1986. Weed hosts of *Meloidogyne,* the root-knot nematodes. In K. Noda and B. L. Mercado, eds., *Weeds and the environment in the Tropics.* Chiang Mai, Thailand: Asian-Pacific Weed Science Society, 101–167.

Bendixen, L. C., D. A. Reynolds, and R. M. Reidel. 1979. *An annotated bibliography of weeds as reservoirs for organisms affecting crops. I. Nematodes.* Wooster, Ohio: Ohio Agric. Res. Dev. Center, 64.

Bos, L. 1981. Wild plants in the ecology of virus diseases. In K. Maramorosch and K. F. Harris, eds., *Plant diseases and vectors: Ecology and epidemiology.* New York: Academic Press, 1–33.

Buntin, G. D., and L. P. Pedigo. 1986. Enhancement of annual weed populations in alfalfa after stubble defoliation by variegated cutworm (Lepidoptera: Noctuidae). *J. Econ. Entomol.* 79:1507–1512.

Duffus, J. E. 1971. Role of weeds in the incidence of virus diseases. *Annu. Rev. Phytophathol.* 9:319–340.

Genung, W. G., and J. R. Orsenigo. 1970. Some insect-weed inter-relationships that a grower should know. *Fla. State Hortic. Soc. Proc.* 83:161–165.

Gibbs, A. J., ed. 1973. *Viruses and invertebrates.* New York: American Elsevier Pub., xvi, 673.

Giblin-Davis, R. M. 1993. Interactions of nematodes with insects. In M. W. Khan, ed., *Nematode interactions.* London; New York: Chapman & Hall, 302–344.

Harris, K. F., and K. Maramorosch, eds. 1980. *Vectors of plant pathogens.* New York: Academic Press, xiv, 467.

Jacobsen, L. A. 1945. The effect of stinkbug feeding on wheat. *Can. Entomol.* 77:200.

Khan, M. W., ed. 1993. *Nematode interactions.* London; New York: Chapman & Hall, xi, 377.

Kranz, J., H. Schmutterer, and W. Koch. 1978. *Diseases, pests, and weeds in tropical crops.* Chichester; New York: Wiley, xiv, 666, [32] leaves of plates.

Leath, K. T., and R. A. Byers. 1977. Interaction of Fusarium root rot with pea aphid (*Acyrthosiphon pisum*) and potato leafhopper (*Empoasca fabae*) feeding on forage legumes. *Phytopathology* 67:226–229.

Manuel, J. S., L. E. Bendixen, and R. M. Riedel. 1981. *Weed hosts of* Heterodera glycines: *The soybean cyst nematode.* Wooster, Ohio: Ohio Agric. Res. Dev. Center, 8.

Manuel, J. S., L. E. Bendixen, and R. M. Riedel. 1982. *An annotated bibliography of weeds as reservoirs for organisms affecting crops. Ia. Nematodes.* Wooster, Ohio: Ohio Agric. Res. Dev. Center, Ohio State Univ., 34.

Manuel, J. S., D. A. Reynolds, L. E. Bendixen, and R. M. Riedel. 1980. *Weeds as hosts of Pratylenchus.* Wooster, Ohio: Ohio Agric. Res. Dev. Center, 25.

Maramorosch, K., and K. F. Harris, eds. 1979. *Leafhopper vectors and plant disease agents.* New York: Academic Press, xvi, 654.

Maramorosch, K., and K. F. Harris. 1982. *Pathogens, vectors, and plant diseases: Approaches to control.* New York: Academic Press, xii, 310.

Norris, R. F., and M. Kogan. 2000. Interactions between weeds, arthropod pests and their natural enemies in managed ecosystems. *Weed Sci.* 48:94–158.

Pickett, C. H., and R. L. Bugg, eds. 1998. *Enhancing biological control: Habitat management to promote natural enemies of agricultural pests.* Berkeley, Calif.: University of California Press, 422.

Richardson, M. J., and M. J. M. D. Noble. 1979. *An annotated list of seed-borne diseases.* Kew, U.K. Zurich, Switzerland: Commonwealth Mycological Institute; International Seed Testing Association, 320.

Speight, M. R., and J. H. Lawton. 1976. The influence of weed cover on the mortality imposed on artificial prey by predatory ground beetles in cereal fields. *Oecologia* 23:211–223.

Sutic, D. D., R. E. Ford, and M. T. Tosic. 1999. *Handbook of plant virus diseases.* Boca Raton, Fla.: CRC Press, xxiii, 553.

Tamaki, G., H. R. Moffit, and J. E. Turner. 1975. The influence of perennial weeds on the abundance of the redback cutworm on asparagus. *Environ. Entomol.* 4:274–276.

Tamaki, G., and D. Olsen. 1979. Evaluation of orchard weed hosts of green peach aphid and the production of winged migrants. *Environ. Entomol.* 8:314–317.

Taylor, C. E., and D. J. F. Brown. 1997. *Nematode vectors of plant viruses.* Wallingford, Oxon England; New York: CAB International, xi, 286.

Thresh, J. M., ed. 1981. *Pests, pathogens and vegetation.* London, UK: Pitman Books Limited, 517.

Waldrep, T. W., D. S. Chamblee, D. Daniel, W. A. Cope, and T. A. Busbice. 1969. Damage to alfalfa by the alfalfa weevil as related to infestation by henbit (*Lamium amplexicaule* L.). *Crop Sci.* 9:388.

Weischer, B., and D. J. F. Brown. 2000. *An introduction to nematodes: General nematology; A student's textbook.* Sofia, Bulgaria: Pensoft, xiv, 187.

# 7

# ECOSYSTEM BIODIVERSITY AND IPM

## CHAPTER OUTLINE

## INTRODUCTION

Biological diversity, or biodiversity, may be considered as a concept, a measurable entity, or a sociopolitical construct (Gaston, 1996). Biodiversity has to be assessed at three overlapping hierarchical levels of resolution:

1. Genetic diversity
2. Taxonomic diversity
3. Ecosystem diversity

Efforts to define the influence of biodiversity on pest management require a wide range of research approaches and attention to interacting elements of the system. Considerations of plant biodiversity and pest biodiversity interact to define the options available for pest management in any agroecosystem. This chapter discusses the implications of taxonomic and ecosystem biodiversity to IPM; the role of genetic diversity in IPM is discussed in Chapter 17 in relation to host-plant resistance.

> "Biological diversity refers to the variety and variability among living organisms and the ecological complexes in which they occur . . ." (U.S. Congress, Office of Technology Assessment, 1987; Heywood and Watson, 1995; Gaston and Spicer, 1998).

## BIODIVERSITY AND AGRICULTURE

As humans shifted from hunting and gathering to arable farming, agroecosystems were created that had less biotic complexity than the preexisting natural ecosystems. The rate of biodiversity decrease was initially fairly slow, but it

**Monoculture is a cropping system where a single crop species is grown over large areas to the exclusion of other vegetation.**

started to accelerate about 300 years ago, and has been most rapid in the twentieth century. In most industrialized countries today, the majority of staple cash crops are grown as monocultures.

It is appropriate at this point to examine why monocultures (Figures 7–1a and b) have become so prevalent in agriculture. Human labor is a major reason. Weeding and harvesting are the two processes in crop production that require the highest human labor input in nonmechanized agricultural systems (Figure 2–4). In the industrialized countries, machines were developed to reduce the need for hand labor to harvest most staple crops. Similarly, weed management changed from human labor to the use of animal-drawn cultivators, and in the last century to the use of tractor-pulled machinery. The efficient utilization of these machines requires growing a single crop species. The machinery required to deal with different crops may be quite varied, so that growing two or more crop plant species mixed together often precludes the use of mechanization for weed control or harvesting.

Because of the limitation imposed by human labor requirements, the within-field biodiversity of agroecosystems in industrialized countries is likely to remain relatively low. This seems particularly true given that most of the population in developed countries is not involved in farming. In this situation crop production utilizes farm machinery in place of labor. It has been suggested that the nonindustrialized countries will also shift toward increased use of monoculture crop-production systems as their population changes from rural to industrial.

(a)

(b)

(c)

**FIGURE 7–1**    Different levels of cropping system complexity. (a) A monoculture crop of cotton in Central Valley, California; (b) monoculture plantation of bananas in Honduras; and (c) polycrop system of corn and soybeans. Sources: photographs (a) and (b) by Robert Norris; (c) by Marcos Kogan.

What are the ecological implications of the practice of growing crops as monocultures to the management of pests? Ecological theory predicts that natural ecosystems with higher species diversity are inherently more stable than those with a limited number of species. From a pest management perspective, this theory has been used to argue that a diverse ecosystem is less likely to suffer catastrophic pest attack than a monoculture. Not all ecologists accept the theory that ecosystem diversity leads to stability (Goodman, 1975; Grime, 1997; Zeide, 1998; McCann, 2000), but it is still widely cited for arthropod IPM, and there is good evidence supporting the concept.

> **Diversity is considered to be the key to the stability of natural ecosystems.**

It should be emphasized that the diversity/stability concept is largely theoretical. It is difficult to design experiments that are large enough to adequately test the theory. Many of the ideas that have arisen from the diversity/stability hypothesis are based on natural ecosystems, and agricultural ecosystems are unstable due to the many activities carried out by humans. Given that agroecosystems are in a constant state of flux, it is likely that they do not operate like a natural system with respect to diversity and stability; although it is important to point out that the scale at which this theory is considered influences assessments of both diversity and stability.

## Scales of Diversity

The significance of the benefits of biodiversity is spatially, and temporally, scale dependent. The scale levels that are important are as follows:

### Spatial Scales

**Global Biodiversity**   Global biodiversity is important to human activity at the level of world ecosystem processes, and is important to IPM in the sense that it provides genetic resources used for breeding resistant crop varieties, and sources of biological control agents. Host-plant resistance to pests is considered in Chapter 17, and biological control in Chapter 13.

**Regional Biodiversity**   Regional biodiversity encompasses farms and associated towns at a landscape level. It is sometimes discussed as a mosaic of "islands" of habitats between which organisms can move (Figures 7–1c and 7–2). The size of the so-called islands, and the distance between them, is an issue in relation to movement of organisms between them. The IPM impact is that pest and beneficial organisms can move between different ecosystems present in the region. It is useful to think of regional biodiversity at the following two levels, because of the difference in who manages the system(s):

a. Level of a single farm, which is under the control of a single person or organization
b. Whole region in which the farm is situated, which is under the control of multiple agencies or persons with varied interests

**Within Field Biodiversity**   Within field biodiversity impacts many aspects of IPM and is addressed in this chapter in greater detail.

### Temporal Scale   Different crop genotypes or different crop species can be grown in an ecosystem over time, and can provide benefits based on increased diversity in a temporal framework. Temporal diversity is the ecological principle

(a)

(b)

(c)

**FIGURE 7–2**   Landscape mosaic of crop fields mixed with noncrop areas showing woodlands, various refugia areas, and potential corridors for organism movement between ecosystem types. (a) Aerial view of mixed vegetation landscape in southern England; (b) small fields with different crops, hedges, woodlands, and other refugia in a European setting; and (c) small-scale farming in India juxtaposed with undeveloped land.

Sources: photographs (a) and (b) by Robert Norris; (c) by Marcos Kogan.

underlying the use of rotations as a means to manage pests; this topic is discussed in depth in Chapter 16.

## Benefits of Biodiversity

There are numerous ecological reasons why high biodiversity is supposedly beneficial; some of these potentially impact integrated pest management. Major reasons for maintaining ecosystem biodiversity include the following, with each presented in association with the scale at which biodiversity is significant.

1. Conservation of genetic resources—global scale
2. Increased sustainability/stability of the ecosystem—global and regional scales
3. Dilution of the effect of resource concentration—regional scale
4. Enhancement of ecosystem processes (oxygen production, runoff control, erosion control) that humans deem useful—all scales (there are also negative processes, from the human perspective, such as pest epidemics)
5. Preservation of ecosystem aesthetics—regional scale
6. Support for wildlife populations—wildlife may depend on pest organisms for food (weed seeds, insects in noncropped areas, vertebrate pests)—within-field and regional scales

7. Source of beneficial insects (the main reason to foster biodiversity from an arthropod management point of view)—within-field and regional scales
8. Reducing the chance of disease epidemics—regional scale, also possibly within-field scale
9. Improvement in overall soil "health"—within-field scale
10. Increased yield from mixtures of plants—within-field scale

When discussing the significance of biodiversity to IPM it is essential that the spatial and temporal scales involved be clearly defined. What may be useful at one scale may be detrimental at another. Pest management considerations are mainly at the regional and within-farm scales. The effects of biodiversity at the global scale have less direct impact on IPM, but are important in terms of conservation of genetic resources that have potential for use in breeding pest-resistant crops, or for importing natural enemies for biological control. We restrict discussion here to agroecosystems (including gardens, etc.), and do not address pest management in natural ecosystems, such as encroachment by invading species, where the significance of biodiversity is different. The implications of diversity extend beyond pest management, and include factors such as wildlife preservation, soil erosion, aesthetics of landscapes, and maintenance of genetic resources.

As noted earlier, one of the major assumptions about biodiversity is that there is a link between ecosystem biodiversity and stability. From an IPM viewpoint it is also important to establish if there is a link between (biodiversity) stability and yield, or economic return. A farmer or a pest control advisor may perceive ecological principles not translated to economics to be of little value; however, part of the IPM philosophy includes the concept of benefits beyond simple economic return. Most work on biodiversity in relation to IPM does not address net return, and ecological benefits are even more difficult to ascertain. The difficulty of establishing the importance of biodiversity to IPM is further compounded by the fact that most IPM programs often only address management of a single pest species or pest category.

Enhancement of biological control is a major ecological mechanism through which increasing biodiversity functions. We discuss biological control in Chapter 13. The astute reader may note the connection between pests that are subject to biological control and the utility of biodiversity to their management. Many arthropod pests experience extensive biological control, and increasing biodiversity can have a major influence. In some cases, biological control of pathogens and nematodes may also benefit from increasing biodiversity, particularly in the soil. Because weeds are generally not well controlled by biological means in agricultural systems, they typically do not show positive response to increased ecosystem biodiversity.

Drawing broad conclusions about the impact of biodiversity on IPM has proved difficult. Here we briefly review the ecological literature on biodiversity and ecosystem processes. We attempt to present a balanced view of the benefits and pitfalls associated with biodiversity and IPM.

## BIODIVERSITY EXPLAINED

Many of the concepts surrounding the significance of biodiversity are not well defined. A massive volume, *Global Biodiversity Assessment* (Heywood and Watson, 1995), provides much information that helps explain some of the ambiguities of the concept.

Three categories of ecosystem biodiversity were proposed by Whittaker (1972) in an attempt to include consideration of scale in assessment of plant biodiversity:

1. Alpha diversity. This represents the species richness within a designated, local (bounded) habitat.
2. Beta diversity. This biodiversity within a region results from exchanges and turnovers of species among habitats (ecosystem islands).
3. Gamma diversity. This total biodiversity includes all habitat types within an area.

These terms are generally used with respect to plant ecology.

Alpha diversity is quantifiable as increased species richness in a community or specified land area. It can be measured by determining the number of permanent species present in an ecosystem, and the relative proportion of each that is present. These measured values are then used to construct a mathematical index for a community that reflects its biodiversity. These indices range from simple relative dominance to various formulae, such as the Shannon-Weiner index, the Simpson-Yule index, and the Berger-Parker dominance index. No one index is universally applicable or accepted.

The indices are applicable to natural ecosystems, but there is considerable difficulty assessing alpha diversity in disturbed agricultural systems. Within-field biodiversity in arable ecosystems varies temporally because at times in the crop cycle there are no plants present, and thus biodiversity of all other organisms is impacted. Due to the use of monoculture cropping in industrialized annual-cropping systems, long-term modification of biodiversity has to be accomplished at the regional level.

Beta diversity is harder to measure because it is a rate or flux. Beta diversity is assessed by comparing the species composition of two areas using alpha diversity indices.

Scale can be considered at different levels:

1. Specific site. There is a large difference between considering biodiversity in the soil versus aerial environments. The individual manager has reasonable control of some aspects of the soil environment (e.g., nutrient status, moisture, and pH), but often has little control over many aspects of the aerial environment. Developments in precision agriculture will allow for more site-specific management.
2. Within-field scale. The area of consideration ranges from a few square feet to hundreds of acres, but is essentially managed as a single entity. The manager has control over cultural factors such as what is planted, the time of plowing, seeding, and harvest.
3. Farm level scale. The possibility exists for controlling the local mosaic of which crops are planted where, and the vegetational complexity of their surroundings. At this level, biodiversity is still under a single management program.
4. Township (village) local regional scale. The biodiversity of agroecosystems results from the sum decisions made by individual farmers and local land management authorities; these parties may all have different objectives. It is at this level that the concept of area-wide IPM is significant, and its implementation poses challenges due to conflicting interests of different segments of the human community.

5. Region. This is much the same as the township, but on a larger scale. The complexity of management is further increased, because regional government agencies are involved and the increased number of ecosystem variables that have to be considered.
6. Country. The significance is now at the level of how overall government policy is established that either fosters or hinders maintenance of biodiversity. Multiple agencies are typically involved, often with conflicting goals.
7. Global. Issues at this level require that different governments work together. National interests and goals may override considerations that impact biodiversity.

There are several approaches to the manipulation of biodiversity within fields and within regions.

1. Variety blends. Growing multiple genotypes of the same crop species within a field or region increases the genetic diversity within the crop. An example would be a mixture of resistance alleles that would avoid the selection of resistance-breaking pests within a field. Such blending of resistant and susceptible varieties has been proposed to delay the development of pest resistance in genetically modified crops (see Chapter 17).
2. Crop rotation. Diversity can be increased by growing multiple genotypes or different crops in time, a process called crop rotation (see Chapter 16).
3. Intercropping. This approach involves growing two crops, or more (when it is called polycropping), in the same field at the same time (Figure 7–1c). The presence of more than one plant species increases the within-field biodiversity, and may also increase total yield. Intercropping has the drawback that it makes weed management and mechanization of farming operations more difficult. The system only works well where hand labor for weed management and harvesting is readily available.
4. Establishment of within-crop alleys. Selected annual plants can be sown in the spaces between rows of the main crop. These plants provide nectar and pollen necessary for certain natural enemies of arthropod pests, and also aid in nitrogen recycling. If properly chosen, such plants do not compete with the main crop and do not become weeds themselves. Intercropping and cover crops are further discussed in Chapter 16.
5. Establishment of refugia. Defined areas are designated to grow specific vegetation that provides a benefit to the surrounding areas. In most agricultural settings, this means land that could potentially be used to grow crops is instead used to maintain ecosystem biodiversity. Examples of refugia include areas of native vegetation, wetlands, groves of small trees, and woodland. Hedgerows, fence lines, ditch banks, and roadsides can also be managed as refugia. These areas of mixed vegetation form a mosaic structure intermingled with fields of different crops (Figure 7–2) between which different organisms can migrate. From a pest management perspective, the distance between refugia becomes an issue in relation to the mobility of organisms. Refugia serve as islands of suitable habitat in the region when crops are not present, and may harbor both beneficials and pests.

Any manipulation of biodiversity has an associated cost. The economic benefits of increased biodiversity therefore need to be compared with the economic

**Economic cost-benefit analyses must be conducted in relation to biodiversity for IPM purposes.**

costs associated with attaining the increased biodiversity. Such formal, economic cost-benefit analyses are an important component of decision making in IPM. When within-field biodiversity has economic benefit for pest management, such as has been demonstrated in some tree crops, then the farmer/pest control advisor will use such biodiversity.

How society pays for maintaining or increasing biodiversity relative to IPM has not been established. With respect to increasing biodiversity, costs cannot be borne by the farmer alone, as the importance of biodiversity at regional and global scales extends beyond individual farms. When biodiversity requires construction and/or maintenance of refugia at the regional level, the cost should be borne by the greater community. The cost of maintaining global biodiversity is an international issue.

# SIGNIFICANCE OF BIODIVERSITY TO IPM

Most current research on biodiversity in relation to IPM has evaluated beneficial arthropods (e.g., Pickett and Bugg, 1998; Landis and Marino, 1999; Landis et al., 2000), and recommendations are usually based on an assessment of arthropod pest problems. This is a narrow view of pest management. The impacts of biodiversity on management of nonarthropod pests are usually not discussed. Such omission makes assessing the utility of biodiversity to IPM difficult. Here the importance of biodiversity to each category of pests is addressed.

## Pathogen Management

The literature on general plant pathology typically does not address the topic of biodiversity in relation to disease management. The concept that pathogens spread, and hence severity of disease is increased by monoculture of crops, was established in the late nineteenth century when the coffee rust epidemic swept through coffee plantations in Ceylon (now Sri Lanka). Epidemics of Southern corn leaf blight, and repeated incidence of rust diseases in cereals, serve as additional examples of how pathogens can rapidly spread through monocultured crops. In most of these cases, however, not only were the crops grown as monocultures over large areas, but they also were genetically uniform. Loss of intraspecific genetic diversity was at the root of the epiphytotics.

Kranz (1990) reviewed disease spread in monoculture and multispecies communities, and found no evidence that disease was less in multispecies communities. Research on the spread of rust disease in sunflowers confirms this conclusion (Alexander, 1991). The evidence based on the spread of both Dutch elm disease on a worldwide basis and chestnut blight in the United States also suggests that ecosystem biodiversity did not stop the spread of these invading pathogens. The host plants in both cases were always present only as part of diverse ecosystems. At best, plant biodiversity appears to only have decreased the rate of pathogen spread, but not to have reduced the final impact.

There is, however, strong evidence that plant species biodiversity can be extremely detrimental to management of numerous pathogens. Many plants can serve as alternative reservoir hosts for viruses and other pathogens (see Chapter 6) and thus elevated biodiversity can exacerbate disease problems caused by these pathogens. Decreasing biodiversity leads to reduction in severity of many diseases. Systematic removal of vegetation that serves as alternate hosts for pathogens is used as a management tool for many pathogens. The classic historical example was the removal of barberry plants to control wheat rust. We provide the additional example of *lettuce mosaic* management in Chapter 18. Altering biodiversity to manage pathogens is something that can be implemented at the farm level for some pathogens, but in many cases it is a regional problem. In the latter situation the responsibility for carrying out the program is often problematic.

Suppressive soils are those in which pathogens do not thrive and so cause only limited plant disease. Crops grown in soils supporting diverse microorganisms are considered less likely to experience catastrophic disease outbreaks, because increased biodiversity is thought to contribute to suppressive soils. Pest managers have yet to manipulate this form of biodiversity for IPM purposes.

## Weed Management

There is no compelling evidence that permitting a more diverse weed flora will offer any advantage over a flora with a more restricted number of species. There is a hazard that the more diverse weed flora will contain some species that can tolerate a particular set of management practices, and their population will increase if such practices are utilized. We thus find it difficult to see any advantages of weed biodiversity per se to weed management due to the trophic dynamics position of weeds in the food web.

Genetic variation within crop plant species as a tool for weed management is the subject of current research, and is discussed in Chapter 17. Simultaneously growing multiple crops in the same field, and temporal variation in crop diversity (i.e., rotations) have value for weed management, and are discussed in Chapter 16. Attempts are being made to replace competitive weeds with less competitive weeds, and such schemes are discussed in Chapter 16.

Proposals to leave weeds in fields for insect management have been made (see biodiversity; arthropod section). Such proposals imply that weeds are economically less important than insects, but there is only limited evidence on which the relative importance of these pests can be judged.

## Nematode Management

Many plants can serve as alternative hosts for nematodes. Increasing biodiversity thus has the potential to lead to increased populations, even when the host crop is not present. There are, however, some plants that are capable of reducing nematode numbers, such as several species of African marigolds in the genus *Tagetes* (see Figure 13–5). Using such plants to increase diversity has the potential to improve management of certain nematodes. A possible difficulty with using African marigolds for nematode management is that they may be excellent hosts for whiteflies. It is therefore essential that the host status of plants for different pest categories be known in order to use biodiversity for nematode management.

Nematode suppressive soils, similar to those suppressing other pathogens, may be enhanced by increased biodiversity. High biodiversity of soil organisms is considered beneficial for nematode management, because this is indicative of high biological control potential.

## Mollusk Management

The predominant evidence is that increased plant biodiversity makes mollusk management more difficult, as the additional vegetation provides both food and shelter. The development of no-till agricultural systems has demonstrated that the added vegetation and cover leads to increased mollusk problems. Large-scale conventional agriculture rarely experiences mollusk problems; this is attributed to the lack of vegetation and cover that occurs between crops.

The presence of additional acceptable food plants may divert slugs from attacking the seedling crop (Cook et al., 1996). However, it is difficult to ensure that the plants present (a) are more palatable than the crop, and (b) do not compete with the crop. As for nematodes, noncrop vegetation has the potential to increase slug and snail populations, and increased biodiversity is likely to make mollusk problems worse. Inclusion of noncropped vegetation corridors in cereal fields in Europe leads to increased mollusk damage on each side of the strip (Frank, 1996).

## Arthropod Management

The main premise that diverse systems are less likely to experience outbreaks of insect pests is based on a combination of the resource concentration hypothesis at the within-field level, and on increased maintenance of beneficials at the within-field and regional levels. Conversely, densities of beneficials may actually be higher in less diverse systems because their prey may be more concentrated in such systems.

Extensive literature reviews by Andow (1990, 1991a, 1991b) show that diverse systems decreased phytophagous insects in 148 cases, had no effect or increased pest insects in 71 cases, and in 46 cases the response varied. Analyses based purely on numbers of insects may be misleading, because the use of increased biodiversity to maintain a beneficial may also support pests. The outcome of increased biodiversity then depends on the efficiency of the particular natural enemy as a biological control agent, and on whether the arthropod is a key or an incidental pest. To define the value of increased biodiversity, it is necessary to establish the economic costs and benefits resulting from the relative enhancement of the beneficial or the pest population. Even from the limited viewpoint of arthropod management, neither Andow (1990, 1991a, 1991b) nor Norris and Kogan (2000) could reach a broad conclusion about the utility of biodiversity as a management tactic. An additional problem occurs with enhancement of botanical diversity within cropping systems, because the plants added to the system may compete with the crop plants. Each situation must be judged on its own ecological and economic merits.

There are several points that need to be considered in relation to biodiversity and integrated pest management:

1. Increased within-field biodiversity can be achieved by growing multiple genotypes or plant species within the crop (mixed cropping) or through intercropping or polycropping. There are several ways that this type of increase in biodiversity can be accomplished:

   1.1. Intercropping or polycropping achieves increased biodiversity by growing two or more crops together (Figure 7–1c). There are inherent yield advantages to these systems in regions where hand labor is readily available to carry out weeding and harvesting. The mixture of crop species supports a more diverse arthropod fauna, leading to reduced arthropod pest problems in some situations.

1.2. Leaving weeds within the field can provide increased biodiversity. Increased vegetational diversity leads to increased difficulty of weed management when the weeds used to achieve the biodiversity are interspersed within the crop. This increased difficulty in weed management must be allowed for in any arthropod management program that utilizes biodiversity. The magnitude of the problem is relatively low where weeds are managed by hand. The level of difficulty increases when extensive cultivation is used for weed management. If herbicides are used extensively it is difficult to leave weeds because of selectivity problems. In most situations the use of vegetationally diverse systems precludes the use of herbicides for weed management, and places increased reliance on hand weeding.

1.3. Uncultivated and unsprayed strips can be left within the crop to increase biodiversity. These are sometimes referred to as corridors, and the term "beetle banks" has been used. Although some predator populations are increased or maintained in such strips, there is little evidence that the pest numbers in the crop are altered (Wratten et al., 1998). There is a hazard that generalist predators may be detrimental, because they may prey on other beneficials. Another problem is that the strip may support an increase in numbers of organisms that are a pest to another crop in the region.

1.4. Increased biodiversity at the local farm level can be achieved through use of conservation headlands (Figure 7–3). These are nonsprayed edges of fields that grow selected cover crops or weeds. This system works well in a set-aside government program that pays for the lost production (Ovenden et al., 1998), and such programs are one means for society to pay for the maintenance of biodiversity. It is also especially useful because it provides general habitat for wildlife, such as pheasants and partridges.

**FIGURE 7–3**
Example of increased diversity in an agricultural system. Sugar beets are to the right, native vegetation strip in the middle, and *Phacelia* planted on the left.
Source: photograph by Robert Norris.

Some work has raised questions about the utility of vegetational biodiversity within the crop for insect management. In California, Costello and Daane (1998) showed that generalist predators such as spiders stayed in the vegetation strips in vineyards and did not move into the vines. Thus, although the predator population was higher in the system with increased biodiversity, there was no impact on the pests in the crop. A similar finding with weed strips in cereal crops in Europe suggests that carabid beetles tend to remain in the strips rather than move to the adjacent crop. Work in U.S.A. soybeans suggests that weed strips may in fact attract beneficials out of the crop and to the weeds (Kemp and Barrett, 1989). In these cases, diverse vegetational strips, islands, or hedges—instead of acting as sources of beneficial arthropods—behave as sinks. These results indicate that careful analysis of the use of weed strips for pest management is essential.

It is of paramount importance (see later in the chapter) that the plants used to increase within-field biodiversity are not themselves weeds in the adjacent crops.

2. Increased regional biodiversity can be achieved in two ways.

**Growing different crops in a mosaic of small fields can increase local biodiversity.**

2.1. At the local level, biodiversity can be increased through use of small fields planted to different crops. This traditional peasant farming system is still used in many regions of the world. Many argue that it works well for arthropod management, but the system is difficult to mechanize. Strip cropping is a way to achieve the same goal but still permit use of farm machinery. Strip-cropping systems present problems in irrigated agroecosystems because of the varied water requirements of different crops. A problem with strip cropping from an IPM viewpoint is that, to our knowledge, no one has demonstrated an economic benefit.

**Refugia serve as areas of increased biodiversity within agroecosystems.**

2.2. At the regional level, increased biodiversity can be achieved through the use of various forms of refugia, such as hedges, woodland, ditch banks, or other areas set aside from crop land that grow native vegetation to maintain beneficials and wildlife (Figures 7–2 and 7–3). The costs of maintaining refugia, and the loss of production, must be weighed against gains in insect management. The potential for harboring crop pathogens and vertebrate pests must also be evaluated.

Way (1977) summarized the impacts of biodiversity on arthropod management and emphasized the possible types of interactions that can result from changes in biodiversity. The important points are as follows:

1. Decreased biodiversity can disrupt the life cycle of the pest arthropod, or decrease damage to the crop.
    1.1. The arthropod pest is deprived of alternate food supply or refuge when the crop is absent, or population increase is reduced in absence of alternate food supply.
    1.2. Arthropod pest attack is diluted by abundance of a preferred host plant.
    1.3. Large uniform units of monoculture suffer less overall damage from pest arthropods that are concentrated on field margins.
    1.4. Standardizing the sowing date of annual crops can put the crop out of phase with the arthropod pest when there are no other host (weed) plants available.

1.5. Damage caused by the arthropod pest is better tolerated when the crop is not competing with weeds.
2. Decreased biodiversity can benefit natural enemies.
   2.1. Decrease in biodiversity of natural enemy species can decrease harm from interfering natural enemies and hyperparasites.
   2.2. Decreased weeds decrease the hazards to natural enemies searching for their prey on crop plants.
   2.3. Decreased prey available on weeds decreases the likelihood that key natural enemies are diverted from the crop.
   2.4. Decreased biodiversity resulting from continuous sequence of monocultured annual crops (weed free) creates opportunities for equilibria between natural enemies and their host/prey equivalent to those obtained in perennial crops.
3. Appropriate increases in biodiversity might be expected to decrease pest arthropod attack by the following methods:
   3.1. Providing camouflage, making the at-risk crop less visible to the pest arthropod (decreased apparency)
   3.2. Acting as a barrier or hazard to the pest arthropod
   3.3. Providing alternative hosts, which divert the arthropod pest from the at-risk crop
   3.4. Benefiting natural enemy action through the following:
      3.4.1. Providing food for nonparasitic/nonpredatory life stages of natural enemies
      3.4.2. Providing food for alternate prey/hosts needed to perpetuate predatory or parasitic life stages of beneficials
      3.4.3. Providing refuges for natural enemies

Way (1977) concluded that the general statement that vegetation diversity and stability are related must be examined in relation to any specific agricultural situation. Reviews of the literature since then agree with this conclusion. Interactions between pests are sufficiently species specific that there is considerable danger in extrapolating from one ecosystem to another. Way (1977) also saw the need for increased knowledge on the impact of vegetation diversity on area-wide spatial distribution, and how this could be used to manage arthropod pests. This conclusion also remains valid, and development of area-wide IPM programs is the manifestation of attempts to use such information.

If the same plants that host beneficial insects also host pest insects, then the losses from the pest may outweigh the gain from all the beneficials. Papers showing increased levels of beneficials associated with biodiversity, such as along field borders, rarely present information about the associated pest species. In the absence of this information it is not possible to accurately determine the advantages provided by the beneficials.

## Vertebrate Pest Management

The role of biodiversity in the management of vertebrate pests is usually not discussed. We have found no evidence that biodiversity enhances control of vertebrate pests. It is feasible that in many situations it may in fact be detrimental, as additional plants can provide shelter and varied food (see Chapter 6). In light of the fact that increased biodiversity is considered beneficial for maintenance of wildlife populations it would seem, from an ecological perspective, that conditions suitable for enhancing wildlife would most likely enhance pest species as well. One of the standard recommendations for controlling pests such as ground squirrels and voles is to remove the habitat that supports the population (see cultural manipulation in Chapter 16). This creates a dilemma for the pest manager who may have to decide between reducing habitat for a pest species and attempting to support wildlife populations.

# USING BIODIVERSITY IN IPM SYSTEMS

## The Cost-Benefit Approach

A cost-benefit analysis must be applied in utilizing biodiversity for IPM. This implies the benefits derived must be balanced against losses before attempting changes in biodiversity. The cost-benefit assessment must include all categories of pests, and potential environmental impacts.

The cost-benefit approach identifies potential problems with the practice of leaving plants as a means to maintain beneficial insects. If vegetation complexity at either the region or within-field scale is to be used to manipulate beneficial organisms, the following pest hexagon (see Figure 6–1) questions need to be addressed.

1. Are the additional plants, at either the within-field or regional scales, alternative hosts for plant pathogenic organisms (e.g., fungi, bacteria, or viruses) that also cause diseases in crops in the region?
2. Are the additional plants within the field also alternative hosts for plant pathogenic nematodes in any of the crops grown in rotation?
3. Can the additional plants, either within the field or directly adjacent, serve as cover and resources for vertebrate or mollusk pests? The latter problem was demonstrated in Germany where slugs in untilled wild-flower strips destroyed wheat up to 1 to 2 m from the strip (Frank, 1996).
4. Are the additional plants weeds? If so, the following two questions must be addressed:
    4.1. Will the weeds cause direct yield or other losses to the crop?
    4.2. Can the weeds attain reproductive status and contribute to increases in either seedbank (annuals/perennials) or budbank (perennials), and thus increase weed management costs in the subsequent rotational crops?

**The manipulation of biodiversity as a management tactic must be based on consideration of all types of pests.**

If the answer to any of these questions is yes, then the use of such vegetation for pest management purposes must be judged in relation to the economic impact of the other pest organisms that will be altered. The impact of weeds serving as alternative hosts for pathogenic organisms and vertebrates is essentially always negative from a pest management perspective. Leaving such weeds for economically unverified beneficial insect management is dubious, or even detrimental, in a pest management sense. If increased plant biodiversity is to be used for IPM purposes, it should not include weeds, but should rather identify plant species with the characteristics that are required, and then sow such species. The use of *Phacelia tanacetifolia* (which is rarely a weed) to supply pollen for syrphid fly adults is an example.

## Special Considerations/Applicability

Certain differences between cropping systems make the use of within-field biodiversity more or less likely to be useful for IPM.

1. Arable annual crops having either of the following:
    1.1. High value, fresh market produce. Biodiversity is unlikely to provide adequate control due to the cosmetic standards for arthropod pests (see Chapter 19), and due to importance of other pests that may be exacerbated by biodiversity. The organic produce market, however, is changing the cosmetic standards, and may open opportunities to implement biodiversity enhancements that are economically as well as ecologically justifiable.
    1.2. Lower value, processed commodities, or animal feed. Due to the intense disturbance in these systems, it seems unlikely that within-field biodiversity can be utilized, but regional biodiversity may be extremely important in some systems (e.g., management of bollworms in cotton).
2. Arable perennial herbaceous crops (e.g., alfalfa) are similar to 1.2.
3. Pastures have the potential to be used to improve regional biodiversity, but direct benefits to the pasture are uncertain.
4. Turf biodiversity is not feasible in highly managed turf due to aesthetics; diverse vegetation is potentially detrimental if flowering plants attract bees and wasps.
5. Perennial tree crops use of within-crop vegetation on the orchard floor is feasible and offers IPM benefits in some situations, and regional-level biodiversity may have potentially useful aspects.

Our analysis of the importance of biodiversity for pest management is that all pests, not just arthropods, must be considered. This chapter and Chapter 6 should serve as a warning: Beware of general statements about IPM that ignore some or most of the underlying complexity of agroecosystems. We also caution that extrapolating from one system to another may be dangerous from an IPM viewpoint. What works well for cereals with a set-aside program may have no relevance even to another grass crop such as rice, let alone a crop such as lettuce or strawberries. Andow (1991a) summed up the situation when he stated, "While some of the major twists in the Gordian knot of vegetational diversity can be perceived, we are a long way from unraveling its complexity."

## SOURCES AND RECOMMENDED READING

The best current general references on biodiversity and agriculture are the books edited by Collins and Qualset (1999) and Wood and Lenné (1999). The review compiled by Tilman et al. (1999) is also useful additional reading on the benefits of biodiversity. Chapter 13 in Southwood (1992) is a good discussion of the methodology for measuring diversity of animal (arthropod) systems, and the edited book by Pickett and Bugg (1998) presents extensive discussion of the impacts of habitat management as a means to enhance natural enemies of agricultural pest arthropods. The reader should be aware that these books do not discuss the full IPM impacts, as they do not consider any of the limitations we have discussed in relation to management of nonarthropod pests. Paoletti et al. (1992) placed the agricultural impacts of biodiversity into a global conservation perspective. *Global Biodiversity Assessment* (Heywood and Watson, 1995) again provides an exhaustive source of information on the general topic of biodiversity. Books edited by Solbrig et al. (1994) and Reaka-Kudla et al. (1997) also serve as good general references on biodiversity.

Alexander, H. M. 1991. Plant population heterogeneity and pathogen and herbivore levels: A field experiment. *Oecologia* 86:125–131.

Andow, D. A. 1990. Control of arthropods using crop diversity. In D. Pimentel, ed., *CRC handbook of pest management in agriculture.* Boca Raton, Fla.: CRC Press, 257–285.

Andow, D. A. 1991a. Vegetational diversity and arthropod population response. *Annu. Rev. Entomol.* 36:561–586.

Andow, D. A. 1991b. Yield loss to arthropods in vegetationally diverse agroecosystems. *Environ. Entomol.* 20:1228–1235.

Collins, W. W., and C. O. Qualset. 1999. *Biodiversity in agroecosystems.* Boca Raton, Fla.: CRC Press, 334.

Cook, R. T., S. E. R. Bailey, and C. R. McCrohan. 1996. The potential for common weeds to reduce slug damage to winter wheat. In I. F. Henderson, ed., *Slug & snail pests in agriculture.* Farnham, UK: Brit. Crop. Prot. Council Symp. Proc., 297–304.

Costello, M. J., and K. M. Daane. 1998. Influence of ground cover on spider populations in a table grape vineyard. *Ecol. Entomol.* 23:33–40.

Frank, T. 1996. Sown wildflower strips in arable land in relation to slug density and slug damage in rape and wheat. In I. F. Henderson, ed., *Slug & snail pests in agriculture.* Farnham, UK: Brit. Crop. Prot. Council Symp. Proc. 66, 289–296.

Gaston, K. J. 1996. *Biodiversity: A biology of numbers and difference.* Oxford; Cambridge, Mass.: Blackwell Science, x, 396.

Gaston, K. J., and J. I. Spicer. 1998. *Biodiversity: An introduction.* Oxford; Malden, Mass.: Blackwell Science, x, 113.

Goodman, D. 1975. The theory of diversity-stability relationships in ecology. *Quarterly Review of Biology* 50:237–266.

Grime, J. P. 1997. Biodiversity and ecosystem function: The debate deepens. *Science* 277:1260–1261.

Heywood, V. H., and R. T. Watson. 1995. *Global biodiversity assessment.* Cambridge; New York: Cambridge University Press, x, 1140.

Kemp, J. C., and G. W. Barrett. 1989. Spatial patterning: Impact of uncultivated corridors on arthropod populations within soybean agroecosystems. *Ecology* 70:114–128.

Kranz, J. 1990. Tansley review no. 28. Fungal diseases in multispecies plant communities. *New Phytol.* 116:383–405.

Landis, A. D., F. D. Menalled, J. C. Lee, D. M. Carmona, and A. Pérez-Valdéz. 2000. Habitat management to enhance biological control in IPM. In G. C. Kennedy and T. B. Sutton, eds., *Emerging technologies for integrated pest management.* St. Paul, Minn.: APS Press, American Phytopathological Society, 226–239.

Landis, D. A., and P. C. Marino. 1999. Landscape structure and extra-field processes: Impact on management of pests and beneficials. In J. R. Ruberson, ed., *Handbook of pest management.* New York: Marcel Dekker, Inc., 79–104.

McCann, K. S. 2000. The diversity-stability debate. *Nature* 405:228–233.

Norris, R. F., and M. Kogan. 2000. Interactions between weeds, arthropod pests and their natural enemies in managed ecosystems. *Weed Sci.* 48:94–158.

Ovenden, G. N., A. R. H. Swash, and D. Smallshire. 1998. Agri-environment schemes and their contribution to the conservation of biodiversity in England. *J. Appl. Ecol.* 35:955–960.

Paoletti, M. G., D. Pimentel, B. R. Stinner, and D. Stinner. 1992. Agroecosystem biodiversity: Matching production and conservation biology. *Agr. Ecosyst. Environ.* 40:3–23.

Pickett, C. H., and R. L. Bugg, eds. 1998. *Enhancing biological control: Habitat management to promote natural enemies of agricultural pests.* Berkeley, Calif.: University of California Press, 422.

Reaka-Kudla, M. L., D. E. Wilson, and E. O. Wilson, eds. 1997. *Biodiversity II: Understanding and protecting our biological resources.* Washington, D.C.: Joseph Henry Press, v, 551.

Solbrig, O. T., H. M. van Emden, and P. G. W. J. van Oordt, eds. 1994. *Biodiversity and global change.* Wallingford, Oxon, UK: CAB International, in association with the International Union of Biological Sciences, vi, 227.

Southwood, R. 1992. *Ecological methods: With particular reference to the study of insect populations.* London; New York: Chapman & Hall, xxiv, 524.

Tilman, D. C., D. N. Duvick, S. B. Brush, R. J. Cook, G. C. Daily, G. M. Heal, S. Naeem, and D. Notter. 1999. *Benefits of biodiversity.* Ames, Ia.: Council for Agricultural Science and Technology, 31.

United States Congress, Office of Technology Assessment. 1987. *Technologies to maintain biological diversity.* Washington, D.C.: Congress of the U.S. Office of Technology Assessment. For sale by the Supt. of Docs. U.S. G.P.O., vi, 334.

Way, M. J. 1977. Pest and disease status in mixed stands vs. monocultures; the relevance of ecosystem stability. In J. M. Cherrett and G. R. Sagar, eds., *Origins of pest parasite disease and weed problems.* 18th Sypm. Brit. Ecol. So., Oxford, UK: Blackwell Scientific Publications, 127–138.

Whittaker, R. H. 1972. Evolution and measurement of species diversity. *Taxon.* 21:213–251.

Wood, D., and J. M. Lenné, eds. 1999. *Agrobiodiversity: Characterization, Utilization, and management.* Wallingford, Oxon; New York: CABI Pub., xiii, 490.

Wratten, S. D., H. F. van Emden, and M. B. Thomas. 1998. Within-field and border refugia for the enhancement of natural enemies. In C. H. Pickett and R. L. Bugg, eds., *Enhancing biological control: Habitat management to promote natural enemies of agricultural pests.* Berkeley, Calif.: University of California Press.

Zeide, B. 1998. Biodiversity: A mixed blessing. *Bull. Ecol. Soc. Am.* 79:215–216.

# *8*

# PEST MANAGEMENT DECISIONS

## CHAPTER OUTLINE

▶ Introduction
▶ Diagnosis of the problem
▶ Techniques for assessing pest populations
▶ Thresholds
▶ Other factors that affect pest management decisions
▶ Summary

## INTRODUCTION

The development of a sound IPM program depends on several interacting pieces of information, which are jointly used to make management decisions. The basic information that is essential for an IPM decision-support system, is diagrammed in Figure 8–1 in what may be called "decision steps" a staircase metaphor that conveys the notion that information must be obtained from the bottom to the top. Incorrect information or absence of information at any step leads to poor or incorrect decisions. Effective decisions in IPM require that information relevant to all steps be obtained before a decision is made. The lack of appreciation of the importance of a step leads to mistakes. The major steps are as follows:

1. The pest species must be correctly identified, as noted in Chapter 4. If the identity is not correctly established, then the information on the biology and ecology of the pest used to make decisions will be incorrect. The result is that incorrect identification leads to the choice of a strategy that does not control the actual pest, or can result in unnecessary or ineffective actions being taken. In the worst-case scenario, a pesticide may be applied that does not kill the target pest but disrupts beneficials. Or, a beneficial insect may be released but the pest is not suitable prey or host. As noted earlier, pest identification is a complex process and is not addressed in

**Pests must be correctly identified to implement effective IPM.**

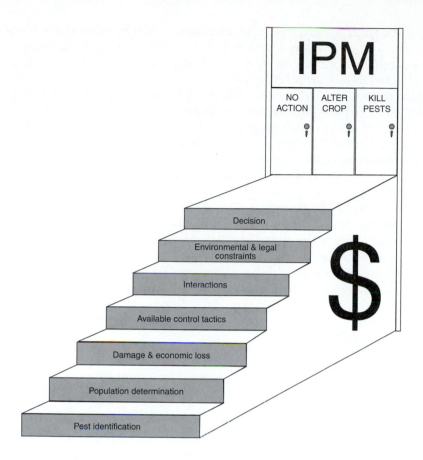

**FIGURE 8–1** The decision staircase. This diagram represents the important types of information that are part of an informed pest management decision. It should be read from bottom to top. The $ sign in the staircase implies that there are costs involved with each step.

this book. If assistance is needed in identification, consult the local university or government experts.

2. The second step in the decision-making process is to determine the pest and crop biological parameters, including pest population size, pest distribution, stage of pest development, identity, distribution and number of beneficials, crop host status, and economics of the crop. The general characteristics of the population biology of the target pest should be reviewed, as this can affect the decision regarding which control strategy to choose.

3. The third step is to evaluate the potential damage sustained by the crop relative to the pest density. This involves consideration of action, economic, or damage thresholds where they are available.

4. The fourth step is to review all the tactics that are available to manage the target pest. The cost of implementing each control tactic must be evaluated relative to the expected economic returns that will be provided by the crop. At this stage it is thus essential to take into account the economics of the crop, in particular the current and projected market value.

5. The fifth step is to consider the possible interactions among the target pest and other pests and beneficials that are present in the agroecosystem. Is the target pest really a key pest? Is it possible to use a tactic that will control the key pest and also provide control of secondary pests? Will the

**Monitoring, or "walking the field," is the essential second step in IPM.**

use of a tactic against the target adversely influence other elements of the agroecosystem? This is a critical step in implementing *integrated* pest management, and it is an element that is, unfortunately, all too often neglected in industrialized agriculture.

6. Once the desired optimum strategy has been identified, it is necessary to evaluate local and regional environmental and societal regulatory restrictions, and to evaluate potential for interactions with other aspects of crop production. In many situations, the restrictions may limit the choice of tactic (pesticides, for example, cannot be used near schools or wetlands).

7. The last step is to make a decision. Following are four general possibilities.

    7.1. No action. The damage caused by the pest is not judged sufficient to warrant action, or that action will adversely influence other pests or beneficials in the agroecosystem, or that restrictions mean that there is no acceptable control tactic.

    7.2. Reduce crop susceptibility to damage. By changing some aspect of the agroecosystem it is possible to limit the damage to an acceptable level.

    7.3. Reduce pest population size. An action is available and recommended that directly reduces the pest population to an acceptable level.

    7.4. Combination of 7.2 and 7.3.

Finally, and importantly, all of the preceding steps must be placed into an economic framework. If the manager cannot carry out any of the steps without losing money, then usually the management strategy is considered economically unjustifiable. However, this may not always be true, as there are cases when economic losses will be accepted for a short time in order to bring a problem pest under control. This requires that the sustainability of the management procedures over multiple seasons be considered.

Because there are costs involved in every step of the decision process, shortcuts are sometimes necessary or steps are completely omitted. Such shortcuts lead to the situation in which pests are controlled but less than the IPM ideal is achieved. Problems often arise because the required information or biology has not been developed or is ignored. Historically, when some of the steps, especially 5 and 6, were omitted, problems such as resistance, resurgence, environmental contamination, and illegal pesticide residues occurred.

This chapter explores the concepts and ideas behind the second and third steps of the decision-making process. Chapters 9 through 17 cover the fourth step and present the various tactics that can be used to suppress pests. Chapters 18 and 19 address the fifth and sixth steps, and cover IPM programs and the various limitations that are placed on pest management because of ecological considerations or laws and regulations intended to protect the welfare of the environment, animals, and humans.

# DIAGNOSIS OF THE PROBLEM

## Pest Identity

Correct identification of the pest species present in the agroecosystem is the most important aspect of pest management. If identification is incorrect, then the choice of tactic is often wrong. Incorrect pest identification may occur when one known species is confused with another known species. This type of error can usually be corrected by consultation with experts in taxonomy of the pest group. Incorrect identification may also occur when a previously unknown species or variant, such as a race or pathotype, of an existing species is classed as a known species. This type of error is more difficult to deal with, and requires scientific research to address the nuances of distinguishing the new pest from the known pest.

A few examples of the types of identification questions that must be addressed include the following:

1. Diseases. Once a symptom is observed it is necessary to first decide if it is caused by abiotic factors or by a pathogen. In the latter case, it must be determined whether it is caused by a bacterium, a virus, a phytoplasma, or a fungus. Often, the organism must be grown in culture to make an accurate identification, which requires special laboratory facilities. If pathovar or *forma specialis* are important, they must be determined. Accurate identification of pathogens is often difficult, although diagnostic approaches using DNA can be applied to some species. Many universities have diagnostic clinics to assist growers and consultants with their identification needs.

2. Weeds. It is necessary to be able to identify seedlings and flowering plants. Seedling identification is probably more important than knowing the flowering plant, because most control actions are taken when weeds are still in seedling growth stages. In some instances, it is also necessary to be able to identify weed seeds, to determine their presence in the soil seedbank or their presence in crop seed.

3. Nematodes. Some symptoms of nematode infection are quite distinct, such as the galls caused by root-knot nematode (see Figure 2–7f and g) or symptoms of red-ring disease in coconut (see Figure 6–14). However, the symptoms of nematode infection may not be as unique, and may resemble nutrient deficiency. Accurate identification of a nematode disease involves collection of soil or plant samples, extraction of nematodes from the samples, and identification and enumeration of the nematodes encountered. Identification of nematode genera is possible by examining the juvenile stages. Identification to the species level is usually made by examining the detailed morphological characteristics of adult females, but may also require examination of other stages as well. Diagnostic approaches are being developed that are based on DNA information. In some cases, it is necessary to conduct a differential host range bioassay to define the nematode pathotype or race. For example, within particular species of root-knot nematodes some races can parasitize cotton while others do not. Accordingly, it may be necessary to determine the root-knot nematode race for nematode management in cotton.

4. Arthropods. It may be necessary to identify adults, the immature stages, and eggs of pest species. The immature stage is often important because the arthropod is most susceptible to control tactics at this stage. Because identification of the immature stage may be difficult, it is often necessary to rear the insect through to the adult stage. Extension publications provide guides to the proper way to collect and preserve specimens for identification. It is necessary to identify beneficial arthropod species as

well, otherwise it is possible to end up spraying a beneficial mistakenly identified as a pest.

The difficulties of accurate pest identification exemplify why the scientific disciplines of taxonomy and systematics are important, and why specific training, or access to resources, in identification of each pest category is necessary for pest managers.

## Monitoring

**Informed choices between pest management options require detailed knowledge of pest biology, ecology, population dynamics, and status.**

Monitoring pest populations requires a sampling plan, can be physically challenging, and may be expensive. Different types of information must be collected and analyzed, and the specifics of the required information depend on the pest category, because some information is pest specific.

To determine pest population status it is necessary to monitor, or measure, pest densities or numbers. There are two very different requirements for pest management in relation to monitoring pests: the pest stages present and the pest density.

1. Phenological events. Phenology describes the growth stages through which the pest passes in relation to time, or a measure of physiological time such as accumulated degree-days. Timing of phenological events is used to establish the first appearance of the pest at the beginning of the season, or the appearance of a particular stage, such as adult moths, during the season. This type of information is used to time control actions, as the susceptibility of many pests to specific tactics varies with phenological stage. Flowering weeds, for example, are usually much harder to control than seedlings, and the hard shell stage (adult) of scale insects is much less susceptible to insecticides than the crawler stage (immature). Assessment of phenological development does not require that the absolute size of the population be determined, but rather that a particular growth stage is present.

2. Population size determination. Population size monitoring involves counting the number of pests. This information is needed prior to using threshold information, because thresholds are usually defined in terms of pest numbers in relation to a unit of measure. Population size determination is usually more difficult, and therefore more costly to conduct, than establishing timing of phenological events. The population size can be expressed on two different units of measurement, which require different approaches to sampling the population:

    2.1.1. Absolute. The number of pests is expressed as a number per unit of land area, such as 2,000 seeds/m$^2$, 50,000 aphids/A, or two elephants/km$^2$. If the pest number is expressed relative to the host or part of the host rather than land area, such as scale insects per tree, lygus bugs per cotton plant, mites per leaf, or nematodes per gram of root, the measure is density, or sometimes referred to as population intensity. Nematode population size may also be expressed as density on the basis of number of eggs or cysts per volume or mass of soil.

    2.1.2. Relative. Sampling methods are said to give a relative estimate of the population if the numbers are expressed per unit of sampling effort, such as the number of rootworm beetles per ten sweeps of a sweep net or numbers of leaf-roller larvae on an apple tree observed in five minutes.

Relative measures have no relevance to weeds, although the proportion of weed and crop leaf canopy may provide a technique for estimating losses based on weed presence to unit crop presence (see later in this chapter).

Regardless of whether population assessment is absolute or relative, it may be possible to use presence/absence sampling, which is technically referred to as binomial sampling. This variation in population assessment involves a sample unit, such as a leaf, fruit, or a quadrat, that is evaluated only for the presence of the target pest. The pest is scored as present or absent, rather than counting the actual number of individuals present. With adequate samples scored this way and the use of appropriate statistical tables, it is possible to estimate the size of the population.

## Spatial Pattern of Pests

Pest organisms occupy different spatial positions within agroecosystems. The spatial pattern of pests within a particular area of the system may be significant to management and may determine the location from which samples should be taken. The pattern of pest organisms in space may be described using three basic frequency distributions (Figure 8–2):

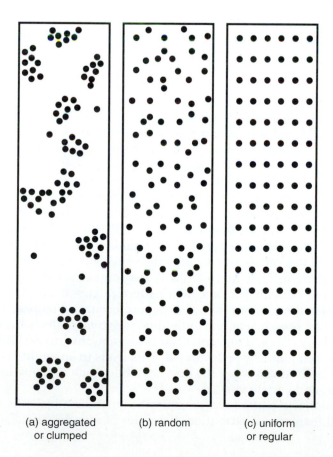

(a) aggregated or clumped  (b) random  (c) uniform or regular

**FIGURE 8–2** The three major possible theoretical natural spatial arrangements for organisms. (a) Aggregated or clumped, (b) random, and (c) uniform or regular.

1. Aggregated pattern. This pattern is also referred to as clumped or contagious, and describes pests that occur in distinct patches or clumps, with areas between the aggregations in which there are few or no pests (Figure 8–2a). Most of the aggregates typically contain relatively low pest densities, but a few contain high numbers. In a contagious spatial pattern, organisms may be considered as attracting each other, so that if one organism is present, the probability of another organism being nearby is increased. There are several reasons why organisms aggregate into groups, including:
   1.1. Organisms remain close to where they originated, so multiple progeny reside in the same place. This is typical of many weeds, nematodes, and other pests with limited mobility.
   1.2. Resources, or suitable environment, are distributed in patches, and nematodes and insects are attracted to a resource or suitable habitat, whereas weeds may thrive in an area of elevated resources or locally suitable habitat.
   1.3. Mobile organisms may be attracted to each other via mating cues; herding or flocking behavior for defense may lead to clustering.
   1.4. Social insect groups have an inherent behavior that results in the social units, such as a termite mound or ant colony.
   The aggregated pattern is common for many pests, especially when the populations are relatively low. This pattern may be described mathematically by several frequency distributions, but the most common is the negative binomial. Aggregation poses a problem for sampling, as the size of the sampling unit relative to the size of the clump determines the apparent nature of the distribution. In a clumped spatial pattern, finding one individual increases the probability of finding another individual of the same species in the immediate vicinity. If a series of samples are taken from a population, an aggregated spatial pattern is suggested if the variance is greater than the statistical mean of the samples.

2. Random pattern. In this case, individuals occur over the field with variable spacing between adjacent pest locations, and no obvious pattern among the locations (Figure 8–2b). Random pattern is typical of many well-established pests when numbers are relatively high, and individuals do not interact, such as occurs with weeds and pathogens. If individuals in a population are randomly distributed, then finding one individual at a location does not alter the probability of finding another individual nearby.
   The random pattern of pest occurrence can be described mathematically by the Poisson frequency distribution. If a series of samples are taken from a population, a random spatial distribution is suggested if the variance is equal to the statistical mean of the samples.

3. Uniform pattern. This implies that all pest organisms are spaced equally distant from each other in a regular pattern (Figure 8–2c), such as once every meter along a crop row. This pattern occurs when pest densities are high and the presence of one individual negatively affects the individuals in its vicinity. Thus, if one organism is present, the probability of another organism being nearby is decreased. This leads to an equidistant spacing between individuals when the pest is mobile, such as many insects, mollusks, and vertebrates.
   The uniform pattern may be described mathematically by the positive binomial frequency distribution. If a series of samples are taken from a

population, a uniform spatial pattern is suggested if the variance is less than the statistical mean of the samples.

In many situations, human activity modifies the aforementioned spatial patterns of pests (Figure 8–3). This is most important for nonmobile pest species such as weeds, nematodes, and soil-borne pathogens. If the spatial pattern has been characterized, that will assist in deciding the best approach for monitoring a population and diagnosing the cause of the problem.

Following are examples of human-influenced patterns.

1. Point source infection. A pest is introduced at a single point in the field and spreads from that point; it typically appears as a single aggregate. Multiple-point source infections can occur providing a multiple aggregate pattern (Figure 8–3a)

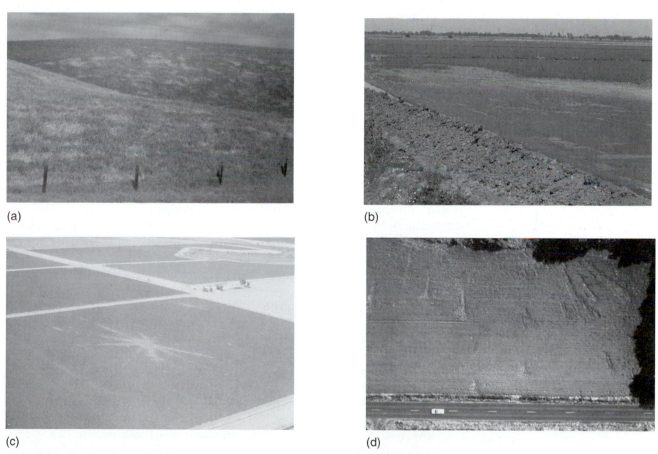

(a)

(b)

(c)

(d)

**FIGURE 8–3**    Examples of different spatial pest patterns. (a) Random point infections of pathogen (seen as light areas), here *barley yellow dwarf virus* disease in barley; (b) increased damage along edge of a field adjacent to levee (foreground), here due to rats in rice; (c) star-shaped pattern of sugar-beet cyst nematode symptoms due to land plane movement of nematodes from a single point infestation; (d) lines of *Ditylenchus,* the stem and bulb nematode, symptoms along the slope of an alfalfa field; the field is higher at the top of the picture and lower at the road end. The symptoms demonstrate the dissemination of the nematode with water runoff.

Sources: photograph (a) by Lee Jackson, with permission; (b) by Terrel Salmon, with permission; (c) by Ivan Thomason, with permission; and (d) by G. Caubel, with permission.

2. Edge of field versus center. Many mobile pests and beneficials invade fields from hedgerows, fence lines, ditch banks, and other surrounding areas (Figure 8–3b). The pests are typically present earlier in the season and at higher density in the field margins than in the interior of the field. If this situation occurs, the manager will know that the field margins should be monitored more intensively and carefully, because assessing populations only in the field center might not detect important initial infestations

3. Prevailing wind and windbreaks. Some pests are blown into fields on the prevailing winds, leading to higher pest density on the upwind side of the field. This area should be monitored more carefully as it may provide earlier evidence of pest arrival. Trees and structures such as fences can act as windbreaks, creating an eddy in the wind currents on their lee side. Some wind-borne pests may be deposited in such areas at higher densities. Special attention should be given to monitoring such areas for pests. Knowing the dispersal mechanisms of pests will help develop the most effective sampling program.

4. Mechanical patterns due to machinery. Soil-borne pests are often carried in soil on equipment resulting in patterns that reflect the movement of the equipment, such as lines across fields or longitudinal stripes. The star-shaped pattern in Figure 8–3c is the result of land-leveling land planes being pulled through a single-point source infestation of sugar beet cyst nematodes.

5. Irrigation patterns. Surface irrigation water can carry pest organisms and is particularly important for organisms such as bacteria, fungi, nematodes, and weed seeds that have little innate capacity for long-distance movement. The pattern of water distribution may be reflected in the incidence of pests, such as increased disease incidence of *Sclerotium rolfsii* in furrows below a site of infection, or nematodes being carried down the field in the natural water flow (Figure 8–3d). Flooding can be a serious problem as it can carry new pests into an area, but usually a particular pattern does not result.

6. Soil conditions. The following soil conditions influence the spatial patterns of pests.

    6.1. Soil moisture status. Many soil-borne pests are sensitive to soil moisture status, and thus their distribution may reflect soil water content. If soils become saturated, then the oxygen concentration may fall, leading to anaerobic conditions (particularly in fine-textured soils), and some nematodes and pathogens are sensitive to low oxygen concentration. On the other hand, many soil-borne pathogens, such as *Phytophthora*, require free water for infection to occur; diseases due to these organisms are more likely to be prevalent in low, wet areas in a field. Many nematode distributions also change with soil moisture status.

    6.2. Soil pH. Soils typically range in pH from about pH 4.0 to 9.0. Most soil-borne pest organisms have an optimal pH range. Knowledge of variations in field pH can suggest what organisms to search for in a particular location.

    6.3. Soil compaction. Some weed species are very well adapted to compacted soil conditions and can grow well where most crops cannot. Most soil-dwelling vertebrate pests avoid compacted soil.

6.4. Soil type. Soil texture may influence aeration and the previous soil factors, and thus exerts an effect on the spatial distribution of nematode populations. It is worth considering however, that although soil type is often cited as altering pest populations, the effects may be mediated by one of the first three mechanisms.

The experienced manager learns where pests are likely to occur in each field, and will usually carefully monitor those areas. The remaining areas in the field should also be checked, but they may be monitored at a lower intensity.

## Statistics of Sampling

What constitutes an adequate sample? For pest management purposes, this usually has to be a compromise between what is considered statistically reliable versus what is economically feasible. Sampling for research often requires greater precision than sampling for pest management decision making.

Statisticians distinguish between two similar terms: *accuracy* and *precision* (or reliability). Accuracy refers to how well the sample result reflects the true population value. Precision or reliability refers to the variability that is inherent in the sampling procedure, or whether a sampling method provides repeatable results. Although one would aspire to achieve highly accurate sampling, in most cases for pest management purposes, reasonably precise or reliable samples are actually more important. If the sampling method is inaccurate, reliability means that the method always has the same bias, allowing for an estimate of percent error from the sample. Absolute population estimates with high accuracy levels are usually necessary in research for establishing population parameters.

The precision of the population estimate is almost always greater with a larger sample size and a greater number of individual samples collected. Practical considerations usually require a trade-off between precision of the estimate gained by sampling and the cost of procuring the samples, but there is no simple rule for all categories of pests and situations. The aim of sampling for pest management purposes is to determine the average, or mean, number of the pests present in a particular system.

Some definitions used in basic statistics of sampling follow, based on those presented by Ruesink (1980).

Statisticians and biologists have developed a technical vocabulary to facilitate communication about sampling programs. Unfortunately, many of the words in this vocabulary also have a nontechnical meaning, and some authors have been ambiguous in their use. Some of the most commonly used technical terms are presented here. This summary is not an exhaustive discussion of statistics as applied to sampling. For more information see Southwood and Henderson (2000), Kogan and Herzog (1980), and Pedigo and Buntin (1994).

A sample consists of a small collection drawn from a larger population, about which information is desired. It is the sample that is observed, but it is the population that is studied. The number of observations (designated as $n$) in the sample is referred to as sample size, while the physical composition and magnitude of a single sample is called the sample unit. For example, the sample unit might be 10 sweeps with a sweep net; a sample size of $n = 20$ would then imply 20 sets of 10 sweeps each.

The arithmetic mean, usually designated as $\bar{x}$ is the most commonly used measure of central tendency and is computed by

$$\bar{x} = \frac{1}{n}\Sigma x_i$$

where $x_i$ represents the $i$th of the $n$ observations of pest $x$. The common measure of spread among the observations is the standard deviation (designated as $s$), which is computed by

$$s = \sqrt{\frac{\Sigma x_i^2 - \frac{1}{n}(\Sigma x_i)^2}{n - 1}}$$

The variance ($s^2$) is simply the square of the standard deviation, whereas the standard error of the mean ($S_{\bar{x}}$) is

$$S_{\bar{x}} = \frac{s}{\sqrt{n}}$$

The standard deviation is an expression of the spread among observations, and the standard error of the mean is a measure of the distance of $\bar{x}$ to the true mean ($\mu$) of the population being sampled. Several ratios that are in use compare the spread among observations relative to the observed mean. The coefficient of variation (CV) is defined as

$$CV = {}^s/_{\bar{x}}$$

A more useful ratio of $S$ to the mean, which entomologists usually refer to as relative variation (RV):

$$RV = (S_{\bar{x}}/\bar{x})(100)$$

These ratios are popular because for many sampling programs $s$ increases as $\bar{x}$ increases, and they are unitless and so can be compared regardless of the sample unit used.

The reliability or precision of a sample, as defined here, is the parameter that is dealt with in pest management. Precision can be measured using relative variation or formal probabilistic statements and confidence intervals. A confidence interval brackets the estimate, usually symmetrically, and states that the true population value lies within the interval with a specified probability, or confidence. When $n$ is large, such as 30 or more, or for some other reason normality can be assumed, a confidence interval for the mean can be written as

$$\bar{x} - t_a\,s/\sqrt{n} \leq \mu \leq \bar{x} + t_a s/\sqrt{n}$$

where $t_a$ is the value of the t distribution with probability level $a$ and degrees of freedom (df) as in $s$. The degrees of freedom are $n - 1$, if $s$ is estimated from the $n$ samples; if added information is used to better estimate $s$, the degrees of freedom will exceed $n$. When sampling pest populations, the errors are seldom normally distributed for small sample sizes (low $n$), and a symmetrical confidence interval is then not appropriate, necessitating computation of asymmetric confidence intervals. Procedures for computing asymmetrical intervals to account for the underlying statistical distribution are available, but because of the computational complexity they have found very little application. An introduction to the literature of asymmetrical intervals appears in Bliss (1967, pp. 199–203).

| TABLE 8–1 | One-Sided Confidence Interval for Three Common Statistical Distributions Given That No Individuals were Found in $n$ Samples |
|---|---|

| Distribution | Upper limit for the mean (individuals/sample) | |
| | General | Example form ($n = 10$, $\alpha = 0.05$, $k = 2$) |
|---|---|---|
| Binomial | $1 - \alpha^{1/n}$ | 0.26 |
| Poisson | $-(\lambda n \alpha)/n$ | 0.30 |
| Negative binomial | $k(\alpha^{(-1/nk)} - 1)$ | 0.32 |

$k$ is the parameter of the negative binomial and $\alpha$ is the confidence level desired.

A relatively uncomplicated method is available for one special case of an asymmetrical interval. This relates to a sampling result in which every one of the $n$ samples produces zero individuals. If there is justification for assuming that the underlying distribution is binomial, Poisson, or negative binomial, it is possible to compute a one-sided confidence interval and state that the population mean is less than some upper limit with probability level $\alpha$. The general formulas in Table 8–1 were derived from concepts of basic probability theory, but the examples indicate that the upper limit is greater for the more clumped distributions.

When the choice between two possible sampling programs depends on costs (either dollars or effort), our knowledge about reliability must be combined with the cost of obtaining the information. One program is said to be more efficient than another if, under specified conditions, it provides more reliable results per unit cost. Relative net precision (RNP) is one measure of efficiency and is defined as

$$RNP = \frac{100}{(RV)(C_s)}$$

where $C_s$ is the total cost of the $n$ samples used to compute RV.

What used to involve tedious computations is easily obtained using spreadsheet or statistical software. Built-in functions allow the direct analysis of sampling data, and some data can be obtained remotely using in-field data loggers.

# TECHNIQUES FOR ASSESSING PEST POPULATIONS

## Pest Monitoring

Monitoring is the process by which the numbers, and life stages, of pest organisms present in a location are established. The population size and the level of activity of beneficial organisms must also be determined for arthropod management; monitoring beneficials is generally not as important for other pest categories.

The process of pest monitoring is often referred to as scouting.

The following point is very important, and is often ignored in pest management discussions. There is no single monitoring technique that works for all categories of pests, and even within a pest category the best techniques differ due to pest biology and ecology. For example, direct counting of nematodes in soil would require the use of different techniques than counting tobacco hornworm caterpillars on a tomato plant.

## Overall Monitoring Considerations

Several important considerations influence the way in which a pest population is monitored. These include:

1. Data required. The effort and equipment involved differ depending on whether actual pest numbers are needed versus presence/absence information, or just phenological event timing.
2. The time of day at which samples are collected. This can alter the efficiency of collection due to differing diurnal activity of the pest, particularly for insects and some vertebrates. Night-flying insects cannot, for example, be sampled accurately during the day.
3. Weather conditions. Within seasons, the weather at the time of sample collection can substantially alter the reliability of the sampling. For example, windy, cool conditions may drive insects out of the top of a plant canopy.
4. Soil conditions. If a field is wet, sampling is difficult. Extracting nematodes and arthropods from wet soils can yield poor results.
5. Organism phenological development. The growth stage of organisms at the time of sampling can alter sampling feasibility or efficiency. The first and second instar larvae of the alfalfa weevil, for instance, are small and feed within the crop terminals, and are thus not easy to dislodge with a sweep net. This can lead to sweep-net counts that underestimate the actual population during the early stage of population development (Figure 8–4). Another example might be the difference between determining how many mature pigweed plants are present versus estimating the number of seeds in the soil. Seasonal weather changes are sometimes reflective of pest activity, and sampling at the wrong time of year may yield poor results.
6. Pest location. Where the pest lives determines where the samples must be collected, and may dictate the type of sampling device.
   6.1. Pests living in soil require the use of different sampling devices than those living in the tree canopy. It can also make a difference whether the pest typically is in the top of the canopy or stays close to the ground.
   6.2. Sometimes it may be necessary to check weeds for pests. For example, the alfalfa weevil lays eggs in the hollow stems of alfalfa (see Figure 5–1k) and also in hollow-stemmed weeds. Assessing egg numbers only in the alfalfa underestimates the actual number of eggs laid.
   6.3. Sampling may be required outside of the crop field proper when a pest population overwinters on vegetation in the surrounding areas. Assessment of lygus bugs in riparian areas can, for example, provide an indication of the magnitude of the problem that can be expected later in adjacent crop areas (Figure 8–5). Some pest control advisors routinely monitor fields adjacent to their target field for the same reason.

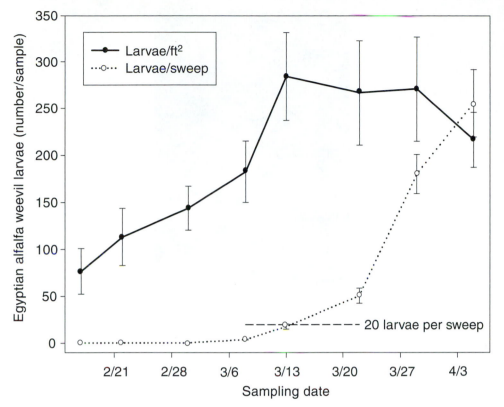

**FIGURE 8–4**
Comparison of Egyptian alfalfa weevil larvae estimated by absolute sampling as total larvae per square foot of alfalfa versus relative sampling with a sweep net. Data are mean ± standard deviation.

Source: redrawn from Cothran and Summers, 1972.

7. Research versus commercial pest management. The statistical accuracy of data collected for research purposes is paramount, often requiring greater numbers of samples. Although data used for pest management purposes should be reliable, they are typically acceptable at lower reliability levels than data required for research samples, and thus less intensive sampling is needed. Sampling techniques that are acceptable for research purposes may not be applicable to practical pest management monitoring due to time involved. Collecting samples for monitoring purposes is an economic cost that has to be considered in the cost-benefit analysis of implementing a management strategy.

## Specific Monitoring Devices and Techniques

***Direct Pest Observation and/or Counting***  A direct count of the organisms in an entire field can be made, such as low densities of weeds or populations of infield vertebrate pests. It is also common to use direct counts of insects per linear length of row when plants are at the seedling stage and the observer can easily see the insects. As the pest populations increase it is not possible to count all the pests in the field, so some form of sampling is necessary. Similarly, as the size of the crop plants increases, and the leaf area to be examined for pests increases, some form of estimation based on sampling must be used. Some methods for sampling fields follow.

**Quadrats**    Using quadrats is probably the simplest form for estimating populations of nonmobile organisms. In this approach, all organisms present within a small, limited area, called a quadrat (e.g., 1 ft$^2$ or 1 m$^2$), are counted. Quadrats at several different areas in the field are usually sampled. The size of the quadrat that must be used for sampling is related to the spatial pattern of organisms in the area. This technique is widely used for estimating weed seedling populations (Figure 8–6a).

**Plant Samples**    Plant part samples, such as stems, leaves, or roots, are collected according to a specific pattern. The amount of damage present, or the number of organisms on the samples, is then determined. Examples of this technique include counting mites per leaf in various crops, counting eggs of various insects such as imported cabbageworm or tomato fruit worm, estimating leaf damage caused by a fungal infection, or estimating the percent of a root system that is galled by root-knot nematodes.

**Knockdown**    Certain insect pests are sampled by dislodging them from the host plant onto a collecting surface or device, and the dislodged insects are subsequently counted. The mechanism used to dislodge the pest varies with pest species, and includes techniques such as using mechanical devices, heat, elevated $CO_2$, or fumigants with aerosol bombs. Large drop cloths, called beat cloths, can, for example, be placed under a grapevine, cotton, or soybean; the canopies are then shaken and the pests and predators that fall from the crop are counted (Figure 8–6b). The beat cloth is not effective for sampling fast-flying insects.

**Sweep Nets**    Nets have been used to collect butterflies for many years. Similar nets can be used to collect samples of pest insects, but the nets are stronger than those used for butterflies. The opening of the net is typically about 15 in. diameter and is reinforced. The net is swept through the canopy of the crop to be evaluated, hence the name sweep net. Insects present in the foliage fall into the net, and are then counted. The technique is used to assess populations of pest

**FIGURE 8–6**   Examples of various monitoring devices and techniques. (a) Counting weeds in alfalfa in a 1 m² quadrat; (b) using a beating cloth to estimate insects in soybeans; (c) sweep net being used for collecting insects in soybeans; (d) suction sampler for spores of aerial pathogens (here powdery mildew of grapes); (e) a portable suction sampler for collecting insects, known as a D-Vac; and (f) various tools used for collecting soil samples; left to right are a shovel, auger, Viehmeyer tube, and an Oakfield tube.

Sources: photograph (a) by Robert Norris; (b) and (c) by Marcos Kogan; (d) by Florent Tromillas, with permission; (e) by Scott Bauer, USDA/ARS; and (f) by Howard Ferris, with permission.

and beneficial insects in alfalfa, cotton, soybeans, and many other crops (Figure 8–6c). The technique has no relevance to pests other than insects. The efficacy of sweep-net sampling varies considerably with the ability and strength of the person doing the sampling. To determine population trends over successive sampling days for comparative purposes, it is desirable that the sweeping procedure be standardized and the sampling be conducted by the same person.

**Suction Air Samplers**   Such mechanical devices draw air over a collector (e.g., a filter or a sticky surface), and are used for certain pathogen spores (Figure 8–6d) and for mobile insects. The suction devices can be fixed or mobile. The movable insect sampling device has a gasoline engine driving a fan that creates a vacuum and is carried as a backpack (Figure 8–6e). It is used mainly as a research tool.

***Damage Evaluation Techniques***    Sometimes observation of symptoms or injury is the only way to determine if a pest is present. Various techniques can be used depending on the pest and the type of symptoms or injury. Beyond a low level of damage, the utility of symptom/injury evaluation techniques may be limited because the damage, and hence at least some loss, may have already occurred. Symptom and injury evaluation techniques cannot be used in situations where cosmetic damage is important (see later in this chapter and in Chapter 19), such as in many vegetable crops, because once the damage occurs it is too late to initiate control tactics. Injury evaluation is of limited value if there is no curative management method available.

**Signs and/or Symptoms**    The manifestations of the pest, or of pest activity, are provided through signs and symptoms.

Signs include evidence such as mounds of soil pushed up by gophers, rabbit fecal pellets, insect frass (feces), slime trails of slugs and snails, sclerotia of fungal pathogens such as lettuce drop and rice stem rot, or holes along the seed line where birds have removed seeds. Such indicators reveal that a particular type of pest is present even if the pest itself is not observed.

Symptoms are typical responses by many host plants to nematode, pathogen, and insect attack. Observation of the symptoms alerts the manager to the presence of the pest, and the severity of the symptoms can often be used as a measure of the pest population density. Amount of diseased area on leaves can be used as a criterion for initiating treatment action for some diseases, for example, the leaf area diseased assessment for powdery mildew of sugar beets. The level of defoliation of cotton and soybean plants is used with actual pest counts to make control decisions. Examples of symptoms include:

Stunted plants—caused by all pest categories and abiotic conditions; most viruses cause general stunting of the entire plant.
Mottled and yellowed leaves—caused by many viruses, leafhoppers, mites, nematodes, or nutrient deficiencies (abiotic).
Curled and red leaves on peaches in the spring—peach leaf curl.
Curled leaves and misshapen apples—rosy apple aphid.
Necrotic spots and dead areas on leaves—caused by numerous pathogens, insect leaf miners.
Wilted plants in hot weather despite adequate water—results from attack by vascular wilt diseases, nematodes, or insect stem borers.
Swollen and galled roots—clubroot of cabbage, crown gall, and root-knot nematodes.
Proliferation or reduction of fibrous roots—caused by some nematodes, such as the sugar beet cyst nematode, bacteria, and some viruses.
Fewer, smaller flowers or delayed flowering—results from some ring stunt, and cyst nematodes.
Root and stem rots—the fungal-like oomycete *Pythium* can cause seed rot, seedling damping-off, and root rots on many types of plants. *Phytophthora* also causes root and lower stem rots on many plant species. Fungi such as *Rhizoctonia* and *Armillaria* also cause these symptoms.
Cankers—localized dead, sunken areas in tree bark that may be caused by fungi, bacteria, and some viruses.
Scabs and warts—an overgrowth of tissue that can be caused by fungi, fungal-like organisms, and bacteria.

Physical injury—observation of injury to the host plant may be the first clue that a pest is present, such as seedlings severed by cutworm attack, holes in leaves from first instar of alfalfa weevil, or leaf damage from slug or snail feeding. Once the injury is noted it is often possible to locate the causal pest, although the pest is not readily observed at first.

Electrical resistance—a different approach to assessing symptoms is based on changes in electrical resistance in diseased tissue (especially wood) as a means of assessing attack by a pathogen. A device called a shigometer has been used in forestry to determine the presence of rotted wood, but has not proved very accurate in diagnosing attack.

**Remote Sensing**   The technique of remote sensing relies on changes in the absorbance or reflectance of plants in response to pest attack. An instrument sensitive to specific wave lengths of radiation can be used to detect such changes. It is remote because there is no contact between the sensing device and the target. Remote sensing is often carried out using equipment mounted in aircraft. The following types of remote sensing are being used or are being developed.

1. Full-color photography. Photographs can be examined or scanned and assessed for chlorosis or other symptoms.
2. Infrared (IR) wavelengths. False-color photography and remote thermometers are used to measure changes in leaf temperature in relation to differential moisture content induced by pest activity (Figure 8–3d is actually an IR photograph).
3. Multiband spectrometers. These are used to measure reflectance at several specific wavelengths. The measurements are typically obtained from aircraft or even satellites, and can be used to detect different types of vegetation. An example of the use of this technique would be to monitor the invasion of weeds in rangelands.

Remote sensing requires some degree of infield assessment (ground truthing) to backup or confirm the remote diagnosis. Uses of remote sensing for pest management include:

1. Pathogens. Infrared remote sensing thermometers can be used to detect some pathogen infections by changes in host moisture status before visually detectable symptoms are evident. Changes in the crop color, or level of stress, can be detected by photographic imaging, indicating where pest infestations have developed.
2. Weeds. Full-color, and false-color infrared, photography is used to record weed patch locations, percent of field infested, and spread over time. Multiband spectrometry is being evaluated to monitor specific species (such as broom snakeweed in rangeland).
3. Nematodes. Remote sensing is being used to detect nematode-induced stress in host plants to assess the pattern of field infestations (Figure 8–3d).
4. Insects. The presence of some insects in crops can be assessed by aerial photography. For example, corn leaf aphid populations were estimated from the sooty mold development using photographs taken from about 6,000 ft. Different areas and levels of infestation were successfully determined using photographic enhancement and computerized area-estimation methods. The assessment of migratory grasshopper breeding areas has been attempted using satellite-based photography with mixed results.

**Leaf Area Index**   The competition between weeds and crops results in crop loss, and researchers are attempting to predict crop loss on the basis of weed leaf area relative to crop leaf area early in the growing season. This approach is being investigated because weed numbers have poor predictive capacity due to the inherent variability in weed size. The relative leaf area index may provide more accurate prediction than simple numerical counts of weeds, but leaf area index techniques are not yet used for routine management in the field.

***Trapping***   Relying on pest mobility, trapping is widely used for insects but has no relevance to soil-borne pathogens or weeds, and is not practical for nematodes. Traps are primarily used to determine if the target pest is present, but are typically not reliable for estimating the actual size of the population. Traps are therefore normally used to monitor population activity. Trapping can also be used to determine the frequency of insect vectors that are carrying pathogens or nematodes. There are several types of traps (Figure 8–7), but all require an attractant, such as a pheromone, to lure the target pest to the trap. The pests are caught (trapped), and are later identified and counted.

Monitoring traps do not reduce the population of the pest, and are thus not suitable as a control technique. Trapping as a control tactic is discussed in Chapters 14 and 15.

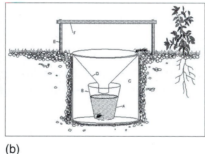

(a)                              (b)                                    (c)

**FIGURE 8–7**   Examples of traps used for monitoring insect populations. (a) Yellow pan trap for monitoring green peach aphids; (b) pitfall trap for estimating activity of ground-dwelling insects; (c) pheromone wing trap for monitoring male moths, here for codling moth in pears; and (d) base of pheromone trap showing rubber cap containing sex pheromone, and codling moths caught in the sticky coating; and (e) blacklight trap for attracting night-flying insects.

Sources: photographs (a) and (e) by Larry Godfrey, with permission; (c) and (d) by Robert Norris, (b) by Marcos Kogan.

(d)                              (e)

**Visual**   Visual cues can be used to lure pests, the attractant being a color. Yellow, for instance, attracts some species of aphids, whiteflies, and leaf miners; a green ball attracts walnut husk flies; and a red ball attracts apple maggot flies. In the latter two cases, the insects become stuck on the ball's sticky surface. Yellow-colored pans full of water or oil are used to monitor green peach aphid populations (Figure 8–7a).

**Bait Traps**   These traps attract the pest to a food source and are used for monitoring some insects, such as almond bait for naval orange worm moths, and grain for some vertebrates, such as squirrels. The utility of these traps is often short lived due to molding of the bait. The bait has to be changed frequently for the trap to remain active.

**Pitfall Traps**   These nonbaited, cup-shaped traps are used to monitor insects that travel on the soil surface, such as carabid beetles. The traps, made from a cylindrical tin can or plastic cup, are installed in a hole in the ground with the lip flush with the soil surface (Figure 8–7b). Insects walk into the trap but cannot escape. The collecting vessel (trap) is usually half filled with soapy water or ethylene glycol.

**Pheromone Traps**   Many male insects and nematodes locate females by a volatile chemical sex attractant called a sex pheromone. Most pheromones are mixtures of chemicals and each mixture is typically species specific, although some may have attraction for other species. Pheromones have now been isolated, identified, and synthesized for many insects. Examples of insects for which pheromone traps are used for monitoring include codling moth (Figure 8–7c), pink bollworm, oriental fruit moth, and many other moths. By placing the synthetic pheromone in a dispenser, called a cap (Figure 8–7d), inside a sticky trap, the male insects are then attracted to the trap. The date of the first appearance of males in the trap can be used as the biofix, that is, the date used to initialize predictive models that are based on degree-day accumulation. The absolute numbers caught in a trap are usually not useful or used, although research is continuing on certain species to establish correlations between trap catches and actual population levels.

**Blacklight**   These traps emit ultraviolet radiation (blacklight) which attracts certain night-flying insects, particularly moths (Figure 8–7e).

**Sticky Traps**   Nonspecific traps with a sticky surface can be used for sampling some types of pathogen spores. Such traps are similar to vacuum air samplers, but the traps collect spores from the air without additional airflow. Sticky traps made of transparent acrylic plastic are used to detect movement of insects in and out of fields or border areas.

***Soil Sampling***   Assessment of soil-borne pests, such as nematodes, some pathogens (e.g., those that form sclerotia), weed seeds, and some insects (e.g., wireworms), requires the use of soil sampling procedures that are typically laborious and time consuming. A shovel is the simplest soil sampling device (Figure 8–6f). There are also various collecting devices that are pushed into the soil, to remove an intact soil core, such as an Oakfield tube (Figure 8–6f). After collecting a soil sample, a technician measures a subsample of known volume and suspends the soil in water. Because soil particles sink more quickly than the organisms, the suspension can be poured through screens or sieves to extract the organic debris and pests. The pests are recovered from the screens and counted. The number of organisms present are expressed relative to the unit of soil, such as number of eggs per gram of soil or root material. The process is difficult and relatively slow.

### Biological Assays

**Baiting**   A bioassay called baiting is used to detect some soil-borne pathogens such as *Phytophthora*. A plant part or fruit sensitive to the target pest is exposed to an aqueous soil suspension. If infection occurs, the pathogen is readily detected by visual inspection of the bioassay tissue.

**Indexing**   A bioassay technique called indexing is used to detect viruses in plants. The plant tissue that is to be tested for viruses is grafted onto a sensitive (index) variety that will display characteristic symptoms if the virus is present. Indexing is used in perennial crops such as grapes, and is a difficult process that typically requires two years. This technique is not used for field diagnosis, but primarily as a means to establish that planting stock of perennial crops such as grapes are free of serious virus diseases. Molecular techniques are being developed as a means to replace indexing. Indexing is not used for other pest types.

**Host Range Bioassays**   The detection of variants of some pests, such as host-specific races of root-knot nematodes, is accomplished by using a differential host range bioassay. The host range is typically determined by inoculating the pest of interest onto an appropriate range of host-plant species, and determining those hosts that support pest reproduction.

**Plating**   Culturing fungi and bacteria on agar media plates permits identification of pathogens. Plant tissue is generally surface sterilized and then placed on an appropriate artificial medium. Pathogens present in the tissue grow out onto the medium and can be identified. Likewise, aqueous soil suspensions can be plated onto an artificial medium, allowing pathogens present in the soil to grow.

### Biochemical Assays

**Immunosorbent Assays**   Identifying and quantifying plant pathogens has traditionally been very difficult. Various immunoassays involving the use of antibodies to detect antigens that are specific to particular pests have been developed. The most well known is the enzyme-linked immunosorbent assay (ELISA). This research tool has been used since the early 1980s, and is increasing in importance for practical pathogen detection and enumeration, such as for *Phytophthora* and many viruses. Special ELISA kits are available for field diagnosis of a number of pathogens, and they can provide accurate identification without the lengthy laboratory procedures that were previously required.

**Isoenzyme Assays**   Enzymes are involved in catalyzing biochemical reactions. There are often different forms of enzymes that catalyze the same reaction, and are called isomers of the enyzmes or isozymes. For example, in assessing variants of isozymes, usually malate dehydrogenase and esterases are used to distinguish among common root-knot nematode species.

**Molecular Techniques**   The polymerase chain reaction (PCR) is a technological development that has revolutionized molecular genetics. PCR is a powerful tool, because it allows the amplification of small quantities of DNA. It is being used to improve methods of pest identification and detection. Techniques based on PCR, such as random amplified polymorphic DNA (often referred to as RAPD), restriction fragment length polymorphism (RFLP) analyses, and others are being used to detect differences in DNA among pest species to facilitate identification. Molecular techniques are also increasingly useful in separating closely related (sibling) species of pests or different populations or biotypes of a pest within a single species. These techniques are used mainly at the research level for pests, but are increasingly being used to address field diagnosis.

**Physiological Techniques**   Multiwell plates with each well containing a different chemical substrate are inoculated with unidentified bacteria to determine their physiological capacity to metabolize the various substrates. The array of substrates that can be used is termed a metabolic profile. The metabolic profile of the unidentified bacterial isolates may be compared with the profiles of known species as an aid to identification.

### Yield Monitoring and GPS/GIS Systems

Crop yield has been used as a summary measure of overall crop health, but a change in yield does not reveal what, from among many possible causes, was responsible. Monitoring yield cannot provide direct information on pest populations, but may indicate the location in a field where action will be needed in future years. Long-term decrease in yields in an area may indicate slow population increase of pests that are not highly virulent. Yield monitoring techniques have greater relevance for problems resulting from soil-borne pests (weeds, nematodes, some plant pathogens, and some insects) than from foliar pests, because soil-borne pest problems will reappear each year at the same place in the field.

The accuracy of yield monitoring is increased by the use of precision farming equipment that can relate yield to a specific location within the field using global positioning system (GPS) and geographical information system (GIS) technologies. The GPS/GIS technology is also being used to relate spatial data on pest distributions to IPM decision making.

We cannot overemphasize that a monitoring technique must be easy to use, produce accurate information, and be relatively inexpensive. As noted, there is a difference between what is done for research and what a pest control advisor or a grower can do. It is unlikely that a monitoring technique will be used if it requires more than a few minutes per sample, because then it will be too time consuming. The cost of monitoring must always be balanced against the benefits derived from monitoring. The crop value is central to such cost-benefit analysis. In low-value crops, such as cereal grains, only minimal monitoring is economically justifiable. More in-depth monitoring is justified in high-value crops such as tomatoes, strawberries, and cut flowers. Because of the costs associated with monitoring, many pest management decisions are based on the history of a given field and the experience of the grower, or by limited monitoring such as presence/absence assessments.

> The cost of monitoring must be balanced by the benefits obtained from such monitoring.

## Sampling Considerations

Pest biology dictates the timing of sampling and locations where samples are collected; no general a priori rule can be applied for all pests. Appropriate sampling strategies should be implemented for each pest that needs to be monitored.

Previous experience may indicate specific locations within the area to be sampled, such as a field, where a particular pest species is likely to occur because it has been there in the past. Examples include weed patches or localized nematode infestations. Sampling early in the growing season and concentrating sampling in areas where pests have historically been a problem may allow detection of pests before their densities increase to damaging levels, or the infestation spreads to a larger area. The approach to sampling depends on many factors, including those listed here:

1. Field size. A greater number of samples are required to reach a given level of accuracy in large fields, and the cost of sampling may dictate the sampling intensity. In addition, large fields are typically planted to lower-value crops, and sampling intensity per acre is usually lower than it is for high-value vegetable crops.

2. Economic value of crop. The increased economic return that results from the management tactic that is used as a result of monitoring must outweigh the costs of scouting. This means that intensive monitoring is not feasible in low-value crops such as wheat and barley.

3. Sample location. The type of crop will determine the location of sample collection. In row crops, sampling is often associated with the row, and samples relate to the number of plants or distance along the row. In orchards, the tree, or part of it, becomes the sampling unit. In broadcast or uniformly spaced crops, unit area determinations such as per square foot, square meter, or other appropriate measures are used. If soil samples are being collected to detect pathogens or nematodes, samples are collected from near the edge of the affected area where pest densities are likely to be highest.

4. Pest density. Densities can range from a few per acre to thousands per plant, or millions per acre. Low pest density requires more extensive sampling to obtain reliable estimates than is required with high pest densities. Sampling intensity usually needs to be greater when the pest populations are low, because accurate estimates are needed so that management tactics can be implemented before pest populations increase.

5. Sample size. Certain formulas allow decisions regarding the minimum number of samples necessary to achieve a desired level of accuracy. To use them, the parameters of the mathematical distribution, which describe the population spatial pattern, are required. For detailed information on the methods and formulas used, see Ruesink (1980). To reduce the number of samples needed to reach a certain level of accuracy, an approach known as sequential sampling has found some use in IPM strategies (see below).

6. Goal of manager. Farm management philosophy and goals can dictate the techniques used and extent of sampling. If the management is lax, or the crop is of limited value, then the necessity for sampling may be limited. If thresholds (see later in this chapter) are being used to guide pest management decision making, then accurate monitoring is essential. If the tactic of pest exclusion is to be used, then presence or absence monitoring is sufficient. Intensive sampling may be required to detect the pest at minimal densities, but quantification of pest numbers is not required.

7. Mobility of pest. The location of pests having low mobility will suggest where monitoring should be conducted in the field. The way that pests get into a field and the direction from which they come may indicate the best place in a field for initial monitoring.

8. Timing of sampling. Monitoring must coincide with the time that the pest is likely to be present, in a detectable form, and in densities that are sufficient to be detected.

9. Frequency of sampling. The frequency of sampling will depend on the rapidity of population development and the rate at which injury to the crop is increasing. During cooler weather, less frequent sampling may be adequate in comparison with the same pest in warmer weather. Sampling frequency often must be increased near harvest, especially for fresh market fruits and vegetables subject to cosmetic damage, as the crop can progress from no, or tolerable, damage to unacceptable damage within a few days.

## Sequential Sampling

The term sequential sampling is used for two different purposes. As a method to improve the precision of arthropod sampling, it is used to indicate repeated sampling at a single time to ascertain the current population. The term is also applied to repeated sampling over time to better track population development.

***Sequential Sampling for Statistical Reliability***   This form of sequential sampling is used to estimate the pest population densities. If the sample statistics do not meet predetermined accuracy requirements, then additional samples are obtained. The process is repeated until the sample precision meets the criteria established. Details of sequential sampling are beyond the scope of this book; see Binns, Nyrop, and van der Werf (2000, Chapter 5).

***Temporal Sequential Sampling***   Pest populations change over time. It is often necessary to know not only the size of the population, but also the rate of change to establish the best management strategy. It is therefore necessary to estimate population growth over time. Knowledge of the population growth rate is important for most pests, particularly those that have multiple generations within the crop season, because within-season population increase the needs to be estimated. Sequential sampling requires repeated, or sequential, monitoring of traps or measuring devices. The sampling interval might be hours for bacteria, days for mites and aphids, weeks for various multivoltine insects and polycyclic pathogens, and years for weeds, univoltine arthropods, nematodes, and single-cycle diseases. To be useful, the samples taken must use the same techniques and protocol at each sampling time. If different sampling techniques or protocols are used over time, it may not be possible to correctly compare the results obtained, thus leading to erroneous decisions.

## Crop Sampling

Tolerance of a crop to pest damage changes with crop phenology. This leads to the concept of plant mapping, which is a numerical index to indicate the growth stage of the plant. To some extent this concept is used for all crops in that a judgment will be made in relation to crop development and the pests present. Plant mapping has been widely accepted for pest management in cotton (described in Chapter 18), soybean, wheat, and other crops.

Correct analysis of the impact of a pest also requires evaluation of the status of the crop.

## Sampling Patterns

Following are possible ways to sample a field for the presence of pests.

1. Random. Truly random sampling will produce the most reliable estimate of the mean population and its variance, regardless of the pest spatial pattern. This assumes that the sample is truly random and is sufficiently large. Attempting to visually select random samples is difficult, and may result in nonrepresentative samples; for example, randomly collecting leaf samples may inadvertently result in only large leaves being selected. Some form of sampling pattern, such as a grid, should be imposed on the field, and then random samples based on grid coordinates can be collected. For most pest management purposes, truly random sampling is either too costly or simply cannot be achieved.

2. Systematic. In most pest management situations, some form of systematic, predetermined sampling will provide the best combination of reliability and cost. In this situation a sampling protocol is designed based on previous knowledge of the pest and the field. It is not possible to state that one pattern is better than another, but samples might be collected on a repeating zigzag or a W pattern across the field (Figure 8–8a), or plants might be selected in a regular way, such as the fifth tree in every fourth row (Figure 8–8b).

3. Directed systematic. This sampling pattern is similar to systematic patterns, but is modified to utilize information about where the pest is most likely to occur.

Many books on management for a specific crop provide suggested sampling patterns.

**Record keeping is an essential component of a sound pest management program.**

## Record Keeping

Accurate record keeping is an important aspect of monitoring for pests. Accurate records provide a history of the managed system. Over time, good records

**FIGURE 8–8**
Examples of two possible field sampling patterns. (a) W pattern in an arable field with x indicating a sample location; and (b) every fifth tree in every fourth row of an orchard.

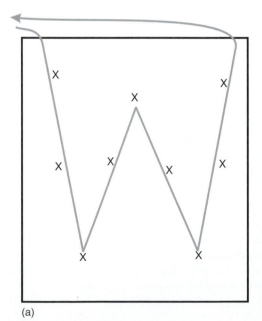

(a)

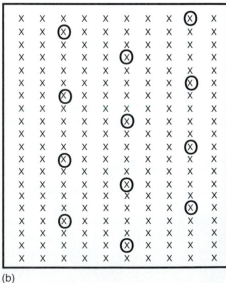

(b)

offer an accurate picture of the pest profile for the site, and make the prediction of future problems much more reliable. For example, remembering details, such as first emergence date of weeds or first appearance date for eggs of cotton bollworm, for many fields over several years is not possible. Record keeping will be discussed again in Chapter 18 as part of developing IPM programs.

Many standard record-keeping forms have been developed for each of the pest categories. Examples of the types of information that should be included are as follows:

1. Field location
2. Identity of the pest and beneficials (who identified them, if appropriate)
3. Density of pests present in units appropriate to the pest
4. Phenological stage(s) of the pest present, and their numbers (expressed in appropriate units)
5. Within-field distribution of the pest
6. Date of observations, possibly even time of day for some organisms
7. Weather conditions at time of sampling, including soil factors (if appropriate)
8. Crop phenological stage and other relevant information
9. Relevant cultural operations
10. Special considerations

## Training

Accurate and reliable assessment of pest populations can be difficult, and requires skill and training. For pest management personnel, a college degree in biology, ecology, agriculture, or similar subject is now considered essential. Training usually also requires at least an additional year of infield experience beyond academic training before a person is capable of making reliable, unsupervised decisions about pest populations. New scouts must be trained to accurately recognize the pest species in all their growth stages and must be shown how to conduct specific monitoring techniques.

# THRESHOLDS

Reaching a pest management decision requires orderly consideration of management objectives and monitoring of pest status. Once the identity of the pest has been established, the pest density determined, and the pest and crop phenological stages defined, the information can be used to decide on a management strategy. The need for implementing control tactics will depend on the pest density relative to the thresholds appropriate for the crop phenological stage. This section discusses the use of the threshold concept in the management of all categories of pests.

## The Damage Concept

When pests are absent, plants will grow and yield at their genetic potential to the extent that the environmental conditions of temperature, moisture, and nutrients allow (see Figure 1–4). This may be considered the optimal yield for the plant growing under those environmental conditions. When pests are present, they injure the plant by disrupting physiological processes and by using plant energy,

**The reduction in yield caused by pests, compared with plants in the same environment without the pest, is called crop loss.**

**FIGURE 8–9**
Generalized theoretical relationship between increasing damage and amount of yield response in the absence or presence of cosmetic standards. "EIL" is the economic injury level, or that pest density at which damage becomes unacceptable.

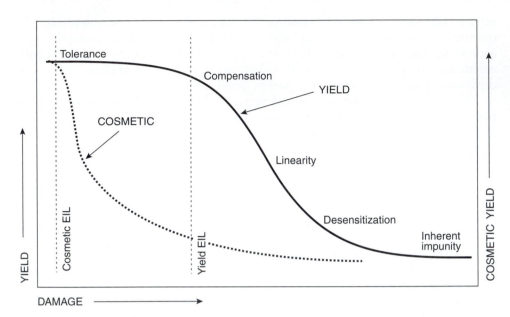

thus depriving the plant of the energy that would otherwise be used for growth and reproduction. When plants in the field are considered together, they constitute the crop, and the reduction in yield caused by pests compared with plants in the same environment without the pest is termed crop loss. Generally speaking, as the density of a pest increases, the crop loss increases (see Chapter 4).

When a pest causes little injury to the crop, the impact on yield (or harvestable product) is usually low (Figure 8–9); however, if the pest causes cosmetic damage this may not be true. Many plants can tolerate low densities of pests without suffering damage. At low densities, the plant is able to compensate for damage caused by the pest, resulting in little crop loss. In some cases, the small amount of damage caused by low densities of pests may result in increased plant growth and yield. This phenomenon is called overcompensation, but is not well understood. It is suspected that as the pest damages the plant, the physiological processes that govern homeostasis are altered so that additional energy is allocated to plant parts that represent the marketable commodity.

The necessity of defining the acceptable amount of injury has led to the concept of thresholds. Once a pest population has increased to the point where the injury/yield relationship is linear (Figure 8–9), the corresponding yield loss is high enough that control tactics should be implemented to manage the pest. At the higher pest population densities, the crop loss eventually becomes independent of any further increases in pest density.

**The concept that low densities of some pests cause little or no injury, and hence no economic crop loss, is central to the use of economic thresholds.**

The concept that low levels of pests cause low levels of injury, and that low levels of injury can be tolerated depending on the value of the crop and plant part that is attacked, is central to much of arthropod and nematode management. The extent to which this concept can be applied to other types of pests is not clear, although the value of thresholds has been recognized for vertebrate management.

The concept that the yield loss per unit pest is more or less constant near threshold densities is central to the thinking that underlies the development of thresholds based on numeric density. The idea of thresholds thus works well for organisms that have essentially a fixed body size for each growth stage (insects, nematodes, vertebrates, slugs, snails), but does not work as well for organisms with variable body size (weeds, fungi). There is a more or less linear correlation between pest numbers and injury for fixed body size organisms versus variable relationships for those with variable body size. At higher pest densities, the linear relationship between density and yield loss breaks down due to intraspecific competition, but at such pest densities the economic thresholds discussed below have been exceeded.

## Economic Injury Level

The economic injury level (EIL) is central to the concepts of thresholds applied to arthropod management and is defined as the pest density at which the economic value of the crop loss prevented by the control action is equal to the cost of that control action (Figure 8–10). The EIL is derived from the mathematical equation:

$$C = V \times P \times I \times D$$

where $C$ = management costs (i.e., costs of control)
$V$ = market value per unit of production
$P$ = pest population expressed as density (e.g., number of pests/A)
$D$ = damage per unit injury
$I$ = injury per pest equivalent

Rearranging this equation to:

$$P = C/(V \times I \times D)$$

**The EIL is a pest density at which the economic value of the commodity lost to pest damage equals the cost of the desired control tactic.**

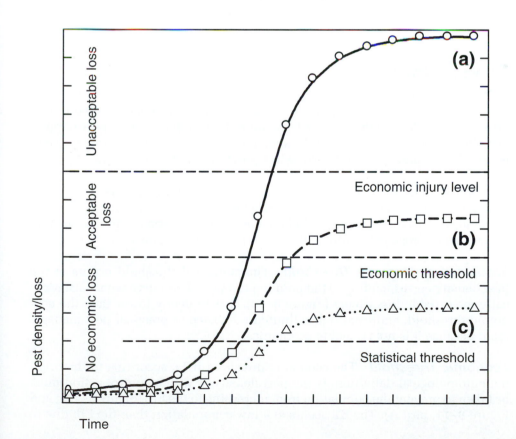

**FIGURE 8–10**
Theoretical time relationships between statistical threshold, economic threshold, and economic injury level for three different pest populations. a = high, b = intermediate, and c = low; or a single pest in three different crops a, b, and c.

**FIGURE 8–11**
Theoretical population development of two pests in relation to economic thresholds and control actions. GEP is the general equilibrium position of the pest. Pest A has a low GEP and pest B has a high GEP.

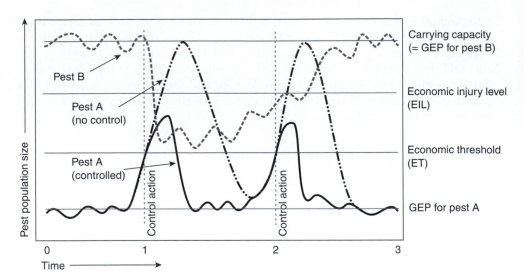

where *P* is the pest density at which damage equals the cost of control, which is the EIL density. An additional parameter must be added to this equation when some damage has to be accepted, as control is not 100%. The revised equation is:

$$P\,(\text{EIL}) = C/(V \times I \times D \times K)$$

where *K* is the proportion of the damage that must be tolerated. This concept of EIL is then utilized as the basis for several different types of thresholds.

## Types of Thresholds

Various types of thresholds have been proposed and are used to varying degrees. How they may be applied to different pest categories is presented in Figures 8–10 and 8–11, which show how these definitions relate to pest population dynamics. The acronym GEP stands for general equilibrium position, and is used by entomologists in referring to equilibrium population density.

***Threshold***   The unqualified use of the word *threshold* appears throughout IPM literature. The term should be qualified when used in relation to pest management. Unless the type of threshold is specified, the term simply means that a stimulus has reached a sufficient level to provoke a response.

***Statistical or Damage Threshold***   The statistical threshold occurs at the pest density corresponding to the point on the yield loss curve where a statistically measurable loss occurs (Figure 8–10). It is typically lower than the economic threshold, and usually has limited relevance to practical pest management because it is not related to economics.

***Economic Threshold***   The economic threshold (ET), according to the original entomological definition, is the pest density at which control action must be taken to prevent the population from increasing to the EIL (Figure 8–10a; and Figure 8–11, pest A). The ET occurs at a lower population than the EIL due to population/damage increase that can still occur after the control action has been taken. Weed science has adopted a definition of ET that is the same as the entomological definition of EIL.

**Nominal Threshold**  The nominal threshold is used by most farmers in the absence of a numerical threshold. Cousens (1987) called this a visual threshold, when applied to weed management. Simply stated, the nominal threshold only requires an assessment of whether the field appears to be in acceptable condition, or if unacceptable crop loss appears likely.

**Comprehensive Thresholds**  The comprehensive threshold is based on assessment of the crop loss caused by combined biotic (all pest categories) and abiotic stresses. Some such thresholds are under development, but none are currently available, because little quantitative information exists on combined stresses relative to crop loss. Farmers and pest managers generally use nominal comprehensive thresholds in dealing with the real-world complexity of multiple pests in the field.

The preceding four thresholds are, by definition, for a single crop, which equates to a single season or year in most cases. The two thresholds mentioned in the following paragraphs are attempts to deal with the multiyear population dynamics of weeds.

**Economic Optimum Threshold**  The economic optimum threshold (EOT) is a multiyear economic threshold which has been proposed to deal with weeds and other organisms that have propagules acting as seedbanks. These seedbanks, in effect, lead to reproduction and thus a modification of population dynamics over several years (Cousens, 1987). The EOT considers the implications to subsequent crops of not controlling pests when they occur at or below the ET. There are no actual field data to support using EOT, but computer simulation models suggest that the EOT may be 5- to 10-fold lower than the single-season ET.

**No-Seed Threshold**  The no-seed threshold (NST), proposed by Norris (1999), argues that many weeds should not be permitted to achieve reproductive status, because the seeds produced by plant densities that are below the ET maintain the seedbank. This concept is based on the enormous potential for seed production and the problems of seedbank longevity. This threshold has not been validated by research, but is used in some form by farmers who recognize the long-term nature of weed populations.

Regardless of the type of threshold being employed under certain circumstance, it is generally better to treat a small pest population than to attempt reduction of large populations at a later date. This concept is especially true of an invading species and for several types of vertebrates that reinfest fields each year, but it does not apply to most endemic insect and nematode pest species. The concept is central to the discussions in Chapter 10.

**Other Types of Thresholds**  Certain types of pests require the establishment of specific thresholds. For instance, aesthetic thresholds are used in decisions related to pest management of ornamental crops. A nuisance threshold is useful when assessing the prevention of pests, such as flies and mosquitoes, which annoy or disturb humans and farm animals.

## Examples of Thresholds

The following are examples of thresholds and how they are expressed. In each case, management action is recommended if the threshold level is exceeded. These examples have been taken from the University of California's IPM website, unless otherwise noted.

- Tomato fruit worm. The economic threshold is 8 eggs/60 leaves below terminal flower cluster (this level is increased if the parasitic wasp *Trichogramma* is present).
- Cotton bollworm in California. The threshold is 20 small larvae/100 plants.
- Cabbage aphid on cole crops. A total of 1% or 2% of plants are infested.
- Sugar beet cyst nematode. The damage threshold in California is 1 egg/1 g of soil.
- Alfalfa butterfly. A total of 10 nonparasitized or nondiseased caterpillars per sweep.
- Velvetleaf in maize in Italy's Po Valley. This is an example of the no-seed threshold (Zanin and Sattin, 1988).
- Web-spinning mites in almonds. The threshold is 45% leaf infestation using presence absence sampling when predators are present.
- Instead of thresholds, disease risk forecasting systems based on inoculum presence, rainfall, temperature, and humidity are used for pathogens. Forecasters that predict disease increases based on environmental conditions are available for potato late blight, hop downy mildew, grapevine downy mildew, blue mold, and *Sclerotinia* stem rot.
- Meadow mice (voles) in artichokes. Threshold is reached as soon as signs are observed.

## Limitations of Thresholds

In several situations economic thresholds are less useful or not appropriate. These include the following:

1. Environmental conditions, especially temperature and moisture, may alter threshold levels, leading to difficulties in extending thresholds from year to year, or from one geographic region to another. For example, the damage threshold for cyst nematodes in the Mediterranean climate of California is 1 egg/1 g of soil, whereas in the temperate climate of the UK it is as high as 15 eggs/1 g of soil.
2. When the economic threshold is immeasurably low, it is not feasible to utilize thresholds for decision making. Typically the action decision has to be made as soon as the pest is observed. This situation arises for management of several pests, including those that directly impact quality and cosmetic appearance to fresh market fruit and vegetables, insects that are vectors for pathogens (bacteria and viruses), or where rapid population growth can occur (vertebrate pests such as voles, and weeds).
3. There is no economically practical way of sampling the pest to determine its population accurately. It is, for example, very difficult to determine the number of spores present for many pathogens, such as those that cause peach leaf curl and late blight of potatoes.

4. There is no effective or economically acceptable tactic that can be implemented once the pest population has been detected and assessed. There are two reasons for this.

   4.1. It may not be biologically feasible to control the organism once it has reached a phenological stage at which its population can be determined.

   4.2. No control tactic is available that can be used economically after the pest population is assessed. Virus diseases of plants cannot be cured. Many fungicides do not have the ability to control an existing infection, and thus must be applied before the population develops. A similar situation occurs for many weeds when there is no herbicide available that offers postemergence control (see Chapter 11). This situation requires using preplant incorporated or preemergence herbicides before the weeds germinate. In both situations the treatments must be made before the pest presence and density can be established.

5. When populations are sufficiently high that they are normally above the economic threshold (Figure 8–11, pest B), the use economic thresholds for decision making is not applicable. This happens with many weeds and many pathogens.

6. If the body size of the pest organism is not fixed, and therefore the damage is not well correlated with numerical density, thresholds based on simple enumeration do not provide accurate predictions. This problem limits the utility of thresholds for management of many pathogens and most weeds. Determining the age structure of the population does not resolve the problem because of the plasticity of growth relative to chronological age.

7. Having inadequate data to establish the damage/yield loss curves that are needed to develop the EIL is a problem for many pests; in the absence of such data, the nominal comprehensive threshold is usually applied, and management decisions are then based on experience.

8. Thresholds are difficult to use in cases of multiple pest attack, where the combined impacts of different pests have not been quantified. The alfalfa data in Figure 6–3 illustrates this problem. Based on thresholds for either the leafhoppers or the *Fusarium* root rot alone, the decision would be that the threshold had not been exceeded and thus no action is recommended. Due to the combined impact of the leafhoppers and the Fusarium root rot, such a decision would be incorrect.

# OTHER FACTORS THAT AFFECT PEST MANAGEMENT DECISIONS

## Cropping History

Cropping history is important to pest management decisions for two reasons. It can provide the manager with valuable clues regarding the pests that are likely to be present. This concept applies to virtually all soil-borne pests, such as all nematodes, many pathogens (e.g., oakroot fungus, verticillium wilt), most weeds, and some insects (e.g., wireworms and cutworms). Cropping history can also alert the manager to carryover problems resulting from previous pesticide use. Pesticide carryover can result in restrictions regarding crops that can be planted (as discussed in Chapter 11).

## Field Location and Size

Field location can place severe restraints on the choice of control tactics. It is thus necessary, as part of monitoring, to record the presence of factors that may limit this choice. The implications are discussed in Chapters 11 and 18. The type of factors that need to be considered include the following:

1. Surrounding crops:
   1.1. Surrounding crops may be a source of a pest problem if the pest is a mobile species; lygus bugs may, for instance, migrate from safflower to cotton as the safflower matures.
   1.2. Adjacent crops may be especially sensitive to spray drift, which may limit the choice of pesticides.
2. Ecologically sensitive areas in the vicinity:
   2.1. Presence of wildlife habitats or refuges
   2.2. Presence of endangered species in the region
   2.3. Presence of public waterways and rivers
   2.4. Presence of soil/geological conditions that may increase the risk of groundwater contamination by pesticides
3. The locations of houses and schools and other structures must be considered.
4. Field size/cropping pattern:
   4.1. In certain parts of the world one can find solid plantings of a single crop over thousands of acres. Such large plantings are difficult to sample, and if a threshold is reached it is difficult to mobilize the equipment in time to treat the entire area, even when spraying by airplane.
   4.2. The thresholds concept may be difficult to apply to small fields that are planted to mixed crops, because thresholds for mixed crops are not available.

## Weather Monitoring

On a short-term basis, weather can dictate the choice of tactics. For example, high winds will limit the application of many pesticides, wet soil can limit the use of heavy machinery in fields, and rain following cultivation can negate the utility of cultivation for weed management. Rain immediately after a foliar pesticide application may wash the pesticide off the treated leaves before it has time to act.

For pests that develop rapidly once infection has occurred, but which have long incubation periods (such as apple scab or late blight of potatoes), monitoring for the visible symptoms is not useful. For such pathogens, it is preferable to monitor the environment and detect conditions, such as appropriate temperature and humidity, that are suitable for infection.

Weather monitoring is also important to pest management as it affects pest development rates. Except for mammals, all pests are cold-blooded (poikilothermic) and thus temperature regulates developmental rates (see heat summation discussion in Chapter 5). Predicting rates of population development is becoming increasingly important for IPM decision making, and so weather

**FIGURE 8–12**
Stephenson screen containing maximum and minimum thermometer and other weather measuring equipment set up in an orchard.
Source: photograph by Robert Norris.

monitoring thus becomes an integral part of developing IPM programs. Data for weather conditions can be obtained from recording equipment set up in the crop (Figure 8–12). Such weather monitoring equipment is usually housed inside a structure called a Stephenson screen or other suitable protection from direct exposure to sunlight or rain. Increasingly, weather data can also be obtained by accessing information available via the internet; such real-time weather data are available in many regions.

## Models for Forecasting Events

The ability to use weather data for predicting pest population development and outbreaks has been greatly improved through the development of computer models. These models can combine knowledge of pest biology with specific weather information to predict when management activity will be needed. The models do not recommend the actual tactic that should be employed. The use of computer modeling for improving IPM is discussed further in Chapters 18 and 20. Weather data are currently used for the following:

1. Degree-day-based insect development models that use data from pheromone traps to define the biofix (set or start) point
2. Temperature and moisture (duration of wetness) models that predict pathogen infection periods

## Cosmetic Standards

Cosmetic standards are set, in part, due to public attitudes toward the presence of pest organisms in food or the damage and blemishes caused by pests, for certain commodities even if yield and nutritional quality are not compromised. Cosmetic standards exist because consumers are unwilling to purchase products that are infested with insects, or that show even small blemishes. Cosmetic standards result in unacceptable losses at much lower injury levels than those causing actual yield loss (Figure 8–9, cosmetic). Such low tolerance means that monitoring techniques cannot be used because unacceptable cosmetic damage will have occurred by the time the pest is detected. In such cases, control tactics are implemented before pests are detected (e.g., by spraying on a calendar schedule), or at the first observation of pests or damage. The former is referred to as prophylactic treatment. The pest management implications of cosmetic standards are discussed further in Chapters 18 and 19.

## Risk Assessment/Safety

The risk assessment dilemma for pest management is demonstrated in Figure 8–10. This graph shows a single pest species in three different crops: In crop a, the pest population increased rapidly; in crop b, it increased moderately; and in crop c, increase was minimal. The graph could also be thought of as three different pests in a single crop, with pest a building up rapidly to high levels, pest b building up less rapidly, and pest c building up to only low levels. The situation for population c is relatively easy to assess; no action is necessary and monitoring will show that although the statistical, or damage threshold was exceeded, the population stopped increasing before reaching the ET level. Pest situation a is also relatively easy to assess because the population increases rapidly at the ET level, and projections would indicate that the EIL would be exceeded. The pest situation b is difficult for decision making, and probably reflects the situation for most pests. At the time that the ET is exceeded, the manager must decide whether to initiate action. In the theoretical example in Figure 8–10 b, no action is required, but that conclusion is not obvious. At the time the ET is reached, it is not clear whether the population is going to continue to increase. Experience plays a major role in this type of situation. Many managers would find the risk of forgoing treatment unacceptable, especially in a high-value vegetable crop. The alfalfa data in Figure 6–3 and the *Fusarium*/root-knot nematode interaction shown in Figure 8–4 reveal another aspect to the risk perception problem. Without knowing about pest interactions, a manager takes a risk in assuming no interactions.

## Economics

There is considerable difficulty in defining the economic losses and gains that result from monitoring and the use of thresholds. The situation often arises in which theoretical thresholds developed under the controlled conditions of research prove difficult to implement in the field. Economic reasons for this difficulty are as follows:

1. If the cost of monitoring is the same as the treatment cost, then the treatment will probably be made without monitoring. This is very much the case in low-value crops, where the treatments that can be made are relatively low in cost, such as spraying a phenoxy herbicide for broadleaf weed control in cereal crops.
2. Uncertainty of the economic losses caused by multiple pest attack is a problem. In this situation, many of the single-species economic analyses that are available may not represent the crop loss that the grower will actually experience. This problem slows the adoption of IPM in many agroecosystems.

3. The value of the commodity may change between the time that a pest management decision is made and the commodity is harvested. The anticipated profit from a crop will dictate how much investment in pest control is acceptable. This is especially true with fresh market vegetable crops, where market volatility can change prices rapidly. For example, a decision not to treat a lettuce crop this week may prove to be disastrous if the crop value doubles at harvest 10 days later. There is no easy answer to this problem, and most growers and pest control advisors will err on the side of caution. This problem also slows the adoption of IPM.

In all cases discussed above, costs and benefits to the environment are not considered. One goal of advanced IPM systems is to incorporate costs and benefits to the environment and to society into the EIL equation.

## SUMMARY

Pest population monitoring is an essential prerequisite to making sound IPM decisions. Numerous techniques and devices are available to monitor pests, but many techniques are specific to a pest category. The person conducting the pest monitoring must be familiar with the appropriate technique for the pest being assessed. Data on pest population development are assessed in relation to economic thresholds prior to implementing a control tactic. Economic thresholds are not universally applicable; or the necessary data are not available, and the biology of some pest categories limits the utility of the economic threshold concept. Although there are limitations to the use of population monitoring and utilization of economic thresholds, these two concepts have become an integral component of IPM programs.

## SOURCES AND RECOMMENDED READING

Chapters 6 and 7 in *IPM in Practice,* by Flint and Gouveia (2001), provide expanded details of many topics discussed in this chapter. Several other books and/or articles provide additional material on the topic of monitoring and decision making. Skerritt and Appels (1995) reviewed general diagnostic methods in the crop sciences; and Binns, Nyrop, and van der Werf (2000) provide detailed discussions of sampling and monitoring in crop protection. Economic thresholds are reviewed by Higley and Pedigo (1996; 1999). A good discussion of thresholds for vertebrate pests is found in the book edited by Singleton (1999). The general concepts of sampling, mainly for insects, are detailed by Kogan and Herzog (1980), Pedigo and Buntin (1994), Binns and Nyrop (1992), and Southwood and Henderson (2000). These sources provide good lists of the relevant literature. The use of GPS/GIS for IPM is reviewed by Ellsbury et al. (2000), and the role of weather by Russo (2000). Riley (1989) offers a good review of applications of remote sensing in insect studies, including IPM. Henkens et al. (2000) provide a review of DNA-based techniques for diagnosis, and the book edited by Schots, Dewey, and Oliver (1994) has many specific examples of the use of ELISA and molecular DNA assay techniques for diagnosis of plant pathogens.

Binns, M. R., and J. P. Nyrop. 1992. Sampling insect populations for the purpose of IPM decision making. *Annu. Rev. Entomol.* 37:427–453.

Binns, M. R., J. P. Nyrop, and W. van der Werf. 2000. *Sampling and monitoring in crop protection: The theoretical basis for developing practical decision guides.* Wallingford, Oxon, UK; New York: CAB International, xi, 284.

Bliss, C. I. 1967. *Statistics in biology; Statistical methods for research in the natural sciences.* New York: McGraw-Hill, v, 199–203.

Cothran, W. R., and C. G. Summers. 1972. Sampling for the Egyptian alfalfa weevil: A comment on the sweep-net method. *J. Econ. Entomol.* 65:689–691.

Cousens, R. 1987. Theory and reality of weed control thresholds. *Plant Protection Quarterly* 2:13–20.

Ellsbury, M. M., S. A. Clay, S. J. Fleischer, L. D. Chandler, and S. M. Schneider. 2000. Use of GIS/GPS systems in IPM: Progress and reality. In G. C. Kennedy and T. B. Sutton, eds., *Emerging technologies for integrated pest management.* St. Paul, Minn.: APS Press, American Phytopathological Society, 419–438.

Flint, M. L., and P. Gouveia. 2001. *IPM in practice; Principles and methods of integrated pest management,* Publication 3418. Oakland, Calif.: University of California, Division of Agriculture and Natural Resources, xii, 296.

Henkens, R., C. Bonaventura, V. Kanzantseva, M. Moreno, J. O'Daly, R. Sundseth, S. Wegner, and M. Wojciechowski. 2000. Use of DNA technologies in diagnostics. In G. C. Kennedy and T. B. Sutton, eds., *Emerging technologies for integrated pest management.* St. Paul, Minn.: APS Press, American Phytopathological Society, 52–66.

Higley, L. G., and L. P. Pedigo, eds. 1996. *Economic thresholds for integrated pest management.* Lincoln, Nebr.: The University of Nebraska Press, 327.

Higley, L. G., and L. P. Pedigo. 1999. Decision thresholds in pest management. In J. R. Ruberson, ed., *Handbook of pest management.* New York: Marcel Dekker, Inc., 741–763.

Kogan, M., and D. C. Herzog, eds. 1980. *Sampling methods in soybean entomology. Springer series in experimental entomology.* New York: Springer Verlag, xxiii, 587.

Norris, R. F. 1999. Ecological implications of using thresholds for weed management. In D. D. Buhler, ed., *Expanding the context of weed management.* New York: Food Products Press, The Haworth Press Inc., 31–58.

Pedigo, L. P., and G. D. Buntin. 1994. *Handbook of sampling methods for arthropods in agriculture.* Boca Raton, Fla.: CRC Press, xiv, 714.

Riley, J. R. 1989. Remote sensing in entomology. *Annu. Rev. Entomol.* 34:247–271.

Ruesink, W. G. 1980. Introduction to sampling theory. In M. Kogan and D. C. Herzog, eds., *Sampling methods in soybean entomology.* New York: Springer Verlag, 61–78.

Russo, J. M. 2000. Weather forecasting for IPM. In G. C. Kennedy and T. B. Sutton, eds., *Emerging technologies for integrated pest management.* St. Paul, Minn.: APS Press, American Phytopathological Society, 453–473.

Schots, A., F. M. Dewey, and R. P. Oliver, eds. 1994. *Modern assays for plant pathogenic fungi: Identification, detection and quantification.* Wallingford, Oxford: CAB International, xii, 267.

Singleton, G. R., ed. 1999. *Ecologically-based management of rodent pests.* ACIAR monograph series no. 59. Canberra: Australian Centre for International Agricultural Research, 494.

Skerritt, J. H., and R. Appels. 1995. An overview of the development and application of diagnostic methods in crop sciences. In J. H. Skerritt and R. Appels, eds., *New diagnostics in crop sciences.* Wallingford, UK; Phoenix: CAB International, 1–32.

Southwood, R., and P. A. Henderson. 2000. *Ecological methods.* Oxford, UK; Malden, Mass.: Blackwell Science, xv, 575.

Zanin, G., and M. Sattin. 1988. Threshold level and seed production of velvetleaf (*Abutilon theophrasti* Medicus) in maize. *Weed Res.* 28:347–352.

C  H  A  P  T  E  R

# 9

# INTRODUCTION TO STRATEGIES AND TACTICS FOR IPM

## CHAPTER OUTLINE

▶ Introduction
▶ Major IPM strategies
▶ IPM tactics
▶ Additional concepts

## INTRODUCTION

Since the early days of IPM, the terms *strategy* and *tactic* have been used to describe the two fundamental components of any IPM system. In IPM, the tactics are the methods available for pest control and the strategies are the various ways tactics are deployed. Effective management is the goal of an IPM strategy, and dictates not only the reduction or elimination of the economic impact of pests, but also the preservation of environmental integrity and the welfare of society.

**Strategy: A plan for successful action based on the objectives of the crop production system, and on the biology and ecology of pests.**

**Tactics: The methods available for pest control.**

## MAJOR IPM STRATEGIES

There are five principal strategies used for managing pests.

1. Prevention. This strategy is intended to prevent the arrival or establishment of pests in areas currently not infested. The size of the area involved may be a large geographic region, such as a continent, or a small area, such as a field. Prevention is the strategy of choice for a pest that has not yet become established in an area or region. Although prevention is a strategy that may involve several tactics, prevention can also be a tactic used to achieve control of a specific pest (see tactics).

2. Temporary alleviation. This strategy uses specific control tactics on an emergency basis to provide a temporary limitation of localized pest outbreaks. The scale is usually limited to areas smaller than a field.

3. Management of within-field populations. Management occurs at the within-field spatial scale on a continuing basis because the pest is well established in an area, is a recurring problem, and has progressed beyond the point where temporary alleviation provides adequate control. This is the standard strategy of most current IPM programs.

4. Area-wide pest management. Most pest problems are handled at the within-field level. For some pests, management must be extended to the regional level to achieve population regulation, especially for many virus diseases and for some mobile insects. This strategy is referred to as area-wide pest management, and requires the cooperation of people throughout the range of the pest.

5. Eradication. The elimination of the entire pest population from an area, eradication is normally only attempted under the gravest of circumstances, such as invasion of the Mediterranean fruit fly into California or of the parasitic weed *Striga* into the Carolinas. If a pest has become established, eradication is generally not possible. It has been argued, however, that eradication is not an IPM strategy because its ecological foundation is considered deficient. Perkins (1982) called eradication "total pest management"—a paradigm departing significantly from the basic tenets of IPM.

## IPM TACTICS

The following chapters (10 through 17) present the control tactics that are actually used for managing pests, regardless of the strategy being followed. Each tactic is discussed separately, but it should be recognized that IPM utilizes a combination of all suitable pest suppression tactics (Figure 9–1). There are three fundamentally different approaches for managing pests.

1. Manipulation of the pest organisms. This approach uses tactics that either directly influence the pest organism or alter its behavior so that it no longer causes unacceptable losses.

2. Manipulation of the host plants. The tactics used here either increase crop tolerance to pest attack or change the crop so the pest no longer attacks it. The tactics employed affect the consumer pest population indirectly through its food source.

3. Manipulation of the environment. These tactics alter the environment such that pest populations do not increase to damaging levels. The environment is made less suitable for the pest, more suitable for the host, or more favorable to natural enemies of the pest.

Each of these overall approaches has several different possible tactics that can be used.

### Pest Manipulation

Three approaches that are used to directly manipulate pest organisms are prevention, pesticides, and nonpesticidal tactics.

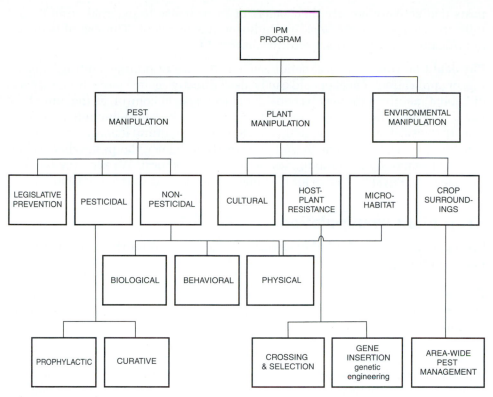

**FIGURE 9–1**    Diagrammatic representation of the control tactics considered to be components of an IPM program.
Source: redrawn and modified after Waibel and Zadoks, 1998.

***Prevention***    Prevention is a key component in IPM programs and the topic is discussed in detail in Chapters 10 (national and regional levels) and 16 (on-farm level). The movement of pests into new areas can be avoided by the enactment of laws at the national or regional level to prevent the spread of particular pests, but pest prevention can also be practiced at the farm level without necessarily requiring the application of legislative rules and regulations.

***Pesticides***    Pesticides are chemicals that have direct toxic influence on the pest. Most are lethal to the target pest(s); however, nonlethal growth-regulating chemicals are also classified as pesticides. The use and regulation of pesticides is discussed in Chapter 11, and societal concerns about pesticides are discussed in Chapter 19.

***Nonpesticidal Tactics***    Three nonpesticidal tactics can achieve direct manipulation of pests.

**Biological Control**    Biological control uses one organism to control another organism. The beneficial, or antagonist, organism reduces the pest population density so that economic loss does not occur. The definition is often broadly interpreted. The topic is discussed in Chapter 13.

**Behavioral Control**    Pest behavior can be modified so that the pest does not cause unacceptable crop losses. Behavior modification is only feasible with pests that actively modify their behavior in response to external stimuli, and thus are not relevant for weed or pathogen management. The use of behavior modification in IPM is discussed in Chapter 14.

**Physical Controls**    Physical action such as the use of pulling, cutting, squashing, or application of heat or cold can be used directly against pests. Some aspects of habitat modification may be considered as physical control, as the aim of the tactic is to alter the physical environment (e.g., temperature, humidity, or wind speed). Physical tactics are sometimes referred to as cultural management, but we consider physical tactics as those that directly influence the pest, whereas cultural tactics modify the crop to indirectly impact the pest. We keep the two tactics separate. Physical tactics for pest management are discussed in Chapter 15.

## Crop Plant Manipulation

Crop plants can be modified to effect pest management either through cultural tactics or host-plant resistance.

*Cultural Tactics*    These involve modification of the practices that are used to grow the crop, to either decrease pest success or increase the crop capacity to tolerate the pest. The topic is addressed in Chapter 16.

*Host-Plant Resistance*    This involves changing the crop genotype to enable it to withstand damage or deter pest organism reproduction through genetic alteration of plant characteristics. Host-plant resistance as a management tactic is discussed in Chapter 17.

## Environmental Manipulation

Environmental manipulation can be achieved at two levels. Microhabitat, such as the humidity within the crop canopy, can be modified so that the pest organism is less able to develop; this involves physical processes, and so is considered under the topic of physical tactics (Chapter 15). Environmental manipulation can also be accomplished over larger geographic areas, through the alteration of habitats both within and surrounding the field (region). These topics have been discussed in Chapters 6 and 7, and are considered further in Chapters 16 and 18.

## ADDITIONAL CONCEPTS

IPM uses all of the above tactics in appropriate combinations, with economic viability and minimization of ecosystem disruption as objectives. An important consideration must also be noted: A tactic, or combination of tactics, that works well for one category of pest under one particular set of environmental conditions may have little relevance to management under other conditions or for another category of pest.

For clarity of presentation, we have presented tactics as if they are mutually exclusive, but in many instances, at least two tactics overlap; for example, cultural practices, such as alteration of crop density, may actually work by changing the environment.

Two phenomena need to be emphasized in relation to all tactics, as they define how different tactics are applied to different pest categories. They are based on the concepts of pest resistance and specificity of control.

## Pest Resistance

The adoption of, and reliance on, one control tactic to the exclusion of all others will eventually lead to decreased efficacy of the control, because the target pest will develop resistance to that tactic. Examples include the widespread repeated use of some pesticides, or the widespread deployment of single-gene-based resistant plants (including genetically engineered plants). Regardless of the particular tactic used for pest management, it cannot be overemphasized that overreliance on a single tactic is a practice that can limit the long-term utility of a tactic. The adoption and continued use of multiple tactics within an IPM framework remains the logical approach to managing pest resistance. The topic of pest resistance and its management is covered in Chapter 12.

> Repeated use of a single tactic to the exclusion of all others will result in selection of resistance within the pest population to that tactic.

## Selectivity of Control

A control tactic that affects only the target species, and does not harm other nontarget species, is selective or specific. The method of application can be used to achieve selectivity, and is essential for management of animal pests. Ideally, the control tactic is applied so that it does not affect any other organism except the pest, because any tactic that affects animals, such as nematodes, insects, and particularly warm-blooded animals, also has great potential to directly affect humans. An additional reason why selectivity is critical for arthropod management in many situations is that it is desirable to kill the target pest but leave beneficial arthropods unharmed. Examples of beneficial arthropods include parasites and predators of pests that are used for biological control, or bees that are used to pollinate crops. The situation is similar for pathogen management. Thus, for management of all categories of animal pests and pathogens, the ideal tactic will affect only the target pest, and will leave all other organisms unharmed.

> Selective IPM tactics can control the target pest without affecting other nontarget organisms.

> The ideal pest management tactic will affect only the target pest or pests, and will leave all other organisms unharmed.

The concept of selectivity applies to weed management in a different way. In a typical weed management situation, several to many weed species must be controlled and the crop selectively left unharmed. In most situations, controlling a single weed species will not provide adequate weed control. Thus, for management of weeds, the ideal tactic will control many weed species but not harm the crop.

## SOURCES AND RECOMMENDED READING

Perkins, J. H. 1982. *Insects, experts, and the insecticide crisis: The quest for new pest management strategies*. New York: Plenum, 304.

Waibel, H., and J. C. Zadoks. 1998. Economic aspects of biotechnology in crop protection. In D. Jordens-Rottger, ed., *Biotechnology for crop protection—Its potential for developing countries*. Feldafing, Germany: Deutsche Stiftung fur Intertnationale Entwicklung, 149–168.

# *10*

# PEST INVASIONS AND LEGISLATIVE PREVENTION

## CHAPTER OUTLINE

▶ Introduction
▶ Historical perspectives
▶ Invasion and introduction mechanisms
▶ The regulatory premise
▶ Legal aspects of preventing invasions
▶ Pest risk assessment
▶ Exclusion
▶ Early detection
▶ Containment, control, or eradication
▶ Summary

## INTRODUCTION

**If a pest does not occur in an area, concerted efforts should be made to keep the pest out.**

Stopping introduction is the first line of defense against pests. Insufficient importance is usually attached to this aspect of pest management. Obviously, if a pest does not occur in a given region, it is unnecessary to manage the pest in that area, which is the optimal situation.

**An invasive pest is an alien species whose introduction will, or is likely to, cause economic or environmental harm.**

Society continues to pay dearly because humans have moved organisms around the world (Figure 10–1; Table 10–1). The problems that result from invasive non-indigenous species are part of the primary struggles facing humans in the coming century. Invasion biology is the study of the introduction, colonization, and spread of non-indigenous organisms into new regions. The study of invasion biology aims at understanding the economic, medical, and environmental impact of non-indigenous invasive species. This chapter considers the enormous extent to which humans have created their own pest problems and describes the legislative actions that have been put in place by society in its often vain attempts to limit the spread of pests.

## HISTORICAL PERSPECTIVES

**Today's pest is frequently yesterday's exotic, introduced species.**

Many of today's important pests originated as the exotic, introduced organisms of yesterday. The description of the spread of wheat rust by human

**FIGURE 10–1**
Surface of Lake Chapala in Mexico covered by the introduced exotic weed, water hyacinth.
Source: photograph by Lars Anderson, USDA/ARS.

| TABLE 10–1 | Estimated Annual Losses, in Millions of Dollars, Due to Non-indigenous Species Introduced into the United States | | |
|---|---|---|---|
| | Loss due to damage | Control costs | Total costs |
| Type of introduced organism | Dollar amount (in millions) | | |
| Pathogens (crops, amenity,[1] forests) | 23,100 | 2,600 | 25,700 |
| Weeds (crops, pastures, amenity, aquatic) | 24,410 | 9,648 | 34,058 |
| Mollusks (zebra mussel and clams) | 1000 | NA[2] | 1100 |
| Arthropods (crops, amenity, forests) | 16,044 | 2,011 | 18,055 |
| Reptiles and amphibians (brown tree snake) | 1 | 4.6 | 5.6 |
| Fishes | 1,000 | NA | 1,000 |
| Birds (pigeons, starlings, etc.) | 1,900 | NA | 1,900 |
| Mammals (feral pigs, mongooses, rats, etc.) | 19,850 | NA | 19,850 |
| Nonagricultural (human and animal diseases, ants and termites, crabs, cats and dogs, etc.) | 28,104 | 6,900 | 35,004 |
| Total | | | 136,630 |

[1]Amenity includes turf, gardens, golf courses, etc.
[2]NA indicates no data are available.
Modified from Pimentel et al., 2000.

activity demonstrates the importance and long-term ramifications of pest introductions.

## Wheat Rust Story

The historical perspective on stem rust of wheat (see box) is from the description by Horsfall and Cowling (1977). Stem rust of wheat is a fungal pathogen that now occurs throughout the wheat growing areas of the world. The fungus infects the aerial parts of wheat, and related cereals, causing reductions of tillers, grain yield, and food value of the kernels. The life cycle of the pathogen involves cereals and an obligate alternate host, the common barberry (see Figure 5–19). In

## Wheat Rust and Barberry

Wheat originated in the Near East and so did the barberry (*Berberis vulgaris*), the nemesis of wheat. As all (plant) pathologists know, *Puccinia graminis,* the cause of wheat rust, goes through the sexual stage on barberries. Therefore, the barberries provide sources of inoculum to start the epidemic anew each season. Barberries and wheat rust have produced epidemics and subsequent famine in the Near East since prehistoric times.

In Roman times, traders carried wheat by ship from the Mediterranean Sea through the Pillars of Hercules to northern and western Europe, but they did not take the barberry. As a result wheat flourished there for centuries, free of rust.

Elsewhere, movements were afoot that would one day introduce rust into the western wheat. Mohammed encouraged his followers to spread the faith of Islam. Some did so by the sword, when necessary. For example, the Saracens rode north, east, south, and west. They spread all across northern Africa and on into Spain at Gibraltar. Conquering Spain, they flowed into France, where they were met and finally defeated by Charles Martel at the battle of Tours. They fled back across the Pyrenees to Spain.

Sometime during this great push of the Saracens, one of them in Italy, homesick for his homeland, imported a barberry bush, "and the fat was in the fire." Birds ate the delicious red berries and spread the seeds north and west. Peasants liked the pretty bush and planted it in their door yards. The missing link was now provided and rust for the first time began to destroy European wheat.

The European farmers discovered the connection between barberries and wheat rust by 1660. By then the idea was so firmly established that the French lawmakers passed a law requiring the eradication of barberries. We think it can safely be assumed that English farmers, at least the observant ones, knew of the relationship. For that reason the English colonists who moved to the United States in the early part of that century had a golden opportunity to move wheat once again across the water without moving its deadly enemy, the barberry. But they bungled it. Whether out of ignorance or stupidity, someone brought barberries to New England and the chance was lost. Wheat rust now raged on in the new land. Man had encouraged his own epidemic. The New Englanders caught up a century later. In 1726 Connecticut passed a barberry eradication law as did Massachusetts a decade later.

It is strange that despite the knowledge of barberries displayed in the colonial laws, the New Englanders took their barberries along with their wheat seed when they moved west—another bungled opportunity.

(Horsfall and Cowling, pages 27–28 in volume II)

regions with cold winters, *Puccinia graminis,* the fungus that causes rust disease of wheat, overwinters as teliospores on infected wheat debris. The teliospores produce basidiospores in the spring, and these spores infect barberry leaves. On barberry, the fungus reproduces sexually with production of aeciospores, which are released in late spring and carried by wind to infect nearby wheat plants. In the wheat plant, the fungus produces uredia from which windblown uredospores are released to infect other wheat plants. When wheat plants become severely infected or begin to mature, the uredia produce overwintering teliospores, and the fungal life cycle continues.

The historical example of the dispersion of wheat rust with human activities illustrates how humans have spread crop pests and pathogens since the beginning of farming. Movement of pests around the world continues today, and opportunities to limit pest spread are still being lost, as this chapter will show.

**Human activities have introduced many of the more serious pest problems to new locations.**

## Overview of Exotic Pests

As the wheat rust example illustrates, people have been moving pests around since the dawn of agriculture. Such movement was relatively slow until the age

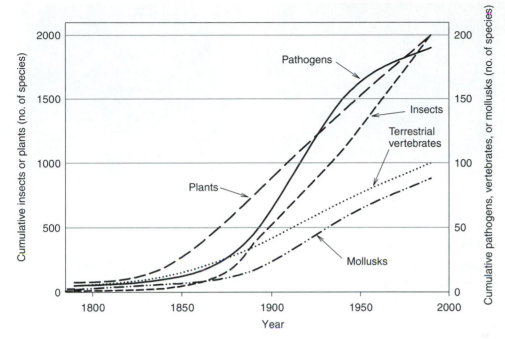

**FIGURE 10–2**

Non-indigenous species introduced into the United States since 1800.

Source: redrawn from OTA report (U.S. Congress, 1993). Note: Scale for plants and insects differs from that for other pest species.

of global exploration which started in the fourteenth century, but has been increasing ever since (estimates for numbers of alien species in the United States are presented in Figure 10–2). The ease of modern travel coupled with the public attitude (more later) has resulted in a vast increase in the number of species that have been transported around the world. Introduction of species from one place to another involves considerable risks; most introduced species are probably harmless, but some have the potential to cause enormous losses if they become established.

The colonial countries, as a part of global exploration, established botanical gardens around the world. The explorers collected plants and brought them back to these botanic gardens. Many crop plant species were introduced to new areas in this way, but so were many pests. It was not until the nineteenth century that people began to realize the serious problems associated with unrestricted importation of organisms. As recently as 1862, one of the charges of the USDA was to introduce new plants. By the end of the nineteenth century, the problems of unrestricted transportation of foreign organisms became sufficiently severe that regulatory actions were taken. By the start of the twentieth century, laws were instituted to regulate transport of organisms through human activities so that invasion of pests could be slowed or stopped. Such laws, and the best means of implementing them, are still in debate.

Halting, limiting, or eliminating the spread of pests requires considerable foresight and long-term planning, which is difficult for a number of reasons. Humans, and the necessary political process, react quickly to emergencies, but are generally poor at long-term planning. Dealing with a problem that does not exist in a particular area may seem to be an unnecessary infringement of people's freedom. People might not see the need for regulations, because the problem does not yet exist in a given area. Accordingly, getting the public to recognize potential problems and cooperate with regulations may be difficult.

**FIGURE 10–3**
Mediterranean fruit
fly (medfly). (a) Adult
male, and (b) close-up
of a cherry showing
destruction of the
tissue and the
maggots.

Sources: photograph
(a) by Scott Bauer,
USDA/ARS; and (b) by
J. Clark, University of
California statewide IPM
program.

(a)

(b)

## Extent and Costs

Plants and animals in their native range are typically kept in check by powerful forces of competition, predation, and diseases. When moved to a new region, a species may be freed from these constraints and spread without restriction, with the potential to negatively impact the ecology and agriculture in the new region. An invasive species becomes a pest if it interferes with human goals and needs.

It has been estimated that there are perhaps as many as 50,000 non-indigenous species in the United States. Many of these, such as crop plants and many animals, provide most of the agricultural production, and many serve as ornamentals or are beneficial organisms. Some, however, have become serious pests. The Office of Technology Assessment (U.S. Congress, 1993) estimated that 79 species have resulted in approximately $97 billion in losses since their introduction. A more comprehensive analysis of losses in the United States (Pimentel et al., 2000), including the direct losses, control costs, and environmental damage from non-indigenous species, estimated that the total losses are approximately $137 billion per year (Table 10–1); of this total, non-indigenous species were estimated to cost plant agriculture $93 billion per year. Non-indigenous species thus cost approximately $400 per year for each man, woman, and child living in the United States.

Estimates of the benefits from excluding a pest also provide another way of considering the costs of non-indigenous species. Excluding the Mediterranean fruit fly from invading California is estimated to save $210 million per year in direct losses, and up to $732 million per year in pesticide treatment costs that would be incurred if the insect became established. Forty-two countries have quarantine laws designed to prevent importation of the medfly (Figure 10–3); and embargoes on produce from infested areas would substantially increase the economic costs due to loss of markets. On a smaller scale, it was estimated that exclusion of the leaf miner *Liriomyza trifolii* from Finland provided between a 3:1 to 13:1 benefit ratio between costs of control versus costs of exclusion. Averting the disruption of biocontrol caused by insecticide use is another benefit of exclusion.

Another measure of the magnitude of biological invasions in the United States can be judged by the 1,567,000 interceptions of plant material and 48,480 interceptions of pests (assume the latter is mainly arthropods and pathogens), by USA agricultural border inspectors during 1998. That same year there were

**FIGURE 10–4**
A single day's contraband intercepted at JFK International Airport in New York by Jackpot 1, a member of the USDA/APHIS beagle brigade.
Source: photograph courtesy USDA/APHIS.

20,700 civil penalties levied, with a total fine value of $1,080,000. The magnitude of this problem is portrayed by the contraband intercepted on a single day at a major international airport (Figure 10–4).

Environmental costs due to non-indigenous species are much harder to estimate. Such species may alter ecosystem processes, but it is usually not possible to place a cost on such an alteration. Purple loosestrife, for example, changes the functioning of wetlands for all resident species. Tamarisk lowers the water table in desert regions, causing springs to dry up, with catastrophic results to organisms using such water sources. Drying of the everglades and loss of sawgrass swamps is attributed, in part, to invasion of the Melaleuca tree. Hardwood forests in the eastern USA changed dramatically when the chestnut trees died following the invasion of chestnut blight. The brown tree snake has eliminated several native bird species in Guam. Introduced animals, such as rabbits, contribute to soil erosion but the impacts are to a great extent unknown. It has been estimated that 50% of endangered species in the United States are endangered because of non-indigenous exotic species.

The following serve as examples of introduced pests, and of the impacts associated with their introduction.

***Pathogens***    The history of pathogen introductions is difficult to ascertain, because many pathogens are relatively cryptic and not easily recognized and identified. Downy mildew of grapes was transported from the USA to France in the 1860s with the result that the French wine grape industry was almost destroyed. White pine blister rust was moved to the USA from Germany in 1906. Citrus canker was brought to the USA before 1914, and resulted in a massive eradication program to save the citrus industry. Dutch elm disease was first observed in the USA in Ohio in 1931 and has now eliminated most elm trees in North America. It also devastated elm trees in England (Figure 10–5). Coffee rust was introduced into Sri Lanka (Ceylon) in 1868 resulting in the destruction of the coffee growing industry. Coffee rust has now crossed the Atlantic from Africa to South America, and it was detected in Brazil in 1970. Additional examples of introduced diseases include chestnut blight, which was introduced to the USA

**FIGURE 10–5** Elm trees dying after infection by the introduced fungus that causes Dutch elm disease; center trees infected while trees to extreme right and left are still healthy.
Source: photograph by Robert Norris.

in the early 1900s from China and has devastated native chestnut trees. Rhizomania, or crazy-root disease of sugar beets, was first detected in the Po Valley of Italy in 1954, and then spread over the sugar beet growing regions of the world in the 1970s and 1980s.

***Weeds***   Many plants were originally imported for agricultural purposes, or as ornamentals, and then 'escaped' from agricultural fields to become weeds. It has been estimated that approximately 60% of the 500 plant species considered as weeds in the United States were introduced by human activity. In the irrigated Southwest, virtually all weeds in warm-season summer crops are introduced plant species. Examples of common agricultural weeds that are not native to North America include nutsedge, johnsongrass, Canada thistle, pigweed species, barnyardgrass, nightshade species, lambsquarters, velvetleaf, and foxtails. Wild oat was probably introduced as a contaminant of animal feed in the early 1600s. These weeds must now be controlled on an annual basis at enormous expense to farmers.

Witchweed, a root parasite on corn and other grass crops, was introduced into North Carolina in 1956. An eradication program costing $225 million has been waged since that time to stop the weed spreading into the Corn Belt where its impact would be devastating.

Many weeds of North American rangeland and wildland (including riparian) habitats have also been introduced. Examples include yellow starthistle (1860s), leafy spurge, medusahead, spotted and Russian knapweeds, Scotch thistle, tamarisk, Scotch broom, and gorse.

Kudzu was first introduced to North America as an ornamental in 1876 from Japan. It was also widely planted in the 1930s for erosion control. Kudzu is a vine that can smother structures and essentially any other vegetation (Figure 10–6). Kudzu is now one of the most widely distributed invasive plants in the southern and eastern USA.

Wetlands are susceptible to invasion by exotic plants, although these are typically not agricultural weeds. Examples include purple loosestrife, Melaleuca, and giant reed.

**FIGURE 10–6**
Introduced vine kudzu is capable of smothering trees and structures; here it covers abandoned farm buildings in Georgia.
Source: photograph by J. Anderson, with permission.

Numerous aquatic weeds have been introduced, often as ornamentals, including water hyacinth (Figure 10–1), hydrilla, Eurasian water milfoil, and giant salvinia. The latter, a fern, is one of the most serious aquatic weeds on a worldwide basis, and was first found in the United States in 1995. African nations alone spend an estimated $60 million per year on control of such introduced aquatic weeds as water hyacinth and water lettuce.

***Nematodes***   Recognition and detection of nematode introductions is difficult. The potato cyst nematodes are thought to have coevolved with potato in South America. Potato was brought to Europe in the 1500s, and over time came to be grown in many areas of Europe. The potato late blight epidemics of 1845 to 1846 caused famine and hardship in Ireland and parts of Europe. In an effort to find resistance to blight, expeditions to Central and South America were undertaken to bring wild and cultivated potato tubers back to Europe. The original introductions of potato cyst nematode to Europe are considered to have occurred by cysts in soil adhering to potato breeding material. The potato cyst nematode has since spread to almost all parts of the world where potatoes are grown. In the United States, the spread of potato cyst nematode (also called the golden nematode) has been limited to Long Island by strong quarantine measures.

The soybean cyst nematode was reported from Manchuria in 1880, Japan in 1915, Korea in 1936, Taiwan in 1958, and Indonesia in 1984. The soybean cyst nematode was first detected in North America in 1954, in an area of North Carolina that had a history of growing flower bulbs imported from Japan. The sugar beet cyst nematode was first observed in Germany in 1859, and it has since been spread to all sugar beet growing regions of the world, probably in soil particles contaminating beet seed.

The pinewood nematode, a pathogen of pine that is transmitted by cerambycid beetles, has devastated pine forests in southern Japan, and has also moved into China. Symptoms of pinewood nematode damage were first reported in Japan in 1913, but the role of the nematode in pine-wilt disease in Japan was not determined until 1971. Although the pinewood nematode occurs in most forested areas of the United States, pine-wilt disease is rare on native pines, leading to the suggestion that the nematode is endemic to the USA and has spread from here to other locations.

***Mollusks***    The rosy wolfsnail was imported into Hawaii in 1955 as a biological control agent and has since eliminated several native snail species. The giant African snail was introduced into Florida in 1969 as a pet, and it is now a pest in several situations. The golden apple snail was introduced into Southeast Asia in 1980 as a potential protein food source; but it thrives in the rice agroecosystem and has now become a serious pest in that crop in the region. In the Philippines it is estimated that the golden apple snail causes losses between $425 million and $1.2 billion annually.

***Arthropods***    A few examples of arthropods introduced into the United States include the following: Hessian fly of wheat in 1779, San Jose scale of fruit trees in 1879, cotton boll weevil in 1892, and oriental fruit moth prior to 1912. The imported fire ant was first recorded from Mobile, Alabama, in 1918, and the alfalfa weevil has been introduced at three different times in the twentieth century. The Mediterranean fruit fly (commonly called the medfly) (Figure 10–3), which is a serious pest of all soft fruits, was first introduced in 1929 and many times since. More recent arthropod introductions include the spotted alfalfa aphid in 1954 and the Russian wheat aphid in 1986. The gypsy moth, which defoliates hardwood forest trees, was being investigated in laboratory studies for possible breeding with silk moths. In 1889 some of the moths escaped, became established, and now cause losses of about $764 million per year.

Phylloxera of grapes was moved from the United States to France in the early 1860s and nearly destroyed the French grape industry. The Colorado potato beetle (Figure 10–7) has been introduced into Europe from North America on several occasions.

***Vertebrates***    Rabbits are the classic example of an introduced vertebrate pest (see Figure 2–14e; Figure 10–8). They were imported into New Zealand in 1838 and Australia in 1859. Farmers in these countries rued the day ever since, because rabbits in Australia are estimated to cause $600 million direct losses per year, plus other additional environmental damage. Pigs and goats introduced

**FIGURE 10–7**
Colorado potato beetle is an example of an insect that has been introduced from North America to Europe.
Source: photograph by Scott Bauer, USDA/ARS.

into Hawaii in 1778 from Europe now damage crops and degrade natural environments. A Shakespeare fanatic, who wanted to have all the birds that are mentioned in Shakespeare's plays in the United States, introduced starlings into New York in 1890. Starlings now cause an estimated $800 million in losses per year. Rats are not native to many regions in the world, but were probably unintentional introductions on early trading ships.

***Others*** The Nature Conservancy prepared a list of the 12 most important invasive species in natural ecosystems in the United States in the late 1990s (Stein and Flack, 1996), and they are listed in the accompanying box.

## The Twelve Most Important Invasive Species

| Common name | Latin name | Type of damage |
| --- | --- | --- |
| Zebra mussel | *Dreissena polymorpha* | Aquatic ecosystems |
| Purple loosestrife | *Lythrum salicaria* | Wetlands and riverbank ecology |
| Flathead catfish | *Pylodictis olivaris* | Native fish displacement |
| Tamarisk | *Tamarix spp.* | Desert ecosystem water tables |
| Rosy wolfsnail | *Euglandina rosea* | Killing native snails |
| Leafy spurge | *Euphoribia esula* | Displacing rangeland vegetation |
| Green crab | *Carvuras maenas* | Coastal ecosystems |
| Hydrilla | *Hydrilla verticillata* | Blocks aquatic ecosystems |
| Balsam woolly adelgid | *Adelges piceae* | Kills Fraser fir in eastern United States |
| Miconia | *Miconia calvescens* | Tree taking over ecosystems in Hawaii |
| Chinese tallow | *Sapium seboferum* | Displacing native vegetation |
| Brown tree snake | *Boiga irregularis* | Kills birds and other native animals |

**FIGURE 10–8** Rabbit damage to a broccoli field. Rabbits came from the riparian area on the left; the only crop remaining is the darker area in the top right of the picture.
Source: photograph by John Marcroft, with permission.

# INVASION AND INTRODUCTION MECHANISMS

## Ecological Perspective

The ability of an organism to spread is a natural part of its life history, as pointed out in Chapter 5. Natural mechanisms of long-distance transport include the ocean streams (aquatic conveyor belts), the jet streams in the atmosphere (aerial conveyor belts), high winds (gales, hurricanes, tornados), land bridges formed during periods of drought, and connection between watersheds, rivers, and lakes during periods of floods. Spread mediated by humans provided a qualitatively new mechanism that increased the distances and speed of spread enormously.

There are two ways in which human-mediated pest introductions occur. Organisms are intentionally introduced by people for a variety of reasons, such as ornamental plants, and game or food animals. Organisms can also be introduced accidentally by human activity. The causes of some introductions are unknown and probably result from the natural mechanisms of spread as noted.

## Intentional Introductions

Most intentional introductions involve plants that later became weeds, or animals that later became a problem. That an organism became a problem following its intentional introduction implies that it was imported without sufficient understanding of, or consideration for, its capacity to reproduce, spread, and compete with local organisms.

***New crop plants***    Originally, a plant is introduced as a new crop, but either its utility has not been adequately determined and it was abandoned, or it was too vigorous under the new conditions and was able to establish as a problem. What this means is that the ecology and invasive potential of the plant was not understood or appreciated, and when placed in the new environment it grew, reproduced, and spread without further human intervention. There are many examples of plants that became weeds in this way (Williams, 1980). Johnsongrass was introduced into the United States as a potential forage crop in the 1800s. Bermudagrass was introduced as a forage crop and as a turf grass, for which it is still used, but it also grows where it is not wanted and is a very serious weed. In Australia, ryegrass was introduced as a pasture grass, but has also become one of the most serious weed problems in cereal crops.

***New ornamental plant***    Similar to crops, many plants have been introduced to new regions of the world as ornamentals and then they escape. Many types of weeds were introduced as ornamentals. Water hyacinth is an example of such a weed (see Figure 2–4a; Figure 10–1). It is now a worldwide problem in lakes and aquatic systems, yet it is still being sold in the nursery trade. Other examples of ornamentals include pampasgrass, morning glory, Miconia, Scotch broom, prickly pear in Australia (used as a fence), and many others.

***New animal food source***    A few organisms have been introduced as a protein source and have later become pests. The introduction of the golden apple snail into the Philippines in the 1980s is such a case, and the snail is now a pest in rice production areas.

**Erosion control**   Several plants have been introduced for erosion control, but have then proved too aggressive and have become weeds. Kudzu is probably the most notorious example in the USA (Figure 10–6); another is salt cedar or tamarisk.

**Biological control agent**   Some organisms introduced as biological control agents have become a problem themselves. An example is the mongoose in Puerto Rico. It was introduced to control rats, but instead fed on other animals and birds. The cactus moth was collected in South America and released around the world to control the weedy cacti of the genus *Opuntia.* In 1989, the moth appeared on cacti in the Florida Keys, and now threatens native cacti, many of which are also used as ornamentals.

The rosy wolfsnail was introduced into Hawaii for biological control of pest snails, but the rosy wolfsnail is now decimating populations of several native snails. Likewise, numerous examples of fruit fly parasitoids once introduced into Hawaii are now attacking native tephritid species that play an important role in the ecology of Hawaiian natural plant communities. It is now considered likely that the tachinid flies introduced to control the gypsy moth have eliminated most, if not all, of the large moths native to eastern USA hardwood forests.

**Misguided or lack of knowledge**   Most people have insufficient knowledge of botany, ecology, or pest management to understand the implications of transporting plants from region to region. Yet many people believe that they have the right to transport any plant material they wish with impunity. Often the short-term financial gain of a few creates long-term problems faced by the whole of society. Witness the few in the United States who ignored the warnings about importing dead wood from diseased elm trees. In exchange for the ability to make furniture for a few years, the whole of the elm tree population in the United States was sacrificed.

**Discarding unwanted organisms**   When people become tired or bored of taking care of an organism, they may choose to discard it. The organisms involved may or may not be killed in the process. Organisms that are not killed have the potential to establish themselves wherever they were discarded. The repeated introductions of the noxious aquatic weed *Hydrilla,* which is sold in the aquarium trade, are attributed to people dumping aquariums into local streams or ponds. The choking weed has also been introduced into the Potomac River in Washington, D.C. This and other aquatic weeds block irrigation and drainage canals, enhance sedimentation in flood control reservoirs, interfere with public water supplies, impede navigation, and generally restrict water uses. An estimated $100 million is spent each year in the United States to control aquatic weeds, most of which are non-native.

There is also potential that pet organisms can become pests in a similar fashion. For example, in 1989, six pet rabbits were released by their owner at the Haleakala National Park in Hawaii. An emergency eradication program was initiated, and by 1991, 100 rabbits had been destroyed at a cost of $15,000 to the park.

**Malicious intent**   Although no documented evidence exists of intentional introduction of harmful organisms, this concept is the basis of biological warfare, and should be a matter of grave concern to responsible governments because some biological weapons could have devastating ecological effects. A proposal to use a fungal pathogen to destroy Colombian coca plantations in the war against drugs (cocaine) was submitted to the USA government. Many scientists expressed concern that the pathogen would potentially have negative effects on crops and desirable native plants. Decision makers should be aware of the serious repercussions that can occur from the release of organisms intended to damage plants.

## Accidental Introduction

Accidental introduction occurs mainly with arthropods, pathogens, and nematodes. The movement of plant material (including soil adhering to roots), and the soil in which plants are growing, may inadvertently introduce pests. Such introductions are hard to detect until the organism has started to become established, which may take many years. Only at this later time is the hidden, insidious problem revealed. For microscopic soil-dwelling organisms, such as pathogens and nematodes, this may be especially problematic. The current regulatory stance is that soil should not be moved from region to region, and that all plant introductions should go through established quarantine centers to determine if the plants are hosts of pests (more later in the chapter). Nearly all accidental introductions result from ignorance or unwillingness to recognize the facts and fully consider the possible dangers of unregulated importation.

***Produce or human food***   Produce transported from one area where a pest is endemic to an area where it is not present has the potential to introduce a pest. This is almost certainly the way that many of our crop arthropod and pathogen pests have traveled around the globe. The medfly is probably the most notorious. It travels as larvae in ripe fruit (Figure 10–3b). California has spent nearly $235 million between 1975 and 1999 to eradicate repeated introductions of this pest. The European corn borer, one of the most important pests of corn and many other crops in the United States, was probably introduced in broom corn imported from Hungary or Italy between 1909 and 1914. From an original pocket of infestation in some Eastern states, the corn borer spread into the midwestern Corn Belt.

Produce is necessarily extended to include wood and wood products. Dutch elm disease came to the United States in lumber imported from Europe for furniture making. The lumber was obtained from trees killed by the disease. Even now, lumber is imported to the United States from Russia and China, risking the introduction of dangerous pests such as the Asian gypsy moth and the Asian longhorn beetle. The longhorn beetle attacks many kinds of living hardwood trees and has the potential to devastate North America's eastern hardwood forests. Thus far, its range in North America is restricted to the vicinity of New York City and Chicago, where it was unintentionally introduced in wooden crates from China. People know the dangers of these importations, but some still act irresponsibly.

***Contaminant of crop seeds/planting stock***   When crop seeds or propagation material is transported, pests associated with these materials can be inadvertently transported. Seeds may harbor pathogens on the seed coat; and some nematodes (see Figures 2–7d and e) and pathogens, especially viruses, can actually be present in the seed and carried in that form. In the past, the seeds of some crop plants were collected from the ground, and soil particles with associated pathogens and nematodes were picked up as well. There is little possibility of visually detecting such contamination. Weed seeds can be present as a contaminant of crop seeds; however, these can be visually detected. Transport of noncertified crop seed has great potential for pest introduction. Prevention of pest spread was a major reason for the development of seed certification programs discussed in Chapter 16.

The use of clonal material for vegetative propagation of crops (e.g., grapes and potatoes) is especially problematic. Any fastidious pathogen requires a living host, and transport of cuttings, tubers, and bulbs provides such a host. Transport of clonal materials that are not certified as pest free at the site of origin is especially dangerous. Unregulated transport of grape cuttings has, for instance, moved pathogens, especially viruses, between France and North America, and vice versa. The imported crucifer weevil was discovered in horseradish fields in southern Illinois in the mid-1970s. It has been speculated that someone brought rootstocks from Europe without submitting them to quarantine procedures, and thus inadvertently transported the weevils.

**Contaminant of feed for animals**   Food and feed for animals may be contaminated with pests. Infested hay is particularly important in this respect, as it often harbors many pests. It is believed that wild oat was introduced into California in the 1700s by contaminated hay used to feed horses.

**On or in live animals**   Pests can be on, or even in, animals that are being transported, resulting in inadvertent importation of the pest. For example, it has been shown that weed seeds and some nematodes will survive passage through the alimentary canal of many vertebrates The introduction of the weed *Parthenium hysterophorus* into Sri Lanka in the late 1980s was attributed to seeds in the dung of goats imported from India by a military peace-keeping mission.

**Contaminated soil**   Many soil-borne pathogens, nematodes, and weed seeds can be transported in soil without any knowledge of the people involved. For this reason most regions now ban transfer of field soil. Only sterilized potting soil can be shipped commercially, but the threat of contamination remains. Even short-distance transportation of pests in soil is of great concern, but it cannot be managed by interstate regulations. Local growers, however, adopt their own so-called quarantine procedures. For example, with the resurgence of the grape phylloxera in vineyards in California and Oregon, managers required farm workers to dip their boots in lime before entering fields in order to prevent the spread of the pest from contaminated areas.

**Irrigation water**   Spread by irrigation water is important at the local and regional levels. Pest organisms such as weed seeds, some pathogen propagules, and some nematodes can float in water and be spread within fields or moved into noninfested fields.

**Transportation vehicles**   Organisms are carried in, or become attached to, vehicles used for transportation. It is well recognized that new pest introductions typically occur around mass transport facilities such as ports, along railroad rights of way, and more recently around airports and along major roads.

The transportation of the gypsy moth on the outside of airplanes, campers, and even furniture shipped across the United States is a classic example. Car tires can carry soil containing pathogens, nematodes, and weed seeds. The ballast water in ocean-going freighters has transported organisms from port to port; the introduction of the zebra mussel to North America is attributed to this mechanism of spread.

***Farm machinery***    This specialized type of transportation is important for pest invasion and spread at the regional and within-farm levels. The topic is discussed in greater depth under cultural management in Chapter 16.

***Military activity***    Numerous organisms have been introduced accidentally because of actions taken by military forces. The actual mechanisms are a combination of those cited previously. The brown tree snake was probably introduced to Guam on military equipment after the end of World War II. The snake has eliminated eight bird species and two gecko species from Guam. To prevent the snake from reaching other vulnerable locations, such as Hawaii, an extensive monitoring and interception program has had to be established at all ports on Guam.

## Introductions by Unknown Mechanism

Sometimes it is not possible to determine how an introduction occurred. Examples include the Egyptian alfalfa weevil, the cereal leaf beetle, and rhizomania of sugar beets.

## THE REGULATORY PREMISE

Regulatory or legislative prevention is based on the idea that the best pest management is exclusion. Costs of exclusion are usually relatively low compared with the cost of control after the pest has invaded a new area (Figure 10–9), and early control (eradication immediately after invasion) is usually much less costly than late control.

It has been argued that society should use the wildfire paradigm in relation to noxious weed invasion (Dewey et al., 1995), and the analogy applies equally well to all pest categories. This paradigm holds that fire prevention is better than fighting fires once they start (fires equates to invasion), because the outcome of a fire is uncertain and potentially devastating. The wildfire paradigm also dictates that it is better to extinguish fires while they are still small (equates to stop the spread while the population is small). This implies that if a pest does invade, efforts should be undertaken to eradicate it as soon as the problem is recognized. Large wildfires are costly to fight and cause enormous losses. Established invading pests can result in elimination of profitable crop cultivation in a region, or can cause farmers to apply environmentally undesirable control measures year after year. The latter often involves increased use of pesticides. The wildfire paradigm also includes public education and awareness as a part of the management strategy. Pest management has done little to enhance public awareness of the serious nature of pest introduction. The wildfire paradigm is a useful analogy relative to the management of invading pests.

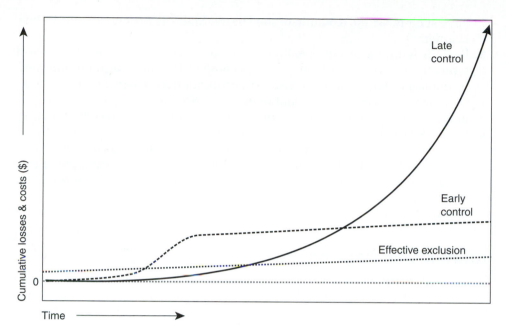

**FIGURE 10–9**
Theoretical cumulative damage over an extended period using three different control strategies for an introduced exotic pest.
Source: modified after Naylor, 2000.

Use of economic thresholds generally should be withheld when applied to invasive pest species as they are detected in a new area. If feasible, the new pest should be eradicated from the area as quickly as possible. Waiting until the invasive pest reaches an economic threshold may allow it to become so well established that eradication is impossible, and the best that can be hoped for is continuous control action (Figure 10–9; late control); at worst it can mean that a crop is no longer grown in the region (e.g., coffee in Sri Lanka following introduction of coffee rust in the mid-1800s).

# LEGAL ASPECTS OF PREVENTING INVASIONS

The realization that human societies could not tolerate the uncontrolled global spread of organisms led to legislation to restrict the movement of organisms that have "pest potential." The scale at which legislation is enacted may be at the level of international treaties, laws within countries, and even laws in regions within countries. For example, many states in the United States have specific quarantine laws.

**Legislative mandate (laws and regulations) can limit the spread of pests into areas not currently infested.**

## Legislation to Regulate Pest Transport

Legislative regulation occurs at different geopolitical levels and requires differing levels of cooperation between interested parties. These levels are as follows:

1. International global. Due to the shipment of goods in commerce between countries, it became apparent that internationally recognized standards regarding the transportation and movement of potential pest organisms were required.
2. Countries. Individual countries enact regulations for protection from invasion of exotic pest organisms. Most major industrialized countries

have such laws; the example of the United States is presented later in this chapter.

3. **Regions within a country (states or provinces).** Pest regulation at this level typically occurs when the region is isolated so that pest introduction can be managed. For example, Hawaii is surrounded by ocean, and California is isolated by ocean, desert, and mountains. Regional regulatory activity may also result from the limits imposed by politically derived boundaries.

4. **Local.** A city or a county can mandate that certain organisms cannot be imported, or restrictions on movement may be imposed. Local regulation may be addressed at the individual farm level, such as the example of the grape phylloxera noted earlier in this chapter. Farm-level regulation of pest invasion is discussed in greater depth in Chapter 16 (see "Prevention").

## International Regulations

Overall, international laws regulating transport of nonindigenous species are weak, because pest regulation must take place in several countries simultaneously, and thus requires development of international treaties and agreements. The International Plant Protection Convention (IPPC) was established in 1951 to foster the necessary cooperation. It is a multilateral treaty organized within the framework of the Food and Agriculture Organization (FAO) of the United Nations. One hundred and thirteen countries were signatories to the convention as of January 4, 2000. A major function of the IPPC is to establish international phytosanitary principles that are acceptable to the participating countries, but do not infringe on their sovereign rights. To this end the IPPC establishes international standards for phytosanitary measures (also called ISPMs). Signatories to the convention agree to abide by these standards. A major component of this process is pest risk analysis (see later in this chapter), as risk analysis establishes the necessity for phytosanitary measures.

In 1995, the World Trade Organization enacted an agreement on the Application of Sanitary and Phytosanitary Measures (the SPS agreement) as a further attempt to regulate the spread of pests due to international trade. The implications of this agreement and the extent to which it will regulate invasion of exotic species is not yet clear.

Clusters of neighboring countries with similar interests and needs in relation to plant quarantine and phytosanitary procedures are organized into regional commissions or organizations (see box).

## Regional Plant Protection Organizations

Asia and Pacific Plant Protection Commission (APPPC)

Caribbean Plant Protection Commission (CPPC)

Comité Regional de Sanidad Vegetal para el Cono Sur (COSAVE)

Comunidad Andina (CAN)

European and Mediterranean Plant Protection Organization (EPPO)

Inter-African Phytosanitary Council (IAPSC)

North American Plant Protection Organization (NAPPO)

Organism Internacional Regional de Sanidad Agropecuaria (OIRSA)

Pacific Plant Protection Organization (PPPO)

# National Regulations

All countries have laws restricting movement of plants, soil, and products that could be carrying pests. Following are three examples. The Brazilian Enterprise for Agricultural Research (EMBRAPA) has established a national center for research on the environment (Centro Nacional de Pesquisa do Meio Ambiente) that monitors the risk of invasive species, establishes quarantine of biological materials imported into the country, and develops legislation to regulate movement of materials with the potential to introduce new pests. The Australian Quarantine and Inspection Service (AQIS) performs similar functions in Australia. In England, the regulation of exotic pest organisms is within the Ministry of Agriculture, Food and Fisheries.

# Major Laws in the USA

United States laws are written by Congress in the form of acts, which are signed by the president. These acts are then promulgated in the form of regulations, which become part of the legal code of the country. The mandate of United States laws concerning pest importation and spread is carried out under the jurisdiction of the United States Department of Agriculture (USDA) and the Commerce Department.

In the United States, the following major acts were intended to limit the introduction and spread of pests and to provide the legal basis on which plant pests are regulated.

- The Plant Quarantine Act (1912) formed the historic basis for establishing quarantines against domestic and foreign plants and plant products. The act provided the authority to quarantine any area to prevent the spread of a plant pest. The act also provided the authority to restrict and, to the extent necessary, prohibit entry into the United States and interstate movement of plants and plant products, based on the need to restrict pest introductions. Although the name was Plant Quarantine Act, its targets were nonplant pests such as arthropods, pathogens, and nematodes. The act was amended many times; the most recent was in 1994. The Plant Protection Act of 2000 has now superceded it.
- The Federal Seed Act (1939) regulates interstate and foreign commerce of seeds and addressed "noxious weed seeds" present in commercially sold crop seeds.
- The Organic Act (1944) provided the authority for the federal government agencies to cooperate with states, farmers, associations, and Mexico to carry out operations to control or eradicate pests that pose a significant economic hazard. This authority was extended by amendment to the act in 1976 to grant authority to cooperate with nations of the Western Hemisphere in the control of pests that threaten the United States. Most aspects of this act have now been superceded by the Plant Protection Act of 2000.
- The Federal Plant Pest Act (1957) provided the authority for the secretary of agriculture to prevent the spread of plant pests. One section gave the secretary authority to declare an extraordinary emergency. Another section provided the authority to take emergency action to seize, treat, or destroy articles or products related to plant pests new to or not known to be widely prevalent in the United States. The Plant Protection Act of 2000 has now superceded this act.

- The Federal Noxious Weed Act (1975) provided authority to a regulatory system designed to prevent the introduction into or through the United States of noxious weeds from foreign countries. The secretary of agriculture was authorized to initiate control and eradication actions against incipient infestations of noxious weeds that were introduced into the USA. This act had the same intention as the Plant Quarantine Act of 1912, except that it specifically regulated importation of plants that may become weeds. It was amended several times. The act has now been superceded by the Plant Protection Act of 2000.
- The Plant Protection Act (2000), or PPA, replaced most of the preceding acts. It is part of the Agricultural Risk Protection Act, and combines the previous activities under a single act and attempts to avoid duplication of effort and to clarify some regulatory activities. The PPA consolidates authority to regulate plants, plant products, certain biological control organisms, noxious weeds, and plant pests. This is the first time that biological control agents are defined and regulated specifically.

The earlier acts, and now the PPA, are the basis of regulations to restrict movement of pests and to provide both civil and criminal penalties for violations. The regulations are developed by various government agencies based on the acts, of which the USDA is the most important. Within the USDA, the major branch involved with pest regulation is the Animal and Plant Health Inspection Service (APHIS). APHIS has Plant Protection and Quarantine (PPQ) inspectors to enforce policies.

Many states have exclusion and quarantine laws. Such state agencies have jurisdiction to control movement of pest organisms into and within the state.

# PEST RISK ASSESSMENT

## Definition Difficulties

A major difficulty arises in determining pest status and risks if an introduction occurs. The definition of pest is anthropocentric (Chapter 1), and people in different parts of the world, and segments of society within a country, may view an invading or potentially invasive species from different perspectives.

Some ornamental plants have great potential to become established as weeds. Weeds in the genus *Ipomoea* (morning glories) are a serious problem for cotton farmers, yet several species are sold as ornamentals in garden seed catalogs. Organic gardening catalogs sell a plant called chufa, because it produces nutty flavored tubers; it is the same plant from which the national drink of Spain, *horchata,* is prepared. This plant is yellow nutsedge, which is considered to be one of the world's worst weeds. Such dichotomies make regulation and determination of pest status difficult.

## Pest Risk Analysis

The regulatory agency must conduct a pest risk analysis in relation to a potential pest that could be transported from one country or region to another. A similar analysis must be made if an introduced pest is discovered in a new area. Invasion biology is the underlying science of pest risk analysis. Following are types of considerations to be included in pest risk analysis.

***Likelihood of becoming a pest***   To assess the probability that an organism will become a pest requires analysis of the biology and ecology of the organism, how it behaves in its current range, and how it might behave in the new environment (from which regulators hope to exclude it).

***Pathways of spread***   Knowledge of how an organism spreads gives regulators an idea of how rapidly it will invade new areas, and provides a base to assess how easy it will be to regulate its movement.

***Amount and type of damage***   Knowledge about the potential of a pest to cause economic injury is a key element in decision making regarding invasive species. The damage potential must be assessed relative to climate and crop production in a given region, and should consider the potential damage a pest might cause, ranging from only minor damage to elimination of the production of a particular crop in the region.

***Ease and cost of control***   Cost-benefit analyses are a necessary component of pest risk analysis, because they form the basis for the economic acceptability of management decisions. The analyses include consideration of the intensity and acceptability of management tactics that will be necessary to control the pest. For example, will it be possible to control the new pest with simple cultural practice modification or plant breeding, or will it be necessary to substantially increase pesticide use because no other tactics are effective?

***Effects and costs to the environment***   Costs to the environment must be considered. Here it is necessary to determine how recommended management tactics will influence the environment. Undesirable environmental impacts such as air or water pollution, or danger to wildlife, will eliminate the use of some tactics. At the ecosystem level, the pest might have substantial negative influences on the environment, such as an invading plant displacing native vegetation or the activity of an animal causing soil erosion.

Pest risk analyses follow strict protocols for each pest category, and are developed by the appropriate government agencies in cooperation with the parties that are likely to be impacted. The public is always allowed a certain amount of time to offer comments on proposed regulations, before such regulations are codified as law.

Obviously, many risk assessment questions are difficult to answer definitively, but the risk analysis must lead to classification of the organism according to its risk as a pest. If the assessment of risk is high for an organism that is not present, it is placed on a list of pests that should be excluded. These are the organisms targeted by inspectors at international ports, or on produce prior to international shipment.

## Regulatory Options

Once the potential risk for an organism has been determined, several regulatory actions can be initiated. One of the following actions would be pursued.

1. Exclusion
2. Early detection
3. Containment, control, or eradication

# EXCLUSION

**Whenever prevention or exclusion is possible it is preferable to attempting eradication after pest establishment.**

Risk analyses have determined that an organism would pose an unacceptable risk if introduced into a noninfested region. It is thus judged more cost effective to exclude the organism than it is to attempt other management options. Exclusion can be at the level of countries. Each country will have a list of particular organisms that are the subject of pest exclusion efforts, as determined by assessments described previously. The lists of quarantine pests are routinely posted on the websites of the regulatory agency for each country, and some URLs are provided at the end of the chapter (APHIS, 2000).

Exclusion can be conducted at the regional level. Efforts to stop the cotton pink bollworm from invading the San Joaquin Valley of California serve as a good example of regional exclusion. The insect is not present in the valley, but is present in the Imperial Valley just 200 miles to the south. Monitoring is conducted to detect the presence of adult moths, overwintering habitat is destroyed by mandatory plow down of plant material, and incoming shipments are inspected. A sterile release program in the San Joaquin Valley protects more than 1 million acres of cotton. Millions of moths are raised and sterilized at a special facility in a USDA laboratory in Phoenix (see Chapter 17 for a more detailed explanation of the sterile-insect release tactic). The sterile moths are released from small airplanes over cotton fields where wild moths have been found. Sterile release occurs from May to mid October, when pink bollworms are most active in cotton. For the sterile release method to work, a ratio of 60 sterile moths to 1 wild moth is needed. This high ratio of sterile to fertile moths makes sterile releases impractical in heavily infested areas, but it works well in the limited geographic area of the San Joaquin Valley where small local infestations can be specifically targeted.

## Pest Interception Tactics

Determining if a pest is likely to be present in a particular situation (e.g., in a shipment of plants or in the car of a vacationing family) is very difficult and relies on a variety of approaches. The intent of the laws is met differently for the general traveling public in comparison with commercial shipments.

***Public travel***   The assumption is made that members of the public should not transport plants or soil, and enforcement is based on that assumption. For example, transport of papayas from Hawaii to the U.S. mainland is acceptable only if the papayas are certified pest free at the point of origin. The problem for regulatory agencies and their personnel is how to determine if plants or soil are being transported. The following techniques are used.

1. Residents returning from a foreign country must complete questionnaire forms. The drawback to this technique is that it relies on the knowledge and honesty of persons completing the form. Written questionnaires are typically coupled with some form of verbal questioning.
2. Trained sniffing dogs can detect various types of materials in baggage that is being shipped. The U.S. Plant Protection and Quarantine inspectors of APHIS operate their so-called beagle brigade for the purpose of detecting illegal materials in baggage (Figure 10–10).

**FIGURE 10–10** A USDA/APHIS Plant Protection and Quarantine beagle sniffing baggage. It is a federal offense to interfere with these dogs.
Source: photograph courtesy USDA/APHIS.

3. Visual inspection of baggage helps to locate and identify plants or soil, or even the actual pest. This procedure works only for vertebrates, most plant material (including seeds), and many insects that can be seen with the naked eye.

Detection of nematodes and pathogens remains a problem because they are microscopic. Rapid molecular methods for pathogen identification may make detection of these pests easier in the future, but at present the only logical approach is to restrict importation of host-plant material or soil.

***Commerce***   Commercial produce shipments are subjected to greater regulation than personal travel, so different approaches may be used for pest detection.

1. Inspection at site of origin. Inspectors employed by the receiving country, for example, PPQ in the United States, work in the country of origin to inspect the crop growing conditions and the produce either before or after packing, but prior to shipment. Certificates stating that the produce is free of quarantined pests are issued; these are referred to as phytosanitary certificates. These types of certificates may also be required to ship produce between regions within a country. Authority to conduct phytosanitary inspection and issue certificates is sometimes given to equivalent authorities employed by the shipping country, rather than having personnel from the receiving country present.

2. Visual inspection. Shipments are visually inspected using protocols specifically developed to find target pests known for a commodity. They are much more likely to detect the presence of a pest than are the random inspections used for the general public.

3. Specific testing. Some pests, such as most pathogens, are difficult to detect. Specific techniques have been developed for some, such as indexing for some viruses, but these have the drawback that they typically require a long time and special facilities. Molecular techniques may make detection of some of these organisms much easier in the future.

**Phytosanitary certificates show that produce being shipped is free of pest organisms.**

# Enforcement Actions

If a pest is detected, or if potential host material is discovered, several possible actions can occur. The public must be aware of the potential dangers of transporting plants from one area to another, and the traveling public must realize that transport of pests or their host material is breaking the law.

***Confiscation***    If inspection of vehicles or personal luggage leads to discovery of plant and animal products entering the country (or region in some cases, e.g., California) such products are automatically confiscated (Figure 10–4) and destroyed. The individual or organization involved may be subject to fine (see Fines). The reality is that because of the sheer volume of traffic, everything cannot be inspected. Confiscation typically occurs at ports of entry or at regional border stations.

***Impound shipment/deny entry***    This action is the equivalent of confiscation at the level of commercial shipments that could potentially be carrying pest organisms. The shipment is seized and held until it is deemed pest free, or it is returned to its country of origin.

***Certification programs***    A phytosanitary certificate can be required before a shipment is permitted entry into a country. The phytosanitary certificate travels with the shipment; entry is denied in the absence of the required certificate.

***Quarantines***    It may be necessary to sequester the produce or host organism in a special containment facility until it is deemed pest free. This approach is used for import of pet animals, and for commercial introduction of new plant material. The United States has several quarantine facilities located at major ports of entry around the country. This technique is not applicable to fresh produce that has a short shelf life.

***Fines***    Large fines can be levied against individuals or organizations for willful intent to import a potential pest or its likely host material.

***Treatment***    For some pests it may be possible to apply control tactics to the product before or during shipment. The tactics can be physical or chemical. Physical processes include the use of heat or cold to kill pest organisms. Radiation has been considered in the past but public concerns have withheld its widespread use, despite the fact that it may be one of the safest ways of eliminating potentially hazardous organisms from foodstuff. Among the chemical processes, fumigation with methyl bromide has been used for several decades, but is being phased out (see Chapter 19).

If pests are present in a shipment, and there is no tactic available to eliminate the pests, then the shipment will be rejected by the receiving country. Vessels traveling between countries can be fumigated as a pest control measure; Australia has required fumigation of aircraft prior to arrival.

***Embargo***    This is a blanket form of enforcement in which any form of suspected soil or plant material cannot be imported because it might harbor a pest. Certain types of products thus cannot be shipped into a country. Embargoes can also be enacted between different regions in a country.

# EARLY DETECTION

Unfortunately, exclusion programs are not 100% efficient, and new pests do become established. The reasons are the same as those outlined earlier in this chapter under invasion mechanisms. Part of a pest regulatory program is to detect such newly introduced pests before they become widespread, and also to assess the damage potential of the pest. The lag time between introduction and detection is, however, one of the major problems in dealing with exotic pest organisms. It has been estimated that many non-indigenous weed species may have been in the United States for 30 years or spread to 10,000 acres before they were detected. The lag has been estimated at 25 years for some nematode pests.

Early detection is the key to stopping establishment of an exotic pest, but is difficult. Federal government and state agencies have biologists trained to look for invading exotic species. These biologists use appropriate traps and other monitoring devices (see Chapter 8). A limitation of this type of monitoring is that it is targeted at specific pests that are deemed especially important. Introduction of a pest that is not targeted may go undetected for considerable time. The extent to which effective monitoring can be implemented is limited, and government agencies have to rely on interested and knowledgeable members of the public, including farmers, conservationists, professional pest control advisors, and the public at large to aid in exotic pest detection.

# CONTAINMENT, CONTROL, OR ERADICATION

## Containment

Once a new pest is detected in an area, management decisions must be made. If it is not feasible to eradicate the infestation, containment measures may be used to halt the further spread to noninfested areas. Typically, the responsibility of maintaining such a program is jointly shared between government and the potentially affected parties.

The potato cyst nematode (sometimes called the golden nematode because of its color) in the United States is a good example of containment action. The nematodes, species *Globodera,* were transported from Europe and became epidemic in the Long Island potato growing area in New York State by the mid-twentieth century. Although unpopular with local growers, a strict quarantine was established that denied shipment of potatoes outside the infested region. A court challenge was denied on the grounds that the action was reasonable and necessary. The U.S. Congress in 1948 passed the Golden Nematode Act, which established control action in the infested area. The imposition of quarantine and the controls is credited with stopping the spread of this nematode to other potato growing regions in the United States.

## Control

Control is the major thrust of the following chapters, and is the appropriate action when the decision has been made that society can or will have to live with the new pest. What this means is that the economic, social, or environmental costs of other alternatives are too high. The onus of management at the control level shifts from the public sector to the private sector. This situation is typical for most pests that are well established and constitutes the majority of practical day-to-day IPM.

## Eradication

Once an introduction has occurred, and the organism has been determined to be potentially very damaging, measures are taken to eliminate all pest individuals while the infested area is still relatively small.

Once issued, an eradication order involves certain legal ramifications. Agencies (government or private) can be required to control the pest, and federal and local government funds may become available to help offset costs. If pest control measures are not implemented (lack of compliance), substantial fines and/or jail time may be levied, and control action implemented at the owner's expense and over the owner's objections. In this situation, the public good overrides the rights of individuals, and may create problems because the concept of a pest is anthropocentric and thus not uniformly accepted by all segments of society. In addition, not all segments of society are equally accepting of particular pest management tactics (e.g., some people are opposed to any pesticide use).

Eradication may not be possible for some types of pests. It is essentially impossible to eradicate soil-borne pathogens and nematodes once they have become established within an area, in part because of the cryptic nature of the pests. By the time the pests have been noticed they have likely spread and established themselves over relatively wide areas. An additional reason why eradication is difficult is the problem of delivering a control tactic effectively to the entire soil profile.

Eradication is not undertaken lightly. The success of such a program is never certain, and to a great extent depends on the biology of the pest involved. Eradication programs may be financially costly because of control efforts and crop destruction orders, such as occurred with citrus canker (see below). Eradication programs are also potentially disturbing to the general public, as the need for control is often poorly understood and control tactics may involve large areas.

Introductions often occur in residential areas, because ports of entry are usually nearby, or uninformed members of the public have transported illegal plant material there. This means that such residential areas are of necessity subjected to sprays or other forms of treatment if the eradication program is to be effective. However, if pest establishment results in unacceptable losses (such as elimination of a crop from a region, or the need for routine pesticide use when none was required before the introduction), then eradication may be justified. In some instances, however, decisions to eradicate were made many years after the pest was established. There is an extensive literature on the debate about the wisdom of eradication in these cases. The boll weevil eradication program in the southeastern United States and the fire ant eradication program in the South are examples.

Following are examples of eradication programs that have been undertaken in the United States.

1. Citrus canker is a bacterial disease of citrus that has the potential to eliminate the industry. Citrus canker was introduced into Florida in the early 1900s. The disease was considered eradicated in 1947 after

destruction of 1.5 million planted citrus trees, 1.5 million alternative hosts, and 6.6 million nursery trees, plus enormous monetary cost. It is the only disease considered to have been eradicated. The disease was found again in Florida in 1984 but the new strain was not as serious as that causing the earlier epidemic.

2. Mediterranean fruit fly larvae cause serious damage to all soft fruits (Figure 10–3b). If this insect were allowed to establish in major fruit growing regions such as California and Florida, its impact would be enormous. Numerous introductions of the med fly into these states have occurred, resulting in eradication actions being taken several times. Some of these have not been popular with the general public, such as when it became necessary to treat large areas of Los Angeles and San Jose in California with aerial applications of malathion mixed with bait.

3. Hydrilla is a submerged aquatic weed that has the capability to completely block irrigation canals, marinas, and waterways. It is a serious threat to paddy rice production. It has been introduced into several states repeatedly. Due to its potential to block irrigation canals and to reduce rice production, eradication was carried out following each introduction.

4. Witchweed is a parasitic flowering plant that attaches to the roots of corn and other grass crops. A local infestation was discovered in the Carolinas in the USA. A federal eradication effort there was considered successful.

5. Boll weevil is a potentially devastating insect that attacks cotton. It is currently under an eradication program in the Southern United States. Several states are now considered boll weevil free, which realizes tremendous savings for cotton growers in those regions. Establishment of the program was, however, controversial on ecological grounds.

Unsuccessful eradication efforts in the United States include attempts to control chestnut blight, Dutch elm disease, and the imported fire ant.

## SUMMARY

People dealing with plants, animals, or agriculture need to be aware of the problems that plant pests can cause, and that quarantines to prevent unwanted introductions of plants and animals must be respected and complied with. According to an Office of Technology Assessment report (United States Congress, 1993), most persons involved in animal agriculture support pest exclusion laws and regulations. Few animal breeders would consider importing an animal without observing the mandated quarantine procedures. Unfortunately, some people dealing with plants, particularly ornamental plants, do not observe quarantine regulations applicable to plants or plant parts. The ease of international travel and the great numbers of people traveling, combine with the rapid expansion of international trade to increase the risk of exotic pest invasions. In addition, as Brasier (2001) points out, despite the enormous long-term costs of invasions, openended market forces do not operate in support of effective quarantines.

The importance of stopping pest invasion and spread cannot be overemphasized. The costs to society and the environment include lost production, contaminated products, increased use of control tactics such as pesticides, and dangers to native species due to loss of habitat through competition with exotic species. Horsfall and Cowling (1977), in referring to the ways in which human beings have transported pathogens from one place to another, said:

> 'Man, in his incredible ignorance, displays astonishing ingenuity in messing up the great ecological system of which he is inescapably a part.'

Their sentiment bears repeating as a fitting end to this chapter. Unless the members of society are willing to cooperate in this area of pest management, it seems likely that plant pests will continue to be spread more widely.

## SOURCES AND RECOMMENDED READING

Several books capture the problems of pest invasion and introduction, and the various legal barriers that humans have erected to slow or stop invasions. The magnitude of the problem has been documented extensively (McGregor et al., 1973; Wilson and Graham, 1983; CAST, 1987; U.S. Congress, 1993; Pimentel et al., 2000; Mack and Lonsdale, 2001). The analysis by Pimentel et al. (2000) is particularly useful for the North American situation; see Mooney and Hobbs (2000) for a general discussion of invasive species in several other countries, and Wingfield et al. (2001) for pathogens in tropical forests and the Southern Hemisphere. Three volumes edited by Khan (1989) on arthropods and pathogens provide in-depth coverage of regulatory issues. Assessment and management of plant invasions is the focus of a book edited by Luken and Thieret (1997) and a review by the Council for Agricultural Science and Technology (Mullin et al., 2000). The role of horticulture in the introduction of invasive plant species is the focus of a review by Reichard and White (2001). The significance of seed quarantines in pest management is the subject of an FAO publication (Mathur and Manandhar, 1993) and for seed-borne pathogens by McGee (1997). A useful review of the actual control tactics used for quarantine has been prepared by Sharp and Hallman (1994), and Dahlsten and Garcia (1989) review eradication and provide examples. Campbell (2001) provides a useful review of pest risk analysis in relation to changing trade regulations. Various websites provide useful information in determining current regulatory aspects of pest invasion management (Anonymous, 2000).

APHIS. 2000. WWW URLs for plant pest introduction laws and regulations. http://www.aphis.usda.gov/ppq/

Brasier, C. M. 2001. Rapid evolution of introduced plant pathogens via interspecific hybridization. *Bioscience* 51:123–133.

Campbell, F. T. 2001. The science of risk assessment for phytosanitary regulation and the impact of changing trade regulations. *Bioscience* 51:148–153.

CAST. 1987. *Pests of plants and animals: Their introduction and spread.* Ames, Ia.: Council for Agricultural Science and Technology.

Dahlsten, D. L., and R. Garcia, eds. 1989. *Eradication of exotic pests: Analysis with case histories.* New Haven, Conn.: Yale University Press, vi, 296.

Dewey, S. A., M. J. Jenkins, and R. C. Tonioli. 1995. Wildfire suppression—A paradigm for noxious weed management. *Weed Technol.* 9:621–627.

Horsfall, J. G., and E. B. Cowling. 1977. Some epidemics man has known. In J. G. Horsfall and E. B. Cowling, eds., *Plant disease: An advanced treatise.* New York: Academic Press, 17–32.

Kahn, R. P. 1989. *Plant protection and quarantine.* Boca Raton, Fla.: CRC Press, v, 3.

Luken, J. O., and J. W. Thieret. 1997. *Assessment and management of plant invasions.* New York: Springer, xiv, 324.

Mack, R. N., and W. M. Lonsdale. 2001. Humans as global plant dispersers: Getting more than we bargained for. *Bioscience* 51:95–102.

Mathur, S. B., and H. K. Manandhar. 1993. Quarantine for seed. FAO plant production and protection paper 119. *Proceedings of the Workshop on Quarantine for Seed in the Near East.* International Center for Agricultural Research in the Dry Areas (ICARDA), Aleppo, Syrian Arab Republic, 2 to 9 November 1991; Rome, Food and Agriculture Organization of the United Nations, 296.

McGee, D. C., ed. 1997. *Plant pathogens and the worldwide movement of seeds.* St. Paul, Minn.: APS Press, iii, 109.

McGregor, R. C., F. J. Mulhern, and Import Inspection Task Force. 1973. *The emigrant pests.* A report to Dr. Francis J. Mulhern, administrator, Animal and Plant Health Inspection Service, v, 167.

Mooney, H. A., and R. J. Hobbs. 2000. *Invasive species in a changing world.* Washington, D.C.: Island Press, xv, 457.

Mullin, B. H., L. W. J. Anderson, J. T. DiTomaso, R. M. Eplee, and K. D. Getsinger. 2000. *Invasive plant species.* Ames, Ia.: Council for Agricultural Science and Technology.

Naylor, R. L. 2000. The economics of alien species invasions. In H. A. Mooney and R. J. Hobbs, eds., *Invasive species in a changing world.* Washington, D.C.: Island Press, 241–259.

Pimentel, D., L. Lach, R. Zuniga, and D. Morrison. 2000. Environmental and economic costs of nonindigenous species in the United States. *Bioscience* 50:53–65.

Reichard, S. H., and P. White. 2001. Horticulture as a pathway of invasive plant introductions in the United States. *Bioscience* 51:103–113.

Sharp, J. L., and G. J. Hallman, eds. 1994. *Quarantine treatments for pests of food plants. Studies in insect biology.* Boulder and New Delhi: Westview Press, Oxford & IBH Publishers, ix, 290.

Stein, B. A., and S. R. Flack, eds. 1996. *America's least wanted: Alien species invasion in U.S. ecosystems.* Arlington, Va.: The Nature Conservancy.

United States Congress, Office of Technology Assessment. 1993. *Harmful nonindigenous species in the United States.* OTA-F-565. Washington, D.C.: U.S. Government Printing Office, viii, 391.

Williams, M. C. 1980. Purposely introduced plants that have become noxious or poisonous weeds. *Weed Sci.* 28:300–305.

Wilson, C. L., and C. L. Graham, eds. 1983. *Exotic plant pests and North American agriculture.* New York: Academic Press, xvi, 522.

Wingfield, M. J., B. Slippers, J. Roux, and B. D. Wingfield. 2001. Worldwide movement of exotic forest fungi, especially in the tropics and the Southern Hemisphere. *Bioscience* 51:134–140.

# *11*

# PESTICIDES

## CHAPTER OUTLINE

► Introduction
► Types of pesticides
► Historical aspects
► Pesticide discovery process
► Chemical characteristics
► Application technology
► Environmental considerations
► Interactions between pesticides
► Toxicity of pesticides
► Legal aspects of pesticide use
► Consumer protection

## INTRODUCTION

Pesticides are a recent development in agriculture. A few inorganic chemicals, such as sulfur and arsenicals, have been used for many centuries, but synthetic organic chemicals have only been used as pesticides since the middle of the twentieth century. Prior to that time, pathogens, nematodes, and arthropod pests were controlled by cultural means, selection of resistant varieties, biological control, and use of the few inorganic pesticides that were available. The chemicals developed since the 1940s have provided control that could not be achieved with other methods, in most instances, and more importantly, they provided control where none was previously available. Because of the efficacy and ease of application of such pesticides, they were quickly and widely adopted.

Weed control has been somewhat different from arthropod, nematode, and pathogen control. Because of the consistently devastating effect of weeds on yield, the importance of weed control has always been recognized. Prior to the development of selective organic chemical herbicides, most in-row weed control was achieved by human labor. The availability of selective chemical herbicides dramatically decreased the human labor necessary for weed control, which is probably a principal reason for the widespread adoption of herbicides for this purpose.

Some assessments of pesticide use might lead a reader to conclude that pests can be managed without the use of pesticides. Theoretically, the use of pesticides for pathogen and arthropod management could be abandoned if the resultant increased losses and the damage to harvested products were acceptable. The rapid growth of the world population (Figure 1–3) however, makes this unacceptable. Western, industrialized societies cannot abandon herbicides, because uncontrolled weeds would devastate crop production, and there is insufficient human labor available to control weeds in the absence of herbicides. In the United States and much of Western Europe, less than 2% of the population is involved in agriculture (see Figure 1–1). It is unlikely, except under conditions of catastrophic food shortages, that people in industrialized countries would be willing to hoe weeds for the extended time that would be necessary to support large-scale agriculture. Given the needs for large-scale food and fiber production, pests will likely continue to be managed with the judicious use of pesticides as a component tactic of IPM.

## Advantages of Pesticides

Why have pesticides remained as important components of many IPM systems? There are several reasons, although they do not apply equally to all pesticide categories. Some of the reasons are real, some are perceived.

1. Pesticides provide control of some pests where no other effective tactics were previously available.
2. Pesticides may be inexpensive when compared with the alternative management tactics, especially many herbicides, for which the alternative is hand labor. In California sugar beet production, a cost analysis by Norris (1995) revealed that $50 to $100 per acre of herbicide can provide weed control that is superior to that obtained from $400 to $700 per acre of hand hoeing. Economic cost-benefit analyses are necessary to properly evaluate the actual benefits arising from pesticide use in IPM.
3. Many of the economic benefits derived from the use of pesticides are attributed to increases in yield that have occurred through their use; the improvements in yield since the 1950s in the United States, shown in Figure 1–2, are in part due to the use of pesticides, in conjunction with use of improved varieties and various agronomic practices.
4. Pesticide use may require less energy than other management tactics to achieve a given level of control, especially for herbicides, where the alternatives include tillage that requires fuel for tractors. For insecticides and fungicides, however, plant resistance and biological controls are far more efficient in terms of energy costs, especially if the energy used in the production and transport of the fungicides and insecticides is considered.
5. Pesticides may require less knowledge of pest biology and agroecosystem processes than do alternative tactics.
6. Pesticides often provide rapid remedial action against the target pest, and allow rapid control of an existing pest problem. This, combined with item 3, has been a major reason for the acceptance and use of pesticides.
7. Pesticides decrease the amount of planning required by the pest manager by providing the ability to control pests when they occur or when they reach the economic injury level. Avoiding pest problems requires planning and consideration of unpredictable future events.

8. When properly used under appropriate conditions, pesticides typically provide a relatively predictable level of control. There is often greater uncertainty associated with the use of other tactics.

9. Pesticides allow for greater control over cropping sequences, because pesticides decrease the necessity to rotate crops for pest management reasons. This allows crops with greater economic value to be grown more frequently, resulting in increased profits over time. This is especially true with herbicides used for in-row weed control in row crops.

10. Some pesticides have permitted the development of new cultural practices. In the absence of herbicides most no- or reduced-tillage systems, for example, would not be feasible.

11. Use of pesticides can permit agricultural production in regions where such production would not be feasible without pesticides.

12. Pesticides can reduce the incidence of toxins in food that result from contamination of the food by rotting organisms.

## Disadvantages of Pesticides

Many problems arise with pesticide use and misuse. The magnitude and importance of these problems vary considerably among pesticide categories. As with the advantages, some problems are real but others are only perceived.

1. Effects on nontarget pests. Pesticides may have nontarget effects, which occur at two levels:
   1.1. Within the agroecosystem. Pesticides may harm useful organisms, such as honey bees, or beneficial insects that are critical in the biological control of pest populations. Some pesticides are toxic to wildlife.
   1.2. External to the agroecosystem. Pesticides may move from the place where they were applied, resulting in contamination of surface water or groundwater, and pesticide accumulation in the food chain.

2. Cost of pesticides. Although low cost may be a reason for the use of pesticides, in other situations pesticides are an expensive choice; this is especially true for management of certain insects where biological control is a feasible alternative. In many developing countries, the cost of pesticides is prohibitive. Often cheap labor offers alternatives to pesticides, as, for example, in hand weeding or in hand picking worms on crop plants. Again, economic cost-benefit analyses are essential to assess the value of pesticide use in IPM.

3. Residues and drift. Residues remain in the soil, and on or in harvested produce after the application of a pesticide. Residues may be of particular concern if pesticides are applied incorrectly. Drift occurs when pesticides are applied during unfavorable weather conditions. Wind may carry pesticides to areas adjacent to the crop fields, resulting in damage to neighboring plants and animals.

4. Food contamination. The possibility exists that pesticide residues in food lead to long-term, adverse health consequences for consumers. Residues are of particular concern on foods for infants, as tolerances may be quite different from those for adults. When tests have been done, pesticide residues are typically either not detectable or are within established tolerances.

5. Toxicity. Because pesticides are toxic chemicals, they have the potential to be toxic to humans, livestock, and wildlife. The regulations governing

the production and application of pesticides (see later in this chapter) are designed to minimize such risks.

6. Hazard to farm workers. Due to their toxicity, pesticides have the potential to cause illness in farm workers, especially for those working in hand-harvested fresh market crops. Because these must be free of cosmetic damage, they require pesticide application close to harvest.

7. Create pest problems. Several potential pest problems can develop with repeated use of pesticides:

   7.1. Pesticide resistance. The repeated use of a single pesticide may lead to the selection of pests that are resistant to the pesticide. This is an enormous problem for use of all types of pesticides, but has been particularly important with insecticides, herbicides, fungicides, and bactericides (antibiotics). We devote Chapter 12 to the discussion of resistance and resistance management.

   7.2. Pest resurgence. When the pesticide (usually an insecticide) kills the target pest, but also kills beneficial insects, the pest population often will increase to a level higher than the level preceeding the application. The phenomenon is called pest resurgence. When the survivors of the target population begin to reproduce, their numbers grow exponentially because the beneficials that once limited population growth are no longer present. Pest resurgence often leads to repeated pesticide applications (see item 8).

   7.3. Secondary pest outbreaks. When the pesticide kills the key pest but not a minor (secondary) pest, the minor pest population may increase and become important. This is a problem for both insecticide and herbicide use. Secondary weed outbreaks are also referred to as replacement, because the minor weed replaces the major weed that was the original target.

8. Pesticide treadmill. Misuse of pesticides may lead to more frequent applications and require higher rates of the product needed to control the same pest. This phenomenon has been called the pesticide treadmill and it is the result of severe ecosystem disruption (see items 7.2 and 7.3). The pesticide treadmill (see Chapter 12) has been of particular concern in the excessive use of insecticides.

This chapter reviews synthetic chemical pesticide use in agroecosystems and describes the contributions of pesticides to effective IPM. Pesticides must be used within the IPM philosophy of acceptable environmental, economic, and social consequences.

## Current Use

Although pesticides are widely used, accurate figures on quantities are not readily available. The United States Environmental Protection Agency (EPA) provides data on pesticide use in the United States. The following data are from the 1997 EPA estimates of pesticide use (Aspelin and Grube, 1999).

1. World pesticide production in 1997 was estimated at 5.7 billion pounds, of which 40% were herbicides, 26% were insecticides, 9% were fungicides, and 25% were other pesticides (fumigants, predacides, molluscicides, etc.) (Figure 11–1).

**FIGURE 11–1**
Estimated world
pesticide production
in 1997, shown
according to major
classes of pesticides.
Percentage values
associated with each
bar are the USA
component expressed
as a percentage of the
world total.
Source: data by Aspelin
and Grube, 1999.

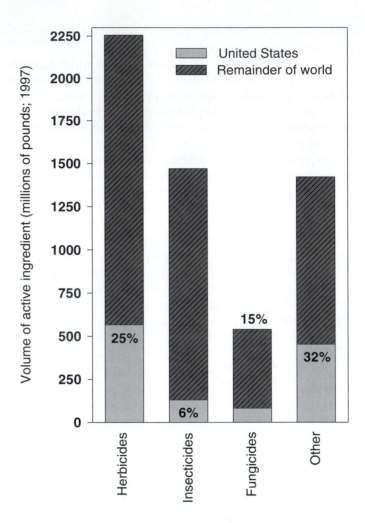

2. Pesticide use in the United States peaked in the early 1980s at about 1.12
   billion pounds, and declined to just under 1 billion pounds in the mid-
   1990s (Figure 11–2). The decline was due primarily to decreased
   nonagricultural uses. Due to increased prices, the dollar value of the
   pesticides that are produced has continued to increase. During the 1980s
   and 1990s, herbicide and fungicide use remained almost constant (Figure
   11–3), insecticide use and other chemical use declined, and use of
   fumigants and other conventional pesticides increased.
3. In 1997, there were about 890 different pesticide chemical structures (not
   products) in use, which was a decrease from a high of about 1,200 in the
   mid-1980s.
4. Approximately 90% of all herbicides and insecticides used in the United
   States are applied on corn, soybeans, cotton, and small grains (including
   sorghum), but agricultural pesticides represent only a small proportion of
   the total pesticide use in the United States (Figure 11–4).
5. Herbicides represent the largest quantity of conventional pesticide use in the
   USA, followed by fumigants, insecticides, and fungicides in that order
   (Figure 11–5). Agriculture accounts for over 80% of all conventional
   pesticide use. The home and garden sector is not permitted to use fumigants.

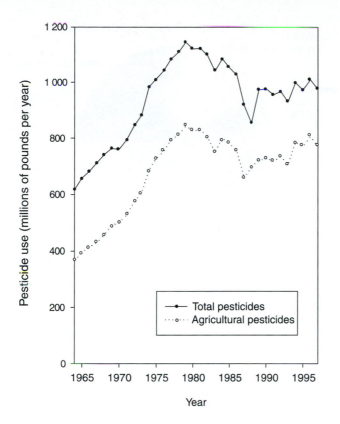

**FIGURE 11–2**
Estimated use of conventional pesticides in the United States from 1964 to 1997. Total pesticides includes both the agricultural and nonagricultural sectors (industrial/commercial/government and home/garden).
Source: data by Aspelin and Grube, 1999.

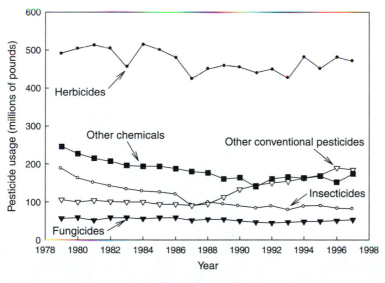

**FIGURE 11–3**
Changes in use of major classes of conventional pesticides, including other chemicals such as growth regulators and desiccants, in the United States between 1978 and 1997.
Source: data by Aspelin and Grube, 1999.

6. In 1997, 16 of the 25 most widely sold agricultural pesticides in the United States were herbicides (top 2), 4 were biocides that are toxic to all pests, 3 were fungicides, and 2 were insecticides.

7. Sodium hypochlorite (chlorine bleach) is used to kill microorganisms in drinking water and swimming pools, and as a disinfectant in hospitals, restaurants, and other public places. It is considered a pesticide and accounts for a large percentage of nonagricultural pesticide use (Figure 11–4).

**FIGURE 11–4**
Proportions of
different types of
pesticides sold in the
United States in 1997;
figures in parentheses
are millions of pounds
per year.
Source: data by Aspelin
and Grube, 1999.

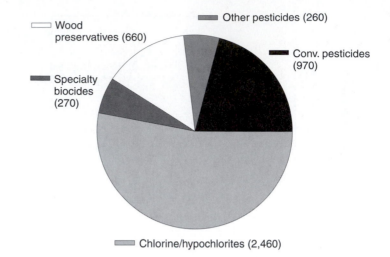

**FIGURE 11–5**
Comparison of the
use of major classes of
conventional
pesticides in the
United States in 1997
according to market
type; percentage
values indicate the
contribution of each
major category of
pesticides to the total.
Source: data by Aspelin
and Grube, 1999.

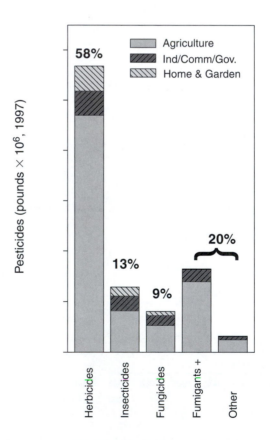

The state of California has accurate data on pesticide use, because the California Environmental Protection Agency (CalEPA) instituted a requirement in 1992 that all commercial pesticide applications must be reported. The data for California pesticide use, up to 1999, are now available on the World Wide Web (California EPA, 2001).

# TYPES OF PESTICIDES

Some may consider a pesticide as a chemical that kills insects. Much like the term *pest* discussed in Chapter 1, this is erroneous, because there are different pesticides that target each pest category. Although one of the roots of the word *pesticide* is *-icide,* which means "to kill," not all pesticides are lethal to their target pests. Some pesticides simply incapacitate the target pest, thereby preventing it from causing damage. Generally, pesticide is used to describe applied substances that have some level of toxicity to organisms defined as pests. Three different origins of pesticides are recognized.

1. Inorganic chemicals. These are pesticides derived from chemical elements other than carbon.
2. Synthetic organic chemicals. This term is used in the chemical sense, and these are carbon-containing compounds typically synthesized and derived from petroleum-based chemicals.
3. Biopesticides. These are pesticides that have a biological origin. The term may refer to chemicals produced by organisms (such as antibiotics or pheromones) or an organism itself (such as a bacterial suspension).

The definition of pesticide does not imply differences due to the original source of the pesticide, or the mode of action of the pesticide. The definition is typically expanded to include chemicals that modify pest growth or behavior, or regulate plant growth. Pesticides therefore include all the chemicals included in the accompanying lists.

**FIFRA defines pesticide as "any substance or mixture of substances, intended for preventing, destroying, or mitigating any pest, or intended for use as a plant growth regulator, defoliant or desiccant."**

| Main Types of Pesticides | |
| --- | --- |
| **Pesticide type** | **Organism(s) affected** |
| Fungicide | Fungi |
| Bactericide | Bacteria |
| Antibiotic | Bacteria |
| Herbicide | Plants (weeds) |
| Silvicides | Trees |
| Mycoherbicide | Plants |
| Algaecide | Algae |
| Slimicide | Slime-forming organisms |
| Nematicide | Nematodes |
| Molluscicide | Slugs & snails |
| Insecticide | Insects |
| Adulticide | Adults |
| Larvaecide | Larvae |
| Ovicide | Eggs |
| Aphicide | Aphids |
| Acaricide | Spiders |
| Miticide | Spider mites |
| Predacide | Vertebrates |
| Rodenticide | Rodents |
| Avicide | Birds |
| Piscicide | Fish |

| Additional Pesticides and Related Chemicals | |
| --- | --- |
| Pesticide chemical | Type of impact |
| Disinfectant (mainly chlorine) | Microorganisms |
| Wood preservatives | Wood rotting organisms |
| Repellents | Keep animal pests away |
| Attractants | Attract animal pests |
| Growth regulators | Modify crop/pest growth |
| Desiccants | Dehydrate foliage |
| Defoliants | Cause plants to shed leaves |
| Adjuvants | Enhance spray characteristics |
| Synergists | Enhance toxic action of pesticide |

# HISTORICAL ASPECTS

The following is a summary history of pesticide development. An extensive list of dates of pesticide-related information is provided in Table 11–1.

## Inorganic Chemicals

Historically, inorganic compounds were some of the earliest chemicals used for pest control. Sulfur was used for fumigating houses in 1000 B.C.E., and arsenic has been used for insect control since about C.E. 900 in China. Lead arsenate was one of the first insecticides that provided reliable control starting in the mid-nineteenth century. About 1880, copper and lime mixtures were introduced for control of mildew on grapes. Borax was used as an insecticide and a herbicide starting in the late nineteenth century. Chlorates were introduced early in the twentieth century for nonselective weed control, but chlorates have the undesirable property of turning dead plants into explosive fire hazards, so they are rarely used now. Pesticides based on mercury compounds were introduced as fungicides early in the twentieth century, but their use has now been discontinued.

A major drawback to using inorganic pesticides is that they are based on chemical elements that do not break down. Thus, repeated use leads to accumulation of the element in the soil. Compounding this is the fact that they were once typically applied at relatively high rates. Some elements (e.g., lead, arsenic, and mercury) are very toxic to a range of organisms, including humans. Although many pesticides based on copper and to a lesser extent zinc are still used, the use of most inorganic pesticides decreased dramatically once the synthetic organic pesticides became available. Many inorganic pesticides were phased out as the environmental/toxicological problems inherent in their use were discovered.

## TABLE 11–1    Historical Record of Events Relating to the Use of Pesticides

### B.C.E.

| | |
|---|---|
| 1200 | Biblical armies salt and ash the fields of the conquered; first reported use of nonselective herbicides. |
| 1000 | Homer refers to sulfur used in fumigation and other forms of pest control. |
| 100 | The Romans apply hellebore for the control of rats, mice, and insects. |
| 25 | Virgil reports seed treatment with "nitre and amurca." |

### C.E.

| | |
|---|---|
| 70 | Pliny the Elder reports pest control practices from Greek literature of the preceding three centuries; most practices based on folklore and superstition. |
| 900 | Chinese use arsenic to control garden insects. |
| 1300 | Marco Polo writes of mineral oil being used against mange of camels. |
| 1649 | Rotenone used to paralyze fish in South America. |
| 1669 | Earliest mention of arsenic as insecticide in Western world, used with honey as ant bait. |
| 1690 | Tobacco extracts used as contact insecticide. |
| 1773 | Nicotine fumigation by heating tobacco and blowing smoke on infested plants. |
| 1787 | Soap mentioned as insecticide. |
| | Turpentine emulsion recommended to repel and kill insects. |
| 1800 | Persian louse powder (pyrethrum) known to the Caucasus. Sprays of lime and sulfur recommended in insect control. Whale oil prescribed as scalecide. |
| 1810 | Dip containing arsenic suggested for sheep scab control. |
| 1820 | Fish oil advocated as insecticide. |
| 1821 | Sulfur reported as fungicide for mildew by John Robertson in England. |
| 1822 | Mixture of mercuric chloride and alcohol recommended for bedbug control. |
| 1825 | Quassia used as insecticide in fly baits. |
| 1842 | Whale-oil soap mentioned as insecticide. |
| 1845 | Phosphorus paste declared as official rodenticide for rats by Prussia; by 1859, it was used in cockroach control. |
| 1848 | Derris (rotenone) reported being used in insect control in Asia. |
| 1851 | Boiled lime-sulfur employed at Versailles by Grison. |
| 1854 | Carbon disulfide tested experimentally as grain fumigant. |
| 1858 | Pyrethrum first used in the United States. |
| 1860 | Mercuric chloride solutions applied to destroy soil-inhabiting forms such as earthworms. |
| 1867 | Paris green used as an insecticide. |
| 1868 | Kerosene emulsions employed as dormant sprays for deciduous fruit trees. |
| 1877 | Hydrogen cyanide (HCN) first used as fumigant, to fumigate museum cases. |
| 1878 | London purple reported as a substitute for Paris green (both are arsenicals). |
| 1880 | Lime-sulfur used in California against San Jose scale. |
| 1882 | Naphthalene cakes used to protect insect collections. |
| 1883 | Millardet discovers the value of Bordeaux mixture in France. |
| 1886 | Hydrogen cyanide used for citrus tree fumigation in California, USA. |
| | Resin fish-oil soap used as scalecide in California. |
| 1890 | Carbolineum, a coal-tar fraction, used in Germany on dormant fruit trees. |
| 1892 | Lead arsenate first prepared and used to control gypsy moth in Massachusetts, USA. First use of a dinitrophenol compound, the potassium salt of 4-6-dinitro-o-cresol, as insecticide. |
| 1896 | Copper sulfate used selectively to kill weeds in grain fields. |
| | British patent refers to inorganic fluorine compounds as insecticides. |
| 1897 | Oil of citronella used as a mosquito repellent. |
| 1902 | The value of lime-sulfur as apple scab control discovered in New York, USA. |
| 1906 | Passage of Federal Food, Drug, and Cosmetic Act (Pure Food Law). Lubricating oil emulsions first applied to citrus trees. |
| 1907 | Calcium arsenate in experimental use as an insecticide. |
| 1909 | First tests with 40% nicotine sulfate made in Colorado, USA. |
| 1910 | Passage of U.S.A. Federal Insecticide Act. |
| 1911 | First publication of the use, outside the Orient, of derris as an insecticide, in British patents. |

*(continued)*

251

C.E. (continued)

| | |
|---|---|
| 1912 | Zinc arsenite first recommended as insecticide. |
| | p-Dichlorobenzene applied in the United States as a moth fumigant on clothes. |
| 1917 | Nicotine sulfate first used in a dry carrier for dusting. |
| 1921 | Airplane first used for spreading insecticide dust for catalpa sphinx at Troy, Ohio, USA. |
| 1922 | Calcium cyanide begins commercial use. |
| | First aerial application of an insecticide to cotton, in Tallulah, Louisiana, USA. |
| 1923 | Geraniol discovered to be attractive to the Japanese beetle. |
| 1924 | Cubb (derris) first tested as insecticide in the United States. |
| | First tests of cryolite against Mexican bean beetle. |
| 1925 | Selenium compounds tested as insecticides. |
| 1927 | Tolerance established for arsenic on apples by United States FDA. |
| | Ethylene dichloride discovered to be useful as a fumigant. |
| 1928 | Pyrethrum culture introduced into Kenya. |
| | Ethylene oxide patented as insect fumigant. |
| 1929 | Alkyl phthalates patented as insect repellents. |
| | n-Butyl carbitol thiocyanate produced commercially as a synthetic contact insecticide. |
| | Cryolite introduced as an insecticide. |
| 1930 | First fixed nicotine compound, nicotine tannate, used as a stomach poison. |
| 1931 | Anabasine isolated from plants and synthesized in the laboratory. |
| | Thiram, first organic sulfur fungicide, discovered. |
| 1932 | Methyl bromide first used as a fumigant, in France. |
| | Ethylene and acetylene discovered to promote flowering in pineapples; first plant growth regulators. |
| 1934 | Nicotine-bentonite combination developed; first dependable nicotine dust. |
| 1936 | Pentachlorophenol introduced as wood preservative against fungi and termites. |
| 1938 | TEPP, first organophosphate insecticide, discovered by Gerhardt Schrader. |
| | Pesticide amendment to Pure Food Law of 1906 passed, intended to prevent contamination of food. |
| | *Bacillus thuringiensis* first tested as microbial insecticide. |
| | DNOC, first dinitrophenol herbicide, introduced to United States from France. |
| 1939 | Rutgers 612 introduced, first good insect repellent. |
| | DDT discovered to be insecticidal by Paul Muller in Switzerland. |
| 1940 | Sesame oil patented as synergist for pyrethrin insecticides. |
| 1941 | Hexachlorocyclohexane (BHC) discovered in France to be insecticidal. |
| | Introduction of aerosol insecticides propelled by liquefied gases. |
| 1942 | First batch of DDT shipped to United States for experimental use. |
| | Introduction of 2,4-D, the first of the hormone (or phenoxy) herbicides. |
| 1943 | First dithiocarbamate fungicide, zineb, introduced commercially. |
| 1944 | Introduction of 2,4,5-T for brush and tree control, and warfarin for rodent control. |
| 1945 | Early synthetic herbicide, ammonium sulfamate, introduced for brush control. Chlordane, the first of the persistent, chlorinated cyclodiene insecticides introduced. |
| | The first carbamate herbicide, propham, became available. |
| 1946 | Organophosphate insecticides TEPP and parathion, developed in Germany, made available to USA producers. |
| | First resistance in houseflies to DDT observed in Sweden. |
| 1947 | Toxaphene insecticide introduced; to become the most heavily used insecticide in USA agricultural history. |
| | Passage of the Federal Insecticide, Fungicide, and Rodenticide Act (FIFRA). |
| 1948 | Aldrin and dieldrin first produced; the best of the persistent soil insecticides. |
| 1949 | Captan, the first of the dicarboximide fungicides, goes into use. |
| | Synthesis of first synthetic pyrethroid, allethrin. |
| 1950 | Malathion introduced, probably the safest organophosphate insecticide. |
| | Maneb fungicide introduced. |
| 1951 | First carbamate insecticides introduced: isolan, dimetan, pyramat, pyrolan. |
| 1952 | Fungicidal properties of captan first described. |

| | |
|---|---|
| 1953 | Insecticidal properties of diazinon described in Germany. |
| | Guthion insecticide introduced. |
| 1954 | Passage of Miller Amendment to Food, Drug, and Cosmetic Act; establish tolerances for all pesticides on raw food and feed products. |
| 1956 | Introduction of carbaryl, the first successful carbamate insecticide. |
| 1957 | Gibberellic acid, plant growth regulator, made available to horticulturists. |
| 1958 | Atrazine, first of the triazine herbicides, and paraquat, first of the bipyridylium herbicides, introduced. |
| | Delaney clause added to FFDCA banning carcinogens in food. |
| 1959 | Cranberries embargoed by USA FDA for excessive residues of the herbicide aminotriazole. |
| | FIFRA (1947) amended to include all economic poisons (e.g., desiccants, nematicides) |
| 1960 | Treflan® herbicide becomes available. |
| | *Bacillus thuringiensis* first registered on lettuce and cole crops. |
| 1961 | Introduction of chlorophacinone rodenticide and mancozeb fungicide. |
| 1962 | Publication of *Silent Spring,* by Dr. Rachel Carson. |
| 1963 | Appearance of Shell No-Pest Strip,® slow-release household fumigant. |
| 1964 | Fungicidal properties of thiabendazole described. |
| 1965 | Development of Temik,® first soil-applied insecticide-nematicide. |
| 1966 | Carboxin, the first systemic fungicide, developed. |
| | Methomyl insecticide and chlordimeform acaricide-ovicide introduced. |
| 1967 | Introduction of the second group of systemic fungicides with benomyl. |
| 1968 | Discovery of tetramethrin, resmethrin, and bioresmethrin, synthetic pyrethroids with greater activity than natural pyrethrins. |
| | First publication of resistance in a weed to a herbicide (common groundsel to atrazine). |
| 1969 | Arizona, USA places a moratorium on the use of DDT in agriculture. |
| | USDA adopts policy on pesticides to avoid use of persistent materials when effective, nonresidual methods of control are available. |
| | Publication of the Mrak Report which laid groundwork for concerted environmental protection, resulting in USA Environmental Protection Agency in 1970. |
| 1970 | Formation of U.S. Environmental Protection Agency (EPA), which assumed responsibility for registration of pesticides (instead of USDA). |
| | All registrations suspended for alkyl-mercury compounds as seed treatments. Authority to establish tolerances for pesticides in foods and feeds transferred from FDA to EPA. |
| 1971 | Glyphosate herbicide first introduced. |
| 1972 | Passage of Federal Environmental Pesticide Control Act (FEPCA or FIFRA amended). |
| | Introduction of first microencapsulated insecticide, Penncap M,® methyl parathion. |
| | California initiates licensing of all pest control advisors. |
| 1973 | Development of first photo-stable synthetic pyrethroid, permethrin. |
| | Cancellation of virtually all uses of DDT by the USA EPA. |
| 1974 | First standards set for worker reentry into pesticide treated fields by EPA (e.g., reentry intervals of 24 or 48 hours dependent on dermal toxicity of pesticide). |
| 1975 | Cancellation of all uses of aldrin and dieldrin, except as termiticides. |
| | Registration of first virus for budworm-bollworm control on cotton. |
| | First insect growth regulator (methoprene) registered with EPA. |
| 1976 | Rebuttable Presumption Against Registration (RPAR) issued for strychnine, endrin, Kepone®, 1080, and BHC in USA. |
| | Passage of Toxic Substances Control Act (TSCA) on October 11. |
| | Most pesticidal uses of mercury compounds canceled by USA EPA. |
| 1977 | Use of dibromochloropropane (DBCP) suspended, and all registered uses for Mirex® canceled by USA EPA. |
| 1978 | USA EPA concludes the first full-scale RPAR of a pesticide, chlorobenzilate. Certification training completed for private and commercial applicators, to apply restricted-use pesticides. |
| | Additional amendments to FIFRA designed to improve pesticide registration process. |
| | First list of restricted-use pesticides issued by EPA. |
| | First registration of a pheromone (gossyplure for pink bollworm) for use on cotton. |
| | Publication of *The Pesticide Conspiracy* by Robert Van den Bosch. |

*(continued)*

**C.E. (continued)**

1979    Most uses of 2,4,5-T and silvex suspended by USA EPA.

1980    Through new legislation Congress assumes responsibility for USA EPA oversight.

1982    Passage of the Comprehensive Environmental Response Compensation and Liability Act (CERCLA or "Superfund") for cleanup of toxic wastes, spills, and dumps.

The Delaney clause is reexamined. No carcinogenic food additives can be used in the processing of foods. Pesticides are exempt from this classification.

Any regulatory action on pesticides by EPA must be preceded by a risk/benefit analysis.

1983    USA EPA cancels most uses of ethylene dibromide (EDB).

1984    USA EPA cancels most registrations for endrin.

1985    USA Congress reauthorizes the Federal Endangered Species Act, originally passed in 1973, and amended in 1978, 1979, and 1982.

First registration of azadirachtin as in insecticide, for nonfood use.

1986    Superfund is amended by USA Congress to include Title 111, The Emergency Planning and Community Right-to-Know Act.

OSHA develops its Hazard Communication Standard, requiring employers to provide Material Safety Data Sheets (MSDS) to employees working with or exposed to hazardous substances.

USA EPA cancels all remaining registrations of DBCP.

USA EPA cancels all agricultural uses of toxaphene.

USA EPA suspends all distribution, sales, and uses of dinoseb.

1987    USA EPA attempts to bring FIFRA into compliance with the Endangered Species Act of 1973, regulating 126 pesticides in 135 U.S. counties.

1988    USA Congress passes amendments to FIFRA referred to as "FIFRA lite."

USA EPA postpones implementation schedule for Endangered Species Act.

Chlordane and heptachlor use as termiticides canceled.

USA EPA announces new risk/benefit "negligible risk" policy for residues of carcinogenic pesticides in processed foods.

USA EPA establishes heavy fee rates for new pesticide registrations, additional sites, and label changes.

1989    USA EPA cancels 20,000 products for nonpayment of maintenance fees.

1990    Azadirachtin uses expanded.

1991    USA EPA cancels 4,500 products for nonpayment of 1990 maintenance fees.

USA EPA issues Final Order stating that the law supports a *de minimis* exception to the Delaney clause for pesticide uses that create at most a *de minimis* risk.

USA EPA cancels most uses for ethyl parathion due to human hazard.

USA EPA and pesticide industry agree to increase registration maintenance fees; Congress enacts proposal.

USA Supreme Court rules that local governments are permitted to enact pesticide legislation more restrictive than FIFRA.

Ecogen granted USA patent for Foil,® genetically enhanced *Bacillus thuringiensis* for control of both beetle and lepidopteran larvae.

Mycogen registers first genetically engineered pesticides with USA EPA, MVPG, and M-Trak,® *Bacillus thuringiensis* encapsulated delta endotoxin.

1992    USA EPA cancels 1,500 products for nonpayment of 1991 maintenance fees.

USA EPA establishes new Worker Pesticide Protection Standard.

USA EPA publishes first complete Pesticide Reregistration "Rainbow Report."

California creates state Environmental Protection Agency (CalEPA), in USA.

1993    USA Congress considers H.R. 967, which requires EPA to expedite minor-use registrations, with waiver of some data requirements.

Neemix® granted USA EPA exemption on food.

1994    California initiates 100% use reporting for all commercial pesticide applications.

1995    USA Federal worker protection standards implemented to Occupational Safety and Health Act of 1970; impacts many aspects of pesticide usage.

1996    USA Federal Food Quality and Protection Act (FQPA) enacted.

This table modified from Ware (1994) with updates.

# Synthetic Organic Chemicals

The use of pesticides based on organic chemicals started in the late nineteenth century. Most of these products were initially discovered as petroleum by-products. From 1900 to the 1930s, use of these types of pesticides increased slowly. Starting in the 1940s, in particular during World War II, the efficacy of many of the pesticides was clearly demonstrated, leading to their widespread use. Foremost among those was the discovery of the insecticidal properties of DDT, rapidly followed by the discovery of selective herbicidal properties of the phenoxy compounds. These chemicals provided spectacular control of pests that routinely caused substantial losses and were difficult to control. The single-tactic chemical control of pests was embraced with almost euphoric optimism, and set in motion the whole concept of the silver bullet approach to pest control. The 1947 presidential address made by C. Lyle to the Entomological Society of America probably summed up the feeling at the time:

> The recent progress in the development of new insecticides and insect repellents has not been equaled in all history . . . at no previous time in history have the achievements of entomologists been of such universal value. . . . The entomologist has become a wizard in the eyes of the uninitiated—and indeed some of the achievements seem little short of magic. (Lyle, 1947)

The development of synthetic organic pesticides rapidly increased during the 1950s and 1960s. The euphoria lasted for about 15 years, during which time some began to question this reliance on chemicals. The publication of *Silent Spring,* by Rachel Carson (1962), dramatically highlighted problems and potential environmental impacts of widespread pesticide use. Resistance of insects to the single-tactic chemical approach started to increase rapidly at this time, and resistance in pathogens and weeds also appeared. The development of increasingly sensitive tools for analytical chemistry during the period 1960 to 1980 also exerted a strong influence on pesticide development, use, and regulation, as these techniques provided improved ability to detect pesticides in food and the environment at very low levels. For these and other reasons the initial euphoria was replaced by reevaluation of the role of pesticides in agriculture, and to their utility in IPM programs. This chapter discusses development, use, and regulation of synthetic organic pesticide chemicals.

# Biopesticides

A diverse group of pesticides that are metabolites derived from biological organisms, or even the organisms themselves, are generically known as biotic pesticides, or biopesticides. Biopesticides include entomopathogenic fungi, bacteria, viruses, and nematodes; plant-derived pesticides (botanicals); and insect pheromones (when used to alter behavior; see Chapter 14). We have adopted the terminology used by Hall and Menn (1999) and Copping (1998) for presentation of this group of pesticides.

**Entomopathogens are microorganisms that cause disease in insects.**

***Living systems***   Some living organisms can be packaged in such a way that they may be sprayed, or applied some other way, to reach the target pest. Many viruses are lethal to insects, but cannot be cultured outside living organisms. For example, a virus that infects *Anticarsia gemmatalis,* an important pest of soybean in North and South America, is obtained by collecting diseased larvae, grinding them in water, and spraying the filtrate onto plants to infect other larvae. The virus is now being produced commercially and can be preserved as dry formulations until ready for use. Certain granulosis viruses, and spores of entomopathogenic fungi, also are produced commercially for use as insecticides. The bacillus agent of the milky disease of the Japanese beetle has been in use for many years. Two fungal pathogens were commercialized for weed management in the late 1970s and others are being evaluated. Two groups of nematodes are used as biopesticides for control of various insects. The first publication on the potential of nematodes for use in biological control was in 1932, when the nematode *Steinernema glaseri* was observed to kill Japanese beetles (*Popillia japonica*). Several species of entomopathogenic nematodes are now commercially available, primarily for control of soil-inhabiting insect pests.

**Fermentation products**    Application of *Bacillus thuringiensis* (generally referred to as Bt) as a biological insecticide was probably the first biopesticide of this type, as its use started in 1938. Many different strains of Bt have now been discovered. Bt produces parasporal crystalline inclusions which are toxic to many insects; the parasporal inclusions are formed by various insecticidal crystal proteins. Most Bt insecticides are produced in large bioreactors, and formulations contain the crystal proteins and some viable spores; in some applications the spores are inactivated.

Some organic chemical pesticides have active ingredients that are microbial metabolites. The metabolites are produced on a commercial scale using large-scale fermentation. The development of such chemicals as insecticides has occurred since the early 1980s, and includes products such as abermectin and spinosad. Antibiotics are toxins produced by some microbes to inhibit other microbes. The most familiar are antibacterial, and have been used to kill pathogenic bacteria since the 1940s. Because some antibiotics can be used to manage bacterial pathogens in agroecosystems, they are considered pesticides.

**Botanical pesticides**    Chemicals derived from plants were some of the first-known pesticides. As plants evolved, the selection pressures exerted by animal pests and by pathogens resulted in the evolution of plant chemical defenses that act to inhibit pest attack. Defensive compounds that plants produce at all times, regardless of whether the pest is present, are termed constitutive defenses. In some plants, defensive chemicals are produced only after pest attack, and those chemicals are referred to as inducible defenses. Plants that produce active constitutive chemical defenses may be grown so that the chemicals can be extracted and applied as pesticides.

Hellebore, a plant in the buttercup family, was used by the Romans to control rats, mice, and insects. Pyrethrum, derived from plants in the genus *Pyrethrum,* and rotenone from the *Derris* plant were in widespread use by the mid-nineteenth century. The use of nicotine for insect control was well established by the mid-eighteenth century. Neem—based on the chemical azadirachtin obtained from the neem tree, *Azadirachta indica*—is an effective insecticide. It has been used for centuries in India where the neem tree is native, but is a relatively recent introduction to the western world. Many plants produce various alkaloids, some of which are applied for vertebrate pest control, such as strychnine. Use of these compounds has continued to the present, and many are considered acceptable to organic farming methods.

***Transgenic plant pesticides*** Some crop plants have been genetically engineered, or transformed, to produce compounds that are not normally present in the plant. Such transgenic plants are often referred to as genetically modified organisms (GMOs). Plants capable of producing the Bt endotoxin are a current example, but others are under experimental development. In 1996, corn was the first crop plant transformed with the Bt endotoxin genes to be used in the field. Cotton and potatoes with Bt endotoxin genes have also been introduced since then. In Europe, transgenic crops (GMOs) have not been widely accepted, and there is some societal resistance to the use of such crops in the USA. The role of this type of pesticidal plants is further discussed under host-plant resistance in Chapter 17, and in Chapter 19 under societal restraints.

# PESTICIDE DISCOVERY PROCESS

The first pesticides were accidental discoveries made through observation by astute people. The phenomenon of finding valuable or agreeable things that were not directly sought is called serendipity. It was not until after the discovery of the insecticidal activity of DDT in 1939 and the herbicidal activity of 2,4-D in 1942 that a formalized discovery process for synthetic organic pesticides was initiated. Since the late 1940s the process has taken the following steps.

**Step 1**

Chemists employed by chemical companies obtain chemicals or synthesize numerous novel compounds. This process has been partially automated since the start of the 1990s with computer-driven robots for adding and mixing chemicals by way of a process called combinatorial chemistry, to generate a wide variety of novel chemical structures that can be tested for biological activity.

**Step 2**

The newly synthesized novel chemicals are subjected to a primary screening process to evaluate biological activity. Historically high rates of the chemical were applied to a group of selected target pest organisms, for example, 10 weeds, 10 insects, and 10 pathogens. If any activity was observed, then further testing occurred. Activity against vertebrates and mollusks was pure chance at this stage, as they were not usually evaluated. The pests included in evaluations were driven by economic considerations. The expected economic return from a pesticide effective against a particular pest would determine whether it was worth including such a pest in the screening process.

Since about 1990, much of the screening process has been automated using robotics to conduct this large-scale testing on a limited number of organisms with extremely low amounts of compound (less than 1 mg per test). This change has been necessitated by the small amounts of compounds synthesized using high throughput synthesis such as combinatorial chemistry, coupled with the need to test more compounds in order to discover a potentially useful one.

Most chemical companies now have some sort of relatively high throughput toxicology and environmental fate screens that are initiated upon discovery of an interesting chemical class. The information from these tests is important, since development of a compound will depend on having good toxicological and environmental fate properties.

### Step 3

If a potential new pesticide has been discovered, then the company will file a patent application.

### Step 4

A secondary screen of compounds that were active in the primary screen is conducted, and involves additional crops, a wider range of appropriate pests, and a range of chemical doses. Companies have also added testing of representative beneficial insects such as honeybees, certain predators, and parasitoids in an attempt to develop selective pesticides to protect natural enemies.

Preliminary toxicology tests are also started at this time, and the development of analytical procedures for residue analysis is initiated. Formulation chemists initiate research on the most appropriate way to package the chemical in a usable form.

During this phase of development, preliminary field tests are conducted at pesticide company research stations.

### Step 5

At this stage, the pesticide company decides whether to pursue development of the chemical. This decision depends on market size, cost of manufacturing the compounds, and initial toxicology information. Some extra lead time is required if a new manufacturing plant must be built.

### Step 6

Acute toxicology studies continue, and long-term feeding studies are initiated to assess possible problems that might arise from use of the chemical, including chronic effects, birth defects, mutagenicity, and carcinogenicity. Toxicology testing is conducted on the parent chemical, formulated product, and all major metabolites. Widespread field tests to establish field efficacy are conducted, still at industry research stations. Further observations on selectivity are made so that the new product can be marketed as compatible with biological control in an IPM program.

### Step 7

The chemical is introduced to university and public agency researchers, often at professional society meetings and through a network of industry technical representatives. New chemicals are typically introduced at this stage as a coded product.

### Step 8

The initial research is completed and all data are compiled. The pesticide company starts putting a marketing force in place.

### Step 9

The proposed label is written. The label is the document that tells the user the crops and pests on which the pesticide can be used, how to safely apply the pesticide, and any safety warnings. A more complete discussion of labels is included later in this chapter. The pesticide company, referred to as the registrant, assembles a packet of information on all aspects of the pesticide and its label. The packet is submitted to the EPA, or appropriate regulatory authority in other countries, requesting that the product be registered as a pesticide under current law. Where required (e.g., in California), the packet must also be submitted for local state registration.

**Step 10**

The governmental agencies consider the information contained in the application packet and decide whether registration is granted and the label approved. A period of public comment is included before the decision to register is completed. After the product has been registered, and the label for use approved, the company can begin to sell the product.

In the United States, the entire development and registration process currently takes six to nine years, and costs about $100 million. The current success rate is about one new product from about 100,000 new chemicals synthesized.

# CHEMICAL CHARACTERISTICS

## Pesticide Nomenclature

The nomenclature associated with each chemical pesticide provides information about the chemical. Examples of the use of these different names are presented in Table 11–2.

***Chemical name***    The chemical name spells out the full nomenclature of the pesticide in currently accepted terms, which are governed by strict international rules. Only chemists and the research community use this name.

***Chemical structure***    A name that denotes the structure of the molecule, including isomeric forms, and contains information important for chemists, toxicologists, and biochemists is called the chemical structure. Chemical activity is often a function of structure.

***Common name***    A common name is assigned to the chemical which may be related to the formal chemical name. Development of acceptable common names follows international rules. The common name is typically much shorter than the chemical name, so it is easier to use. It is used by anyone who needs to refer to the chemical, but does not wish to use the full chemical name or trade name. The use of common names is the accepted form for scientific communication as it does not specify a particular brand name, and it solves the problem that arises when more than one trade name is used for the same parent chemical (see Trade Name). Pesticide common names are the equivalent to the generic names of drugs. We use common names rather than trade names throughout this book, unless a trade name is required for clarity.

***Trade name***    The trade name is a registered brand name and is the property of a particular commercial entity (usually a chemical company). It is used for commercial sale of a formulated pesticide that includes a chemical as its active ingredient. It is based on a name assigned during the process of label acceptance. A trade name is registered by a company and can be used on any product the company so desires as long as it fits within certain parameters. It is not necessarily tied to a patented product; however, the trade name varies by country, and thus a company may sell the same pesticide under different names in different countries. Once the patent on the chemical pesticide has expired

| TABLE 11–2 | Representative examples of pesticides selected from a few of the major families of pesticides. Each compound shows the chemical structure and includes the common chemical name, trade name(s), full chemical name, chemical family and type of pesticide, and comments about typical use and pest(s) controlled. |
| --- | --- |

### Various aliphatic compounds

These are compounds with carbon skeletons not formed into ring structures.

(a) Common name: methyl bromide
Trade name: none, sold as methyl bromide
Chemical name: bromomethane
Type of pesticide: general biocide, fumigant
Type of use: controls most organisms in the soil and in stored products

(b) Common name: fosetyl-Al
Trade name: Aliette®
Chemical name: aluminum tris(-O-ethyl phosphonate)
Type of pesticide: organophosphate, fungicide
Type of use: effective vs. soil-borne oomycetes

(c) Common name: malathion
Trade name: sold under trade name of Malathion and Cythion®
Chemical name: O, O-dimethyl-S-1,2-di(carboethoxy)ethyl phosphorodithioate
Type of pesticide: organophosphate, insecticide
Type of use: contact, many insects, nerve poison, relatively low mammalian toxicity

(d) Common name: glyphosate
Trade name: Roundup®
Chemical name: N-(phosphonomethyl) glycine
Type of pesticide: miscellaneous, herbicide
Type of use: nonselective, translocated, postemergence, inhibits shikimic acid pathway

(e) Common name: aldicarb
Trade name: Temik®
Chemical name: 2-methyl-2-(methylthio) propionaldehyde O-(methylcarbamoyl) oxime
Type of pesticide: carbamate; insecticide and nematicide
Type of use: systemic, soil applied, very toxic to mammals

(f) Common name: EPTC
Trade name: Eptam®
Chemical name: S-ethyl dipropylthiocarbamate
Type of pesticide: thiocarbamate, herbicide
Type of use: preplant incorporated, inhibits germination, selective in beans and alfalfa (corn, but only when used with an antidote)

(g) Common name: maneb
Trade name: sold as Maneb
Chemical name: manganese ethylenebisdithiocarbamate
Type of pesticide: dithiocarbamate, fungicide
Type of use: systemic, probably acts by isothiocyanate production which inhibits sulfhydryl groups on amino acids

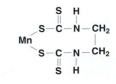

## Compounds with benzene rings (also called aromatic compounds)

The benzene ring consists of six carbon atoms. The hydrogen atom located on each carbon atom has not been shown in the drawings. One or more of the hydrogen atoms are replaced with other atoms or side chains.

(h) Common name: 2,4-D
Trade name: many trade names
Chemical name: 2,4-dichlorophenoxyacetic acid
Type of pesticide: phenoxy, herbicide
Type of use: selective, kills dicots in cereals and other grass crops
(e.g., turf), translocated, postemergence, probably alters translation of genes

(i) Common name: diuron
Trade name: Karmex® and others
Chemical name: 3-(3,4-chlorophenyl)-1,1-dimethyl urea
Type of pesticide: substituted urea, herbicide
Type of use: photosynthetic inhibitor, apoplastic movement, soil applied, limited selectivity

(j) Common name: carbaryl
Trade name: Sevin®
Chemical name: 1-naphthyl methylcarbamate
Type of pesticide: carbamate, insecticide
Type of use: broad-spectrum insect control,
toxic to bees, relatively
low mammalian toxicity

(k) Common name: permethrin
Trade name: Ambush®, Pounce®
Chemical name: m-phenoxybenzyl (∓)-cis,
trans-3-(2,2-dichlorovinyl)-2,
2-dimethylcyclopropanecarboxylate
Type of pesticide: synthetic pyrethroid,
insecticide
Type of use: many insects, low rates, stable
in sunlight

(l) Common name: warfarin
Trade name: numerous
Chemical name: 3-()′-acetonylbenzyl)-4-hydroxycoumarin
Type of pesticide: coumarin, rodenticide
Type of use: anticoagulant, used in baits for rodent
control, inhibits blood clotting

## Compounds with heterocyclic rings

Heterocylic means that the rings making up at least part of the molecule contain mixed atoms, usually carbon and nitrogen in pesticides.

(m) Common name: nicotine
Trade name: numerous
Chemical name: 1-3-(1-methyl-2-pyrrolidyl) pyridine
Type of pesticide: alkaloid, botanical, insecticide
Type of use: contact, nerve poison

*(continued)*

261

| TABLE 11–2 | Representative examples of pesticides selected from a few of the major families of pesticides. Each compound shows the chemical structure and includes the common chemical name, trade name(s), full chemical name, chemical family and type of pesticide, and comments about typical use and pest(s) controlled. *(continued)* |
|---|---|

(n) Common name: captan
Trade name: sold under trade name Captan
Chemical name: N-(trichloromethylthio)-4-cyclohexene-1,2-dicarboximide
Type of pesticide: sulfenimide, fungicide
Type of use: preventative, contact

(o) Common name: benomyl
Trade name: Benlate®
Chemical name: methyl 1-(butylcarbamoyl)
-2-benzimidazolecarbamate
Type of pesticide: carbamate, fungicide
Type of use: systemic

(p) Common name: atrazine
Trade name: Aatrex® and others
Chemical name: 2-chloro-4-(ethylamino)-6-(isopropylamino)
-*s*-triazine
Type of pesticide: triazine, herbicide
Type of use: selective in corn, apoplastic transport,
mainly preemergence, inhibits photosynthesis;
most widely used pesticide in the United States.

(20 years from date of filing), any company can manufacture and market products based on the chemistry under whatever trade name they choose. It is possible to have several trade names being used for the same active ingredient, which can lead to confusion.

In some situations there may be one trade name for agricultural uses and another for home and garden or specialty uses; and there will be different labels for each use. The formulations that are intended for home and garden use have specific labels and are often sold at higher prices. Pesticide applications must be made according to the label, or the application is illegal. Table 11–2 provides the nomenclature in relation to some common pesticides.

## Chemical Relationships

As outlined, the pesticide discovery process leads to the identification of new compounds that may have a previously unknown chemical structure. Chemicals with similar structure may have similar, or related, activity. Once an active chemical is discovered, the chemists synthesize other analogs with similar structures. This process has led to the development of families of pesticides that are all based on a core chemical structure. The following examples show the core of the molecule from which the name is derived. We have used accepted organic chemical structural designations; the letter *R* in these diagrams implies that substitutions of various chemical structures can occur in these places.

***Chlorinated hydrocarbons***    Also known as organochlorine compounds, these pesticides are compounds containing carbon and hydrogen (hydrocarbons) in which one to many hydrogen atoms have been replaced with chlorine. There are several subgroups depending on the complexity of the molecule. These compounds are primarily insecticides, with relatively low mammalian toxicity, but which are generally persistent. Many chlorinated hydrocarbon pesticides also possess the ability to accumulate in the food chain (more later). The oldest and best known insecticide, DDT, is a member of this group; others include aldrin, dieldrin, BHC, lindane, and chlordane. Use of several members of this group has been curtailed due to their persistence and accumulation in the environment. The ones remaining in use, such as methoxychlor and dicofol, are relatively nonpersistent.

***Carbamates***    This family of pesticides includes compounds that are derivatives of carbamic acid; they contain a central carbon linked to two oxygen atoms, and to a nitrogen atom (as shown). One or two of the oxygen atoms can be replaced with sulfur, which results in production of thiocarbamates and dithiocarbamates, respectively. All classes of pesticides are represented in this group. Widely used insecticides and nematicides in this group include carbaryl (Table 11–2j), methomyl, carbofuran, and aldicarb (Table 11–2e). Examples of fungicides in the family are thiram, maneb (Table 11–2g), zineb, and metamsodium. Carbamate herbicides include chlorpropham, EPTC (Table 11–2f), and phenmedipham.

***Organophosphates***    Organophosphates form a diverse family of phosphorus-containing compounds mostly with insecticidal and nematicidal properties, but it also include some important herbicides. There are several subfamilies of OPs.

***Phenoxy***    The earliest selective herbicides, such as 2,4-D (Table 11–2h) and MCPA, are in this chemical family. They have a central phenol structure with various substitutions on the ring and the hydroxyl position. These typically possess strong foliar activity against dicotyledon weeds, including perennials.

***Substituted Urea***    This group of long-lived soil-applied herbicides have urea at the core of the molecule. They are absorbed by the roots and translocated to the foliage where they inhibit photosynthesis. Diuron (Table 11–2i) is probably the best known example; diuron is called DCMU in the biochemical literature, where it is cited because of its ability to block photosynthetic electron flow.

Herbicides based on the sulfonyl-urea core form a subset of the substituted ureas. This chemistry was first developed in the late 1970s. The herbicides have very high activity and are used at low application rates (typically in the range of 0.5 to 1.0 oz/A applied to the soil).

***Triazines***    This large group of broad-spectrum selective soil-applied herbicides is based on a heterocyclic six-member ring of alternating carbon and nitrogen atoms. The single most widely used conventional pesticide in the United States is atrazine (Table 11–2p). Atrazine is used in corn, because corn is capable of biochemically altering the chemical to make it nonphytotoxic. Corn has the ability to replace a chlorine atom on the ring with a hydroxyl group, rendering the molecule inactive; most weeds do not have this ability and are killed.

***Synthetic Pyrethroids***    Chemists were able to determine the structure of the natural botanical insecticide pyrethrin and then to make synthetic insecticides derived from that structure. Permethrin (Table 11–2k) is an example of an insecticide in this group.

***Benzimidazoles***    The benzimidazoles are a group of systemic fungicides with activity against many different pathogenic fungi. Benomyl (Table 11–2o) is probably the most widely used. All exhibit activity through inhibition of tubulin synthesis.

There are numerous other families of pesticides, but we make no attempt here to discuss them all. For a comprehensive review of the current pesticides and their characteristics, see Tomlin (2000).

## Modes of Action of Synthetic Organic Pesticides

**The metabolic (biochemical) process that is affected by the pesticide defines the mode of action of the pesticide.**

Synthetic organic pesticides are toxins that interfere with, or block, a metabolic process or processes in the target organism. Depending on the importance of the process to the organism, the toxin may disrupt growth, paralyze, or kill the pest. For herbicides there are about 20 different known modes of action; for fungicides, nematicides, and insecticides there are about half as many. In some cases the more limited number of modes of actions of insecticides and fungicides has had serious implications for pesticide resistance and its management (see Chapter 12).

The following are some of the more generally recognized pesticides modes of action. They do not apply to all categories of pesticide or pest; in fact, some have no relevance beyond the target category of pest.

### Modes of Action that Exclusively Affect Animals

**Nerve Poisons**    Many insecticides and nematicides act through disruption of the nervous system. Compounds with this mechanism of action have no effect on organisms without a nervous system, and thus are not active in plants or pathogens. They are, however, toxic to all animals, including humans. The use of such insecticides and nematicides thus poses considerable hazard to workers and nontarget animals.

Many of the nerve poisons are primarily acetylcholinesterase inhibitors. Cholinesterase is an enzyme involved in regulating signal transmission between neurons in the nervous system of all animals. When cholinesterase is inhibited, the neurons pass into a state of continuous discharge, resulting in constant muscle contraction, which inhibits movement and breathing, and ultimately causes death. Since nematodes do not breath, nerve-poison nematicides may be referred to as "nematistatic," because instead of killing, they inhibit directed movements. Accordingly, nematodes may recover from exposure to cholinesterase inhibitors. Many organophosphate (e.g., malathion; Table 11–2c) and carbamate (e.g., aldicarb; Table 11–2e) insecticides have inhibition of cholinesterase as the mode of action. An inherent problem with the use of nerve poisons is that they may be highly toxic to mammals.

**Anticoagulant**   These compounds, such as warfarin (Table 11–2l), decrease the ability of blood in warm-blooded animals to clot. Anticoagulants cause the animal to bleed to death. Anticoagulants are only relevant to the control of birds and mammals.

**Juvenile Hormones**   Insect growth regulators (IGRs) and other hormonelike compounds mimic the action of insect hormones that regulate the growth and molting processes. These are highly specific and selective compounds often compatible for use together with biological control agents in IPM systems. Only a few insecticides have this mode of action.

**Acute Muscle Poison**   Alkaloids such as strychnine are acute muscle poisons. These highly toxic chemicals must be used with extreme caution.

**Contraceptive**   Contraceptives are chemicals that block or impede reproduction, and may be useful for vertebrate pest control; the use of immunocontraceptives is experimental.

**Antifeedant**   Metaldehyde is used to control slugs and snails, and the mode of action is to cause the animal to stop feeding. Insecticides based on azadirachtin are also antifeedants.

**Repellent**   Chemicals that cause an animal to move away from the treated area are called repellents. The chemical does not necessarily have a toxic effect.

### Modes of Action that are Specific for Plant Systems

**Photosynthesis Inhibitors**   Inhibition of photosynthesis is the mode of action in several families of herbicides, including the triazines (Table 11–2p) and substituted ureas (Table 11–2i). These herbicides are only toxic to green plants that carry out photosynthesis. Animals do not photosynthesize, so these compounds typically are of little toxicity to them. The mode of action is to block electron flow in the process of transferring light energy from chlorophyll to the system used to convert water and carbon dioxide into sugar precursors.

**Tubulin Inhibitors**   Tubulin is a component of the subcellular organelles called microtubules. These are involved with regulating cell division and cell wall synthesis. Inhibition of tubulin leads to abnormal cell wall production and cessation of cell division. Several herbicides (e.g., dinitroanilines) and fungicides (e.g., benzimidazoles) disrupt tubulin production in plants and fungi, but do not affect tubulin in animal cells. Tubulin inhibitors thus exhibit only low toxicity to animals.

**Inhibition of Aminoacid Biosynthesis**   Several herbicide classes inhibit the action of enzymes involved with synthesis of branched chain and aromatic (essential) amino acids in plants. Animals cannot synthesize all amino acids; those that cannot be synthesized must be ingested and are said to be essential. Herbicides with this mode of action typically have low toxicity to mammals; the herbicide glyphosate (Table 11–2d) is an example.

**Sterol Synthesis Inhibitors**   Several fungicides act through inhibition of the sterol synthesis pathway in fungi.

**Nuclear Regulation**   Herbicides in the phenoxy group, such as 2,4-D (Table 11–2h), and related chemicals are thought to regulate the transcription of DNA-encoded information to RNA and thus modify plant growth. These herbicides are sometimes referred to as synthetic auxins, as they mimic the action of the natural growth-regulating hormone indole-3-acetic acid (IAA, or auxin) when used at nonlethal doses. Inhibition of RNA synthesis is the mode of action of several fungicides in the phenylamide group.

**Safeners**   These chemicals are not themselves phytotoxic, but are added to a herbicide formulation that makes the parent herbicide less toxic to a non-target plant, usually the crop. The safener increases selectivity in relation to the crop.

### Modes of Action that Affect Life Processes

Certain modes of action affect some of the most fundamental life processes and thus are active against all living organisms (biocides).

**Uncoupling of Oxidative Phosphorylation**   Oxidative phosphorylation is the process by which energy is transferred to the carrier ATP molecule during the process of respiration in mitochondria. All organisms respire to release energy from food. Compounds that inhibit this process are therefore toxic to most organisms, and such compounds are true biocides. Many pesticides based on substituted phenols have this mode of action, but most have been phased out of use due to their toxicity.

**Cell Membrane Disruption**   The cells of all organisms are surrounded by a membrane. If this membrane is disrupted or destroyed, the cell is killed. Several pesticides, such as paraquat, and various oils have the ability to damage cell membranes and are thus toxic to most organisms.

## Pesticide Acquisition by Pests

### Contact Pesticides

Such pesticides must actually be sprayed onto a pest, or a surface where a pest may walk over it in the case of mobile pests, so that there is physical contact between the pest and the active ingredient of the pesticide.

### Ingestion Pesticides

These pesticides must penetrate the animal body through the mouth during feeding. The process applies to molluscicides, many insecticides, and predacides, but has no relevance to herbicides and fungicides.

### Translocated Pesticides

**Symplasm: living components of a plant, including cell contents and the phloem. Apoplasm: nonliving component of plants, including cell walls and the xylem (dead at functional maturity).**

Some herbicides are mobile within the plant rather than acting by contact. It means that the herbicide can affect the plant in tissues other than those that directly received the spray. Herbicides applied to the soil can be translocated in the xylem from the roots to the shoots. Such herbicides are said to exhibit apoplastic movement in the plant as they travel in its non-living component. Herbicides applied to the foliage can be moved from the sprayed leaf to the roots in the phloem; they are said to exhibit symplastic movement as they move in the living components of the plant. Chemicals that display apoplastic movement tend to move upward in the plant and accumulate wherever water is being used. Chemicals that display symplastic mobility can be distributed in any part of the plant, but tend to accumulate where carbohydrates are being utilized, such as shoot and root growing points, in developing fruits and seeds, and in rhizomes and tubers.

**Systemic Pesticides**   These pesticides are nonherbicidal pesticides that are mobile within the plant or the pest. They are thus functionally the same as translocated herbicides because the whole organism does not have to be contacted by the pesticide to achieve desired effects.

**Topical Versus Soil Application**   Many pesticides are directly applied to the target pest as a topical, or surface, application. Other pesticides are applied to the soil and are taken up by plant roots or have fumigant (toxicant forms a gas) action that provides contact with the target pest. Topically applied pesticides must be absorbed by the part of the pest that is sprayed. Many herbicides are poorly absorbed by foliage, and many soil-borne pests are not exposed to topical applications. Pesticides that are usually applied topically often have limited activity if applied to the soil.

   Soil-borne pests such as nematodes and pathogens are aquatic organisms, and live in the water film which surrounds soil particles. To be effective, a nematicide or soil-applied fungicide needs to move through the soil, and soil moisture, to the target. Soil-dwelling nematodes that feed on the roots of plants occur in the same soil strata as plant roots, so nematicides must be delivered to appropriate soil depths to be effective. Nematicides can move through the soil as vapor or dissolved in soil water.

**Persistent Versus Nonpersistent Pesticides**   Persistence refers to the duration of the pesticide activity after its release into the environment. It is a relative term; a fungicide or insecticide that remains active for a few weeks is considered to be persistent, but a herbicide active for a few weeks is not. In the former cases, nonpersistence implies that a fungicide or insecticide lasts for a few hours to a day or two, although there is no exact definition. Persistence at the ecosystem level usually implies that the pesticide degrades very slowly and will remain active over a period of a few months to several years (see Persistence in the soil later in this chapter), or will accumulate in the tissues of organisms that consume it.

**Preventative Versus Curative Application**   Pesticides that are applied before the pest is present or has not yet reached damaging population levels are said to be preventative (fungicides and bactericides) or prophylactic (nematicides and insecticides). Preventative fungicides must be applied before infection occurs, as they cannot control or eliminate an existing infection. Pesticides that are effective after the pest is present and has become established are referred to as curative treatments. A curative fungicide treatment, for example, is capable of controlling a pathogen that has invaded a host and established an infection.

**Pesticide Selectivity**   Selectivity refers to the ability of a pesticide (or other treatment) to kill target pest species while leaving other species of the same category unharmed. (For a more thorough discussion, see Toxicity of Pesticides later in the chapter.) Selectivity is crucial to the use of herbicides in crops (Figure 11–6). Ideally the herbicide would kill all weeds but not harm the crop. In reality, something less is usually obtained; the crop is only damaged slightly

*A selective pesticide kills some organisms (target pests) while not harming desirable species.*

**FIGURE 11–6**
Selective postemergence weed control in corn, showing almost complete weed control with no damage to the crop.
Source: photo by Doug Buhler, USDA.

and typically a few weed species are not well controlled. Selectivity is also advantageous for vertebrate management, as control of a specific vertebrate pest is usually required while all other vertebrates, such as people and birds, are unharmed. In reality such selectivity has not been achieved. Selectivity is essential for insect management because of the ability of broad-spectrum insecticides (which have little or no selectivity) to kill not only the target key insect pest, but also beneficial insects and other nontarget organisms. Insecticides that do minimum harm to beneficials are often termed soft insecticides.

The means by which selectivity can be achieved vary with category of pesticide and the target pest. Biochemical differences in the organisms being treated are generally the best basis for achieving selectivity. This means that a genetic tolerance or ability to withstand the chemical occurs in the nontarget organism. Biochemically based selectivity is relatively reliable, but is not always available. Selectivity can be also achieved by factors such as timing of application and placement of the pesticide, and is known as ecological selectivity. Because the latter often depends on variable ecological conditions, this type of selectivity is not as reliable as biochemical selectivity.

Herbicide-resistant crops (or HRCs) are crops that have been genetically manipulated to increase their tolerance to a herbicide. This subject is discussed in more detail under host-plant resistance in Chapter 17.

Nonselective or broad-spectrum pesticides kill most organisms within the pest category. Nonselective herbicides are useful in situations where control of all vegetation is desirable, such as industrial facilities, roadsides, ditch banks, and canals. Many of the first insecticides were broad spectrum, and killed both the target pests and many nontarget beneficials; use of such insecticides is usually considered inappropriate for IPM systems.

***Pesticide Efficacy***    Efficacy is a collective term applied to any pest control measure, and indicates the realized treatment effect on the target pest relative to the desired effect. A pest management tactic is efficacious if it achieves the desired reduction in the pest population and the damage caused by the target pest.

## Classification of Pesticides by Application Timing

Pesticides can be applied to crops at many different stages of crop growth. The application is often named according to a particular stage of growth; this is especially valid for herbicides, but it also applies to certain preventative uses of fungicides, insecticides, and nematicides.

*Seed treatment*   The pesticide is coated onto the surface of seeds prior to planting. This typically involves fungicides and some insecticides used to protect the seed and seedling from pathogen and soil insect attack. Treated seeds cannot be used for food or feed purposes.

*Stale seedbed*   Herbicide is applied to kill existing weeds on planting beds that are formed weeks to months prior to seeding the crop.

*Preplanting*   Pesticide is applied to soil weeks or months before the crop is planted to control pests at or following planting. Soil fumigant insecticides and nematicides, and some herbicides, are applied this way.

*Preplant incorporated (PPI)*   The pesticide is applied to soil immediately prior to planting and is physically mixed into the soil. This technique is used in areas where rainfall is low and surface irrigation techniques are employed. Granular nematicides, insecticides for soil insects, and several herbicides must be applied this way.

*At-planting incorporation*   Some granular nematicides are applied as a band over the planting bed during planting operations. Control of some soil insects in row crops consists of a sidedress application at planting. The insecticide is delivered in a furrow parallel and generally slightly deeper than the seed furrow.

*Preemergence*   Pesticide is applied to the soil surface after planting the crop but before the seedling emerges above the soil. Many herbicides are applied at this time in rain-fed or sprinkler-irrigated systems.

*Cracking*   This stage occurs when the crop is about to emerge from the soil; the shoot pushing up causes the soil to crack, hence the name. Some nonselective herbicides such as paraquat can be applied at this time without damaging the crop.

*Postemergence (POST)*   The pesticide is applied after the crop (and/or the weeds) has emerged above the surface of the soil. This stage is sometimes subdivided into early post and late post in relation to the size of the crop plants.

*Lay-by*   Refers to a row-crop pesticide application using ground equipment at the time that the last cultivation or other operation is feasible prior to canopy closure.

*Preharvest*   Pesticides are applied at crop maturity but prior to harvest. Insecticides or fungicides are often applied at this time to protect the harvested product; however, the pesticide label carries restrictions for the minimum time that must be allowed between application and harvest.

*Postharvest*   Pesticide is applied following harvest of the crop.

*Dormant*   Pesticide application is made to perennial crops in the season when the crop plants are not growing. In dormant herbaceous crops such as alfalfa, herbicides are used at this time. In tree crops, dormant application of insecticides and fungicides is considered essential for many IPM programs.

*Crop phenology*    Many pesticide applications are timed according to the phenological stage of the crop. The terminology used is crop specific, such as "petal fall" in almonds, the details of which are beyond the scope of this text.

*Fallow*    Pesticide application is made at a time when there is no crop present. Herbicides are the type of pesticide most frequently used at this time because of the importance of weed control during fallow periods.

# APPLICATION TECHNOLOGY

Pesticides used in an IPM program must be delivered to the target pest while maintaining worker safety and minimizing impact on nontarget organisms and the environment. Sophisticated delivery systems have been developed to achieve these goals. The necessary components of the delivery system include an appropriately formulated pesticide, a suitable application tool, and correct environmental factors.

## Formulations

**Most pesticides cannot be used in IPM systems in their native chemical state.**

All pesticides must be formulated to produce a commercial product that is suitable for practical use. There are several reasons why different formulations are necessary.

1. Most pesticides must be delivered as fine drops or particles to obtain the necessary coverage of the target pest (contact pesticide), or plant part (systemic or ingestion pesticide). To be dispersed as fine drops or particles, the pesticide must be mixed into a carrier that can achieve the necessary coverage. The most widely used carrier is water, and pesticides suspended in water are often delivered as a fine spray. However, the active ingredients of most pesticides are not water soluble, which makes delivery of the native chemical in water difficult if not impossible. The water-solubility problem is resolved by using an appropriate formulating process. Some pesticides are delivered as fine droplets suspended in air as the carrier, which requires a different formulation than is used for water delivery. A few pesticide uses require that the native chemical be formulated and sold premixed with the carrier (e.g., dusts and baits).

2. In many cases the efficacy of the native chemical is low unless specific additives are used to increase uptake of the pesticide by the target organism. This type of formulation is especially important for contact pesticides where a high level of coverage is required.

3. Many pesticides are sufficiently toxic to people that they must be formulated so that they can be handled and applied safely. In many cases, dry powders can provide adequate control of the target pest, but create an unacceptable hazard to the applicator or wildlife, and so the pesticide is formulated as a liquid instead.

4. Most pesticide formulations include additives to increase shelf life and aid in storage, such as materials to inhibit the caking of the pesticide at the bottom of the can.

Pesticides are generally presented in two basic formulations, either dry or liquid. Some of the liquid pesticides are so volatile that they actually function as a gas. Pesticide formulations are often referred to by letter acronyms (as shown in parentheses in the following sections) rather than the full name. The active ingredient is listed in the formulated product as a proportion (usually a percentage) of the total material of the delivered pesticide.

## Liquids

1. Solution (S or SC). In this formulation, the parent chemical will dissolve in water for application. The formulation contains agents to prolong shelf life.
2. Emulsifiable concentrate (EC). The native compound is dissolved in a suitable organic solvent because it is not soluble in water. The solvent with the dissolved pesticide is then suspended in water as an emulsion for application. EC formulations contain compounds called emulsifiers to aid in the emulsification process. A good emulsion does not separate or settle out, but remains as a milky suspension. The solvents are organic chemicals that can pose some undesirable side effects themselves, such as direct toxicity, or contribute to pollution (if they are volatile compounds).
3. Flowable (FL). The parent compound is not soluble in water, but can be suspended in water as a finely divided powder. This type of formulation needs stabilizers that reduce the tendency of pesticide particles to settle out. Flowable formulations have the disadvantage that the particles are abrasive to spray equipment.
4. Aerosols. The pesticide is formulated so that it is dispensed as a fog of ultrafine droplets. Aerosol formulations are generally only used in confined spaces, such as greenhouses. Aerosols are never used as herbicides because of the hazard of movement to nontarget plants.

## Dry

1. Dusts. The parent compound is mixed into an appropriate finely ground carrier. Dusts have the serious potential problem of drift to nontarget organisms (see later section in this chapter); for this reason herbicides are never formulated as dusts.
2. Granules (G). The native pesticide chemical is either coated onto, or mixed into, an inert carrier such as refined clay that can be formed into granules. Granular formulations are usually referred to in relation to the percentage of concentration of the native chemical. A 10G formulation, for example, contains 10% active ingredient. Both dusts and granules may cause problems because they can be hazardous to applicators and field workers. An economic disadvantage is that the relatively bulky carrier has to be shipped, and the shipping cost makes them relatively expensive.
3. Wettable powder (WP). The parent compound is finely ground and mixed with inert dry materials. The formulation contains wetting agents that permit the powder to wet and mix with water. Wettable powders have the same problem as flowables in that they are abrasive to equipment. They are also prone to settling out in the spray tank unless well agitated.
4. Soluble powder. The parent compound is soluble in water and can be dissolved directly in spray solution for application.

5. Water-dispersible granules. These highly concentrated granules of pesticide disperse when added to water, similar to a wettable powder.

6. Baits. The native pesticide is applied to or mixed into an attractive food bait. The pest eats the bait and so ingests the pesticide. Used widely in vertebrate, mollusk, certain ants, and fruit fly pest control, and to a lesser extent for some soil-dwelling insects (e.g., cutworms). The food material in the bait tends to decay, creating a problem with maintaining the bait in attractive edible form for any length of time. Baits are not used for herbicides and fungicides.

7. Slow release (encapsulated). The native compound is physically encased or entrapped in an inert medium from which it escapes slowly. This is advantageous because the release rate is limited, extending the control achieved from a single application, particularly for compounds that are volatile or that break down rapidly. This formulation can reduce the hazard to applicators.

8. Impregnates. Similar to granules, as pesticide is impregnated into a solid medium from which it escapes slowly. A good example is impregnation of pesticides onto fertilizers; this type of formulation is widely used in the home lawn and garden industry for insecticides and herbicides.

### Gas

**Fumigants**   The native compound is active in the gaseous form. Application can be in the form of a gas from a pressurized cylinder, a liquid that volatilizes, or a solid that releases the gas when it comes in contact with water.

## Adjuvants

Adjuvants are chemicals added to pesticide formulations or to the spray solution to enhance performance. The manufacturer adds them to the formulation when practical, or their use is spelled out on the pesticide label to be added at the time of mixing in the spray tank.

1. Surfactants (**surf**ace **act**ive ag**ents**). These chemicals accumulate at interfaces between liquids and solids and reduce surface tension. They are similar to household detergents. The reduced surface tension that the surfactant creates permits drops to spread out on a surface, such as a leaf, instead of staying as a spherical drop. This permits coverage of a larger surface area with the given amount of liquid, resulting in enhanced uptake of a pesticide. The use of surfactants is most important for topically applied pesticides, and the pesticide label often requires such use.

2. Spreader stickers. These additives are intended to stop spray droplets from rolling off their target.

3. Antifoaming agents. Some additives, especially surfactants, can create foam in the spray tank because of the necessary agitation to stop the pesticide from settling out. Antifoaming agents may be added to the formulation to reduce foaming.

4. Drift-control agents. Drift is a potential problem when pesticides are sprayed. Under some conditions it is necessary to add thickeners to the spray solution to reduce the number of fine drops created at the nozzle. Fewer fine droplets result in less drift.

5. Buffers. Some pesticides have a narrow range of pH at which they either provide optimum performance or at which they do not break down. If pH is critical to the performance or stability of a pesticide, buffers may be added to maintain the spray solution at the correct pH.
6. Performance enhancers. This mixed group of compounds boost performance.
7. Nonphytotoxic oils. Inclusion of small amounts of oil in the spray tank can substantially increase the activity of several pesticides. The label may require their use. Depending on the pesticide and its intended target, the oil may be of mineral or vegetable origin.
8. Fertilizers and other mineral additives. The performance of several herbicides is substantially improved by adding to the spray tank small amounts of fertilizer, such as ammonium sulfate.

# Application Equipment/Technology

Most formulated pesticides are not applied in concentrated form, but have to be diluted in a carrier or delivery medium. Exceptions are aerosols, baits, granules, and dusts, which are usually applied at the concentration in the formulated product without further dilution.

**Most pesticides are not applied in their concentrated form, but rather are diluted in a carrier.**

Water is the most versatile and most frequently used carrier. Air is used where coverage on all surfaces of the target is necessary (upper and lower leaf surfaces; all twigs in a tree) and is used primarily in perennial tree and vine crops. Due to costs, granules and baits are only used when water and air are neither practical nor feasible.

Water-based pesticide application systems rely on some kind of sprayer. The basic components of a sprayer are essentially the same for all kinds of sprayers (Figure 11–7), although numerous variations exist for different specific situations (Figures 11–8 through 11–11).

*Sprayer Components*    Sprayers come in a variety of sizes. Some are small, handheld or backpack, for use in domestic or small commercial vegetable and flower gardens. Large, tractor-pulled or self-propelled units, are capable of dispersing hundreds of gallons of spray.

**Tank**    All sprayers must have a tank to contain the spray solution. Due to the problem of pesticide settling out in the tank, or of nonuniform mixing, there must be some form of agitation of the solution in the tank. The most common form is to use excess pressurized liquid in the form of jet agitation (Figure 11–7), but many systems have built-in paddles for mechanical agitation. Tanks are made of materials that do not corrode easily, such as high-density plastics or stainless steel.

**Filters and Strainers**    Clogging of nozzles is a serious problem; most spray systems therefore have an inline strainer between the tank and pump to remove large particles. Most spray systems also have a series of screens as part of the nozzle assembly to trap particles that are too large to pass through the orifice in the nozzle tip (Figures 11–7 and 11–12).

**Pump**    All systems must have a pump to pressurize the spray solution. The pump ranges from hand-operated pistons for small hand sprayers to various kinds of mechanical centrifugal and positive displacement pumps on field sprayers.

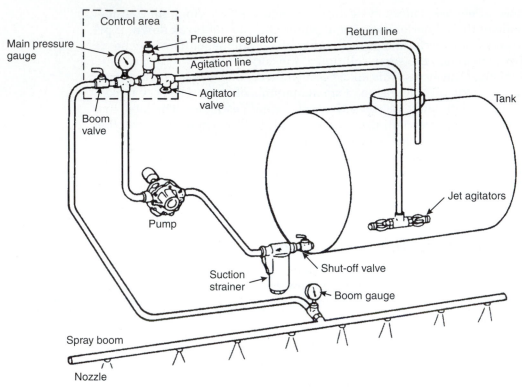

**FIGURE 11–7**   Schematic diagram showing basic components of a pesticide sprayer (modified after Bohmont, 2000). Actual configuration of a sprayer varies by crop and intended use; see photographs for examples of major types of sprayers.

**Controls**   It must be possible to shut the spray system on and off, and to regulate the pressure in the system. Accurate application of pesticides depends in part on having constant pressure in the delivery system. All sprayers, with the exception of handheld sprayers (Figure 11–8a), therefore have some form of adjustable pressure regulator. A pressure gauge is also necessary so that the regulator can be set to the required pressure. An on/off valve is also necessary. These controls must be within easy reach of the operator.

**Boom and Nozzles**   The system must be capable of delivering the spray solution to the desired area by way of a system of nozzles arranged on a boom. The boom delivers the pressurized spray solution to the individual nozzle assemblies. The boom must be sufficiently supported so that its height in relation to the target (soil surface or plant foliage) does not vary along its length. Both the spacing and type of nozzles depend on the type of application. Each nozzle assembly (Figure 11–12) consists of an interchangeable nozzle tip attached to a nozzle body on the boom. A nozzle screen is housed inside the nozzle body. As the pressurized spray solution is forced through the orifice in the nozzle tip, it is reduced to a spray of fine droplets. Different nozzle tips can be used to achieve different spray characteristics, such as flat fan versus hollow cone pattern, or coarse versus fine drops. Of the many nozzle tips available, the proper tip must match the specific requirements of the application.

(b)

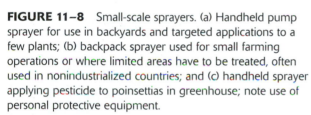

**FIGURE 11–8** Small-scale sprayers. (a) Handheld pump sprayer for use in backyards and targeted applications to a few plants; (b) backpack sprayer used for small farming operations or where limited areas have to be treated, often used in nonindustrialized countries; and (c) handheld sprayer applying pesticide to poinsettias in greenhouse; note use of personal protective equipment.
Sources: photograph (a) by Robert Norris; (b) by R. Faidutti, FAO; and (c) by Peggy Greb, USDA/ARS.

(c)

***Specialized Modifications***  Due to the variety of conditions under which pesticides are applied, there are numerous modifications in the basic application machinery.

**Ground or Air Application**  Pesticides are usually applied using ground equipment (Figure 11–9). In certain situations, use of ground equipment has its limitations:

- Scale: Large areas can be difficult to cover sufficiently and quickly.
- Soil condition: If the soil is too wet, ground equipment may get stuck.
- Cultural practices: It is difficult, for instance, to get equipment into a flooded rice paddy.
- Large crop plants: Tall plants such as maize or trees may interfere with equipment or the equipment may damage the plants.

In these situations, an aircraft may permit spraying when ground application is not possible or appropriate (Figure 11–10). Aircraft application is more expensive than ground application, and generally does not give as uniform distribution of pesticide on the target. The most serious drawback of application by air is the increased likelihood of pesticide drift. Pesticide drift increases with equipment speed during application and the distance of the spray boom from

(a)

(b)

**FIGURE 11–9**    Ground application field spraying equipment. (a) Broadcast boom sprayer for solid planted crops; (b) drop nozzle row-crop sprayer.

Sources: photograph (a) by Ken Giles with permission; (b) by Bill Tarpenning, USDA/ARS.

**FIGURE 11–10**
Fixed-wing aircraft for aerial pesticide application; here applying herbicide to dormant alfalfa.
Source: photograph by Robert Norris.

the target. Aircraft can travel faster than a ground sprayer, and the boom is farther from the target than it is on ground equipment. Helicopters are usually capable of delivering the pesticide to the target with greater precision than fixed-wing aircraft.

**Carrier**    Water, air, granules, dusts, and baits are used as carriers; the choice of carrier has several impacts on pesticide applications. Water sprays tend to hit one side of an object, whereas air carrier applications can achieve more thorough coverage of all surfaces. Air is often used as the carrier in situations where coverage of all surfaces is desired, such as pathogen and insect control in perennial orchard and vine crops (Figure 11–11). When air is used as a carrier, drift is a greater hazard than when using water. Application of granules and baits is usually more expensive than using water or air as the carrier. Use of granules and baits allows more precise targeting of application than can be

**FIGURE 11–11**
Air-blast sprayer used for treating trees in orchards.
Source: photograph by Marcos Kogan.

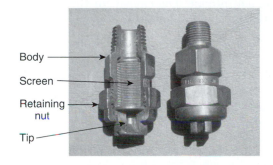

Body

Screen

Retaining nut

Tip

**FIGURE 11–12**
Components of a typical spray nozzle assembly; right is normal, left is a cutaway to show the internal components.
Source: photograph by Robert Norris.

obtained with sprays or air-blast sprayers. Some pesticides for vertebrate control, such as strychnine, can only be used safely as baits placed underground.

**Closed-system Loading**　For worker protection, closed-system loading is desirable, and may be required in some regions, for all class I pesticides (see later). A closed-loading system permits transfer of the formulated pesticide concentrate from its original container into the sprayer tank without exposing personnel to the chemical. Most closed-loading systems for liquid formulations have some form of probe that is inserted into the container through which the pesticide is sucked into the spray tank. Systems for dry formulations involve a special loading tank into which the pesticide is placed inside a water-soluble bag; after the lid has been sealed, the tank is flushed with water to dissolve the bag and transfer the pesticide to the main spray tank.

**Band Versus Broadcast**　For crops that are planted with either narrow rows (less than about 10 in.) or are planted broadcast (e.g., wheat, alfalfa), pesticide applications are also applied broadcast. When crops are planted in rows that are about 20 in. apart or greater (e.g., tomatoes, cotton) it is feasible to apply the pesticide in bands rather than broadcast. Band application is often used with herbicides, where cultivation is used between the crop rows to mechanically kill weeds, and the herbicide is applied over the bed to control in-row weeds. This technique reduces pesticide cost and lowers the quantity of pesticide placed into the environment.

**Directed Spray**    These sprays are made at a time when a crop is present. The nozzles are set so that they spray downward below most of the crop foliage. Directed sprays are mostly used for herbicide applications where crop tolerance is inadequate.

**Drop Nozzles**    This application is similar to directed spray except that there is no effort to keep the spray off the crop foliage. The nozzles hang from pipes below the boom such that they are within the crop canopy. This configuration is most often used for insecticides and fungicides (Figure 11–9b).

**Shielded or Hooded Spray**    The spray nozzles are placed between shields or under hoods so that the spray does not hit the crop. This type of nozzle is primarily used for herbicides.

**Chemigation**    In regions where irrigation is used, it is possible to inject chemicals directly into the irrigation system and have them distributed with the water. This process is called chemigation. No additional equipment is needed other than injection equipment. The caveat for this type of application is that the pesticide distribution is based on water distribution, which may be uneven.

**Soil Incorporation**    In regions of low rainfall, or for pesticides that are volatile, it may be necessary to physically mix the pesticide into the soil to obtain satisfactory efficacy. This is referred to as incorporation, and is achieved by disking for a broadcast pesticide. Where band application is required, sophisticated power tillers are used (Figure 11–13). The pesticide label specifies if incorporation is required.

**Soil Injection**    Fumigant pesticides cannot be sprayed, so specialized equipment is used that directly injects the diluted pesticide solution into the soil several inches below the surface. Where the fumigant is very volatile it is necessary to cover the treated area with polyethylene tarp to prevent the fumigant from moving out of the soil and into the air (Figure 11–14).

**FIGURE 11–13**
Power tiller/planter rig applying preplant incorporated herbicide, injecting insecticide, planting crop, and applying preemergence herbicide in a single operation.
Source: photograph by Robert Norris.

**FIGURE 11–14**
Polyethylene tarped field following methyl bromide application.
Source: photograph by Robert Norris.

**FIGURE 11–15**
Bait station for vertebrate pest control. Bait is placed in central vertical tube with pest animal access through lateral tubes on ground.
Source: photograph by Robert Norris.

**Bait Delivery Systems**   Specialized systems are required to deliver poisonous baits for control of vertebrate and some invertebrate pests. In most situations, the application technique must ensure that nontarget wildlife species are not likely to be able to feed on the bait. Bait stations (Figure 11–15) allow the pest species to climb in and eat the bait while excluding larger, nontarget species. Strychnine is used as the poison for gopher bait, but strychnine baits must be placed below the soil surface to restrict access by nontarget species. The ideal place to put gopher bait is in the burrow system of the target animal(s), which can be done by hand but is time consuming. A mechanical "burrow builder" constructs an artificial burrow into which the bait is placed. When a gopher explores the new "burrow," it encounters and eats the bait. This is an example of technological innovation that decreases environmental hazard. There are insecticide-laced baits for various insects (see Chapter 14).

# ENVIRONMENTAL CONSIDERATIONS

The environment can influence the distribution of a pesticide, altering how much of the pesticide gets to the target. This is of direct concern to pest management practitioners. Another often more important issue is the impact that pesticides can have on nontarget species and the ecosystem in general. Ecological effects of pesticides are significant to pest management practitioners, ecosystems, and society. Understanding interactions between pesticides and the environment is essential if pesticides are to be used effectively and safely.

## Volatility

Every chemical can change physical state from solid (or liquid) to gas phase. The ease with which this occurs is measured as the vapor pressure. The lower the vapor pressure, the lower the volatility. Vapor pressure is measured in much the same way as atmospheric pressure, and is expressed as millimeters (mm) of mercury (Hg). The importance of vapor pressure to pesticides is that it determines the ability, and thus likelihood, of the material changing into the gas form.

Examples of vapor pressure for some pesticides are:

Acrolein = 240 mm Hg at 25°C (It is a gas at normal temperature and pressure, and is used as a fumigant.)

Eptam (EPTC) = $3.4 \times 10^{-2}$ mm Hg at 25°C (It is a liquid at normal temperature and pressure, but volatilizes readily. This creates restrictions on how it can be used; see below.)

Glyphosate = $1.8 \times 10^{-7}$ mm Hg at 25°C (It is a solid at normal temperature and pressure, with low vapor pressure. It is considered to be nonvolatile.)

Fumigants have high vapor pressure and diffuse readily as gases. In the gaseous form, fumigants can fill large volumes, such as in soil or a building. It is assumed that a fumigant is volatile at normal temperature and pressure; thus, to use the pesticide, it must be contained in the area where the activity is needed. Typically fumigants are injected into the soil, and then the soil surface is sealed to prevent the gas escaping. Soil may be sealed by rolling for a compound with moderate volatility, such as metam sodium, or by a polyethylene tarp, as is the case of more volatile products such as methyl bromide (Figure 11–14). Fumigants are also used in enclosed structures to kill pests (mainly insects) present either in the building or on produce placed inside the structure. The former use is referred to as structural pest control, the latter is fumigation as part of the process of postharvest pest management for produce shipped between regions or countries.

For all nonfumigant pesticides, high volatility is a potentially serious problem. The problem is of particular concern with herbicides, but applies to all pesticides. The significance is twofold. Materials with high vapor pressure may escape from the application site and may not provide the intended activity. These highly volatile pesticides must be applied so that vapor loss is minimized, or they should not be used. Several moderately volatile herbicides, such as EPTC, must be mixed into the soil immediately after applying them to reduce the loss due to volatility.

Pesticides in the gaseous form can readily move and have effects distant from the intended target and/or create pollution hazards. Such movement can be a problem with all fumigants if they are not used in strict accordance with their label. Foliar applied chemicals with high volatility, such as esters of phenoxy herbicides (e.g., 2,4-D), can easily move off target and their use is therefore restricted.

# Drift

The intent of an application is to place the pesticide on or near the desired target. A pesticide can, however, move away from the intended target area during the application process. This is called drift. Off-target movement varies from a few inches, which is a problem for band spraying, to as much as several miles, which can be a serious regional problem. Drift can occur in one of three forms:

**Drift is the unintended off-target movement of a pesticide during or following its application.**

1. Fine spray droplets of the carrier, containing the pesticide
2. Fine droplets, or particles, of the native pesticide (This occurs when the carrier evaporates or when dry formulation, such as dusts, are used.)
3. Pesticide vapor, in the case of a volatile compound

Pesticide drift causes the following problems.

1. Lost activity. Pesticide that has drifted away from the target area can no longer provide the intended level of activity. For large fields sprayed broadcast, this is not a serious problem; but for band spraying over the row, it can be very serious as the treated area can be displaced relative to an entire crop row.
2. Off-target impacts. Pesticide drift has several consequences to organisms in the surrounding area.
   2.1. Plants. Herbicide drift can cause damage to surrounding crops (Figure 11–16), ranging from transient symptoms to death. The impact ranges from none to total crop loss. Plant damage is not usually considered for drift of other types of pesticides.
   2.2. Humans. Pesticide drift onto people can cause illness. The problem is particularly important in relation to workers in nearby fields. The significance depends on toxicity of the pesticide in use. This problem may occur with all categories of pesticides.
   2.3. Beneficial insects. Drift of insecticides can disrupt natural enemies resident in vegetation surrounding the target field or in adjacent fields.
   2.4. Wildlife in surrounding areas. Refer to 2.3.
3. Illegal residues in adjacent crops. Pesticide drift onto an adjacent crop has the potential to create an illegal residue on or in the adjacent crop. This occurs if the pesticide that moved off target is not registered on the crop onto which it has drifted. If drift occurs near the harvest date for a neighboring crop, this can also lead to illegal residues, and crop destruction may be required. The residue problem is especially important for organic farming when such farms are adjacent to conventionally sprayed farms, because any pesticide residue results in loss of organic certification.

(a)

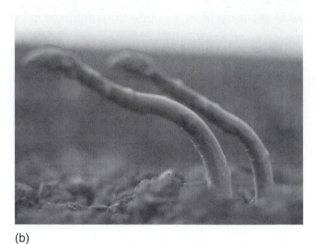

(b)

**FIGURE 11-16** Injury to asparagus due to herbicide drift. A contact herbicide inhibited growth on the upwind side of the spears causing them to bend toward the source of the drift: (a) field view; (b) close-up of damaged spears.
Source: photographs by John Marcroft, with permission.

The magnitude of drift is influenced by several factors. When drift is anticipated, the following variables should be modified to minimize the drift potential.

1. Spray droplet size. Large droplets are less likely to drift than small droplets. The following variables impact droplet size.
    1.1. Pressure. High pressure produces a greater quantity of fine drops than low pressure.
    1.2. Spray volume. Low volume applications typically produce more fine drops than high volume applications, due to the necessity of using nozzle tips with small orifices to achieve adequate coverage.
    1.3. Shear. High wind shear across the nozzles produces more fine drops than low wind shear. Shear is a function of the following two variables.
        1.3.1. Vehicle speed. The faster the vehicle is traveling the greater the wind shear.
        1.3.2. Nozzle angle. Downward-facing nozzles experience greater shear than backward-facing nozzles. For this reason, nozzles on aircraft are often pointed backward to minimize impact of shear on droplet size.
    1.4. Nozzle orifice size. Smaller size nozzles produce more fine particles than larger size nozzles for a given pressure.
    1.5. Nozzle type. Nozzles are designed to produce different patterns and droplet size characteristics in addition to regulating flow rate. Flooding-type nozzles typically produce less fine drops than fan nozzles. The use of swirl plates in jet nozzles typically increases the percentage of fine droplets.
    1.6. Additives. Thickeners, as noted in the earlier adjuvant section, can be added to the spray solution and can reduce the proportion of fine droplets.
2. Wind speed. The degree of drift depends on the rate of airflow, and drift increases rapidly as wind speed increases. Wind speeds in excess of 10 mph

**FIGURE 11–17**
Boom sprayer with hooded boom to minimize spray drift.
Source: photograph by Robert Norris.

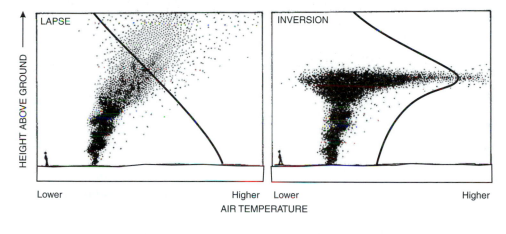

**FIGURE 11–18**
Diagram showing smoke dispersing under lapse conditions and smoke trapped under inversion conditions, demonstrating how fine drops of a pesticide spray can be trapped under inversion conditions. Line graphs represent temperature in relation to height above ground.
Source: modified from Marer, 1988.

are usually too high for most pesticide applications because of increased drift; wind speeds between about 1 to 5 mph are preferred. It should be emphasized that dead calm conditions can also be a problem (see next item). To a certain extent the impact of wind speed on drift can be minimized by the use of hooded boom on ground equipment (Figure 11–17). Here the boom is covered so that the spray pattern is protected from wind.

3. Inversion conditions. Atmospheric inversions occur when there is no wind and the temperature is cooler at the surface than higher in the atmosphere (e.g., 100 ft up) (Figure 11–18). The result is that the cool air is trapped at the soil surface and does not mix with the rest of the atmosphere. If spraying is carried out under these conditions, the stagnant air can be loaded with fine pesticide drops or particles. Later, when the inversion conditions break, the whole air mass can move with its associated pesticide load. Some cases of long-distance herbicide drift have been attributed to aerial spraying under inversion conditions.

4. Boom height. Everything else being equal, the further the boom is above the target the greater the chance for drift.

5. Scale. The larger the area being treated, and the more quickly that the application is made, the greater the potential for drift.

Drift probably occurs with all pesticides, but is very serious for herbicides because symptoms on sensitive plants provide visible evidence of drift; fungicide and insecticide drift can only be detected with specially designed assays. Herbicides in the phenoxy group (e.g., 2,4-D), paraquat, glyphosate, and others are particularly prone to causing drift problems. Drift by herbicides has, in some cases, caused sufficient damage to adjacent crops that the only solution has been to restrict, or even ban, the use of the herbicide under certain conditions.

## Pesticide Behavior in Soil

**Most pesticides end up in the soil.**

Many pesticides are intentionally applied directly to the soil, and much of foliar-applied pesticides eventually reach the soil. Activity of soil-applied pesticides, and the fate of all pesticides in the environment, is thus to a great extent dependent on phenomena that occur in soil.

**Chemicals are attracted to interfaces between dissimilar phases.**

### *Adsorption/Desorption Phenomena*
Dissolved chemicals accumulate at interfaces between two dissimilar phases (e.g., liquid and gas; liquid and solid) due to molecular charge relationships at such interfaces. The types of attracting forces vary from strong ionic bonding (e.g., the positive charge on the herbicide paraquat reacts with the negative charges on soil particles), to hydrogen bonding, and to electrostatic effects such as van der Waals forces. Pesticide molecules may also be physically trapped in the lattice structure of clays. Except for ionic bonding, the attraction is a reversible phenomenon that depends on the chemical nature of the molecules involved, and to their concentration. The degree to which adsorption occurs depends on factors that alter the adsorptive capacity of the soil. The main factors are percentage of clay content and percentage of organic matter content.

Adsorption phenomena cause pesticides in soil to be attracted to the soil/soil water interfaces. This results in two significant impacts on how pesticides are used.

**Use rate of soil-applied pesticides may have to be adjusted in relation to soil type.**

**Pesticide Availability**   As a pesticide is adsorbed to the soil, its availability for activity against the target pest is decreased. The result is reduced activity and hence decreased efficacy. Strongly adsorbed pesticides such as paraquat have no activity in soil, and thus cannot be used as soil-applied pesticides. For many soil-applied pesticides (especially herbicides), the rate used must be adjusted depending on the absorptive capacity of the soil. Clay soils, for instance, require higher use rates than sandy soils. Many soil-applied pesticides cannot be used in muck soils because of their high organic matter content and high adsorptive capacity. Information about rate adjustments in relation to soil type is specified on the pesticide label (Figure 11–19).

**Leaching is the downward movement of chemicals through the soil by water.**

**Leaching**   When a pesticide reaches the soil, it typically accumulates at the soil particle–soil water interface. The pesticide is said to be "bound" to the soil. As pesticide-free water moves through the soil, the bound pesticide is released from adsorption sites and dissolves in the water. Strongly adsorbed pesticides may have to be physically mixed into soil, as they will not move through the soil profile with water (e.g., trifluralin). Pesticide movement into the soil profile is necessary to obtain activity, either through contact with the pest or to permit uptake by plant roots.

**Preemergence:** Apply to the soil surface at planting (behind the planter) or after planting, but before weeds or crop emerge.

Table 3: Bicep II – Preplant Surface, Preplant Incorporated, or Preemergence – Corn

| Soil Texture | Broadcast Rate Per Acre | |
| --- | --- | --- |
| | Less Than 3% Organic Matter | 3% Organic Matter or Greater |
| **COARSE** Sand, loamy sand, sandy loam | 1.5 qts. | 1.8 qts. |
| **MEDIUM** Loam, silt loam, silt | 1.8 qts. | 2.4 qts. |
| **FINE** Sandy clay loam, silty clay loam, clay loam, sandy clay, silty clay, clay | 2.4 qts. | A. 2.4 qts. B. 2.4-3.0 qts.* |
| Muck or peat soils (more than 20% organic matter) | DO NOT USE | |

*For cocklebur, yellow nutsedge, and velvetleaf control on fine-textured soils above 3% organic matter: Apply 3.0 qts. of Bicep II per acre.

**A.** Do not exceed this rate on highly erodible land with less than 30% plant residue cover. Control of certain weeds may be reduced and a tank mix partner or an application of a postemergence herbicide may be needed.

**B.** Use this rate for all other applications.

**FIGURE 11–19**
Portion of a soil-applied herbicide label showing modification of application use rate in relation to soil type.
Source: photograph by Robert Norris.

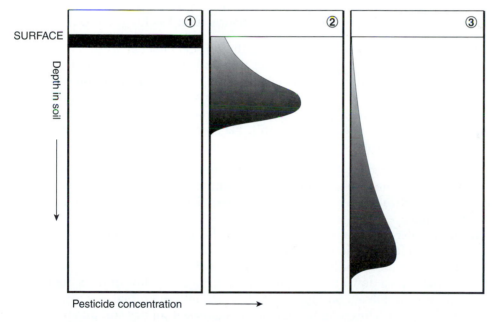

SURFACE

Depth in soil

Pesticide concentration

**FIGURE 11–20**
Diagrammatic graph of pesticide movement into the soil profile by leaching. (1) shows concentration of pesticide at surface at application prior to rain or irrigation; (2) shows pesticide location in soil profile after intermediate amount of rain or irrigation; (3) shows pesticide location in soil after larger quantity of rain or irrigation.

Weakly adsorbed pesticides are released from the soil particles, and move down the soil profile with water (Figure 11–20). Increasing quantities of irrigation water, or rain, move the pesticide deeper into the soil profile. Negatively charged pesticides are the most prone to leaching, as they are repelled from the soil particles, which are also negatively charged. The propensity of a chemical to leach can result in label restrictions on how such pesticides are used in soil. Leaching has resulted in groundwater contamination by pesticides, and has created environmental concerns (see Chapter 19). The extent to which leaching occurs depends on four interacting characteristics, of which adsorptive phenomena are the most important.

1. Adsorption phenomena determine how tightly the pesticide is bound to the soil particles, and is the single most important factor in determining likelihood of leaching. Adsorption phenomena are regulated by:
    1.1. Chemical structure relationships. Factors such as ionic charge, presence of hydrogen atoms, and electromagnetic effects of the interatomic bonds are important. Predicting adsorption based on chemistry has not been very accurate.
    1.2. Soil components. The amount of clay and organic matter present in the soil are prime determinants of adsorption. Pesticide adsorption is strong in soils that are high in clay or organic matter, and weak in sandy soils that are low in organic matter. Pesticide leaching is much faster in sandy soils than in clay or muck soils.
2. Water volume. For a given soil type and pesticide, the larger the quantity of water passing through the soil column the greater the extent of leaching (Figure 11–20).
3. Water solubility. Within a class of chemicals, increasing water solubility can lead to increased leaching. Between classes of chemicals, the chemical structure/activity relationships are usually more important than water solubility.
4. Persistence. The longer a pesticide persists in the environment without being degraded, the greater are the chances that leaching will occur.

***Persistence in the Soil***     Pesticides in the soil degrade chemically or are broken down by the metabolic activity of microorganisms. The degradation rate varies with the specific pesticide and the environmental conditions. Because microbial activity is essential in the conversion and breakdown of pesticides in the soil, factors that impact microbial activity therefore play a major role in determining pesticide persistence in soil. Factors that regulate pesticide breakdown by microbes, and hence persistence, include the following:

1. Chemical structure. The chemical structure of a pesticide molecule determines the overall likelihood of chemical reaction or microbial attack. No generalizations can be made about which structures are more or less persistent.
2. Soil temperature. Temperature regulates the activity of microorganisms and, therefore, pesticide breakdown. Pesticides persist much longer under cool than warm conditions because microbial activity is low in cool weather. Breakdown during winter months is often slow or does not occur.
3. Soil moisture level. Soil moisture also regulates microbial activity. Breakdown is slow, and persistence is increased, in dry soil due to reduced microbial activity. Breakdown is most rapid in warm, moist soils. Waterlogged clay soils tend to become anaerobic, a condition that can increase the longevity of some pesticides (e.g., organochlorines such as DDT) or enhance degradation of others (e.g., dinitroaniline herbicides).
4. Soil organic matter. Soil organic matter serves as food for microbial organisms, and therefore soils rich in organic matter support high microbe populations. Pesticide persistence is longer in soils with low organic matter, because such soils support lower microbial populations; breakdown is more rapid in soils with higher organic matter content.
5. Degree of adsorption. A pesticide that is strongly adsorbed to soil colloids is not available to microbial degradation (e.g., paraquat), and such pesticides have increased longevity in soil.

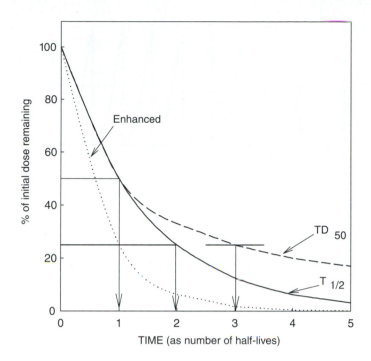

**FIGURE 11–21**
Graph showing theoretical pesticide decline in relation to time, demonstrating the half-life ($T_{1/2}$) concept, with (dotted line) or without (solid line) enhanced breakdown, and the time to 50% dissipation ($DT_{50}$) concept (dashed line).

6. Depth in the soil. The upper layers of the soil have the greatest microbial activity; deeper in the soil there is little microbial activity. Pesticides are thus more persistent deeper in the soil.

The best way to quantify persistence is to determine the time required for 50% of the applied material to be broken down. This value is called the half-life or $T_{1/2}$ (Figure 11–21; solid line) for purely physical phenomena. The quantity of chemical remaining can then be expressed relative to the number of half-lives that have passed; after four half-lives, the amount of chemical remaining would be 1/16 of that initially applied. For pesticides, the concept of time to 50% dissipation, or $DT_{50}$ (Figure 11–21; dashed line), may be more useful than the constant $T_{1/2}$ because the dissipation decreases as the concentration of the remaining pesticide decreases. This results in longer persistence than is predicted using constant $T_{1/2}$ values; in the theoretical figure, the $DT_{50}$ for the second 50% loss is double that for the predicted $T_{1/2}$. The values of $DT_{50}$ vary from minutes for some pesticides, to months or even years in some cases. The half-life of a chemical is not a fixed value, but rather varies over a range depending on the previous six factors.

Many pesticides actually show what is described as multiphasic degradation. That is, initial degradation may be very rapid, followed by slower rates of degradation. The reasons for these biphasic or multiphasic curves are complex, and not always well understood. One possible explanation is that the soil consists of various "compartments," which have different rates of adsorption and desorption. With time, most pesticides move into the compartment where the rate of release is much slower and hence the chemical is less available for degradation and leaching.

Use of the half-life concept shows that the maximum accumulation of a pesticide in the soil can only be twice the application rate if repeat applications are only made at or after the time required to achieve the $T_{1/2}$ (Figure 11–22). It is only when applications are made at times shorter than the first $T_{1/2}$ ($DT_{50}$) that

**FIGURE 11–22**
Graph showing theoretical accumulation of a pesticide in the soil following repeat applications made at the time of achieving 50% dissipation.

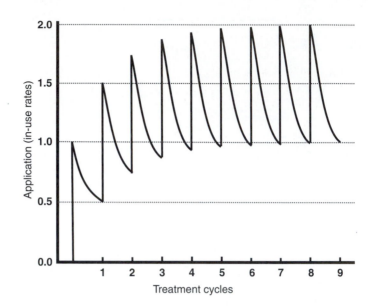

| TABLE 11–3 | Examples of Typical Pesticide Half-life and $DT_{50}$ Times in Soil and Their Relative Persistence |
|---|---|

| Pesticide | Use type | Representative $T_{1/2}$[1] | $DT_{50}$ (days)[2] | Persistence level |
|---|---|---|---|---|
| DDT | Insecticide | 2 to 15 years | 12,419 | Highly persistent |
| Paraquat | Herbicide | 1.5 to 13 years | 500 | Highly persistent |
| Benomyl | Fungicide | 3 months to 1 year | 220 | Persistent |
| Picloram | Herbicide | 3 months to 1 year | 168.5 | Persistent |
| Atrazine | Herbicide | 3 to 6 months | 81.4 | Persistent |
| Chlorothalonil | Fungicide | 1 to 3 months | 49.5 | Moderately persistent |
| Metolachlor | Herbicide | 2 to 8 weeks | 35.9 | Moderately persistent |
| Abamectin | Insecticide | 2 to 8 weeks | | Moderately persistent |
| Terbufos | Insecticide | 2 to 4 weeks | 12.2 | Moderately persistent |
| 2,4-D | Herbicide | Approx. 7 days | 11.7 | Low persistence |
| Metam sodium | Soil fumigant | 1 to 7 days | 5.0 | Low persistence |
| Captan | Fungicide | 1 to 5 days | 3.0 | Low persistence |
| Malathion | Insecticide | 1 or 2 (to 5) days | 1.0 | Low persistence |

[1]The half-life values are average predicted ranges under conditions favorable for breakdown; adverse conditions can substantially increase these times. All information obtained from the ExToxNet website (2000).
[2]$DT_{50}$ values from Gustafson (1993).

a pesticide can accumulate in the soil to levels higher than twice the application rate. No agricultural pesticides in use today are reapplied at a frequency greater than the $DT_{50}$ when label directions are followed, with the exception of the herbicide paraquat, which is so strongly bound to soil that it is not available for microbial degradation.

A selected list of pesticides indicating the possible range of half-lives that exist among pesticides is presented in Table 11–3. Most of today's pesticides have half-lives that range from a day or two to a few months.

The implications of pesticide persistence at the ecosystem level are considered in Chapter 19.

***Plant-Back Restriction***   Persistent pesticides can remain in the soil sufficiently long that significant residues are present when the next crop is planted. The remaining pesticide causes two different problems, which result in so-called plant-back restrictions regarding the length of time between a pesticide application and planting of the next crop.

1. Phytotoxicity to the crop. Persistence is a problem for many herbicides, because residues restrict the crops that can be grown in a rotation. Crop tolerance to herbicide residues is one reason why most herbicides cannot be used in intensive vegetable production systems, and why they should be avoided in backyard vegetable gardens. Many herbicides are subject to plant-back restrictions based on herbicide residues; following are two examples.

    1.1. Picloram is not used in arable cropland because it kills or damages most broadleaf crops planted within two to three years following an application. The use of picloram is restricted to rangeland and noncropped areas because of plant-back restrictions.

    1.2. Atrazine is the most widely used pesticide, but its persistence in the soil limits rotation following corn or sorghum for the next 18 months after application. This is not a limitation in regions where continuous monocultures are common, such as corn after corn, but is a significant limitation in regions where crops such as alfalfa, tomatoes, sugar beets, and kidney beans are grown in rotation with corn.

2. Pesticide residue in rotation crops. When a pesticide remains in the soil following application to one crop, the next crop in the rotation may take it up. If the pesticide is not registered on the rotation crop, any pesticide residue is illegal and crop destruction can be required. To avoid the possibility of illegal residues, plant-back restrictions are required so that the pesticide breaks down before the next crop in the rotation is planted. Plant-back restrictions because of the potential for residues apply to any pesticide that may be used late in the crop-production cycle, but are most frequent for insecticides and fungicides because they are often used near harvest.

Plant-back restrictions are clearly specified on the pesticide label (see later in this chapter), and such restrictions must be followed.

***Problem Soils***   The repeated application of pesticides to soil may lead to reduced pesticide efficacy over time, because microorganisms that are capable of degrading the pesticide are selected by repeated applications. Increased population densities of such microbes result in an enhanced degradation rate for the pesticide (Figure 11–21; dotted line). Soils with such enhanced degradation have been called "problem soils." Examples include the following:

1. Phenoxy herbicides break down in soils much more rapidly if phenoxy herbicides have been applied previously. This fact was demonstrated in the early 1960s, and was the first example of the phenomenon.

2. Repeat applications of carbamate insecticides and herbicides have resulted in decreased activity with subsequent applications, especially in the United States Corn Belt. The result has been almost complete lack of efficacy in severe cases. One approach to solve the problem has been to add compounds to the pesticide formulation that block the action of the enzymes involved in the degradation of the pesticide.

# INTERACTIONS BETWEEN PESTICIDES

Frequently, more than one pesticide is applied to a crop. The pesticides may be applied at different times, or simultaneously. In the latter case, it would seem logical to mix the two pesticides and actually apply them together, but this may not be appropriate. Unexpected interactions can occur between chemicals in the same or different classes of pesticides if they are mixed, and the interactions can decrease the efficacy of the pesticide or result in phytotoxicity to the crop. For example, interactions can occur between two herbicides, two insecticides, or between a herbicide and an insecticide.

## Formulation Incompatibility

Sometimes it is necessary to mix pesticides of differing formulation; however, mixtures of some pesticide formulations in a spray tank may result in a negative reaction. A typical problem is the production of a precipitate, which is manifested as a sludge that clogs the sprayer. One general rule of compatibility is that if two pesticides are to be mixed, one formulated as a wettable powder and the other as an emulsifiable concentrate, mix the powder with water first and then add the emulsifiable formulation. Mixing in the powder second causes the mix to cake up and precipitate, because the oil in the emulsion coats the powder particles and causes them to coalesce. There are two additional rules for mixing pesticides. The most important is that pesticides should not be mixed if the label cautions against doing so. If there is no indication on the label regarding compatibility of mixtures, it is wise to mix a small amount of the two pesticides and observe to determine the reaction.

## Altered Crop Tolerance

Pesticide mixtures can alter the activity of one or both pesticides on target or nontarget organisms, which can be a problem. An example of this type of interaction is the loss of crop tolerance to certain herbicides if they are mixed with insecticides. Rice is normally tolerant of the herbicide propanil, but in the presence of organophosphate insecticides, this tolerance is lost and the herbicide kills the rice (Figure 11–23). This occurs whether the herbicide and the insecticide are mixed in the tank or on the plant by separate applications. A similar loss of selectivity occurs in corn when certain sulfonyl-urea herbicides are used with the insecticide terbufos, and in soybeans when certain insecticides are applied with the soybean herbicide metribuzin. The same rules outlined above for chemical incompatibility apply to this type of interaction.

## Alteration of Efficacy

Mixtures of pesticides can alter the efficacy of one or both chemical components on their target species. Increased activity is useful, but decreased activity creates a problem. In known cases where antagonism of activity occurs, pesticide labels indicate that they should not be mixed.

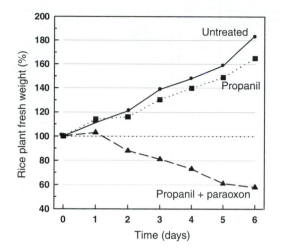

**FIGURE 11–23**
Graph of effect of herbicide propanil on rice in the absence and presence of the organophosphate insecticide paraoxon.
Source: redrawn from Matsunaka, 1968.

## TOXICITY OF PESTICIDES

A pesticide, by definition, is a chemical that kills or injures a pest. This means that all pesticides are toxic, at minimum, to the target pest. Pesticides can also be toxic to nontarget species, including the human applicators or field workers. The study of the action of toxic chemicals is known as toxicology. The next section deals with toxicology fundamentals.

**All pesticides are toxic (otherwise they would not be used).**

## Dose/Response Relationship

"The dose makes the poison" is a well-known statement, and means that toxin concentration and the period of exposure to the toxin determine the effect on an organism. Examples of potentially toxic chemicals that humans may encounter on a routine basis include alcohol, table salt, nicotine, caffeine, aspirin, and gasoline. At high doses all of these chemicals are toxic and can kill, at appropriate intermediate doses they are useful, and at low enough doses they do not have a detectable toxic effect.

Increasing the dose leads to an increased effect (Figure 11–24). When a population of test organisms is exposed to a toxicant, a few individuals respond to relatively low doses, most respond to a medium dose, and a small number require a relatively high dose before they are affected (Figure 11–25). When the response in Figure 11–25 is plotted as cumulative numbers of individuals affected, a sigmoidal response pattern results. The sigmoidal response starts with low level of activity that increases fairly rapidly as dose increases, before slowing to an asymptotic maximum (Figure 11–24). The lowest dose at which an effect is discernable is referred to as the no observable effect level (NOEL).

This view of the dose-response relationship is not universally accepted (Calabrese and Baldwin, 1999), and other responses to a toxin are possible. A phenomenon known as hormesis exists for many chemicals, including pesticides, where instead of acting as a toxin, the chemical enhances growth when applied at very low doses. Practical use of the effect of the impact of low doses of pesticides would require a reevaluation of the NOEL concept, and the application of NOEL to risk analysis for pesticide residues.

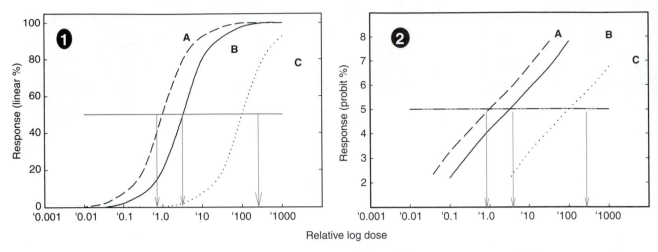

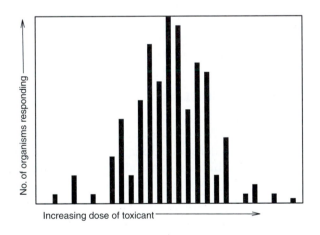

**FIGURE 11–24**   Graph showing theoretical dose-response curves. A, B, and C represent responses of different test species or biotypes to a single toxicant, or a single test species to three different toxicants. Graph (1) plotted using linear scale y-axis; graph (2) plotted using a probit scaled y-axis. Note that both graphs use a logarithmic scaled x-axis.

**FIGURE 11–25**
Theoretical graphical example of the proportion of a population of test organisms responding to increasing doses of a toxicant.

## Acute Toxicity

**Acute toxicity is the response of an organism to a single exposure to a given concentration of a toxicant.**

The toxicity of chemicals, including pesticides, is determined by tests on animals. All data for humans are indirect extrapolation from animal testing, or through insights obtained when people are subject to accidental exposure. The toxicity determined through animal testing varies between organisms of different sizes, species, and even gender. Because different animals have different body sizes, toxicity is usually expressed as units of toxicant per unit of body weight (e.g., mg/kg). Acute toxicity data are expressed relative to the mode exposure to the toxin, through ingestion, inhalation, or absorption through the skin or eyes.

**$LD_{50}$**   The lethal dose 50% ($LD_{50}$) is the single-exposure dose that kills 50% of the organisms in the test population (see horizontal lines in Figure 11–24) when the toxicant is ingested.

Examples of $LD_{50}$ values in rats for chemicals from different pesticide classes are shown in Figure 11–26. The $LD_{50}$ values of several well-known

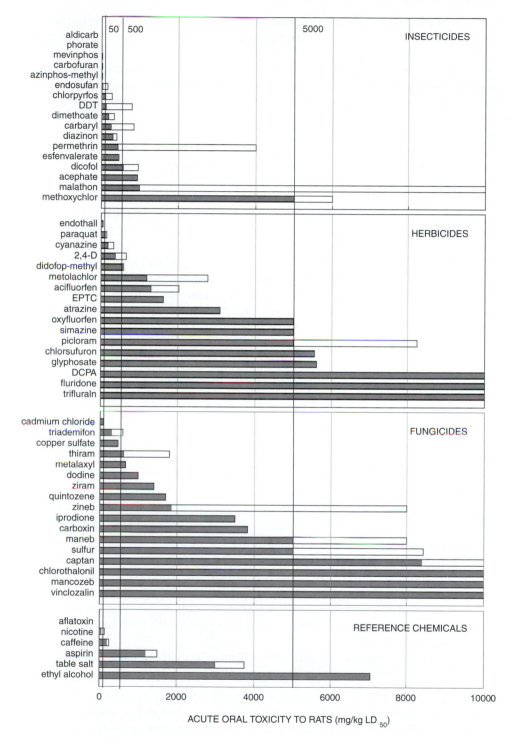

ACUTE ORAL TOXICITY TO RATS (mg/kg LD$_{50}$)

**FIGURE 11–26**

Graphical presentation of comparative single-dose acute toxicology LD$_{50}$ values for examples of selected insecticides, herbicides, and fungicides tested on male rats. Values for representative common household chemicals are provided for reference purposes; aflatoxins are natural by-products of fungal decay universally present in peanut products. Many of the products showing LD$_{50}$ values of 10,000 mg/kg are actually listed as "over 10,000 mg/kg." Open bars represent special formulations.

household chemicals are included for reference. These provide a comparison of the relative toxicity of different pesticide classes. Insecticides and nematicides target animal pests, and so are typically more toxic to animals than are fungicides and herbicides that often target physiological or biochemical systems not present in animals. The data used for Figure 11–26 are derived from tests using the native chemical, and formulations for application may be manipulated to reduce the toxicity of the commercial product in comparison with that of the native chemical. The data presented are from experimental determinations with male rats, as these are most frequently used. Even female rats may not have exactly the same response. Other test animals include mice, rabbits, beagles, fish (e.g., bluegill), and birds (e.g., quail), and all may have different sensitivity to pesticides and dosages. All estimates of toxicity to humans are based on extrapolations from these test organisms.

***LC$_{50}$***    When the route of exposure to a toxicant might be in air or water, the acute toxicity is expressed on the basis of concentration and duration of exposure rather than milligrams per unit body weight. This is the lethal concentration, hence LC, to kill 50% of the test population. For example, an LC$_{50}$ might be 15 parts per million (may also be expressed 15 mg/liter) for 12 hours.

***Selectivity***    Biochemically based selectivity is explained at the physiological level. Difference in the quantity of toxicant required to produce a lethal reaction differs between species and types of organisms. In Figure 11–24, selectivity between species A and B is less than an order of magnitude of the quantity of toxicant, whereas the selectivity between A and C is over two orders of magnitude (all vertical arrows shown in Figure 11–24).

## Chronic Toxicity

**Chronic toxicity describes the long-term effects on an organism of repeated exposure to nonlethal toxicant doses.**

To understand the effects of pesticides on organisms, it is necessary to establish the effects of long-term exposure to the toxicant at doses that do not immediately kill the organism. These are referred to as long-term, or chronic, studies. The consequences of chronic exposure to toxicants can result in different responses in the exposed organisms. The following physiological responses are typically evaluated for pesticides.

1. Carcinogenesis—the production or increase in cancer frequency in the test organisms relative to exposure to a toxicant. This requires lifetime feeding studies, or long-term inhalation studies.
2. Teratogenesis—the change in frequency of birth defects relative to long-term exposure to toxicants. This test requires a minimum of two generations of the test organisms.
3. Oncogenicity—a change in the frequency of noncancerous tumors relative to long-term exposure to a toxicant. This requires lifetime feeding or inhalation studies.
4. Endocrine system disruption—the ability of a toxicant to disrupt the endocrine hormone system in animals is evaluated.

In the United States, establishing chronic toxicity is mandatory for all new pesticides developed since the Federal Insecticide, Fungicide, and Rodenticide Act was passed (see later in this chapter), with mandatory assessment of endocrine disruption added in 1996. The toxicology information developed prior to that time must be upgraded under supervision of the EPA. Long-term toxicology testing is time consuming and expensive. There is discussion within the scientific community regarding the definition of appropriate, accurate experimental protocols for evaluating chronic effects that can be extrapolated meaningfully to humans.

# Types of Exposure

Organisms may be exposed to pesticides through a number of routes, including:

1. Ingestion or oral exposure. The pesticide is either swallowed or eaten. The major concern is contamination of food and drink.
2. Inhalation exposure. Here the pesticide is present in the air and is inhaled into the lungs. The main concern is for field workers, applicators, and people living near sprayed fields, who might breathe in spray mist or dust.
3. Dermal exposure. In this case the pesticide is absorbed through the skin, which can occur very rapidly. Mixing of pesticides with bare hands should be avoided, because dermal uptake of pesticides such as methyl-parathion (no longer in use) through the skin can lead to death. Similarly, walking barefoot through treated foliage is extremely dangerous. Dermal uptake can also occur from contaminated clothes.
4. Ocular exposure. Direct pesticide contact with the eyes can lead to rapid uptake; corneal damage, and in extreme cases blindness or death. Uptake via eyes from pesticide splashes is of greatest concern for pesticide applicators, because they handle pesticide concentrate.

People involved in professional pest management are well aware of the importance of these different exposures that provide routes of pesticide entry into the body. Untrained farm workers, especially in less developed countries, and many homeowners often handle pesticides in an unsafe manner because they do not understand that all pesticides have toxic properties and may be taken up via different routes.

# Hazard

Hazard is not the same as toxicity. A pesticide can be extremely toxic, but if there is no potential for exposure, there is no hazard. If a chemical is not very toxic, then exposure does not entail significant hazard. Talcum powder is a good example. Hazard in relation to pesticides depends on the following:

**For a toxicant to be considered a hazard, there must be the potential for damage and exposure.**

1. Inherent toxicity of the chemical. This hazard is a dose-related problem, and risk analysis assumes that the hazard decreases as the quantity of pesticide decreases. The only direct control over the inherent toxicity as a pesticide hazard is to not use compounds that are considered too toxic.
2. Exposure. Reducing the chance and duration of exposure reduces the hazard. People involved in mixing and loading pesticides for application have the greatest hazard from pesticides because they are the most likely to be exposed. Common sense and the law dictate that pesticides should never be put in drink containers and that impervious gloves, goggles, and shoes be worn when handling pesticide concentrate. Hazard reduction is the objective of a requirement for closed-system loading of category I pesticides. These measures are intended to reduce hazard by lessening the chance of exposure. They are discussed in greater detail under the topic of worker protection later in this chapter.

In the United States, the EPA developed a system for rating the hazards of pesticides based on the types of acute toxicology data discussed in the preceding sections. A summary of the information is presented in Table 11–4. All pesticides are

**TABLE 11–4**    **Oral, Dermal, and Inhalation Ratings of Pesticide Groups Ranked by EPA Toxicity Categories**

|  | EPA Pesticide Categories | | | |
|---|---|---|---|---|
|  | I | II | III | IV |
| Designation | Highly hazardous | Moderately hazardous | Slightly hazardous | Relatively nonhazardous |
| Signal word | Danger—Poison | Warning | Caution | Caution |
| Lethal dose[1] | Few drops to 1 tsp | 1 tsp to 1 oz | 1 oz to 1 pt/lb | Over 1 pt or 1 lb |
| **Hazard indicators** | | | | |
| Oral $LD_{50}$ | Up to and including 50 mg/kg | From 50 through 500 mg/kg | From 500 through 5,000 mg/kg | Greater than 5,000 mg/kg |
| Inhalation $LD_{50}$ | Up to and including 0.2 mg/liter | From 0.2 through 2.0 mg/liter | From 2.0 through 20 mg/liter | Greater than 20 mg/liter |
| Dermal $LD_{50}$ | Up to and including 200 mg/kg | From 200 through 2,000 mg/kg | From 2,000 through 20,000 mg/kg | Greater than 20,000 mg/kg |
| Eye effects | Corrosive; corneal opacity not reversible within 7 days | Corneal opacity reversible within 7 days; irritation persists for 7 days | No corneal opacity; irritation reversible within 7 days | No irritation |
| Skin effects | Corrosive | Severe irritation at 72 hours | Moderate irritation at 72 hours | Mild or slight irritation at 72 hours |

[1]Estimated approximate range for a 155-lb person.

placed into broad categories that provide a readily understood, rapid indication of the relative hazard of compounds without having to know the details of the toxicology. Only one positive hazard indicator is required to place a pesticide in a high hazard category. For example, a native chemical with an oral $LD_{50}$ between 50 and 500 mg/kg (category II) formulated with a solvent that is corrosive to the eyes will be placed in EPA category I. The categorization of pesticides into hazard classes is extremely important in relation to label development and use restrictions.

## Residues

After application, pesticide residues normally decline exponentially over time (Figure 11–21). The decline rate depends on many of the same factors that influence decline in the soil.

> A residue is the portion of an applied chemical that remains on or in the crop or the soil.

1. Pesticide chemistry. Some core molecules are more resistant to microbial breakdown than others; the chlorinated hydrocarbon insecticides are an example of long-lasting pesticides.
2. Target plant. The rate of residue decline varies by plant species, growth stage at application, the part of plant that was exposed to the pesticide, and the part that is harvested.
3. Physiology. The interaction between the pesticide and plant physiology alters the accumulation of the pesticide in plant tissues. A pesticide that moves in the transpiration flow (water) of the plant will accumulate wherever water is being used, such as leaves and large, water-containing fruits.
4. Environmental conditions. Changes in plant condition in response to temperature and water availability can substantially alter residue levels. Residue levels are often higher in drought-stressed plants than in well-watered plants.

Two practical field-level environmental problems associated with residues on plant (or soil) surfaces influence how a pesticide can be used.

1. Liftoff occurs when part of the pesticide residue is released from the target surface, in part due to pesticide volatility. Liftoff can result in unacceptable drift.
2. Dislodged residues are particles of a pesticide that can be dislodged from leaves and fruits by worker activity, such as brushing against the plant. This may be a problem when farm workers reenter a treated field.

Pesticide residues that remain in produce intended for public consumption are also a potentially serious problem, and are considered later in this chapter.

## Tolerance

Pesticide tolerance is legally defined as the quantity of pesticide residue that can remain in the harvested product. A pesticide tolerance is determined by taking the NOEL obtained through acute and chronic testing, and then dividing it by 100 or 1,000 depending on the toxicological characteristics of the pesticide. The tolerance is expressed in parts per million (ppm) or parts per billion (ppb). One part per million is the equivalent of 1 oz of salt (about enough to fill a typical salt shaker) evenly mixed through 12,500 5-lb bags of sugar.

> The pesticide tolerance is the maximum quantity of pesticide residue that can remain on or in the harvested product at the time of sale.

Compliance with tolerances requires the accurate determination of how much residue is present in the plant tissues intended for consumption. Accordingly, pesticide development requires that a reliable and accurate extraction/detection procedure be developed for the native pesticide and all its major metabolites. Produce is then subject to analytical testing to quantify the residue present. If produce is determined to have a residue that exceeds the tolerance, then the produce is "red-tagged" and all sales and uses of that produce are suspended. The produce must be destroyed if residues do not decline to levels below the tolerance.

If a pesticide is not registered for a particular crop, then there is no tolerance level for any detectable amount of pesticide in that crop. With no tolerance, *any* residue of the nonregistered pesticide is illegal. Such illegal residues have important implications in relation to pesticide drift and plant-back restrictions, and for any applications to nonlabeled crops.

## Reentry Interval

After a pesticide has been applied, a certain amount of time must pass before it is safe for workers to reenter the field. Reentry interval is the time required for the surface residues of a pesticide to decline to a safe level following application. Reentry intervals in the United States range from four hours (minimum) to several days, and depend on the toxicity of the pesticide and its half-life. The reentry requirements are specified on the label. The pesticide-treated area must be posted with a warning sign. The warning sign notes what pesticide was applied, the length of the reentry restriction, and the time when the reentry restrictions expire (see later section in this chapter on posting).

## Preharvest Interval

The preharvest interval is the time required between pesticide application and harvest. It is the time that is adequate to permit pesticide residues on or in plant tissues intended for consumption to decline to below tolerance levels. Preharvest intervals are most important for pesticides used on fresh market produce. The preharvest interval varies from none for nontoxic materials such as *Bacillus thuringiensis,* to months for persistent chemicals such as the chlorinated hydrocarbons. The preharvest interval is specified on the label and must be followed.

## LEGAL ASPECTS OF PESTICIDE USE

### Overview of Pesticide Regulations

**Society needs a coherent policy for integrated pest and pesticide management.**

Pesticide policy addresses the concerns of society and the ecosystem constraints (see Chapter 19), and has the following broad goals.

1. Develop quality control of pesticides. Ensuring the quality of pesticides was the main reason for the original laws; this is now a minor issue as earlier sampling showed a high level of compliance.
2. Develop responsible pesticide use policies:
    2.1. Crop protection. Provides assurance that the pesticide would perform as described, which is a major reason for regulations.

2.2. Pesticide handler/applicator protection. Assures people who are working with pesticides that they are not placed at unnecessary risk.

2.3. Field worker protection. Assures farm workers who are not involved with pesticide application that they are also not at risk of pesticide exposure.

2.4. Consumer protection. Ensures that the public is not exposed to hazardous levels of pesticides.

2.5. Environmental protection. Ensures that the impacts of pesticide use on the environment is minimized. This is a major issue.

3. Develop safe international trade:

3.1. Export protection. It is necessary to ensure that one country is not exporting pests or pesticide residues to another country.

3.2. Import protection. Likewise, it is necessary to ensure that a country is not importing pests or pesticide residues from other countries.

4. Stimulate technology. Regulations can drive the development of new technologies for pest management, even though this may not have been the intent. For example, legislation requiring closed-system loading for class I liquid pesticides in California led to development of equipment to achieve the mandate of the new law.

5. Promote pesticide education programs.

6. Foster communication between organizations and agencies.

## History of Pesticide Regulation

Pesticide regulation goes back to the beginning of the twentieth century, at which time the concern was user protection from misrepresentation of efficacy and to ensure product quality. The following dates outline the legislative milestones for the United States; most industrialized countries have followed a similar pattern of legislation.

1910   Federal Insecticide Act. This act was the initial attempt to regulate pesticide quality.

1938   Federal Food Drug and Cosmetic Act (FFDCA). This act provides the vehicle for monitoring of pesticide residues in food, which came about because of excessive arsenic and lead residues in various crops in the 1930s. The act has been amended several times since 1938, but it is still the authority for monitoring pesticide residues in food.

1947   Federal Insecticide, Fungicide, and Rodenticide Act (FIFRA). This act was the start of the modern era as it defines pesticide by "intended use." What this means is that any compound that is to be used to kill any form of pest will be designated as a pesticide and its regulation will fall under the mandates of FIFRA. The year 1947 saw the beginning of synthetic organic chemical pesticides (e.g., chlorinated hydrocarbons such as DDT, phenoxys such as 2,4-D). FIFRA remains the vehicle that provides the legislative mandate to manage registration of pesticides, and it has been amended numerous times since its introduction. The United States Department of Agriculture (USDA) originally carried out all mandates of FIFRA.

1958   Delaney clause added to FFDCA. This amendment required that any cancer-causing compound should not be permitted in food—the so-called zero tolerance. The wording of the clause did not permit any risk-benefit analysis. Improved technology has allowed the detection of smaller and smaller amounts of pesticides, and has made the Delaney clause untenable (see 1996 FQPA).

1970   Environmental Protection Agency (EPA). President Nixon required all mandates under FIFRA to be transferred to the EPA from the USDA. Up to that time, in-state use of pesticides that were not federally regulated were allowed. Such registrations were discontinued under the EPA.

1995   Additional federal worker protection standards. Implemented within the framework of the Occupational Safety and Health Act (1970), many aspects of the standards impacted pesticide application and use.

1996   Food Quality Protection Act (FQPA). This act provided major revision of how pesticide residues in food are evaluated, altered cancer risk assessment to solve the dilemma created by the Delaney clause, created tolerance levels for children, regulated endocrine system disrupters, and developed the concept of "risk cup" for lifetime residue exposure. The risk cup implies that residues of all pesticides with similar mode of action must be considered relative to lifetime exposure, and in aggregate rather than singly.

Other federal acts have been passed that also have direct impact on pesticide regulation and use. These include the Clean Air Act (1970), Occupational Safety and Health Act (1970), Clean Water Act (1977), Safe Drinking Water Act (1974), Resources Conservation and Recovery Act (1976), Endangered Species Act (1973), and Superfund Amendments and Reauthorization Act (1986).

## Codes and Regulations

Laws are enacted by federal or state governments. Laws created by the acts are implemented as part of the code of regulations. In the United States, the federal code for agriculture is under Title 7 (pesticides are also impacted by some other codes). Various agencies carry out the mandates of the relevant parts of the regulations. The federal agencies responsible for regulating pesticides and pest management in the United States, with mention of specific regulatory activities, include the following:

United States Environmental Protection Agency (EPA)—oversees all aspects of pesticide regulation

United States Department of Agriculture (USDA)—oversees regulation of exotic and quarantined pests

Food and Drug Administration (FDA)—monitors acceptable pesticide residue levels in foods

Federal Aviation Administration (FAA)—licenses pilots involved with aerial application of pesticides, and all aspects of agricultural aircraft operation

Department of Transportation (DOT)—regulates interstate shipments of pesticides

United States Geological Survey (USGS)—provides data on pesticide contamination in water

United States Fish and Wildlife Service—involved with regulation of pests and pesticides that impact fish and wildlife

Occupational Health and Safety Administration (OSHA)—regulates aspects of agricultural worker safety

# Major Requirements of FIFRA

FIFRA is divided into many sections; each section defines a particular set of laws. For example, section 3 deals with registration of pesticides, section 18 deals with exemptions of federal and state agencies, and section 24 deals with special state authority (24c deals with "special local needs" for use of particular pesticides that are otherwise unavailable).

Prior to selling a pesticide (whether inorganic, synthetic organic, microbially derived, or a biochemical of biological origin) in the United States, the following must occur.

1. All pesticides must be registered by the EPA before they can be used.
2. Product registration requires specific information so that the product when used according to label directions will:
   2.1. effectively control pests listed on label.
   2.2. not injure humans, crops, livestock, wildlife or damage the environment.
   2.3. not cause illegal residues in food or feed.
3. Pesticides must be classified into general use or restricted use.
4. Restricted-use pesticides can only be applied by a certified applicator.
5. Pesticide production facilities must be registered and inspected by the EPA.
6. Use of any pesticide in a manner inconsistent with the label is prohibited; the approved label is a legal document and it is unlawful to apply a pesticide in a manner contrary to the label.
7. States may register pesticides on a limited basis for local needs when no alternatives are available. These are called special local needs (SLN) registrations; they are promulgated under section 24 of the FIFRA code. This typically extends the use of a pesticide to a crop, or situation, which is currently not included on the label. Pesticide uses can also be allowed on an emergency exemption basis if certain criteria are met and there are no available alternatives. This is allowed under section 18 of FIFRA prior to establishment of a federal tolerance for that food/feed use.
8. Violations of the label specifications are subject to heavy fines and/or imprisonment.
9. The EPA can, under FIFRA, suspend, cancel, or restrict use of previously registered pesticides that have been deemed to pose an unacceptable risk to consumers or the environment.

# Restricted-use Pesticides

The EPA must classify all pesticides as either general use or restricted use. Restricted-use pesticides are considered sufficiently hazardous that only appropriately trained personnel can apply them. There are several "triggers" (specific hazardous characteristics of the pesticide) that are used to classify a pesticide as restricted use.

1. Acute toxicology. Pesticide is sufficiently toxic that there is immediate health hazard. (All category I pesticides with a 50 mg/kg $LD_{50}$ or lower acute toxicity are automatically designated as restricted use only.)
2. Special hazard to applicators or workers (e.g., eye hazard due to toxicity of solvent used).

3. Hazard to nontarget animals, such as honey bees, or to wildlife.
4. Possess environmental problems, such as drift to other crops, or likely to cause groundwater contamination.
5. Problems with persistence in the soil, such that it poses a hazard to subsequent crops.

Restricted-use pesticides can only be applied by a certified applicator who has been trained in appropriate procedures and who has passed a written examination. If farmers wish to apply restricted-use pesticides on their own land (i.e., not for hire), they must have received training equivalent to that of the certified applicator. Restricted-use pesticides are not available for use by the general public. Individual states can impose additional conditions.

## Requirements for Federal Registration of a Pesticide with the EPA

The following information must be provided before a federal pesticide registration will be approved.

1. Product chemistry must be fully documented.
2. Analytical procedures to extract and detect the pesticide, and its metabolites, must be developed, including efficiency of extraction from all harvested plant parts on which the pesticide will be registered.
3. Hazards to the following organisms must be identified:
    3.1. Wildlife, especially birds (e.g., quail used in toxicology studies)
    3.2. Aquatic organisms: fish (e.g., bluegill) and various other test species
    3.3. Humans and other animals (inferred from testing on rats, mice, rabbits and dogs)
    3.4. Phytotoxicity to nontarget plants
4. Environmental fate must be established in air and water.
5. All major metabolites must be identified and their toxicological properties determined.
6. The following toxicological studies must be conducted:
    6.1. Acute toxicity. Establishes $LD_{50}$ and other relevant data for several different classes of organisms.
    6.2. Chronic. Establishment of chronic toxicity requires long-term feeding tests at sublethal concentrations; requires two generations to evaluate reproductive effects (mice, rats, and dogs).
        6.2.1. Mutagenicity (ability to cause genetic alteration; uses the Ames test on bacteria)
        6.2.2. Teratogenicity (ability to cause birth defects)
        6.2.3. Carcinogenicity (ability to induce cancer; requires autopsy of all major organs)
        6.2.4. Endocrine system disruption (ability to induce changes in hormone balance)
7. Data must be obtained to document the efficacy of the pesticide; presently, federal law does not require that these data be submitted for regulation.

The cost of developing the above mentioned information for a new active ingredient is approximately $20 to $40 million. After satisfactorily completing all data requirements, the registrant, usually a major chemical company, submits a summary of the information called a packet of data to the EPA. After reviewing the submitted data, the EPA makes a decision to register the pesticide, to require further data to resolve unanswered questions, or to deny the registration. The registration decision on registering a new active ingredient is published in the *Federal Register* (the official record of government actions in the United States). At the time of registration, the pesticide is given a registration number, which appears on the approved label. The EPA charges a one-time fee for establishing the tolerance levels for a new active ingredient and an annual maintenance fee. Following federal registration and label approval, the states must approve the registration. For many states this process is automatic, but for others, such as California, there is a further exhaustive review before granting a state registration.

Similar registration procedures are now used in most developed countries (Garner et al., 1999). Due to increasing international trade in pesticides and pesticide-treated commodities, it is becoming imperative that the same rigorous registration processes be used on a worldwide basis, but such uniformity has yet to be achieved.

Currently there is uncertainty on the part of the EPA regarding how to register transgenic plants that express genes producing a pest control chemical, such as the *cry* genes from Bt. The agency has been taking the hard line that a chemical that kills pests is a pesticide regardless of origin. This would require that transgenic crops that express a toxin, such as the Bt endotoxins, be registered as a pesticide.

## The Pesticide Label

The registrant writes the label and submits it for approval at the time of requesting registration by the EPA. Once approved, the label can only be altered with the further approval of the EPA. A label requires that certain information be provided. The fictitious DePesto (Figure 11–27) and DeWeed (Figure 11–28) labels depict the type of information required on the label for a restricted-use pesticide (DePesto) and a general-use pesticide (DeWeed). The following section numbers are targeted to the appropriate place on the fictitious labels.

> **The pesticide label is a legal document. Application of a pesticide in a manner inconsistent with the label is a crime.**

1. Brand or registered trade name/trademark (varies by language and country)
2. Type of pesticide (e.g., insecticide, miticide, herbicide)
3. Chemical name of active ingredient(s); referred to as the active ingredient
4. Common chemical name
5. Type of formulation (e.g., wettable powder, emulsifiable concentrate)
6. List of active ingredients
7. Total contents in the container (e.g., 5 gallons, 5 lb)
8. Name and address of the manufacturer
9. Registration number issued by the EPA at the time of label approval
10. Establishment number (indicates the factory where the pesticide was manufactured; may also include a lot number)
11. Use classification (general or restricted)
12. Signal word (indicates the relative hazard of the pesticide, and is based on the hazards presented in Table 11–4):
    12.1. DANGER (PELIGRO); skull and crossbones on package; category I
    12.2. WARNING (AVISO); category II
    12.3. CAUTION: categories III and IV
13. Precautionary statements (provide information on how to use the pesticide safely; Itemize how to reduce hazard to eyes, skin, etc.)

## DIRECTIONS FOR USE CONTINUED (18)

**METHODS FOR APPLICATION:** The minimum gallonage requirement is 10 gal. of finished spray per acre with ground equipment, 2 gal. per acre with aircraft.

**(17) ALFALFA:** Ground or air application - alfalfa and Egyptian alfalfa weevil larvae, pea aphid, and in New York state for snout beetle control. Apply the amount of De Pesto indicated in the chart, when feeding is noticed or insects appear. Alfalfa weevil adult – apply 1-2 pints per acre prior to bloom. Observe the indicated number of days after application before cutting or grazing. Do not apply more than once per season. Apply only to a field planted to pure stands of alfalfa. For waterfowl protection, do not apply in fields in proximity of waterfowl nesting areas or areas where waterfowl repeatedly feed.

(16)

| Pints of De Pesto per acre | Do not cut or graze within: |
|---|---|
| ½ | 7 days |
| 1 | 14 days |
| 2 | 28 days |

(20)

**CORN, FIELD:** Ground application — corn rootworms — use ½ pints of De Pesto per 13,000 linear feet (1 acre with 40 inch spacing). Apply, at planting, as a 7 inch band over the row or injected of the row by mixing with water or liquid fertilizers; mix in the following way to make sure that the mixture is physically compatible. Premix 1 part of De Pesto with 2 parts water. Add this premix to the tank of fertilizer along with rinsings from the premix container. Maintain agitation in the tank after mixing and during application. Do not mix until ready to use.

**SUGARCANE:** Sugarcane borer – apply 1-1½ pints of De Pesto per acre using aerial equipment. Check sugarcane fields weekly, beginning in June and continuing through August. Make first application only after visible joints form and 5% or more of the plants are infested with young larvae feeding in or under the leaf sheath and which have not bored into the stalks. Repeat whenever field checks indicate the infestation exceeds 5%. Do not apply within 17 days of harvest. Do not use in Hawaii.

**(12) ORANGES, LEMONS, GRAPEFRUIT & TANGELOS in Arizona and California:** Air or ground application – Citrus thrips – apply De Pesto at ½ to 1 pint per acre. Use the higher rate on severe infestations of thrips. Use sufficient water to obtain thorough coverage (5 to 15 gallons per acre by air). Apply in the early spring before bloom when the new growth is about 3 to 4 inches long. Make additional applications as needed until the new fruit is walnut size. Application at petal-fall may be critical to prevent scarring of the fruit. Applications during mid-summer to protect new growth on young trees are also recommended. Do not apply within 3 days of harvest. Do not graze livestock in treated orchards for 10 days after treatment.

**POTATO:** Tubeworm, cabbage looper, aphids, and in areas east of the Mississippi river leafhoppers and flea beetles. Apply De Pesto at indicated rates when field checks show that the insect infestation is above 5%. Tubeworm, cabbage looper and aphid – apply ½ to 1 pint per acre. Leafhopper and flea beetles – ½ pint per acre. Do not apply within 14 days of harvest.

---

## RESTRICTED-USE PESTICIDE (11)

**For retail sale and use only by certified applicators or persons under their direct supervision, and only for those uses covered by the certified applicator's certification.**

# DE PESTO (1) (2)

INSECTICIDE
FLOWABLE CONCENTRATE (5)

ACTIVE INGREDIENT: (3)
Pestoff (*trisalicylic-2, 5-dichlorocarbamate*) ......... 43.8% (4)
INERT INGREDIENTS ................................ 56.2% (6)
TOTAL ............................................... 100.0%

THIS PRODUCT CONTAINS 4.0 LBS OF PESTOFF PER GALLON

**KEEP OUT OF REACH OF CHILDREN**

**DANGER  PELIGRO** (12)

**STATEMENT OF PRACTICAL TREATMENT**

**IF SWALLOWED**-Drink 1 or 2 glasses of water and induce vomiting by touching back of throat. Do not give anything by mouth to an unconscious person. Get medical attention.

**IF ON SKIN**-In case of contact remove contaminated clothing and immediately wash skin with soap and water. Get medical attention.

**IF INHALED**-Remove to fresh air. If not breathing, give artificial respiration, preferably by mouth. Get medical attention.

**IF IN EYES**-Immediately flush eyes with plenty of water for at least 15 minutes. Get medical attention.

*NOTE TO PHYSICIAN - Pestoff is a cholinesterase inhibitor. Atropine is antidotal.*

SEE SIDE PANEL FOR ADDITIONAL PRECAUTIONARY STATEMENTS

A-Z Chemicals Inc., (8)
Chemcity, Minnesota 558888
EPA Registration No. 102357-A2 (9)
EPA Est. 102357-MN-1 (10)

**A-Z** LOGO

NET CONTENTS ONE GALLON (7)

---

## PRECAUTIONARY STATEMENTS (13)

### HAZARDS TO HUMANS AND DOMESTIC ANIMALS
**DANGER**
Fatal if swallowed, inhaled or absorbed through the eyes. Do not get in eyes or on clothing. Avoid contact with skin. Do not breathe dust or spray mist.

### PERSONAL PROTECTIVE EQUIPMENT (PPE) (14)
Applicators and other handlers must wear:
• Protective eyewear
• Coveralls over long-sleeved shirt and long pants
• Waterproof gloves
• Socks and shoes
Discard clothing that has been drenched or heavily contaminated with concentrate of this product.

### ENVIRONMENTAL HAZARDS
This pesticide is toxic to wildlife and fish. Use with care when applying in areas frequented by wildlife or adjacent to any body of water. Keep out of lakes, streams or ponds. Do not apply when weather conditions favor drift from target area.

### BENEFICIAL INSECT CAUTION
This pesticide is highly toxic to bees exposed to direct treatment or to residues remaining in the treated area. Do not apply when bees are actively visiting the crop. Do not apply to the crop, or weeds blooming in the treated area. Applications should be timed to provide the maximum interval between treatment and the next period of bee activity.

### PHYSICAL OR CHEMICAL HAZARDS
Flammable! Keep away from heat or open flame.

## DIRECTIONS FOR USE (15)
It is a violation of federal law to use this product in manner inconsistent with this label
### RE-ENTRY STATEMENT (19)
Do not treat areas where unprotected humans or domestic animals are present. Do not allow entry into treated fields within 48 hours of treatment, unless full PPE is worn. Consult appropriate state and local regulatory officials for State Re-Entry Restrictions which take precedence if more restrictive than those stated on this label.

### STORAGE AND DISPOSAL (21)
Do not contaminate water, food or feed by storage, disposal or cleaning of equipment. Pesticide, spray mixture, or rinsate that cannot be used or chemically processed should be disposed of in a landfill approved for pesticides or buried in a safe place away from water supplies. Triple rinse (or equivalent) and dispose in an incinerator or landfill approved for pesticide containers, or bury in a safe place. Consult federal, state, or local disposal authorities for approved alternative procedures.

**FIGURE 11–27** Hypothetical label for a fictitious restricted-use insecticide. The numbers indicate key components of the information required on the label, and are keyed to notes in the text.

**FIGURE 11–28** Hypothetical label for a fictitious general-use herbicide. The numbers indicate key components of the information required on the label, and are keyed to notes in the text.

14. Personal protective equipment (PPE) (indicates the type of protective equipment to wear when applying the pesticide or when entering a treated field prior to expiration of the reentry period)
15. Directions for use (explains use rates and how to use the pesticide to control pests; typically broken into several subsections)
16. List of application rates (This list provides the legal maximum amount of pesticide that can be applied. The pesticide can be used at a lower rate than is listed on the label but it cannot exceed the stated rates. The maximum rate must be stated per season or per year when multiple applications are permitted.)

**It is a violation of federal law to apply a pesticide to a crop for which it is not registered.**

17. List of sites, or crop for most agricultural pesticides, to which the pesticide can be applied (These listed sites (crops) are the only ones to which the pesticide may legally be applied.)
18. List of organisms controlled (provides a list of those organisms expected to be controlled by the pesticide)
19. Reentry statement (indicates the time required before workers can safely return to the site (field) without wearing specified PPE; time varies from four hours to several days depending on pesticide and crop)
20. Preharvest interval (indicates the time that must elapse between application and when the crop is safe to harvest, and is based on the time for the pesticide residue to decline below the established residue tolerance)
21. Storage and disposal (states how the pesticide should be stored, how excess spray mix must be handled, and the proper procedures for cleaning and disposal of the empty container)

Although the label appears to be a set of instructions on how to use the pesticide, it is in fact a legal document that defines the safe and accepted applications of the pesticide. The label clearly states that it is illegal to apply a pesticide in a manner inconsistent with the label; thus, civil, and even criminal, penalties can be levied against persons applying pesticides in a manner contrary to the label. The main reasons for this requirement are that uses not presented on the label can lead to residues in food or feed that have the potential to be dangerous to persons or animals eating the product, it could be dangerous to field workers, and it could lead to contamination of groundwater or other undesirable nontarget effects.

## State and Local Pesticide Regulation

States and other local regions within a country can enact laws that place additional restraints on IPM tactics. In the United States, such local laws cannot be less stringent than those set by the federal government, but states can impose stricter requirements than those imposed by the federal government. California leads the United States—and probably the world—in the development of pesticide regulations aimed to meet the local needs of the region. Some examples of additional California requirements will be provided in the next section of this chapter. Other states and local regions are adopting measures similar to those in California. It is incumbent on any person working in pest management (see next section) to learn not only the federal regulations that affect this area, but also the local laws that may impose additional requirements.

# Pesticide Users

Several different groups of people are involved with pest management decision making and the application of pesticides.

***Farmer/Landowner***   All persons can apply general-use pesticides to crops on land that they own, provided such application is in compliance with the label. Landowners must meet certified applicator requirements to apply a restricted-use pesticide, however. This means that they have been trained and have demonstrated competence in all special requirements for the use of the pesticide.

***Pest Control Advisor***   If pest control advice is provided for a fee, including recommendations for pesticide use, then the person giving such advice is called a pest control advisor (PCA). This title is now widely used, but the levels to which PCAs are regulated vary widely by country, and by state within the United States.

Contingent on local regulations, a well-thought-out written pest management recommendation will include the following types of information.

1. Name of the grower or land manager
2. Name of the PCA making the recommendation
3. Name of the crop to be treated
4. Identity of the pest(s) to be controlled
5. Name of the pesticide(s) to be applied and the rate of application
6. Site location information
7. Alternative ways to solve the problem (likelihood of success, costs, etc.); referred to as mitigation alternatives
8. Location of any environmentally sensitive areas (wildlife refuge, river, etc.) within the area
9. Adjacent crops that might be sensitive to spray drift (e.g., melons to sulfur dust)
10. Public health hazards in the area (e.g., schools, houses, or other development)
11. Reentry intervals for field workers
12. Harvest interval (time from spraying to harvest adequate for dissipation of residues)
13. Special considerations (e.g., PPE, plant-back restrictions)

For greater details, see Chapter 9 of Flint and Gouveia (2001).

***Applicators***   Applicators apply pesticides for hire; they are also referred to as pest control operators, or PCOs. A strictly PCO company makes no recommendations. Any PCO can apply general-use pesticides, but only a licensed certified applicator who has passed the required special exams can apply restricted-use pesticides.

***Pesticide Dealers***   These companies typically only sell pesticides, and are generally licensed by the state.

***Full-service Companies***   These companies carry out the functions of advising, selling, and applying pesticides. Full-service companies thus have PCAs who make recommendations, sell pesticides, and apply them. Several of these companies operate regionally or even nationwide in the United States.

***Homeowners***    Individuals can handle and apply small amounts of pesticides if they are labeled for homeowner use. Pesticide use by homeowners is probably where the greatest pesticide application and disposal abuses occur. For this reason most pesticides available to homeowners are sold in less concentrated formulation than those used in agriculture. Homeowners often do not have the expertise or training to safely apply pesticides, and are often poorly informed as to the potential dangers of inappropriate pesticide use. It is not uncommon to observe even well-educated persons applying pesticides dressed only in a pair of shorts and wearing open-toed thongs. This practice is indicative of the total disregard for safety on the part of many homeowners when applying pesticides, because for even the most innocuous pesticide, the label states that a long-sleeved shirt, long trousers, and socks and shoes should be worn.

## Pesticide Use Reporting

Reporting of pesticide use is not required in most regions of the world, so there is limited information on the exact extent of pesticide usage. In most situations, the data for pesticide use are only estimates. There are no federal requirements for reporting pesticide use in the United States. Since 1993, California has required that commercial pesticide applications be reported to the state via the local agricultural commissioner's office. This is referred to as 100% use reporting. California data are therefore available on agricultural use of pesticides. Other states and countries are adopting similar reporting.

## Protection of Pesticide Users

**Workers must be informed about pesticide use and be protected from exposure.**

It is necessary that workers who apply pesticides, and the workers in agricultural fields, be protected from poisoning by pesticides. When pesticides were first introduced, worker protection standards were low and, hence, many workers were poisoned. In the developed countries this was deemed unacceptable and worker safety standards have been steadily increased to the point that workers are unlikely to be exposed to toxic levels of pesticide when applications are made in compliance with the label (Figure 11–29a). It is a sad statement about human society that in the less developed countries worker protection from pesticide exposure is poor (Figure 11–29b), and many workers are still poisoned every year.

In the United States, right-to-know laws and worker protection and safety laws have changed as attitudes toward worker protection have changed. Following are examples of current United States worker protection standards for agricultural pesticide users.

1. Right to know. All workers must be informed about the hazards involved with pesticides.
2. Training requirements. All workers must be trained in relation to pesticide hazard and what to do in an emergency. Records of such training must be maintained.

(a)                                               (b)

**FIGURE 11–29**  (a) Person wearing full personal protective equipment while applying a pesticide; note use of impervious suit with full hood, gloves, face mask, and respirator. (b) Worker in a nonindustrialized country applying pesticide with lack of proper protective equipment, which is a potential exposure to pesticide poisoning.
Source: photograph (a) by Jack Clark, University of California statewide IPM program; and (b) by Marcos Kogan.

3. Medical records and testing. Baseline cholinesterase levels must be established, and periodic monitoring must be performed for all persons involved in pesticide application.
4. Safety equipment or personal protective equipment (PPE). Such equipment varies with class of pesticide, the requirements of which are spelled out on the label. The PPE ranges from full respirator and impervious clothes (shoes or boots, coveralls, gloves), to regular coveralls, socks, and shoes.
5. Reentry interval and posting. The posted information informs workers when it is safe to reenter the field. The interval varies from four hours for nontoxic materials to several days for more persistent toxic compounds. The reentry period must be posted (Figure 11–30) and observed; anyone entering the field before the reentry period expires must be wearing the PPE designated by the label.
6. Language barriers. Pesticide labels that are printed in a language that is foreign to the agricultural field worker must be interpreted for those workers.

***Mixer/Loaders and Applicators***  Workers involved in loading and mixing pesticides have the highest risk of exposure to concentrated pesticides, and therefore require more stringent safety standards. Guidelines for safety include the following considerations.

1. Appropriate protective clothing/safety equipment must be provided by the employer and must be worn by the employee. The employer must also provide washing and changing facilities.
2. The employer must provide appropriate training for employees, and the training must be documented.
3. The employer must provide baseline and updated cholinesterase testing for persons working with certain types of pesticides.
4. All high toxicity liquid pesticides (category I) should be introduced into the sprayer using a closed loading system, as has been adopted as a requirement in California. The goal of such systems is to allow pesticide transfer without directly exposing the mixer-loader to the pesticide.

**FIGURE 11–30**
Sign, or posting, on perimeter of field showing what pesticide has been applied and when it is safe to reenter the field without designated protective equipment.
Source: photograph by Robert Norris.

***Field Workers***    These workers are typically not exposed to pesticide concentrate, but can come in contact with spray residues over an extended period. Guidelines for protection of field workers are often mandated by law, but should include the following:

1. Safe reentry times, based on toxicity and amounts of residues that may be dislodged, a time at which it is safe to reenter a treated field is established. The minimum reentry time is four hours and can extend to several days for more persistent compounds that are toxic to humans. Long reentry intervals can pose problems for picking of perishable crops that do not store well.
2. Oral and written notice of treated areas is required.
3. Warning signs must be posted around the field indicating the pesticide applied and the duration of the reentry restriction (Figure 11–30).

## Illnesses from Pesticides

In the United States, death due to accidental pesticide poisoning is low. Most pesticide-related deaths are suicides, and a few are attributed to murder. As noted earlier, deaths due to pesticide poisoning in less developed countries are still common.

The extent of illnesses in workers due to pesticide poisoning is difficult to establish because the available data are problematical. The National Poison Control Centers document perhaps as many as 17,000 pesticide-induced illnesses per year in the United States. The World Health Organization estimates that perhaps as many as 1,000,000 people in less developed countries suffer pesticide-related illnesses each year. These figures may be inaccurate, however, because of difficulties in establishing the exact cause of an illness, illnesses are not reported, and workers may not seek medical help.

# CONSUMER PROTECTION

A major goal of pesticide regulation is to ensure that there are no unacceptable pesticide residues in food. To this end, the United States Food and Drug Administration conducts routine sampling of produce to determine the quantity of pesticide residues present and if tolerances have been exceeded; and in California, the California Department of Pesticide Regulation conducts similar testing.

Residue tolerances in produce are based on an acceptable pesticide residue, and provide consumer protection from pesticides in food. Residue is that amount of pesticide remaining in the produce at harvest. Tolerance is the maximum amount of pesticide residue that can legally be present at harvest, expressed in parts per million (ppm) or parts per billion (ppb). The EPA establishes the residue tolerance when the registration label is issued. The residue tolerance is based on the no observed effect level (NOEL) established during toxicology testing. A safety factor set at 100- to 1,000-fold lower than the NOEL is used to establish the residue tolerance. The mechanism by which residue tolerances are established is somewhat controversial as toxicologists believe that testing high doses of pesticides on cancer-prone rodents may not provide data that represent the activity of low doses of the pesticide in humans. Persons fearful of pesticides believe that any indication of a residue is too great. Added to this controversy is the phenomenon of hormesis. Judging the risk and the benefit of pesticides is thus not an easy task, and the tendency on the part of the EPA has been to err on the side of conservative estimates.

The established residue tolerance is used as a pesticide enforcement tool. The residue of a pesticide in the harvested produce (human food or animal feed) cannot legally exceed the residue tolerance level. Any residue in produce from a crop for which a pesticide is not registered, that is, no tolerance level has been established, is illegal.

**The concept of residue is central to pesticide use enforcement.**

**It is illegal to sell a crop with a residue exceeding the tolerance.**

## Surveys of Pesticide Residues in Food

Samples of produce are taken from the field and from produce distribution centers, and are analyzed for pesticide residues. The analyses can be multipesticide marketplace surveillance (many pesticides), or priority pesticide surveys that are targeted at specific high-risk pesticides. If residues are detected that exceed the tolerance levels, the produce is withheld from shipment and cannot be sold, unless the residue declines to below the tolerance; criminal and civil action may be taken against the producer if violative residues (government jargon) are the result of intentional misapplication. There are few violative residues; most pesticide detections are well within the established residue tolerance.

Data for the 1998 FDA and the 1996 (latest available) CDPR surveillance monitoring programs are presented in Table 11–5. The data were obtained from multiresidue screen analysis (over 200 pesticides can be detected), except for the California Priority pesticide program which targeted 26 pesticides of special health interest.

Most of the "at or below tolerance" samples were actually below 10% of allowable residue. For the California marketplace, 0.23% of the violative samples were overtolerance of a registered pesticide, and 1.31% had residues in crops for which the detected pesticide was not registered. The latter are typically well within tolerance for registered crops. Both FDA and California testing show that most imported produce have pesticide residues within the established residue tolerances.

The data in Table 11–5 are typical of what has been found in surveys over the last 30 years at the federal or state level. Residue tolerances are

| TABLE 11–5 | Federal and California Pesticide Residue Analysis Results | | | |
|---|---|---|---|---|
| Level | No. samples | No residues detected | At or below tolerance | Violative |
| | | ——% of samples—— | | |
| Federal (FDA) | 3,597 (domestic) | 64.9 | 34.3 | 0.8 |
| | 3,860 (import) | 68.1 | 28.9 | 3.0 |
| California (CDPR) | 6,097 (marketplace) | 60.4 | 38.1 | 1.5 |
| | 1,472 (priority)[1] | 80.0 | 19.8 | 0.2 |

[1]All priority samples were from treated crops.
FDA data are for 1998 and California Department of Pesticide Regulation data are for 1996.

rarely exceeded, and typically by not over 1 ppm. Overtolerance residues are mainly due to drift onto nonregistered crops, trace residual in soil, or freak environmental conditions, and are rarely from direct misuse.

Regulations have evolved as society has found new ways to use, and abuse, pesticides. Human nature, being what it is, makes it unlikely that society will ever be able to regulate itself to the point where there is no hazard from pesticide application. As society continues to expand its knowledge of pests and ecosystems and the technologies used for pest management, improvements in the laws and regulations will be made.

## SOURCES AND RECOMMENDED READING

There are many books about the use of pesticides. The following are particularly useful. *The Pesticide Manual: A World Compendium* (Tomlin, 2000) is probably the most comprehensive listing of pesticide chemicals and their characteristics; it is updated periodically. *The Pesticide Book* (Ware, 1994) also provides general coverage of all pesticides. *The Standard Pesticide User's Guide* (Bohmont, 2000) and *Agrochemical and Pesticide Safety Handbook* (Waxman, 1998) provide thorough coverage of pesticide use and application, laws, and regulations. *Botanical Pesticides in Agriculture* (Prakash and Rao, 1997) is an extensive compendium of botanical pesticides and their properties. Two compendia provide extensive coverage of the topic of biopesticides (Copping, 1998; Hall and Menn, 1999). For herbicides, the Weed Science Society of America's *Herbicide Handbook* (1994, 1998) is a particularly valuable reference for information on all aspects of chemistry, uses, toxicology, and fate in the environment. Many aspects on the chemistry, mode of action, and use of fungicides is covered in *Fungicides in Crop Protection* (Hewitt, 1998). Information on toxicology and environmental impacts of most pesticides is available at the ExToxNet website (2000). The following books deal with pesticide registration: *Pesticides: State and Federal Regulation* (Anonymous, 1987) and *International Pesticide Product Registration Requirements: The Road to Harmonization* (Garner et al., 1999). The book *Chemical Pesticide Markets, Health Risks and Residues* (Harris, 2000) documents pesticide hazards, with an emphasis on problems of pesticide abuse in nonindustrialized countries. Books of historical interest that addressed concerns with pesticide use are by Carson (1962) and Van den Bosch (1978), and Hardin (1968) provided a classic paper on problems with protecting the environment.

Aspelin, A. L., and A. H. Grube. 1999. *Pesticides industry sales and usage: 1996 and 1997 market estimates.* Washington, D.C.: Biological and Economic Analysis Division, Office of Pesticide Programs, Office of Prevention Pesticides and Toxic Substances, U.S. Environmental Protection Agency, ii, 39.

Bohmont, B. L. 2000. *The standard pesticide user's guide.* Upper Saddle River, N.J.: Prentice Hall Inc., 544.

Bureau of National Affairs. 1987. *Pesticides: State and federal regulation.* Rockville, Md.: Bureau of National Affairs, Inc., 151.

Calabrese, E. J., and L. A. Baldwin. 1999. Reevaluation of the fundamental dose-response relationship. *BioScience* 49:725–732.

California EPA. 2001. California pesticide use summaries database, http://ucipm.ucdavis.edu/PUSE/puse1.html

Carson, R. 1962. *Silent spring.* New York: Fawcett Crest, 304.

Copping, L. G., ed. 1998. *The biopesticide manual.* Farnham, UK: British Crop Protection Council, xxxvii, 333.

ExToxNet. 2000. Pesticide information profiles, http://ace.ace.orst.edu/info/extoxnet/pips/

Flint, M. L., and P. Gouveia. 2001. *IPM in practice; Principles and methods of integrated pest management,* Publication 3418. Oakland, Calif.: University of California, Division of Agriculture and Natural Resources, xii, 296.

Garner, W. Y., P. Royal, and F. Liem. 1999. *International pesticide product registration requirements: The road to harmonization.* Washington, D.C.: American Chemical Society, xi, 322.

Gustafson, D. I. 1993. *Pesticides in drinking water.* New York: Van Nostrand Reinhold, xii, 241.

Hall, F. R., and J. J. Menn, eds. 1999. *Biopesticides: Use and delivery. Methods in biotechnology,* vol. 5. Totowa, N.J.: Humana Press, xiii, 626.

Hardin, G. 1968. The tragedy of the commons. *Science* 162:1243–1248.

Harris, J. 2000. *Chemical pesticide markets, health risks and residues.* Wallingford, Oxon, UK; New York: CABI Publishing vii, 54.

Hewitt, H. G. 1998. *Fungicides in Crop Protection.* Wallingford, Oxon, UK; New York: CABI Publishing, vii, 221.

Lyle, C. 1947. Achievements and possibilities in pest eradication. *J. Econ. Entomol.* 40:1–8.

Marer, P. J., M. L. Flint, and M. W. Stimmann. 1988. *The safe and effective use of pesticides.* Oakland, Calif.: University of California Statewide Integrated Pest Management Project Division of Agriculture and Natural Resources, x, 387.

Matsunaka, S. 1968. Propanil hydrolysis inhibition in rice plants by insecticides. *Science* 160:1360–1361.

Prakash, A., and J. Rao. 1997. *Botanical pesticides in agriculture.* Boca Raton, Fla.: Lewis Publishers, 480.

Tomlin, C., ed. 2000. *The pesticide manual: A world compendium.* Farnham, Surrey, UK: British Crop Protection Council, xxvi, 1250.

Van den Bosch, R. 1978. *The pesticide conspiracy.* Los Angeles, Calif.: The University of California Press, xiv, 226.

Ware, G. W. 1994. *The pesticide book.* Fresno, Calif.: Thompson Publications, 386.

Waxman, M. F. 1998. *Agrochemical and pesticide safety handbook.* Boca Raton, Fla.: Lewis Publishers, 616.

Weed Science Society of America. 1994. *Herbicide handbook.* Champaign, Ill.: Weed Science Society of America, x, 352.

Weed Science Society of America. 1998. *Herbicide Handbook—Supplement to the seventh edition.* Lawrence, Kans.: Weed Science Society of America, vi, 104.

# *12*

# RESISTANCE, RESURGENCE, AND REPLACEMENT

## CHAPTER OUTLINE

▶ Introduction
▶ Resistance
▶ Resurgence
▶ Replacement
▶ Warning about the three Rs

## INTRODUCTION

This chapter explores the ecological responses of pest populations to pest control tactics, with an emphasis on the phenomena of resistance, resurgence, and replacement—the three Rs of IPM—that often occur when pesticides are used. The phenomena described in this chapter represent a practical manifestation of the principle of natural selection. Natural selection, sometimes colloquially referred to as "survival of the fittest," is the differential survival and reproduction of organisms, as determined by inherited characteristics. The principle of natural selection and its role as a mechanism for organismal evolution, or "descent with modification," was first formally presented to the Linnaean Society in 1858 in a joint paper by Charles Darwin and Alfred Russell Wallace. Natural selection is a powerful force, and its practical importance in applied agriculture and medicine cannot be overstated.

Any pest management tactic may act as a strong selection pressure (see margin note), and the net responses of pest populations to selection have been collectively termed "ecological backlash." The response to selection reflects the ability of some individuals within a pest population to recover, or even thrive, following use of a pest control tactic. It is a manifestation of evolution by natural selection.

A single pest management tactic acts as a selection pressure when it is ineffective against some individuals in a population, and those individuals survive and reproduce. Such individuals possess resistance to the tactic. If the application of a tactic results in the elimination of beneficial predators and

**Selection pressure is biotic or abiotic stress that changes the survival and reproduction of individuals of a particular genotype relative to individuals of different genotypes within a species.**

parasites, the surviving pests are able to reproduce at a level that leads to rapid population growth. In such a scenario, the pests are said to show resurgence. A tactic may control the target pest, but allow a minor pest that was not previously causing damage to increase in numbers so that it causes unacceptable crop loss. This phenomenon is referred to as replacement.

The overall subject is presented here under the general topics of resistance, resurgence, and replacement. These three Rs of pest management may be considered to account for the nearly inevitable downfall of all single-tactic approaches to pest management, and are major reasons why it is necessary to manage pests within an IPM framework. Understanding the ecology of the three Rs is essential if pest management tactics are to be sustainable, but is of no value unless the knowledge is applied.

The problem of selection pressure leading to pest resistance is not limited to agriculture. Bacterial pathogens responsible for disease in humans are pests of people—they cause infections, sore throats, pneumonia, and many other ills. Antibiotics are essentially pesticides, used to control these pests. Excessive use of antibiotics has acted as a selection pressure, leading to the evolution of resistance to antibiotics in many bacterial pathogens of humans. The ecological basis of selection is the same as that for agricultural pest resistance; a single control tactic has been exclusively used, and it has acted as a strong selection pressure. Many antibiotics are of limited effectiveness today because of this phenomenon.

Pest management tactics must be considered as selection pressures, and the potential for resistance assessed if pest management tactics are to remain useful over long periods.

# RESISTANCE

When an organism has a genotype that allows it to tolerate or withstand an environmental stress (whether the stress is biotic or abiotic), the organism is said to have resistance to that stress. It is important to emphasize that the term *resistance* is used in two different IPM contexts. In the first context, resistance of pests to pesticides, and other control tactics, is considered undesirable, as discussed in this chapter. In the second context, resistance of plants to pests is highly desirable as an IPM tactic (see Chapter 17). The two uses of resistance should be clearly understood and not confused.

**Any selection pressure applied to a population will favor individuals with the genotype best able to withstand the selection pressure.**

Of the three Rs of pest management, resistance is considered to be the most important. Resistance will almost inevitably occur if a single pest management tactic or method is used exclusively and repeatedly. The result is that the pest population develops the ability to tolerate the control tactic, requiring the use of an alternative tactic if the pest is to be controlled. This phenomenon is particularly well documented for pesticides. When resistance to a pesticide has been recognized, a new pesticide is applied and then used until it, too, does not work. This scenario is referred to as the pesticide treadmill (Figure 12–1). When a specific pesticide is repeatedly applied, it is generally not a case of *if* pesticide resistance will occur, but rather *when* will it occur. Unless actions are taken to decrease the selection pressure imposed by repeated pesticide applications, natural selection dictates that resistance will occur.

**Selection pressure in the form of repeated pesticide application will increase the frequency of genes within a population that confers resistance to that pesticide.**

The problem of pesticide resistance in pests has received special attention since the 1960s; however, the ecological principle of natural selection applies

**FIGURE 12–1**
Graphic
representation of the
pesticide treadmill.
Source: modified from
Thompson, 1997.

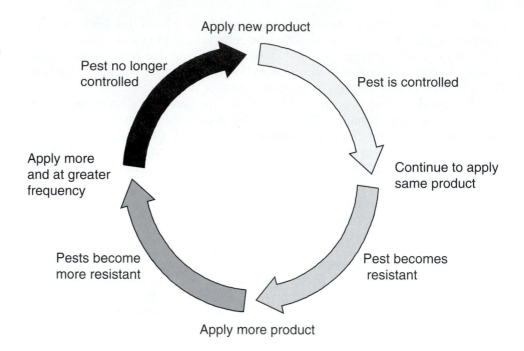

to all management tactics. Some of the earliest examples of pests developing resistance were weeds that adapted to cultural practices.

Pest resistance to pesticides and other control tactics is a major problem for pest management, and indirectly, for society. As pesticides cease to work due to development of resistance in a pest, there must be a constant search for either new pesticides or alternative control tactics. Pest resistance to pesticides may result in the following:

1. Increased pest damage due to inadequate control, resulting in decreased availability of food and fiber as the result of increased crop loss.
2. Increased production costs for the agroecosystem manager if the new pesticide is more expensive, resulting in lowered net revenue.
3. Environmental contamination, if resistance is not recognized and users simply increase application rates in an attempt to regain pest control.
4. Increased commodity cost for the consumer if the costs noted in item 2 are passed on, or if commodity prices increase due to inadequate control (see item 1).
5. If pest resistance becomes widespread, then an effective pest management tactic (the pesticide) has been lost.
6. Decreased sale of an ineffective pesticide with consequent loss of revenue to the manufacturer.
7. Continual investment of time and effort in the development of alternative control tactics.

The importance of pesticide resistance in pests cannot be overemphasized. Growers and pest control practitioners may be confused by pesticide resistance, because the onset may be sudden, and unexpected. This is particularly true when the pesticide has previously worked well. Growers may erroneously conclude that the pesticide was misapplied.

# Historical Development and Extent

Although the phenomenon of resistance to an insecticide may have been observed as far back as 1897, the first documentation of resistance is credited to A. L. Melander, who in 1914 observed San Jose scale still alive under a layer of lime sulfur. It is important to note that resistance in this first example was to an inorganic pesticide, demonstrating that the phenomenon occurs without regard to pesticide type.

Effective new synthetic-organic insecticides, such as the chlorinated hydrocarbons (e.g., DDT) were introduced in the 1940s, and resistance evolved rapidly following their introduction. Since then, almost every new class of insecticides has resulted in development of resistance. There are now estimated to be over 400 insect species resistant to at least one insecticide (Figure 12–2), and many insecticides have been taken off the market because they no longer provide adequate control.

Resistance to a fungicide first appeared in 1940 when *Penicillium* resistant to biphenyl was reported. With the introduction of highly active systemic fungicides in the 1970s, instances of resistance increased rapidly (Figure 12–2). Several fungicides have now been "lost" to resistance.

Resistance to herbicides was the last to appear. Although noted in late 1950s, it was not until 1968 that triazine resistance was demonstrated in common groundsel. During the 1970s, evolution of herbicide resistance was slow, but since the early 1980s there has been rapid evolution of resistance. Over 200 weed species are now resistant to one or more herbicides (Figure 12–2), and several herbicides are no longer used, or their use has been severely restricted, because of lack of efficacy.

Nematode resistance to nematicides has not been documented in the field. Pesticide resistance in vertebrate populations is low, with the only significant example being warfarin resistance in rodents.

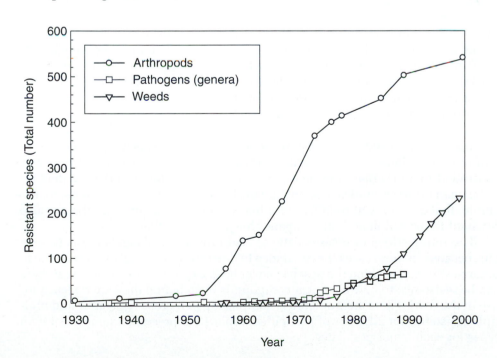

**FIGURE 12–2**

Development of pest resistance to insecticides, fungicides, and herbicides in relation to time. Data for pathogens are by genus, not species; no data could be located after 1990.

Sources: data for insects and pathogens are from Georghiou and Lagunes-Tejeda (1991) and Whalen (2001); data for weeds are from HRAC (2000).

The problem of resistance has become so severe that the chemical industry, in cooperation with the scientific community, has established action committees in an attempt to address the problem. Initially, various specific insect and insecticide "family" committees were formed. In 1981, the Fungicide Resistance Action Committee (FRAC) was formed. This was followed by one for insecticides in 1984, the Insecticide Resistance Action Committee (IRAC). The Herbicide Resistance Action Committee (HRAC) was formed in 1989. All of these action committees have published reviews of resistance in their respective pests, and have developed guidelines that, when implemented, are intended to slow or stop the development of resistance to existing or new pesticides. The HRAC also maintains a running list of all known resistant weeds, and the chemistry of the herbicides to which they are resistant. All action committees have developed websites to provide current information; these are listed at the end of this chapter.

## Resistance Terminology

Several terms are used in relation to pesticide resistance, but none are uniformly accepted.

> **Pesticide resistance is the naturally occurring inheritable ability of some pest biotypes within a given pest population to survive a pesticide treatment that should, under normal use conditions, effectively control that pest population.**

***Resistant***    We have presented a generally accepted definition of resistance in the accompanying margin note. Entomologists have used the term *insensitivity* as a synonym for resistant. The term *tolerant* is variously used in weed science, such as indicating a weed species that is not controlled by a herbicide dose that usually kills most target species. Plant pathology has, however, used tolerant as equivalent to resistant. As noted earlier, resistance is used by plant breeders to indicate a crop variety that can withstand a pest.

***Cross-Resistance***    The term *cross-resistance* is widely used, but also subject to interpretation. A simple definition is resistance of a pest to two or more pesticides due to the same physiological mechanism of resistance. Often resistance to one member of a chemical family results in resistance to other members of the family that have the same mode of action. For example, houseflies resistant to DDT, an organochlorine insecticide, also are resistant to other organochlorine insecticides such as BHC, chlordane, or heptachlor. Cross-resistance can be conferred by single or multiple genes.

***Multiple Resistance***    The term *multiple resistance* means the pest possesses two or more different resistance mechanisms. Multiple resistance is typically controlled by more than one gene (a multigene characteristic) that confers resistance to different modes of action or families of pesticides. Houseflies, for example, that are resistant to DDT and other organochlorine insecticides are also resistant to parathion and other organophosphate insecticides.

The implications of cross-resistance and multiple resistance are very serious, because in these cases the resistance to one pesticide results in resistance to other pesticides to which the pest has never been exposed. This means that there can be resistance to new chemical compounds even before they are released for use in the field. There are numerous examples of both cross-resistance and multiple resistance for all types of pesticides. Table 12–1 presents examples of both types for each major pest category.

| TABLE 12–1 | Some Common Pests and the Pesticide Categories to Which At Least Some Populations Have Developed Resistance | |
|---|---|---|

| Pest type | Pest | Pesticide classes to which pest is resistant |
|---|---|---|
| Pathogen | *Botrytis cinerea* | Benzimidazoles, dicarboximides, anilopyrimidines |
| Pathogen | Late blight | Phenylamides |
| Pathogen | *Erisyphe* powdery mildew | Sterol biosynthesis inhibitors, STAR fungicides |
| Weed | Ryegrass and blackgrass | All ACCase inhibitors, ALS inhibitors, dinitroanilines, glyphosate, amitrole, others |
| Weed | Green foxtail | All ACCase inhibitors, dinitroanilines |
| Weed | Lambsquarters | ALS inhibitors, triazines, auxin type |
| Insect | Colorado potato beetle | Organophosphates, carbamates, pyrethroids, others |
| Insect | Diamondback moth | Organophosphates, carbamates, pyrethroids, Bt, others |
| Insect | Greenhouse whitefly | Organophosphates, carbamates, pyrethroids |
| Arachnid | Two-spotted spider mite | Organophosphates, carbamates, pyrethoids, others |

# Development of Resistance

The development of resistance can be traced to the genetic variability that occurs in any population of living organisms, coupled with the process of natural selection. When an external stress is applied to a population, those individuals with the genetic makeup (genotype) best able to survive and reproduce despite the stress will increase in numbers, while those that cannot tolerate the stress will tend to decrease in frequency. Initially the total population will be reduced if the genotype that cannot tolerate the stress was most frequent, but as the stress-resistant genotype increases in frequency, population numbers will eventually return to the prestress state.

Individuals in a population that cannot tolerate a pesticide are said to be susceptible to that pesticide. They are defined genetically as carrying the susceptible or s-allele(s), for that particular mode of action. Individuals that can tolerate the pesticide are said to be resistant, or carrying the r-allele(s). Individuals with s-alleles or r-alleles are naturally present in all pest populations. In the absence of the pesticide, the s-alleles normally occur at a higher frequency in the population than do r-alleles, because of fitness penalties often associated with the r-allele. Thus, individuals that carry the r-allele are at a disadvantage when competing with s-allele individuals in the absence of the selection pressure. Natural frequencies of r-alleles are typically less than 1:1,000, and are often lower than 1:1,000,000. For example, fungal r-alleles are thought to be as low as 1:1,000,000,000.

Application of a pesticide creates a selection pressure on the target pest population, which reduces the frequency of the s-alleles and increases the frequency of the r-alleles. The change in frequency of the two types of alleles occurs because individuals with the r-allele can reproduce, whereas those with the s-allele are killed or have reduced fecundity. Only a small portion of the wild-type population possesses the r-alleles and can survive the selection pressure (Figure 12–3). If the selection pressure is applied over multiple generations, the population of each successive generation contains a greater proportion of the r-allele and more of those individuals survive and reproduce. After three generations, the pesticide dose A in the theoretical example in Figure 12–3 is no longer effective. The typical reaction has been to use the higher dose B, or to

**FIGURE 12–3**
Theoretical impact of repeated pesticide application on proportion of pest population exhibiting resistance to the pesticide; wild-type susceptible population to left and increasingly resistant populations to right. Line A represents pesticide dose used to control susceptible population; line B represents increased pesticide dose required to kill moderately resistant pests, which is not effective for resistant pests.

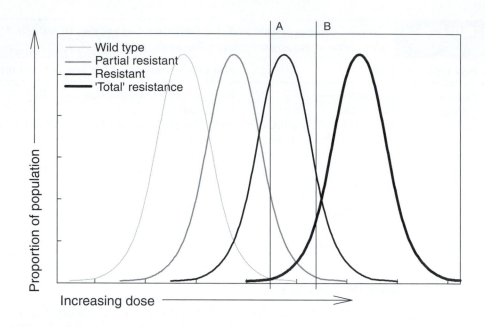

| TABLE 12–2 | Theoretical Appearance of Resistance in a Pest Population, Based on a 99% Control Level | |
|---|---|---|
| **Cycle** | **Apparent population susceptibility** | **Resistance frequency (%)** |
| 1 | Susceptible, control satisfactory | 0.002 |
| 2 | Still susceptible, control satisfactory | 0.02 |
| 3 | Control probably acceptable, but a few escapes noted | 0.2 |
| 4 | Control may still be adequate, but escapes now quite noticeable | 2.0 |
| 5 | Pesticide no longer provides adequate control | 20 |

increase the frequency of applications. The higher rate or more frequent applications result in increased selection pressure, and thus increase the frequency of individuals in the population that carry the r-allele. In effect, the $LD_{50}$ for the population is increasing. Using Figure 11–24, line A represents a susceptible population, line B a moderately resistant population, and line C a highly resistant population. At some point, the cost of control becomes prohibitive or the toxicology and environmental concerns that arise because of higher rates and more frequent applications result in cessation of use of the pesticide.

A theoretical example of the time required for resistance to develop in a population might follow the scenario outlined in Table 12–2. It should be noted that the table uses pest generations rather than years. For pest organisms such as spider mites or pathogens that have multiple generations per season, resistance would be appearing at the end of one season and could be serious after two seasons. For an insect or a pathogen with one generation per year, and little population carryover beyond one year, resistance would appear in five years. Resistance would not appear until after five years for organisms that have an inactive survival stage that endures beyond one year, such as weeds with a seedbank, because dormant weed seeds are not exposed to the selection pressure.

# Fitness

The ecological fitness (or just fitness) of the resistant versus the susceptible bio-types determines their initial proportions in the population prior to exposure to a pesticide, and the rate at which the original balance will reestablish after the selection pressure has been removed. Selection pressures change the relative fitness of different genotypes.

Fitness, as defined in the ecological sense, does not necessarily mean "vigor or strength." In many cases, the r-allele is associated with a reduced fitness, or fitness penalty, that keeps the r-biotype at a low level in the population when there is no selection pressure. For example, common groundsel biotypes that are resistant to atrazine have slightly lower photosynthetic rates than plants with the s-allele, which means that the r-allele plants do not grow as well as the s-allele plants. The green peach aphid offers one of the best examples of de-creased fitness in a resistant insect. The aphid strains with the most resistance to the organophosphate, carbamate, and pyrethroid insecticides are about half as fit as the susceptible biotypes, in the absence of the insecticides, probably be-cause of the relatively large proportion of the body proteins that is directed to the detoxification of the insecticides. In the highly resistant strains, the detoxi-fying esterase enzymes are 3% of the total body protein.

If the r-allele does carry a fitness penalty, the frequency of the r-biotype de-clines once the selection pressure is eliminated. If there is no fitness penalty be-tween biotypes with the r- or s-alleles such that both biotypes are equally fit, then the frequency of r-alleles will not decrease in the absence of selection pres-sure unless extensive immigration occurs. Knowing if the pests with the r-allele are equally or less fit than those with the s-allele can alter the approach to a re-sistance management program; but determining the fitness is difficult.

> **Ecological fitness is the relative intraspecific competitive ability of a genotype, expressed as the average number of surviving offspring relative to that of competing genotypes.**

# Intensity of Resistance

Resistance may vary in intensity among the various biotypes of a species. Inten-sity is measured in terms of how much more tolerant the resistant population is to the pesticide in comparison with that of the wild-type population. In Figure 11–24, the C population is about 100-fold more resistant than the A population, and the $LD_{50}$ has increased by two orders of magnitude. Once the $LD_{50}$ dose has increased 10-fold, then resistance is generally no longer manageable. There are many instances when increases greater than 100-fold have been recorded.

# Rate of Resistance Development

The rate at which resistance develops in a pest population is determined by several interacting factors that are based on the level of pest exposure to the se-lection pressure. The following are some of the most important factors.

1. The frequency of r-allele(s) in the wild-type population and the population biology of the organism which are governed by the following:
    1.1. The number of generations per year affecting how rapidly the population can show a shift in genotype; a higher number of generations results in a faster appearance of resistance.
    1.2. A fitness penalty associated with the r-allele that may keep r-allele(s) at relatively low frequency. The lack of a fitness penalty can lead to a more rapid appearance of resistance.

1.3. The number of r-alleles involved that can alter the rate at which resistance appears. If resistance is controlled by a single gene, it will develop more rapidly than if controlled by multiple genes.

1.4. The ease with which genes can be exchanged between members of the population (gene flow) altering the rate at which r- and s-alleles spread through the population. Higher gene flow among treated and nontreated populations tends to slow the appearance of resistance in the population.

2. Generalizations about how pesticide chemistry and mode of action influence resistance development do not appear possible between different pesticide families. The rapidity with which resistance appears is specific to the pesticide family involved; and within a family, related chemicals often show similar propensity to select for resistance. Some modes of action appear to be much more conducive to selecting for resistance. The acetolactate synthesis pathway in plants seems particularly prone to developing resistance to herbicides that block enzymes in that pathway. Several families of insecticides are cholinesterase inhibitors; resistance to these insecticides usually involves a combination of reduced inhibition rates and final detoxification of the insecticide. Many examples of cross-resistance between organophosphate and carbamate insecticides are the result of selection of biotypes capable of circumventing cholinesterase inhibition effect of the insecticide. Resistance to many of the systemic fungicides developed very rapidly after their introduction (a few years), yet there is still little or no resistance to sulfur or copper fungicides.

3. Longer residual activity and environmental persistence of the pesticide may lead to more rapid appearance of resistance. When the pesticide remains active for longer periods, the selection pressure is increased because it is exerted over a longer time.

4. The higher the dose of pesticide applied, the greater the selection pressure.

5. Multiple use of the same pesticide across several crops, or repeat applications within a season on the same crop, lead to the more rapid appearance of resistance.

6. Resistance will develop most rapidly if all the pest population is treated. For arthropods, pathogens, and vertebrates, the presence of noncropped, nontreated areas that harbor pests with s-alleles can slow the development of resistance, provided the pest can move back into the treated area. For weeds, the dormant seedbank is not exposed to a herbicide, and thus acts to maintain s-alleles in the population.

7. When individuals in a population with the r-alleles for a pesticide can move into nontreated areas, they can cause decreased efficacy of that pesticide, even if it had not been used in that area before. Mobile insects and vertebrates can literally transport resistance genes to new areas. For weeds, the r-alleles can be carried in the pollen for wind-pollinated plants such as kochia and pigweed species, or the whole plant can spread seeds that have r-alleles, such as the Russian thistle (tumbleweeds). For pathogens, the spores of many species are wind borne and can transport resistance genes from one area to another.

Table 12–3 summarizes the importance of several of the factors noted here. Transgenic crops that have been engineered for resistance to pests raise particular concern regarding the selection of pests capable of circumventing the re-

| TABLE 12–3 | Summary of Pesticide Resistance Risk in Relation to Various Pest Management Tactics | | |
|---|---|---|---|

| | Risk of Resistance | | |
| Management option | Low | Medium | High |
|---|---|---|---|
| Pesticide mixture or rotation in cropping system | > two modes of action | two modes of action | one mode of action |
| Use of same mode of action per season | Once | More than once | Many times |
| Cropping system | Multicrop rotation | Limited rotation | No rotation |
| Pest infestation level | Low | Medium | High |
| Pest control in last three cycles | Good | Declining | Poor |
| IPM system | All tactics (biological, cultural, physical, behavioral, chemical) | Pesticide and limited other tactics | Pesticide only |

sistance. When transgenic crops that carry resistance to pests are deployed, there is no rotation of mode of action, and the selection pressure is essentially continuous while the crop is present. Current examples include the transfer of the genes that encode the production of the Bt endotoxin into corn and cotton, and various viruses coat protein genes into cucurbits. Unless very strict techniques for combating resistance are adhered to, the protection that is provided by these genes is likely to be short lived. The deployment of crops containing genes expressing the Bt endotoxins has the potential to severely compromise, or even eliminate, the use of sprayable Bt as a pest management tool.

Great caution should be exercised in the deployment of crops genetically engineered to resist pests, because there is no ecological reason to assume that the technology will provide any more stable solution to pest management than has been provided by the use of pesticides. The ecological risks of pest resistance with transgenic crops are potentially as great as or greater than the use of pesticides. The impact of the pesticides delivered through genetically engineered crops may be irreversible if genes move into wild populations, and if nontransgenic varieties are no longer available.

## Mechanisms of Resistance

Pest resistance to a pesticide can be due to one or several physiological changes in the pest. Following are some major changes that may occur.

1. Uptake or translocation within the pest. Changes in either uptake or translocation mean that the pesticide no longer reaches its target site of action at concentrations that are sufficient for activity.
2. Target enzyme or site of action. Changes in the binding site on an enzyme or other protein can stop the pesticide from exerting its normal inhibitory effect (Figure 12–4). The pesticide simply does not work any more. For

**FIGURE 12–4**
Diagrammatic representation of altered binding site causing reduction in action of a pesticide. (a) Pesticide binds to active sites in enzyme blocking normal substrate access to the site; (b) binding site is altered so that pesticide cannot bind, leaving site open for normal substrate access.

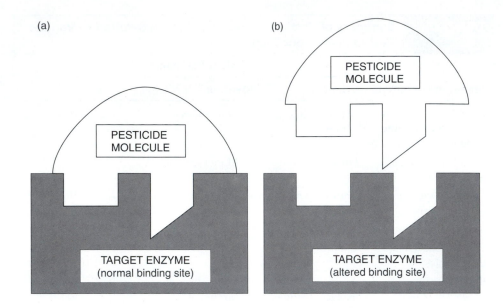

example, the herbicide atrazine, which acts by inhibiting photosynthesis in susceptible plants, does not bind to the active site in the photosynthetic pathway of resistant groundsel. Alteration of the site of action is the most common reason why resistance develops.

3. Increased metabolism of the pesticide in the resistant pest. In this situation the pesticide is rapidly metabolized into an inactive form in the resistant pest.

4. Sequestration of the pesticide. The resistant pest is able to compartmentalize the pesticide in its cells so that the pesticide cannot reach the active site, for example, by sequestering or isolating it in vacuoles. This mechanism occurs with herbicides and fungicides, but is not considered important for insecticides.

5. Behavioral resistance. This implies that the resistant pests can avoid the pesticide due to their own action, and only applies to mobile pests with nervous systems. The behavior of outdoor-dwelling mosquitoes rendered them resistant to DDT because they avoided exposure by not landing on treated buildings. Alternatively, indoor-dwelling mosquitoes were killed by DDT because their behavior included landing on treated building walls. The population shifted to the outdoor-dwelling biotype in response to the selection pressure of routine spraying of interior walls with DDT. Most vertebrates, such as rats and ground squirrels, sickened but not killed by a pesticide-treated bait will never take the bait again; they are said to be "bait shy."

## Measuring Resistance

A fundamental question that arises in relation to resistance is, to detect the on-set of resistance and how to measure its progress in a population. To manage resistance, one must determine the extent to which it is present. Assays to measure resistance are pest-type specific and can be one of the following:

1. Bioassays. Dose/response curves are developed to compare the susceptible population with the suspected resistant population. The curves provide information on both the frequency and intensity of resistance. The actual bioassay used will depend on the type of pest being evaluated. The time required to conduct bioassays presents a problem, as timeliness of information is important for initiation of resistance mitigation measures.

2. Biochemical and immunochemical assays. These assays detect the presence of altered enzymes in the suspected resistant population. Most such assays are used at the research level, but may be of practical future use in the field. There is, however, little economic incentive to develop these more complex assay types at the field level.

3. Molecular genetic techniques. Differences in DNA between susceptible and resistant pest populations can be detected with the use of molecular techniques, and provide information on the genetic differences among populations of a species. Molecular genetic techniques are currently not used at the field level, but are being employed by private labs, academic researchers, and the chemical industry to better quantify the extent of resistance.

## Resistance Management

Resistance management involves tactics that reduce selection pressures, or reduce the proportion of the population that is exposed to a selection pressure. The ultimate goal in managing resistance is to maintain the usefulness of management tactics based on cultural methods, pesticides, or plant resistance genes (more on the latter in Chapter 17). Management of resistance in Bt transgenic crops is of special concern and must be part of any program that recommends the use of such crops.

Entomologists have developed a strategy called integrated resistance management (IRM). Pesticide industry action committees assess pesticide resistance and provide guidelines for managing resistance to their respective pesticide type. Techniques for avoiding or decreasing the development of resistance usually attempt to reverse the factors that contribute to that development. The following steps are part of a resistance management program.

**Strategies to slow the appearance of resistance genes in the population must be developed and implemented, to preserve the efficacy of a control tactic.**

1. Monitor presence of resistant biotypes. The first step in managing resistance is to determine both the frequency of occurrence and the strength of resistance. Early detection is essential if resistance is to be managed before it eliminates the utility of the pesticide. Resistance can be confused with poor control resulting from one of several factors, such as application at the wrong growth stage of the pest, incorrect application rate, poor application resulting in inadequate cover, or even use of defective or denatured chemicals. If resistance appears to be present, assays of the putative resistant population must be made. Assuming that the presence of resistance is confirmed, some of the following actions must be taken.

2. Modify pesticide usage.
   2.1. Alternate or rotate pesticides that have different modes of action. Do not continue applications of the same pesticide or others with the same mode of action. There is a difficulty here, as most users do not

know the modes of action of pesticides. Coding of pesticide containers to show modes of action has been suggested as a means to remedy this difficulty. The lack of effective pesticides with different modes of action may limit the option to rotate pesticides with different modes of action.

2.2. It is essential that the problem of cross/multiple resistance be considered. Rotating to a different pesticide may not accomplish a reduction in selection pressure if the pest is resistant to more than one mode of action.

2.3. Mixtures of active ingredients from different chemical families with different modes of action should be used. This tactic is widely recommended for pathogen and weed management, but is not usually employed for arthropods.

2.4. Using a reduced rate of a pesticide is a standard recommendation for retarding the development of resistance, as it lessens the selection pressure on the target population.

2.5. Repeated low rate treatments of the same pesticide, however, should not be used; this is especially noted for some fungicides.

2.6. Where feasible pesticides with short persistence should be used, short-persistence pesticides reduce the selection pressure as compared with the pressure exerted by use of pesticides with a long residual effect.

2.7. Applications should be as listed on the label, and timed so that they are applied to the most susceptible life stage. This is considered particularly important for insects.

2.8. Development of new pesticides (an action for the industry, not for individual growers) has been a strategy widely used in the past by the pesticide industry. Relying on new pesticides will not provide a long-term solution to problems with resistance because:

2.8.1. Resistance develops against each new pesticide, contributing to the pesticide treadmill (Figure 12–1).

2.8.2. It has become increasingly difficult and costly to bring new pesticides to the market.

2.8.3. Cross-resistance may compromise the activity of even a newly developed pesticide.

3. Use combinations of control tactics. This is referred to as rotating mortality factors and it is one reason for initially advancing the concept of integrated pest management.

3.1. Use alternative pesticides (see item 2.1).

3.2. Use cultural management to reduce pest impact to the extent possible.

3.3. Use mechanical control if feasible, an option particularly relevant to management of resistant weeds. The strategy includes use of hand hoeing to remove weeds that have escaped from a herbicide treatment; if they are physically removed before they set seed they do not pass genes to the next generation.

3.4. Preserve beneficials and implement biological control if feasible. This tactic is particularly important for arthropod pesticide resistance management, but has little relevance to management of resistance in most other categories of pests.

3.4.1. It is important to recognize that resistance to pesticides is a desirable trait in natural enemies. Certain strains of insecticide-resistant predatory mites have been intentionally produced for mass releases to control pest mites in combination with use of a chemical insecticide to control other pests.

4. Rotate crops. Rotating to a different crop often changes the pests that have to be controlled and the pesticides that can be employed. This tactic is very useful for herbicide resistance, as herbicides are usually crop specific due to the selectivity requirements. Rotation to another crop is not as useful for other pesticide classes because the same pesticide is often used in different crops. Rotation is also extremely useful for insect and nematode management when the rotational crop interrupts the life cycle of the pest (see Chapter 16). Rotation is not an option for perennial crops (such as established trees and vines).

5. Preserve susceptible genes. Resistance development can be delayed or even stopped if a proportion of the pest population is not subjected to the selection pressure. The aim of this strategy is to leave at least a portion of the pest population untreated. This can be achieved by leaving portions of the crop untreated, or by setting aside specific areas that are not treated; the latter are referred to as refugia. A difficulty with implementing the preservation of the s-allele strategy is that there are no clear guidelines regarding the size or the proportion of the population that should be left untreated. Current research on conservation biology and application of the concepts of island biogeography may eventually aid in the design of more efficient refugia.

## Implementation Problems

Implementation of tactics to mitigate development of pesticide resistance in pest populations has only been moderately successful in the past. There are many complex reasons for this, and we mention only a few here.

1. Pesticide industry. The pesticide industry probably holds the key to managing resistance. A major difficulty has been obtaining cooperation between competing companies. Development of the chemical industry's resistance action committees has helped to overcome the latter problem.

2. Farmers. Unless farmers are ecologically aware, they will probably utilize the product that appears to offer the greatest immediate return, regardless of the likelihood that resistance will develop in the future.

3. Education/regulatory establishment. The public sector was initially slow to discuss resistance and to convey its importance to farmers. A weed scientist in Idaho noted, "I could have predicted resistance to chlorsulfuron, but I didn't." The academic and regulatory community must continually emphasize the problems that resistance brings to pest management.

4. The public. A portion of annual pesticide use in the United States occurs outside of agriculture (e.g., homeowners, golf courses, parks). Resistance management, generally speaking, is poorly recognized in these areas, and exemplified by the overuse of antibiotics for use in people and the development of resistant bacteria.

## Examples of Resistance/Management

***Pathogens***    Resistance occurs to many of the major fungicide groups, and has necessitated changes in how fungicides are used. To date, however , there is no evidence of resistance to copper- or sulfur-based fungicides, although they have been widely used for over a century. There is also little evidence of resistance to dithiocarbamate, phthalimides, or dinitrophenol fungicides. Resistance magnitudes of 100- to 500-fold as compared with wild-type sensitivity has been recorded for more recently introduced fungicides, many systemic, including:

1. Aminopyrimidines. Resistance is present in many types of pathogens; fungi in the genera *Botrytis* and *Venturia* are the most important.
2. Benzimidazoles. Many fungi are resistant to this class of fungicides, including *Botrytis* in grapes, *Penicillium* rots in citrus, and *Mycosphaerella* in bananas. Cross-resistance occurs between all fungicides in the class; benomyl is the best known example. Some pathogens developed resistance as quickly as 2 to 3 years, but others took 10 to 15 years.
3. Phenylamides. This group includes metalaxyl and related fungicides that are active against *Oomycetes.* This class of fungicides was introduced in 1977, and within three years resistance appeared in downy mildew of cucumbers, late blight of potatoes, and downy mildew of grapes. Mixtures of fungicides having different modes of action have kept resistance to a manageable level. Management of *Oomycete* pathogens is difficult without chemicals.
4. Dicarboximides. The major pathogen of concern for resistance to this group of fungicides is *Botrytis cinerea.* Dicarboximides were introduced in the mid-1970s and resistance appeared within three to five years. Cross-resistance within the dicarboximides is common.
5. Sterol biosynthesis inhibitors (SBIs). The two different groups are composed of dimethylation inhibitors (DMIs) and morpholines. Practical resistance problems developed to many of these fungicides after about 10 years of use following their introduction in the mid-1970s. The most important pathogens are powdery mildews of barley, cucumber, and grapes in the genera *Erisyphe, Venturia,* and others.

There are several additional fungicide types to which resistance is known to occur. For a more in-depth discussion of many aspects of fungicide resistance in pathogens, see Heaney et al. (1994) or visit the FRAC website (2000).

***Weeds***    Herbicide resistance is now widespread in weed populations world-wide, and most major herbicide types are included. Figure 12–5 provides the extent for examples of specific groups. A few specific examples are as follows:

1. Triazines. Atrazine was the first major herbicide against which resistance was confirmed. Now approximately 61 weed species are known to have populations resistant to atrazine and related triazine herbicides. Many exhibit cross-resistance to other triazine herbicides.
2. Branched chain amino acid synthesis inhibitors. These herbicides inhibit the acetolactate synthase (ALS) enzyme. Herbicides in the sulfonyl urea and imidazolinone groups have this mode of action. There are now 63 weed species with populations resistant to these herbicides. Weeds resistant to one herbicide exhibit cross-resistance to all other herbicides in

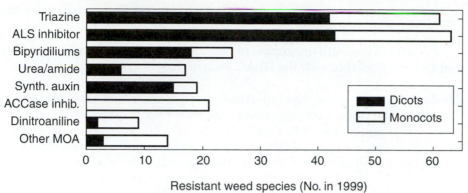

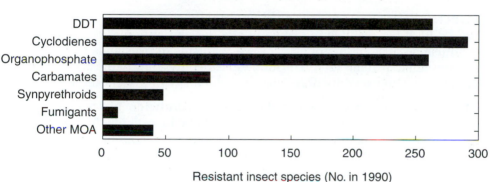

**FIGURE 12–5**
Graphical examples of the numbers of weeds and insects resistant to different classes of herbicides and insecticides.

these groups. Resistance appeared in weed populations within about four to five years of commercial introduction. The problem has become so severe that the use of products has stopped in some instances.

3. Lipid biosynthesis (ACCase) inhibitors. Herbicides with this mode of action inhibit the enzyme acetyl CoA carboxylase, and provide exceptional control of grasses. Ryegrass in Australia has essentially become totally resistant to all grass herbicides, necessitating a complete change in the weed management paradigm. Twenty-one common weeds show resistance to these herbicides, including wild oats, blackgrass, and foxtails.

4. Glyphosate. Resistance to this herbicide has now been confirmed in Australia; and is probably also present in California and Maryland.

5. Herbicide-resistant crops. Similar to crops genetically engineered for resistance to insects, the use of herbicide-resistant crops (see Chapter 17 for more details) poses enormous challenges for resistance management. Widespread use of the same herbicide in different crops will place increased selection pressure on weeds, and increased appearance of resistant weeds should be anticipated.

6. Resistance to cultural practices. If the same cultural practice is used repeatedly there is no ecological reason why resistance to that practice will not develop. Of the several examples of weeds that have become resistant to cultural practices, one of the earliest is false flax in cereal crops, which was selected because the weed seeds are the same size as the crop seeds. Multiple harvestings of alfalfa in California, selected for

prostrate forms of yellow foxtail (see Figure 5–23), that thrive under the frequent cutting regime.

For a more in-depth discussion of herbicide resistance in weeds, see Powles and Holtum (1994) or visit the HRAC website (2000).

**Mollusks**    There is not much pesticide use for control of slugs and snails in agricultural settings, except in some high-value specialty crops such as grass seed production in the U.S. Pacific Northwest. There is no evidence of biochemical resistance to metaldehyde, the most commonly active ingredient used in slug baits. Bait shyness, however, has been noted in some mollusk populations.

**Insects**    Resistance has developed to virtually every type of insecticide (Figure 12–5). It has resulted in many insecticides and miticides being withdrawn from use because of loss of efficacy. Mechanisms of resistance to insecticides are presented in Table 12–4. The following are only a few of the many examples of the extent and severity of resistance problems with insects and mites.

1. Diamondback moth is a pest of cole crops worldwide. In Thailand, this insect is resistant to every available synthetic insecticide. In most regions of the world, diamondback moth is resistant to organophosphates, synthetic pyrethroids, chitin inhibitors, and others.
2. Whiteflies of the genus *Bemisia* are resistant to organochlorines, several organophosphates, certain pyrethroids, and there is evidence of decreased susceptibility to several new insecticides with novel chemical structures and modes of action. Populations showing resistance factors from 360 to over 1,000 have been reported worldwide.
3. In several parts of the world, the green peach aphid is resistant to many insecticides belonging to the principal classes. In the late 1990s, a promising new class of aphicides was introduced—the chloronicotinyls—but low-level resistance already has been detected in some populations.
4. Lepidopterous leaf miners are resistant to many different insecticide types. The field life of a new insecticide in Florida against these pests has typically been three years before resistance occurs.
5. The Heliothini moths include some of the most serious insect pests of major crops. Different species attack cotton, corn, soybean, tobacco, sorghum, and many vegetable crops, especially tomato. The corn earworm (*Helicoverpa zea*) has developed resistance to many insecticides, but it attacks crops, such as soybean, where the insecticide pressure is lower, so resistance is not widespread or as severe as it is for some of the other species. In North America, the tobacco budworm (*Heliothis virescens),* and in parts of Europe, Africa, and Asia the cotton bollworm (*Helicoverpa armigera),* are subject to intensive sprays on cotton, and populations have developed resistance to all insecticides used against them in significant quantities.
6. Some populations of the Colorado potato beetle in the eastern United States are resistant to virtually all insecticides used against them.
7. The corn rootworm, as noted in Chapter 5, has developed a new biotype adapted to the two-year corn/soybean rotation that once controlled it. This serves as a clear example that resistance can occur to management practices in addition to pesticides, and emphasizes that resistance can develop in pest populations in response to any selection pressure.

| TABLE 12–4 | A List of the Principal Mechanisms of Resistance to Insecticides in Arthropods | |
|---|---|---|

| Mechanism | Cross-Resistance Pattern | General Observations |
|---|---|---|
| Penetration | Organophosphates<br>Pyrethroids<br>Pyrethrins<br>Cyclodienes<br>Abamectin<br>DDT<br>Organotins | Generally low levels of protection.<br>May delay onset of symptoms or knockdown. |
| Metabolism | | Resistant insects may show symptoms but then recover. |
| MFO[1] and hydrolase | Insecticides with similar functional group | |
| GSH S-transferase[2] | Preference for methoxy versus ethoxy substituted organophosphates | |
| DDTase | DDT and trichloroethane analogues of DDT | |
| Insensitivity of the nervous system: kdr type | DDT<br>Pyrethroids<br>Pyrethrins | |
| cyclodiene type | Cyclodienes | |
| Altered AchE[3] | Certain organophosphates and carbamates<br>Pattern of cross-resistance depends on the AchE isozyme | |

[1]MFO = mixed function oxidases
[2]GSH S-transferase = Glutathion S-transferase
[3]AchE = acetylcholinesterases
Modified from Scott, 1990.

For a more in-depth discussion of insecticide and miticide resistance in arthropods, see Roush and Tabashnik (1990), Georghiou and Lagunes-Tejeda (1991), or visit the IRAC website (2000).

**Transgenic Crops**    Genes from the soil-dwelling bacterium, *Bacillus thuringiensis,* which encode the production of Bt endotoxins, have been inserted into several crops using genetic engineering technology (see Chapter 17). These crops express the gene(s) and produce the endotoxin, conferring on the plants the ability to kill insects sensitive to the respective Bt endotoxin. Widespread adoption of these transgenic crops will increase selection pressure on populations of the susceptible insects that feed on them. Starting in the mid-1980s, resistance to sprayable Bt had already developed in several places, indicating that resistance to these endotoxins is present in wild populations. Planting a portion of the crop acreage with varieties that lack the Bt endotoxin is a strategy recommended to reduce the selection pressure. However, there is no real enforcement of such mixed plantings, and the ability of these refugia to delay or stop development of resistance has not been tested.

***Vertebrates*** Resistance to pesticides in vertebrate pests is not common. Resistance to warfarin has been recorded in both rats and mice. Resistance to warfarin, an anticoagulant chemical that causes internal bleeding when ingested, may be mediated through enhanced degradation of the anticoagulant, or through alteration of B-complex vitamin K metabolizing enzymes. Of theoretical interest is the fact that some strains of mice have a high level of resistance to DDT.

Bait shyness, in which the target animal modifies behavior to avoid certain baits, is a form of resistance. At present there is little concern about resistance in vertebrates, but monitoring for resistance is probably worthwhile.

# RESURGENCE

**Resurgence occurs when a pest population rapidly recovers from a pesticide application and subsequently increases to densities greater than pretreatment.**

Resurgence is a phenomenon that occurs following pesticide application. The pest population is initially decreased by the pesticide, but recovers to attain a higher population density than was present before the application (Figure 12–6). Resurgence is attributed to several ecological processes that can be perturbed by the pesticide.

1. Reduced biological control. The pesticide kills beneficial organisms that normally regulate the size of the pest population. In the absence of effective biological control, the remaining pest population increases to a higher level.
2. Reduced competition. Theoretically, the pesticide could be differentially more effective against a competing organism, allowing the original pest to recover with lowered competition, and hence attaining a higher population density. There is little evidence supporting involvement of this process in resurgence of pests.
3. Direct stimulation of the pest. Physiological processes of the survivors are enhanced to make them more ecologically fit, such as enhanced egg laying

**FIGURE 12–6**
Theoretical pest population recovery, called resurgence, following a pesticide application that kills the beneficials that were previously keeping the pest population under partial control.

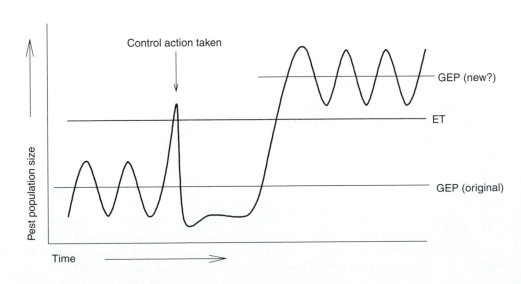

by insects. The phenomenon of hormesis might be involved. The term *hormesis,* or *hormoligosis,* originated in the human pharmacological literature and refers to drugs that have both stimulatory and inhibitory effects depending on doses and timing of treatment.

4. Improved crop plant growth. Application of the treatment decreases the target pest density, and thereby leads to improved plant growth. Reproduction in the surviving pests is effectively stimulated by the increased quality and quantity of host tissue, in effect decreasing intraspecific competition among pests.

Reduced biological control is probably the most important reason for resurgence. Although resurgence is presented here in relation to pesticide use, there is no inherent ecological reason why similar phenomena would not occur with other pest management tactic.

In general, resurgence only occurs in pests where population development is limited by competition or active biological control by organisms of the same type. The possibility of direct stimulation is theoretical, and the role of hormesis in pest resurgences is unclear.

Resurgence has been a serious limitation to management of some arthropod pests. For example, arthropods such as spider mites, and various lepidopteran larvae (caterpillars), have been increased by the use of insecticides that kill the beneficials normally controlling the pest.

Nematicides are generally applied as preplant treatments. After treatment, nematode populations may be observed to increase to densities that are greater than in nontreated areas. Such resurgence in nematode populations is probably related to the phenomenon of improved crop plant growth described previously.

Resurgence is not discussed in relation to weed management, because biological control of weeds by other plants is rare; thus, the main driver for resurgence does not operate. There has been no evidence that the other mechanisms operate on weeds. The regulation of weeds by plant-eating animals, however, has been amply documented in the literature on biological control of weeds (see Chapter 13). Killing or removing the plant-eating animals will result in increased population densities of the weeds, such as occurred when myxomatosis was introduced into Australia for rabbit control.

# REPLACEMENT

Replacement is a term that has been used in arthropod management, but is applicable to all categories of pests, and like resurgence, replacement occurs in response to pesticide application. Entomologists sometimes also use the term *upsurge.*

As shown in Figure 12–7, the population of the target pest (A) is reduced, but a minor pest (B) increases to utilize the resources no longer being used by pest A. This results in pest B becoming a major pest. The population of pest B did not increase prior to the pesticide application, because it was suppressed by competition from pest A and by natural enemies (in the case of arthropod pest replacement). By removing the superior competitor, pest A, and a range of natural enemies, the factors that limited the population growth of pest B are altered and allow it to increase to atypically high densities. Replacement is a potentially serious problem, because the released minor species may cause more crop loss than the original target species. As with all "ecological backlash"

**Pest replacement occurs when pesticide treatment kills the target pest, but the target is replaced by a different species that was formerly a minor pest.**

**FIGURE 12–7**
Theoretical population numbers of two pest species as they might respond during pest replacement. Original pest A was controlled by management tactics that do not control pest B, which is then able to increase to a damaging level, replacing pest A as the problem.

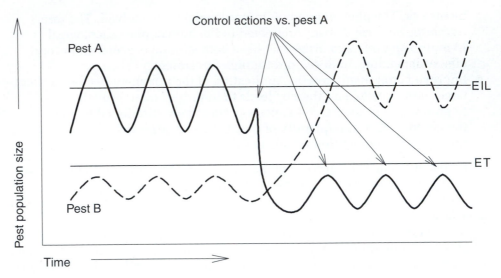

phenomena, replacement has probably been most serious as a result of pesticide use, but it is important to note that it can also occur in response to other pest management tactics or alteration of cultural practices.

***Pathogens***   There is currently no unequivocal evidence for replacement in pathogens, but there is no theoretical reason to think that it does not occur.

***Weeds***   Weed science literature is replete with examples of minor weeds that have become major weeds as a result of herbicide use. Herbicides have selectivity, and those that do not kill the crop usually do not control closely related plant species. Additionally, unrelated species may also escape control through the random quirks that govern selectivity. A classic example of weed replacement was the change from mixed annual weeds, mainly dicots, in cereal crops to mainly grass weeds. This change occurred following the introduction of the phenoxy herbicides that kill dicots but have almost no effect on grasses. The grasses thrived in the absence of competition from the dicots. In a different example, widespread adoption of dinitroaniline herbicides in cotton resulted in a shift in the weed species to those that are not controlled by the herbicides, such as nutsedge and nightshade species. Nightshade species, which are related to tomatoes, only became a problem in processing tomatoes after the adoption of selective herbicides that killed most weeds but not the crop. Species shifts like these are frequent, and must be anticipated with the use of herbicides. IPM programs for weed management must address the phenomenon.

***Nematodes***   Replacement as a direct result of nematicide use is not likely, because nematicides have little specificity. However, it is possible that changes in cultural practices and nematicide use may alter the relative importance of nematodes over time. One possible example concerns plant-parasitic nematodes in Hawaiian pineapple. Pineapple production became an important industry in Hawaii in the late 1800s and early 1900s, and the most important plant-parasitic nematodes were the root-knot nematodes. By the mid-1950s, the reniform nematode was replacing the root-knot nematodes as the primary yield-loss-causing nematode. It is worth noting that effective nematicides were discovered in the 1940s in Hawaii, and that once discovered, nematicide use in pineapple increased over time. However, during this same time period soil acidity became a problem, with many fields having a pH as low as 3.8. The increase in importance of the reniform nematode during the 1950s has been attributed, in part, to increasing soil acidity.

***Arthropods***   Many arthropod outbreaks have occurred as a result of replacement. Perhaps the best documented case of an upsurge of a secondary pest is that of the Heliothini (cotton bollworm and tobacco budworm) in cotton. Since the invasion of the boll weevil in the Lower Rio Grande Valley of Texas, cotton was under a constant spraying program; first with calcium arsenate and later with the new organosynthetic insecticides. By the early 1960s, the boll weevil became resistant to the organochlorine insecticides and growers switched to organophosphates and carbamates. At that time, however, the two Heliothini species, still considered minor secondary pests, began to appear in large numbers and caused severe economic losses. The situation became catastrophic when the tobacco budworm became resistant to all available classes of insecticides. Growers would spray up to 20 times during a growing season, and still suffer severe losses due to the tobacco budworm. The situation only improved when IPM was introduced in the 1970s for boll weevil control. Pesticide pressure was reduced and the bollworms again became the target of recovered populations of natural enemies.

## WARNING ABOUT THE THREE Rs

The key to management of the three Rs is implementation of mitigating strategies *before* resistance, resurgence, or replacement become a problem. Pest management practitioners seem to have great difficulty with this concept, despite the general acceptance that a manager must adopt IPM practices and not rely on a single control method (rotate mortality factors). If all mitigation measures fail, the pesticide is lost. History shows that the long-term sustainability of our pest management resources has frequently been squandered for short-term gain. This problem still exists, at the level of the farmer, the pest control advisor, the pesticide industry, and even the government regulators. The new techniques for engineering pest-resistant plants seem to have led to collective amnesia in these same groups regarding the need for mitigation strategies and the IPM approach to pest management. This attitude cannot continue, or pest management will be in for difficult times in dealing with the inexorability of the three Rs. Unless management of the three Rs is learned, and what is known is implemented, pest managers may eventually run out of options for pest control.

## SOURCES AND RECOMMENDED READING

Two useful general reference texts on the topic of resistance are National Research Council (1986) and Green et al. (1990). There are no current reviews on the topics of resurgence or replacement, but for a good historical account of the cotton bollworm upsurge, see Perkins (1982).

FRAC. 2000. Fungicide Resistance Action Committee homepage, http://PlantProtection.org/FRAC/

Georghiou, G. P., and A. Lagunes-Tejeda. 1991. *The occurrence of resistance to pesticides in arthropods.* Rome: Food and Agriculture Organization of the United Nations, xxii, 318.

Green, M. B., H. M. LeBaron, and W. K. Moberg. 1990. *Managing resistance to agrochemicals: From fundamental research to practical strategies.* Washington, D.C.: American Chemical Society, xiii, 496.

Heaney, S., D. Slawson, D. W. Holloman, M. Smith, P. E. Russel, and D. W. Parry, eds. 1994. *Fungicide resistance.* BCPC monograph 60. Farnham, Surrey, UK: British Crop Protection Council, xii, 418.

HRAC. 2000. Herbicide Resistance Action Committee homepage, http://PlantProtection.org/HRAC/

IRAC. 2000. Insecticide Resistance Action Committee homepage, http://PlantProtection.org/IRAC/

National Research Council, ed. 1986. *Pesticide resistance: Strategies and tactics for management.* Washington, D.C.: National Research Council, National Academy Press, xi, 471.

Perkins, J. H. 1982. *Insects, experts, and the insecticide crisis: The quest for new pest management strategies.* New York: Plenum Press, xviii, 304.

Powles, S. B., and J. A. M. Holtum, eds. 1994. *Herbicide resistance in plants: biology and biochemistry.* Boca Raton, Fla.: Lewis Publishers, 353.

Roush, R. T., and B. E. Tabashnik, eds. 1990. *Pesticide resistance in arthropods.* New York: Chapman and Hall, ix, 303.

# CHAPTER

# *13*

# BIOLOGICAL CONTROL

---

*"The enemy of my enemy is my friend."*

---

## CHAPTER OUTLINE

▶ Introduction
▶ Why biological control?
▶ Biological control concepts
▶ Types of biological control and implementation
▶ Examples of biocontrol using various agents
▶ Summary

## INTRODUCTION

The numbers of animals and plants in natural populations usually fluctuate between upper and lower limits due to the combined effects of abiotic forces, such as climate, soil conditions, and natural disasters and biotic forces, such as food, intra- and interspecific competition, predators, and diseases. Most animal populations fluctuate locally, with extinctions or explosions to uncontrollable numbers rarely occurring over the entire geographic range of the species. Although local extinctions may occur under certain conditions, those areas are eventually recolonized (see also Chapter 5). Ecologists have argued about the relative importance of the components of the environment that act to regulate population numbers. Now, however, it is accepted that ecosystem characteristics determine which of the abiotic or the biological forces exert the strongest influence in population regulation, and when those factors operate. There is general agreement that major fluctuations are the result of local short- or long-term disturbances affecting the regulatory forces.

Within any agroecosystem, the crop provides the foundation of the food web. A food web is the set of feeding interactions among the organisms within the system, and is the relationships that describe "who eats whom," as discussed in Chapter 4. Ecologists have long studied the feeding interactions that occur between species in a food web in an attempt to understand how the interactions determine the population dynamics of the interacting species. The interactions considered include competition, herbivory, predation, and parasitism, all of which are involved in natural control in agroecosystems. Biological control thus focuses on the manipulation of these interactions to reduce pest numbers and limit crop damage.

Agriculture, as practiced today, is one of the most pervasive factors in the disturbance of ecosystems. The occurrence of pests is frequently the direct result of that disturbance. Certain potential pests, however, are kept under natural control by the set of endemic natural enemies that occur in an area. Evidence for the effectiveness of natural control agents is provided in cases of resurgence and replacement as discussed in Chapter 12. Pest outbreaks on all major crops worldwide, on the other hand, document the fact that pests can become devastating in the absence of control tactics.

The fact that some organisms are designated "pest" by humans is an indication that natural control factors have failed to sufficiently limit the organisms numbers once systems have been disturbed. In addition, the natural, dynamic equilibria that existed among organisms in an ecosystem no longer function or may not sustain system characteristics desired by humans. If biological control, or other natural factors regulating population growth, keeps the numbers of an organism below the point at which it interferes with human activity, the organism will not be considered a pest. Alternatively, in certain cases, natural control factors do not keep pests at sufficiently low levels, although conditions might appear optimal for such pest regulation. Another reason that natural control factors might be insufficient in agroecosystems is that human activity has perturbed the system to the point where population regulation by natural enemies can no longer stop the organism from interfering with human needs.

In western industrialized countries, current agricultural practices are intensive, and typical production and cropping practices limit the biodiversity that might play a role in natural pest regulation. On an ecological/evolutionary basis, it is unlikely that most of the widely distributed endemic organisms considered as major pests will be controlled biologically to the point where no other management tactic is required. This is because human activity has modified ecosystems to the extent that natural controls cannot provide the level of pest suppression dictated by human needs. It is essential that management tactics, including pesticide use, be integrated in ways that foster the contribution of biological controls to the cumulative limitation of pest populations. In the future, it is conceivable that agricultural systems will be designed to reduce disturbance in order to enhance biological pest control, with the constraint that adequate weed management is maintained, thus allowing production that satisfies human needs.

Introduced exotic or invasive species are different than endemic species, and many important pests in agroecosystems are introduced species, as discussed in Chapter 10. Introduced species may become pests because the new habitat has food sufficient to support prolific reproduction, or because the natural enemies that regulate the pest population in the native habitat are absent in the new environment. Searching for the natural enemies that suppress the introduced pest in its native habitat can yield organisms that feed on it, and thus regulate its population size. The process of searching for natural enemies in the countries of origin of a foreign pest, the importation of the natural enemies, and their release in the new environment is known in entomology as "classical biological control."

It is unlikely that most species of weeds will be controlled through biological means, because most plants that are considered weeds are weeds in all regions of the world. This means that everywhere in the world where a particular weed grows there are not enough organisms feeding on it to stop it from attaining the status of weed pest. Although some of the perennial weeds that have been introduced into new ecosystems have been subject to successful biological control, such examples should not be considered representative of what can be achieved for all weed species.

Biological control is an attempt to reestablish some level of natural population regulation in agroecosystems that are maintained in an inherently unstable state. Under classical biological control, natural enemies of pests are found, and they are introduced into regions the pest has colonized. There are many examples of successful biological control of the major pests of many crops around the world. The cottony cushion scale on citrus in North America (California) is the oldest and best known example of successful classical biological control (see Figure 3–7). Cottony cushion scale nearly decimated the citrus industry in the late nineteenth century, but it is now controlled so well biologically that it is no longer considered to be a major pest.

A caveat is appropriate about the ecological safety of biological control. Entomologists generally have held that biological control is a most desirable tactic in a pest management program. This view, however, needs to be critically evaluated in relation to other pest categories. All pest control tactics have drawbacks and limitations in addition to benefits, and in this regard, biological control, including the importation and release of exotic beneficial arthropods, is no different than other pest management tactics. The use of biological control may have unintended, undesirable negative effects on nontarget organisms in the ecosystem. It should be remembered that nature may have unanticipated reactions to human manipulations.

# WHY BIOLOGICAL CONTROL?

Biological control is an important tactic in IPM systems, and should be utilized wherever it is feasible. There are several advantages to the use of biological control in comparison with other pest control tactics.

1. If populations of effective biological control agents can be established in an area, there is essentially no further cost. Successful biological control is thus relatively inexpensive once the initial costs of establishing the natural enemies are met.
2. For a successfully established biological control program, the pest never exceeds economic injury levels. If the pest population starts to increase, the biological control agent increases and reduces the pest population growth rate in a density-dependent fashion.
3. Biological control leaves no pesticide residues capable of contaminating the crop or the environment.
4. Biological control can be very effective in extensive permanent ecosystems such as perennial crops of alfalfa, many tree and vine crops,

and rangeland. In the rangeland system, biological control may be the only economically viable tactic for management of many pests.

5. Unlike pesticides and physical tactics, biological control typically does not disrupt other pest control practices.

6. Biological control tactics have traditionally been considered nondisruptive to ecosystems. In some instances, however, there may be unintended nontarget effects of releasing nonnative natural enemies of pests in new areas (see later in this chapter). Through research, scientists involved with evaluation of biological control agents seek to minimize or eliminate such nontarget effects.

# BIOLOGICAL CONTROL CONCEPTS

## Definitions

**Biological control is the use of parasitoid, predator, pathogen, antagonist, or competitor populations to suppress a pest population, making the pest less abundant and less damaging than it would be in the absence of the biocontrol agents.**

There are many variations on the definition of what constitutes biological control. We have selected the definition in the margin note, which can be paraphrased as the use of one organism (beneficial) that either feeds on, or competes with, another organism (the pest) to reduce the population growth rate of, and the damage caused by, the pest. The original entomological concept of biological control was based on multitrophic-level interactions, involving parasitoid or predator natural enemies that used the pest (usually called host or prey) as a food source. In plant pathology and vertebrate pest management, the concept also includes competitive exclusion. The reduction of plant-pathogenic organisms in this way is often referred to as disease suppression, and the term *suppressive soil* refers to soils in which diseases are suppressed because of microorganisms that are antagonistic to soil-borne pathogens and nematodes. In some texts a much broader definition of biological control is accepted which includes cultural practices and host-plant resistance; such a broad definition is not adopted in this book.

In IPM terminology, the agent that is providing biological control is generally referred to as a beneficial, a natural enemy, or an antagonist. Biological control is often shortened to biocontrol.

## The Concept

**Biological control does not cause immediate reduction in target pest population.**

**Biological control only achieves partial suppression of the target pest, as a residual pest population is necessary to maintain natural enemies.**

After the arrival or introduction of a biological control agent (point A in Figure 13–1) in a field or area, its population gradually increases. Eventually the natural enemy population increases to the point at which it starts to reduce the population growth rate of the target pest, causing its decline. This is a description of the density-dependent activity of the natural enemy. Density dependence is a term that describes the feedback mechanism between the population densities of the natural enemy and the pest. The numbers of the natural enemy increase after the target host increases (a positive feedback). As the numbers of the natural enemy increase, the host population densities are suppressed (negative feedback) and a decline in natural enemy numbers will eventually follow. The density-dependent parasitism or predation of the natural enemy will, in due course, cause a decline in the pest population to a lower general equilibrium density, which is operationally referred to in the entomological literature as the general equilibrium position (GEP). Popula-

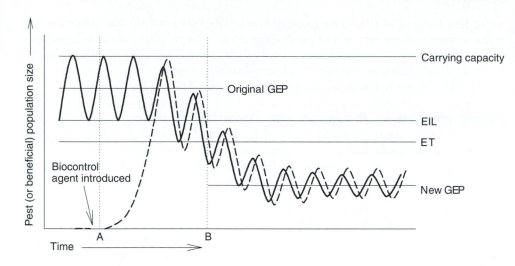

**FIGURE 13–1**
Idealized theoretical population dynamics of a pest and a biological control agent in relation to general equilibrium position, economic threshold, and economic injury level.

tions at the GEP are in balance with the biological control agent. For effective biological control that does not require further intervention, the new GEP must be below the economic injury level for the pest (Figure 13–1).

Figure 13–1 also demonstrates several important attributes of biological control.

1. There is an inevitable time delay (or lag) between the introduction of the biological control agent (point A) and the time at which effective pest reduction is attained (point B). The length of this delay varies from a week or two for organisms with short generation times (e.g., spider mites), to several years for organisms with long generation times (such as a perennial weed). In most situations, biological control cannot rapidly reduce pest population densities.

2. Biological control does not achieve 100% reduction of the pest, but rather reduces the pest population until it is in balance with the biological control agent. In some situations, the new GEP may still be above the economic threshold. In an effective biological control program, residual populations of both pest and beneficial should remain on or in the crop.

3. For a biological control system to function effectively over longer periods, the natural enemy must be able to increase its population density once the pest population starts to increase (Figure 13–1; time after point B). This increase by the beneficial must occur quickly in relation to the change in pest densities, and the increase must be in synchrony with the pest, otherwise the pest population may exceed the economic injury level.

4. The carrying capacity for the pest has not been changed. Any perturbation of the system that reduces the efficacy of the biological control agent will result in the pest population returning to a higher level.

## Trophic Relations Revisited

The basic concept underlying most biological control is that the natural enemy feeds on the pest, resulting in decreased numbers of the pest so that it no longer causes economic loss. In biological control based on competitive exclusion, resources needed by the pest are less available because they are used by a nonpest

organism feeding at the same trophic level. The competition for resources results in reduced impact of the pest. Exploitation of trophic relationships provides the basis for manipulation of biological control agents as a management
tactic in IPM. The pest management terminology used for organisms at different trophic levels in the food chain was presented in Chapter 4.

***Plants, Herbivores, Predators, and Parasites***    Predation, parasitism, and
diseases are the most common types of interactions involved in the biological
control of arthropods and weeds. They also apply to biological control of mollusks, nematodes, and vertebrates. Biological control of plant pathogens involves
competition and microbial antibiosis, in addition to parasitism/predation.

**Carnivores and Herbivores**    Natural enemies, or beneficial organisms, are usually predator or parasite second-order consumers that feed on herbivores (primary consumers) (see Figure 4–8, chains A, B, and D). The exceptions are organisms such as nematodes and mites that do not attack herbivores, but rather prey
on fungal or bacterial pathogens or decomposers (fungivores or bacterivores, respectively). For weeds, the biological control agents are herbivores; they are thus
primary consumers rather than carnivores (see Figure 4–8, see "weed" chain).

**Secondary Predators and Hyperparasitoids**    There are predators and parasites that attack beneficial organisms, and these higher trophic-level organisms
can limit the effectiveness of biological control agents. Such higher-order organisms are secondary predators, called hyperparasites and hyperparasitoids
(see Figure 4–8, chain B), when they attack beneficial arthropods. For biological control of weeds, the higher-order trophic levels are the primary predators
that feed on herbivores (in this case, the herbivore is the natural enemy of the
weed), and are thus members of the same guild as the beneficial for arthropod
management (see Figure 4–8, see 'weed' chain). Trophic interactions involving
hyperparasitoids therefore hamper managed biological control of arthropods,
but they are positive for the biological control of weeds, as will be discussed
later in this chapter.

Ɂ   The terminology of higher-order interactions must be used carefully relative
to biological control. For example, the term *hyperparasite* does not apply only
to parasites of beneficials. In plant pathology terminology, there are hyperparasites of pathogenic primary parasites of plants. Biological control of some plant-
parasitic fungi (e.g., *Rhizoctonia solani*) can result from hyperparasitism by soil
fungi such as *Trichoderma harzianum.*

**Competition within a Trophic Level**    Many beneficial arthropods (e.g., praying
mantis and many spiders) are relatively nonselective in what they will eat. Two
different natural enemy species may share a pest prey, and the two may also attack
each other or other beneficial species. The later phenomenon is referred to as intraguild predation. Some natural enemies are even cannibalistic and thus eat their
own species (e.g., lacewing larvae). Intraguild predation and cannibalism are
shown in the food web in Figure 4–8, chain A and Figure 4–8, chain B for between
species, and the circular arrow in Figure 4–8, chain B for within species. These
types of trophic interactions are negative from the standpoint of biological control,
and can limit the efficacy and utility of such generalist biological control agents.

Ɂ   Complex interactions among species of natural enemies that occupy similar
ecological niches also have a bearing on biological control. When natural enemy
species enter into direct competition, competitive displacement may occur,

wherein one of the species in the complex is displaced by the other. The displaced species may disappear completely from the area where the two species previously coexisted. Competitive displacement has occurred in California where *Aphytis lignanensis* was introduced to control the California red scale; a later introduction of *Aphytis melinus* led to the disappearance of *A. lignanensis*. A different example of competitive displacement or competitive exclusion is discussed later in the chapter under the topic of biological control of plant pathogens.

**Tritrophic and Multitrophic Feeding**    Tritrophic interactions refer to processes that occur across three trophic levels, for example, the interactions among plant, pest herbivore, and primary predator. In some situations, the pest can acquire chemical substances from the plants on which they feed, and those substances may be passed on to beneficials that feed on the pest. Such effects involve three trophic levels. The beneficial in chain D of Figure 4–8 experiences such an effect; the arrow in pest D indicates that a substance is passed on from the crop. Tritrophic interactions appear to be particularly important in the biological control of aphids, but they may affect even generalist predators such as mantids. Tritrophic interactions can be detrimental from a biological control viewpoint when the substance passed on by the herbivore inhibits the carnivore. Other tritrophic interactions may be favorable for the carnivore when plant cues such as odors or color are used to help locate the prey.

Multitrophic interactions are quite frequent in arthropod biological control. Many beneficial arthropods feed at different trophic levels during the different stages in their life cycle. Typically it is the larval stage of natural enemies that actually feed on a pest species and thus limit pest population growth, while the adult stage either does not feed at all or feeds on plant products such as pollen or nectar. Many parasitic wasps, for example, require nectar or honeydew (from aphids) to lay large numbers of eggs. Because such beneficial organisms are feeding both as a carnivore and herbivore, they are referred to as being multitrophic. A multitrophic relationship leads to more complex habitat requirements, as both suitable prey and suitable plants must be present, and thus explains why ecosystem diversity is desirable for arthropod biological control (see Chapter 7).

**Food Source Stability**    The life cycle of most biological control agents must be synchronized to some degree in time and space with their primary food source, the target pest. For a natural enemy that relies completely on the pest as a food source, it needs to be in a life stage that allows survival without food (a resting stage), such as spores, eggs, diapausing larvae, or pupae, when the pest is not present. Some parasitic fungi, of nematodes for example, are also capable of growing saprophytically when they are not parasitizing nematode hosts. Some beneficial arthropods lack a resting stage and so require the constant food source of suitable host(s); thus, an alternative host must be available when the primary target pest is absent. This is represented diagrammatically in Figure 4–8, where the beneficial in chain D also uses a host living on plants outside the managed ecosystem. *Anagrus epos,* the parasite of the grape leafhopper (see Figure 6–10), provides an example of this type of survival mechanism.

**_Competition and Microbial Antibiosis_**    A special case of competitive exclusion involves use of resources by a nonpest organism at the same trophic level as the pest. This concept of biological control is particularly important in the control of plant pathogens. The root-gall-forming bacterial pathogen

*Agrobacterium tumefaciens* can be controlled by dipping seedlings or cuttings in a suspension of an appropriate strain of *Agrobacterium radiobacter.* The *A. radiobacter* strain colonizes roots and produces an antibiotic, agrocin 84, which inhibits most pathogenic Agrobacteria that arrive at the root subsequently. The biological control agent is often referred to as an antagonist, and is depicted in Figure 4–6 (horizontal arrow). The same concept can be applied to weed management, in that a nonweedy plant, usually a crop, uses the same resources needed by the weed. In this case, the tactic involves the crop plant itself, and so as with host-plant resistance, it is not typically considered biological control, but is designated as a form of cultural management, and is so presented in this book.

## Additional Concepts and Terms

Humans, as noted in Chapter 10, have transported many pests into new regions of the world and continue to do so. The movement of exotic organisms to new geographic locations is a problem of significance to world agriculture, conservation biology, biodiversity, and ecology. Once an organism becomes established in a new area, it is extremely difficult, if not impossible, to eliminate it. Introduced organisms may experience rapid population growth in a new area, because the climate is suitable, they compete effectively with local organisms, and they are usually introduced without their associated predators and parasites.

Parasites are organisms that live in or on another organism (the host) for the majority of their life, obtaining their nutrition from the host and causing real damage to it. Parasites can be internal (endoparasites), or external (ectoparasites). Pathogen and nematode pests of plants are parasites. Parasites, as exemplified by internal parasites of mammals, usually coexist with their host for a long time. Parasite survival depends on the survival of the host, although the host usually is debilitated as a result of high densities of parasites. Some parasites eventually kill their hosts, and parasites that become associated with new or atypical host species may be lethal to those hosts.

Because true parasites do not typically kill their hosts, they are generally limited in their use in biological control efforts. There are important exceptions, however, such as in cases where parasites effectively eliminate reproduction by the target pest. Parasites that kill their hosts as part of their life cycle are termed by entomologists as being a *parasitoid,* meaning "parasite like." Insect parasitoids kill the host as a requirement for their normal development from egg to adult. Historically, parasitoid has been used only in reference to insects. However, fungal parasites of nematodes, insects, and arthropods also kill their hosts as part of their life cycles, and so they, too, would accurately be described as parasitoids. Endoparasites and endoparasitoids consume their host from the inside; many beneficial wasps and flies have larvae that are internal parasitoids, and many fungi penetrate their host and kill it in the process of producing new spores. Ectoparasites are attached to the outside of their host, and are able to suck the contents from the host. Ectoparasitism also occurs among wasps, parasitic flies, and beetles. Endoparasitism appears to be more frequent on insect hosts that are exposed, such as foliage-feeding caterpillars or eggs; and endoparasitism of soil-dwelling nematodes is relatively common. Ectoparasitism seems to be more common on insect hosts that live within protected spaces, such as leaf rollers, stem borers, and gall makers or are protected by a cocoon or puparium.

When dealing with pathogens it is necessary to differentiate between obligate parasites and facultative parasites. Obligate parasites require a living host at all times, whereas facultative parasites can also obtain nutrition by saprophagy, or feeding on nonliving decaying matter. Conservation of facultative parasites is easier than obligate parasites.

Obligate arthropod parasites require an alternate prey to maintain the beneficial when the primary host is not present. *Anagrus epos,* a parasite of the grape leafhopper, requires a suitable host at all times because it does not enter diapause (see Figure 6–10). Grapes are deciduous and thus there are no grape leafhoppers to be parasitized in the winter. *Anagrus* survives by parasitizing a leafhopper on blackberries as an alternative host during the winter. In the spring, *Anagrus* returns to vineyards and grape leafhoppers are available for it to parasitize. Vineyards near riparian areas where blackberries grow experience higher levels of biological control than do vineyards more remote from such areas. Alternate prey can also serve to increase the population of a beneficial insect prior to emergence of the target pest. Syrphid flies can, for example, utilize the sow thistle aphid before moving to crop aphids.

Predators are organisms that physically ingest part or all of their prey, usually resulting in the death of the prey. They are usually not physically attached to the prey and, as a rule, they attack multiple prey to complete their development. This is a difference from the parasites and parasitoids that are typically associated with a single host for most of their lives.

The use of pesticides in IPM programs needs to be carefully considered, because pesticides can disrupt biological control. Pesticide use, especially insecticides, should be avoided in any system where biological control agents are established and effective, because broad-spectrum insecticides can kill many beneficial arthropods. In California, the use of broad-spectrum insecticides nearly eliminated the biological control of the cottony cushion scale in citrus by the vedalia beetle in the late 1950s. Only by avoiding broad-spectrum insecticide applications was biological control reestablished.

If insecticides are used, it is preferred that those having low toxicity to beneficials are applied. These are the selective insecticides discussed in Chapter 11. Because selective pesticides are more compatible with environmental integrity, they are frequently referred to as "soft" pesticides. Chemical toxins that occur naturally in plants and fungi have been used in the development of biopesticides (termed "biorational pesticides"). Some biopesticides exhibit high levels of selectivity, and have minimal direct effect on beneficials. An approach to minimize the influence of insecticides on beneficials is the artificial selection of beneficial variants, or strains, of natural enemies that are resistant to insecticides. There have been attempts to produce strains of predatory mites with high levels of insecticide resistance.

Fungicides and nematicides also have the potential to reduce or eliminate biological control of insects and weeds by pathogens and nematodes. Similarly, those chemicals can interfere with biological control of pathogens and nematodes. Herbicides do not seem to directly impact biological control in most cases, but the use of herbicides can indirectly alter resources for biological control agents by changing the floristic composition of ecosystems.

## Fundamental Principles

Several principles must be followed to establish a biological control program. Some are described in the next section.

***Accurate Identification of the Target Pest***    Correct taxonomic identification of the target pest is essential, otherwise the beneficial organisms selected for use may not feed on the target pest. Many biological control agents are relatively host specific, and incorrect identification of the target pest may lead to failure of introduced exotic biological control agents. The coffee root mealybug in Kenya is an example. Misidentification of the mealybug led to the importation of natural enemies that did not feed on the target mealybug species. Only after a correct identification was made were appropriate biological control agents imported and the mealybug brought under control. Importation of herbivores against the wrong species of the aquatic fern *Salvinia* is another example of the importance of taxonomy. It also is critical to obtain the correct identification of the natural enemies employed for the same reason. Differences in the biology of closely related species may determine success or failure in a biological control program.

***Source of Beneficial Organism***    A suitable natural enemy must be used to implement a biological control program. Currently, agents are either exotic or endemic. In the future, natural enemies may possibly be enhanced or produced with the aid of genetic engineering technology.

**Exotic Agents**    Foreign exploration is conducted in regions where the target pest naturally occurs, and organisms that prey on or parasitize the pest are identified. Because many of the pests that are targeted for biological control were originally introduced (see Chapter 10), foreign exploration is usually carried out in the geographic area where the pest originated. Once a potential natural enemy has been identified, it is subjected to research to determine its potential efficacy, host specificity, and safety to nontarget organisms before it can be imported for biological control purposes.

**Endemic Agents**    If endemic beneficial organisms are already present in the system, they do not need to be imported. The drawback to using endemic beneficials is that they typically have their own natural enemies or higher-order organisms that feed on them. Those natural enemies generally limit the population growth rate of the desired beneficial so that they cannot provide adequate control of the target pest.

***Specificity of Agent***    The host range of the biocontrol agent must be determined. A fundamental consideration for biological control is that the agent must not attack any nontarget organism useful to humans, or organisms native to natural ecosystems.

Selectivity is a significant problem for biological control of weeds, as the biocontrol agents for weeds are herbivores, or plant pathogens. If such an agent attacks the target weed and a crop plant, other desirable plants, or even native plants, then it can become a serious problem that is difficult if not impossible to reverse. If biological control agents of weeds switch host plants, the result may potentially have a greater economic impact than if an insect switches to another host insect. However, switches of both kinds may have profound ecological consequences. Several biocontrol agents imported into Hawaii for the management of arthropod pests have switched to feed on native species in the same family as the pest, resulting in an extinction risk to the native species. Such local extinctions reduce biodiversity and can bring about ecological imbalances (see further discussion of the Hawaiian case below). In addition, an introduced arthropod biological control agent must not attack bees that are important pollinators.

Ideally, a beneficial agent will have a very limited host range (the agent is oligophagous), to the point that it will die when the host is unavailable, or it must go into an inactive resting stage when the target host is not present. Obviously, if no hosts are available, natural selection will act to favor any variants of the natural enemy that have some capacity to survive by utilizing alternative hosts. The release of natural enemies to new areas is approved only when the agent has been tested and deemed no threat to a wide range of nontarget organisms. In reality, it is impossible to check the activity of a natural enemy against all possible nontarget species. The release of any exotic biological control agent is thus a calculated risk.

***Culture, Rearing, and Delivery of Biocontrol Agents***   If the natural enemies are going to be released in large numbers, there must be a satisfactory method for rearing the agent under artificial conditions. Mass rearing is necessary so that there are enough organisms for research purposes, and large numbers of organisms can be produced for release once the biocontrol agent has been approved. Efficient rearing procedures are especially critical for natural enemies that are mass produced for inundative releases (see below). Rearing can be problematic. For example, the obligate bacterial nematode parasite *Pasteuria penetrans* shows great promise for the biological control of important plant-parasitic nematodes, including the root-knot nematodes, but there is currently no method for culturing sufficient numbers of the agent. The expense involved in raising the fungal pathogen *Colletotrichum gloeosporoides* for control of northern jointvetch is one of the reasons given why the biological control pathogen is no longer commercially available.

Not only must the biological control agent be capable of being reared or cultured, but mechanisms must also be available for appropriate delivery and dispersal of the agent into the target ecosystem without loss of fitness. If the agent is targeting a soil-borne pest, delivery and dispersal into the soil are extremely difficult. Mass releases of beneficials above ground are not unlike spraying a pesticide, except that the active ingredient must be kept alive and vigorous so that they are able to function normally in the new environment. For a relatively low number of mobile arthropods, dispersal can be as simple as opening a container and letting them fly away. Where large numbers are needed for inundative approaches, sophisticated delivery systems have been devised.

***Ecosystem Constraints***   A classical biological control agent will only succeed when the ecosystem conditions are suitable for its rapid population increase. Both environmental conditions and habitat suitability must be properly evaluated prior to introduction and release of a potential biological control agent. If conditions are inappropriate, the natural enemy will not control the pest in the desired way, or may not become established in the desired habitat.

**Environment**   The temperature range in the environment where the agent is going to be released must be suitable. Low temperatures limit the survival of biological control agents in many systems. For example, *Pediobius foveolatus,* a parasitic wasp native to tropical India, was imported for the control of the Mexican bean beetle on soybean in temperate zones of North America. The parasitoid is effective during the summer growing season but cannot survive the winter. Overwintering cultures are maintained in the laboratory and the wasps are re-released each year in small nursery plots for multiplication and natural dispersal into the soybean fields. Rainfall and day length can limit the range of some agents. Humidity is very important for establishment of pathogens as biological control agents. Low humidity, for example, severely limits the utility of pathogens for biological control in regions with arid climates.

**Habitat Suitability**    Most biological control agents require a relatively stable habitat for population growth and survival. Lack of stability is difficult in most agroecosystems because of periodic tillage and harvesting operations. If an agent cannot become established in an area, it must be reintroduced at periodic intervals as necessary. Greater ecosystem stability is a major reason why biological control often is more successful in orchards and extensive nontilled systems than in short season, arable row crops.

### Freedom from Hyperparasites

If a biological control agent is to be introduced to an area, it is essential that parasites or predators of the agent not be introduced concomitantly. The utility of the puncturevine seed weevil is, for example, limited by a parasite that stops it from attaining high population levels. Higher-order organisms must be evaluated for the extent of possible hyperparasitism prior to release of a biocontrol agent. Of course, it is generally difficult to assess all possibilities in this regard, and hyperparasites have limited the utility of several introduced arthropod agents (see later section on arthropods as agents).

**Monitoring for the presence and activity of beneficials, in addition to pests, is an essential component of IPM.**

### Monitoring

The process of monitoring the level of biological control present in an agroecosystem is an integral part of IPM. Management programs for many insects stress the need to monitor the pest species and their biocontrol agents, as well as their levels of activity. This permits estimation of the proportion of the pest population that is under biological control. For parasites, such estimates are expressed as the percentage of the pest population that is attacked by a beneficial and the percentage that is not attacked, while for predators the ratio of pest numbers to beneficial numbers is used. If biocontrol activity is sufficiently high, then it may not be necessary to apply other treatments to control the target pest, or it may be possible to predict that biological control will eventually increase to an adequate level such that other treatments are unnecessary.

The numerical relationship between pest abundance and natural enemy abundance allows establishment of what has been defined as the "inaction threshold." Some decision thresholds include assessment of biological control; if the economic threshold has been reached, but there are sufficient natural enemies in the system, then it may be possible to delay control action to see if biocontrol increases to an acceptable level. This only applies to crops where cosmetic damage is not an issue. For crops subject to cosmetic standards the damage that occurs during the "wait and see" period will probably be unacceptable. A standard practice for determining the need to treat mites in almonds is to determine the ratio of phytophagous (pest) to predatory mites and six-spotted thrips (beneficials). Control of the alfalfa butterfly is another example; the recommendation is to not apply an insecticide until there are over 10 nonparasitized larvae per sweep-net sample. The decision is based on the abundance of nonparasitized larvae, because if there is adequate biological control, then a pesticide treatment is unnecessary and may even aggravate the situation, as the possibility of pest resurgence is always present.

# Constraints to Biocontrol

Several factors can limit the success of biological control, and many of these factors are essentially the inverse of the principles listed in the preceding section.

***Value to Humans***   Because the concept of pest is anthropocentric, people working in different groups perceive weeds differently and assign different values to the same plant. What is a weed to one group may be valued as a useful plant to another group. Yellow starthistle is such a weed; it is viewed as useful by beekeepers, but is an enormous problem in the rangeland of western United States. Attempts to introduce biocontrol agents against this weed are viewed with concern by beekeepers, because yellow starthistle provides an excellent nectar and pollen source at times when there are few alternative flowers. A similar example of such a dichotomy occurred in Australia when attempts to introduce biological control of *Echium plantagineum* were met with consternation or hope, depending on one's point of view. The plant is commonly known as 'salvation Jane' by beekeepers because it is a good source of nectar, but as 'Patterson's curse' by ranchers because the plant is toxic to grazing animals. Such dichotomy in the assignment of value to biological pest control has not arisen for pests other than weeds.

***Ecosystem Stability***   Agroecosystems experience regular disturbance, and the lack of stability is considered a serious limitation to the success of biological control in arable agricultural systems. Increased ecosystem stability is a major reason for suggestions to increase diversity in agroecosystems, as discussed in Chapter 7. There are now attempts to increase stability by establishing refugia on the borders of fields, and even weed strips within fields. In most instances there are few biological or economic data to substantiate the utility of such habitat modification for biocontrol purposes. In the absence of such data, the role of habitat modification must be carefully assessed from an IPM viewpoint due to the potential for detrimental interactions (see Chapter 7).

***Lack of Adequate Control***   As noted earlier, it is not ecologically likely that biological control can attain 100% suppression of any target pest. For fresh market crops where cosmetic appearance plays a role in purchase, less than 100% pest control may be unacceptable, because any damage to the produce can make it unmarketable (see Chapter 19 for problems of cosmetic appearance of produce).

The length of time that is necessary for natural enemies to reduce target pest populations may be problematic because biological control may be relatively slow in achieving adequate pest suppression. The population of the beneficial can only increase after that of the pest has increased, at which time cosmetic or other damage may already have occurred. This means that the beneficial may not build up sufficiently or quickly enough to provide adequate control in a short season, fresh market crop such as lettuce. For example, the polyhedrosis viruses used for insect control eventually kills the target pest. The difficulty is that once infected, the insect continues feeding and causing crop damage for several to many days before it succumbs to the virus. Some delay between application of a natural enemy and an effect on the target pest is of limited consequence to long season crops such as sugar beets, or soybeans, where a *Baculovirus* has been used with great success, but it can be catastrophic to short season crops such as lettuce or broccoli.

***Compatibility with Other IPM Tactics***    In IPM systems that use pesticides, there are problems of fungicide and insecticide incompatibility relative to bio-control agents. Such chemicals can disrupt development or kill the agents. Biological control is usually specific to a given target pest species, so that the other pests that are present will still have to be managed. If the management includes the use of a pesticide, compatibility must be considered.

***Contamination of Harvested Product***    One advantage to biological control is that it does not leave pesticide residues in the product; however, the residual bio-control agents can themselves be a problem in the harvested product. Most consumers in western societies do not want to eat beneficial insects, such as lacewing eggs (Figure 13–2a) or damsel bugs (Figure 13–2b), any more than they do a pest insect. This imposes a severe limitation on the ability to utilize biological control for IPM in fresh fruit and vegetable crops, because such insects are likely to be found in the produce. In this situation, agent residues can preclude the use of biological control. The biological control by polyhedrosis viruses can be very effective, but the worm cadavers remain in the produce for many days (Figure 13–2c). The presence of dead caterpillars in broccoli, or mummies of parasitized aphids in lettuce are generally objectionable to consumers.

***Agent Access to Pest Organism***    Biological control of some pests can be limited by the ability of the agent to find its prey (the pest). Host-finding ability is a critical component in the biological control equation. The rosy apple aphid injects a toxin into apple leaves which causes the leaf to curl. The aphid lives inside the curled leaves and is hidden from parasites such as wasps, and consequently escapes predation or parasitism. Lack of accessibility by beneficials is a major problem with several insect pests of vegetables such as broccoli and cauliflower (the pest is hidden in the head), and Brussels sprouts (cabbage aphids are inside the sprouts and are completely hidden).

Establishing biological control in soil is difficult because of the inability of biological control agents to disperse and locate their prey. Similarly, adding biological control agents to soil in a way that ensures distribution of the agent throughout the soil profile is difficult, because of the semisolid nature of soil.

***Host Race Specificity***    The same ecological phenomena that select for pesticide resistance in pest populations are also operating in relation to biological control agents. The biological control agent essentially exerts continuous selection pressure on the host pest, and it is thus inevitable that host biotypes that best tolerate or evade the natural enemy survive and reproduce. Such selection may limit the long-term effectiveness of biological control programs when the target host can develop resistance (i.e., a new race). This phenomenon has been observed for biological control of skeleton weed in Australia using the rust fungus *Puccinia chondrillina*. The level of control has decreased since the rust was first introduced, and this decline in efficacy has been attributed to selection of rust-tolerant races of the weed.

Perhaps the best documented case of resistance in arthropods to a biological control agent is that of various pest species against the entomopathogen *Bacillus thuringiensis* (Bt). Resistance was first recorded against Bt formulations in populations of the diamondback moth in Hawaii, which has raised serious questions about the stability of the genetically modified crop varieties incorporating the genes for the Bt toxin (see discussion in Chapters 12 and 17).

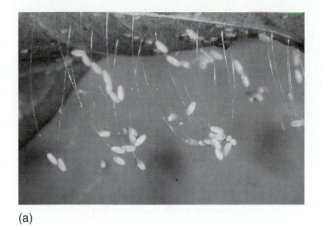
(a)

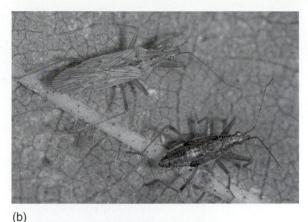
(b)

**FIGURE 13–2**   Examples of biological control agents that can contaminate fresh produce. (a) Lacewing eggs; (b) damsel bugs; and (c) cadaver of an armyworm killed by a polyhedrosis virus.

Sources: photograph (a) by Marcos Kogan; (b) by Jack Clark, University of California statewide IPM program; and (c) by Robert Norris.

(c)

Highly host-specific parasitoids also exert considerable selection pressure on their hosts such that the development of resistance becomes a possibility; however, there are few records of such development. The ichneumonid wasp was introduced into North America to control the larch sawfly. After 27 years of exposure, the sawfly developed immunity to the ichneumonid. The resistant sawfly strain had the physiological capacity to encapsulate the endoparasitoid larvae, thus impeding the larvae from further feeding in the host. There is no record of a prey developing resistance to a predator, mainly because predators have more generalist feeding habits.

***Host Specificity***   Although host specificity is considered essential for biological weed control, it is also a serious impediment to widespread adoption of biological weed control in annual row crops. Annual row crops have many weed associates typical of the crop. Removal of a single species, which is the inevitable result of adoption of a species-specific biological control agent, may simply result in a secondary weed becoming dominant, and does not resolve the overall weed management needs for that crop. The weeds not subjected to the biological control continue to grow normally, or perhaps even better than normal. The biocontrol agent effectively induces replacement, as discussed in Chapter 12. It is unlikely that biological control will be able to resolve weed management problems in annual row crops.

***Lack of Host Selectivity***   The biological control agent may switch to another host. Host switches have been of some concern for biological control. Specificity, however, is a major concern for any organism released for weed control, because biological control agents for weeds are herbivores and might switch to crops or other useful plants. The problem of host specificity in biological control agents has become a major issue among many ecologists concerned about conservation and biodiversity. This is discussed under the next topic heading, and in Chapter 19.

***Ecological Disturbance***   During the 1990s, concerns mounted about unpredictable long-term ecological disturbance/perturbation resulting from the introduction of biological control agents (see later under societal/environmental concerns in Chapter 19). As mentioned, biological control using introduced nonnative agents must include consideration of possible nontarget effects. This is particularly important, because the long-term ecological consequences have the potential of being more serious than those caused by pesticides and, thus, biocontrol agents cannot readily be reversed or cancelled if undesirable nontarget effects develop. Ecological disturbance results from incomplete assessment of the potential for a biological control agent to attack organisms other than the target pest (i.e., its host specificity was not sufficiently high). The following are presented as examples of this problem.

The small Indian mongoose was introduced into the West Indies, Hawaiian Islands, Mauritius, and Fiji to control rats in agricultural fields. The anticipated rat control was never achieved as the mongoose is diurnal, and rats are primarily nocturnal. The mongoose then turned to other prey, and is now considered to be a pest because it contributes to the decline of native birds in those areas. Similarly, ferrets were introduced into New Zealand to control rats, but like the mongoose, have now become a pest.

Fish have been introduced into several aquatic systems for aquatic weed management. Although highly effective for killing the target weeds, this strategy has resulted in major damage to ecosystems because of the almost complete destruction of all aquatic vegetation. It is argued that every fish introduced for aquatic weed control has had major detrimental effects on nontarget species.

With a rich agriculture based on the production of high-value crops, Hawaii has been in the forefront of classical biological control since early in the twentieth century. Many of its major crops were exotic species brought to Hawaii from North and South America; thus, they were obvious targets for importation of natural enemies. The southern green stinkbug was accidentally introduced into Hawaii in 1961. Following that introduction, two major natural enemies of stinkbugs were imported into the islands: the egg parasitoid *Trissolcus basalis* and *Trichopoda pilipes,* a fly parasitoid of late nymphs and adults. In the years following the introduction of the parasitoids, populations of the nontarget native species of koa bug, as well as a few native predaceous stinkbug species, declined noticeably, and those of the nonnative minor pest harlequin bug virtually disappeared. Similarly, the rich, nonpest fruit fly (Tephritidae) fauna native to Hawaii was gravely impacted after importation of parasitoids for the control of four major fruit fly pests that invaded the islands in the late 1800s.

Biological control agents introduced to the northeastern United States for gypsy moth control probably provide the most insidious example of effects on nontarget native insects. In 2001, researchers noted that the tachinid flies introduced to control gypsy moth larvae have probably eliminated all large lepidoptera from the hardwood forests in this region.

In all these cases, hindsight suggests that more caution should have been exerted before releasing the biocontrol agents into new environments, and close monitoring should have been performed after the releases. Could the dangers have been foreseen? Follet and Duan (2000) Simberloff and Stiling (1996), and Wajnberg et al. (2001) provide in-depth discussions of this topic.

# TYPES OF BIOLOGICAL CONTROL AND IMPLEMENTATION

Described in this section are several different classes of biological control programs.

## Classical Biological Control

Classical biological control is usually targeted at an imported alien pest species. It employs a natural enemy introduced from the region where the pest species is native. Once the biological control agent is introduced and established, classical biological control usually is self-sustaining.

Classical biological control is usually implemented by regional or national government agencies in cooperation with research institutions. The individual pest manager does not normally manipulate classical biological control.

## Inoculative Biological Control

Inoculative biocontrol involves periodic releases and reestablishment of a biological control agent that dies out each year, but which can rapidly expand its population when conditions are suitable. The annual reintroduction of the larval parasitoid for Mexican bean beetle control in common beans described earlier (see Chapter 13, page 347) is an example of an inoculative biological control program.

As inoculative biological control programs are typically regional in nature, they are administered through local government agencies or specially formed task forces. Like classical biocontrol, individual managers do not normally perform inoculative releases, although individual growers may implement certain programs using releases of predatory mites.

## Augmentative Biological Control

Augmentative biocontrol involves the periodic release of a natural enemy, usually endemic, that is already present (e.g., lady beetles in vegetable and fruit crops) but does not build up its population sufficiently or quickly enough to effectively control the pest. Figure 13–3 shows the population dynamics of the phytophagous two-spotted spider mite on strawberries in the absence and the

**FIGURE 13–3**
Graph of numbers of two-spotted mites in field-grown strawberries in the absence of, and after release of, the predatory mite *Phytoseiulus persimilis.*
Source: graph drawn from data in Oatman et al., 1977.

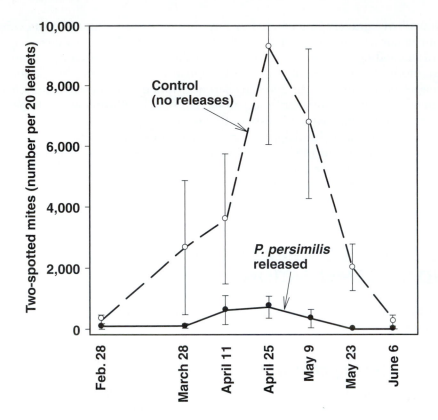

presence of releases of the predacious beneficial mite *Phytoseiulus persimilis.* This example demonstrates how effective an augmentative release of a biological control agent can be.

The individual manager usually carries out augmentative biological control, with the required natural enemies purchased from a commercial supplier.

## Inundative Biological Control

The mass release of the biological control agent that cannot reproduce and thus cannot attain adequate population size without human intervention is called inundative biocontrol. The mass releases of *Trichogramma* egg parasitoids to control lepidopterous pests in annual crops such as corn and cotton are examples of this type of biological control. Certain programs use releases of hundreds of thousands of these wasps per acre to achieve the desired level of control. In these cases, the biocontrol agent is considered and treated as a biotic pesticide.

Inundative tactics can be at the regional level and thus carried out by government agencies, or at the level of the individual crop by a pest manager. In either situation, the biological control agent must be artificially reared or mass produced, making this strategy viable in supporting commercial products.

## Conservation Biological Control

The conservation approach to biological control attempts to maintain endemic biocontrol by avoiding disruption of ecosystem function. It applies mainly to arthropod management, and it has been widely discussed (e.g., van Emden,

1990; Pickett and Bugg, 1998; Collins and Qualset, 1999). Although it is difficult to obtain experimental proof of the benefits of conservation biological control, the outbreaks of secondary pests due to the disruption of natural control agents, described in Chapter 12, provide strong indirect evidence.

Conservation biological control is accomplished through maintenance of habitat that can sustain beneficial organisms, and is implemented at both the farm and regional levels. Implementation of conservation biological control requires that suitable host plants are grown, and that special measures are taken not to disrupt the existing natural enemies, rather than direct purchases and introductions of biological control agents.

## Competitive Exclusion

The competitive exclusion approach applies mainly to pathogens. The ecological concept is different from that of feeding by a higher tropic-level organism limiting the population development of its prey. Competitive exclusion may involve the interaction of organisms at the same trophic level. The control of crown gall of fruit trees with nonpathogenic strains of the bacterium is an example (see pages 343–344).

The individual manager implements competitive exclusion. As with inundative biocontrol, the agents are produced commercially and purchased by the manager.

## Induction of Suppressive Soils

It is necessary to distinguish between the soil ecosystem and the aerial environment in relation to biological control of plant pathogens. As noted, biological control in soil is subject to different constraints than biological control in the aerial environment. The increase in soil microorganisms that reduce nematodes and disease-causing pathogens leads to suppressive soils. Suppressive soils are typically identified as those in which the pathogen or nematode is present, but plant disease and crop losses do not occur. Suppressive soils may be induced by continuous monoculture of particular plants, or by cultivation of plant species that support antagonists. In the case of continued monoculture of a particular crop plant, the crop allows pest pathogen or nematode populations to increase in numbers, and simultaneously supports the increase of antagonists of those pests. When monoculture is maintained for some time, the pests increase sufficiently to support high densities of their endemic natural enemies, and thereby stimulate a density-dependent decrease in the pest population. Eventually, the soil is sufficiently suppressive that even though the pest is present, the crop can be grown without substantial damage.

## EXAMPLES OF BIOCONTROL USING VARIOUS AGENTS

Arthropod feeding is what most people think of when they consider biological control. In reality, any organism has potential as a biological control agent if it attacks a pest.

## Pathogenic Parasites as Natural Enemies

Pathogenic organisms, primarily the fungi, bacteria, and viruses, disrupt the normal function of their host, resulting in reduced growth or even death. Pathogens that use pest organisms as hosts therefore have the potential to provide biological control of those hosts (Figure 13–4).

***Pathogens versus Pathogens***    Microorganisms that are considered beneficial can be used to control other microorganisms that are considered pests. A range of trophic interactions are involved in this type of control, including competition, antibiosis, and parasitism. Competitive exclusion occurs when the beneficial organism competes with the pest at the same trophic level and preempts resources required by the pest for normal growth and reproduction. The basic idea is that the "infection court" or physical area on the host which is required by the pest pathogen for infection is occupied by a nonpathogenic strain of a microorganism (the "natural enemy" of the pathogen). This type of competition may involve the production of antibiotics by the natural enemy.

There are fungal hyperparasites that parasitize plant-pathogenic fungi. Some of the best known hyperparasitic fungi are in the genus *Trichoderma*. For example, *T. harzianum* has been shown to parasitize *Rhizoctonia solani* in

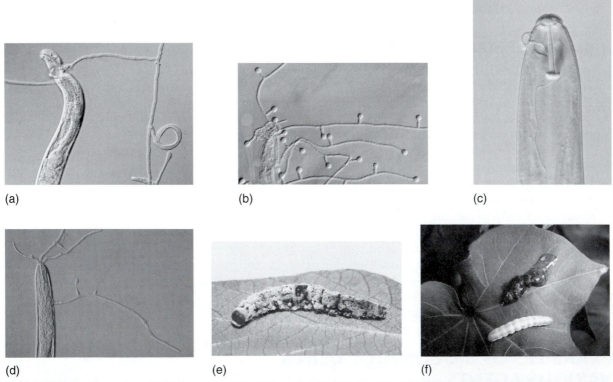

(a)    (b)    (c)

(d)    (e)    (f)

**FIGURE 13–4**    Examples of pathogens for biological control. (a) Nematode trapping fungus that "lassos" its prey; (b) nematode trapping fungus *Monacrosporium* which has sticky knobs; (c) spore of fungus *Hirsutella* penetrating a nematode larva; (d) *Hirsutella* growing from an infected dead nematode larva; (e) lepidopterous worm larva killed by an entomopathogen; and (f) cotton bollworm killed (top) by virus entomopathogen.

Sources: photographs (a), (b), (c), and (d) by Bruce Jaffee, with permission; (e) by Marcos Kogan; and (f) by David Nance, USDA/ARS.

vitro. Experiments in suppressive soil indicate that the *T. harzianum* also parasitizes *R. solani* in a density-dependent manner. The fungus *Coniothyrium minitans* parasitizes sclerotia (resting structures) of several plant-pathogenic fungi, and it appears that such hyperparasitism decreases plant disease caused by the target pathogens. Fungal hyperparasites are thought to be involved in the observed disease suppressiveness of many soils, which is an active research area in plant pathology.

Examples of successful biological control of pathogens include the following:

1. Crown gall is a serious bacterial disease of several orchard crops caused by the bacterium *Agrobacterium tumefaciens.* Biological control of this pathogen is achieved by dipping bare-root transplants into a suspension of a nonpathogenic strain of the bacterium *Agrobacterium radiobacter.*
2. Fire blight of pear trees is caused by the bacterium *Erwinia amylovora.* Biological control with nonpathogenic strains of *Pseudomonas syringae* bacterium is being used commercially to manage this disease.
3. Attempts are being made to inoculate sweet potatoes with a nonpathogenic strain of *Fusarium* fungus to protect them from attack by the pathogenic form.
4. *Gliocladium, Trichoderma,* and *Penicillium* species and strains of these fungi are being investigated as antagonists of *Botrytis cinerea* on various crops such as grapes, strawberries, rubber, and onions.
5. Suppressive soils involve the increase of microbial populations that are antagonistic to pathogens and nematodes, resulting in decreases in disease severity even though a pathogen is present.
6. Bacteriophages, which are viruses of bacteria, have been evaluated as biological control agents against bacteria with essentially no success.

With a few important exceptions, managed biological control of plant pathogens is limited; but with more research, the possibilities are increasing. A major limitation is that in many situations there is no known higher trophic-level organism actively feeding on the pathogen. Several reviews (Sutton and Peng, 1993; Backman et al., 1997; Mehrotra et al., 1997; Rush and Sherwood, 1997) provide more detail on biological control of plant pathogens.

Many microorganisms act through antibiosis, which is the production of toxins. This forms the basis of using microorganisms as biopesticides, which is the second approach for control of pathogens with other pathogens.

**Pathogens versus Weeds**     Many pathogens attack plants, and when the attacked plants are weeds, the pathogen can reduce the vigor of the weed and make it less competitive. Pathogens usually do not kill the weed. There are several examples of successful biological weed control programs using pathogens. The concern, however, is that a pathogen that attacks weeds might alter its host range to include crop plants related to the weed; this concern has slowed development of this type of biocontrol. Another major difficulty of applying pathogens for biological weed control is the difficulty of obtaining reliable infection of the target weed species under field conditions.

The use of a rust fungus (*Puccinia condrilina*) to control rush skeletonweed is an example of classic biological control applied to weed management using a pathogen. Control of the weed has been effective in Australia and in the western United States, but problems of race specificity and resistance limit effectiveness. Another serious problem is that the fungus performs poorly under dry climate conditions. Numerous other pathogens are being investigated as control agents for other weeds, such as rust on mallows and smut on grasses.

There are two examples of inundative release of pathogens for biological weed control. The term "mycoherbicides" has been coined in relation to such use of a pathogen, as it is sprayed in much the same fashion as a herbicide. The commercial product Collego® (based on the fungus *Colletotrichum gloeosporoides*) was developed for control of northern jointvetch in rice in the southern United States. The production of Collego® has been stopped, based on expense and lack of compatibility with fungicide use in the crop. A product called DeVine® (based on *Phytophthora palmivora*) was used for control of strangler vine in Florida citrus. The pathogen has been very effective, with a single application lasting many years. Unfortunately, because repeated applications were not necessary, the economic return for commercial production of the fungal agent was poor, and it is no longer commercially available.

***Pathogens versus Nematodes***   Several different types of parasitic pathogens attack nematodes and can provide substantial levels of biological control.

**Fungi**   Many nematophagous fungi prey on nematodes and play a role in causing decline of nematode populations in soil. Some fungi have different types of morphological adaptations which allow them to efficiently capture their nematode prey (Figures 13–4a and b), while others parasitize nematode hosts (Figures 13–4c and d). There are different types of adhesive or nonadhesive traps produced by "nematode-trapping" fungi, including simple traps such as adhesive mycelia, hyphal branches or knobs, and three-dimensional networks that serve to trap nematodes by adhesion. Some fungal species have highly developed mycelia which form rings to trap nematodes by rapidly contracting upon nematode contact, and thus constricting any nematode within the ring (Figure 13–4a). Other fungi, such as *Dactylella candida,* produce adhesive knobs and noncontracting ring traps on the same mycelium.

An example of a soil suppressive to nematodes involves the cereal cyst nematode and susceptible cereals in England. Cereals were grown for more than 10 years, and although the cyst nematode numbers increased during the first 4 years of cereal monoculture, over the next 10 years the numbers of cyst nematode declined dramatically. The suppression of the cyst nematode over time was caused by the increasing densities of four species of nematode-parasitic fungi, including *Nematophthora gynophila* and *Verticillium chlamydosporium.* The *N. gynophila* parasitize the cyst nematode females and completely destroy them within a week.

**Bacteria**   The unusual mycelial and endospore-forming bacterial genus *Pasteuria* includes species that are endoparasites of plant-parasitic nematodes. The root-knot nematodes are parasitized by *Pasteuria,* which produces spores that adhere to the cuticle of second-stage juvenile nematodes. The nematodes contact spores as they move through soil, seeking host-plant roots. About a week after a spore contacts and adheres to a juvenile nematode, the spore germinates and penetrates the nematode cuticle. A vegetative colony of bacterial cells results, and the nematode body contents are consumed. The end result is that the nematode reproductive capacity is destroyed. Eventually sporangia and endospores fill the cadaver of the nematode host. During the period of bacterial growth, the nematode continues its feeding and development, but eventually the nematode is killed, with adult female nematodes eventually filled with as many as 2 million bacterial spores. Later, the roots and nematode bodies decompose and the spores are released into the soil.

**Pathogens versus Mollusks**   Although some pathogens attack mollusks, according to Godan (1983) these are not likely to become important for use as biological control agents that can be manipulated.

**Pathogens versus Insects**   Several different types of parasitic pathogens infect arthropods and can provide substantial levels of biological control. An epidemic of a pathogen against arthropods (and other animals) is referred to as an epizootic. Under favorable conditions, an epizootic can kill all susceptible individuals in a population, resulting in a population crash of the target pest.

**Fungi**   Many fungi attack insects. Species in the genus *Entomophthora* are probably the best known. When an epizootic occurs, *Entomophthora* can provide very high levels of control of many insects, such as aphids, lepidopterous caterpillars, and grasshoppers (Figure 13–4e). The fungus *Nomuraea rileyi* is probably the single most important mortality factor regulating populations of the velvetbean caterpillar in temperate and subtropical areas where soybean is grown in North and South America. Onset of *Nomuraea* epizootics depends on adequate moisture, so under dry conditions delays may occur that allow pest populations to explode. The pathogen *Verticillium lecanii* provides biological control of aphids on chrysanthemums in England. All these pathogens are endemic and conditions that lead to an epizootic are not well understood. Biological control afforded by these pathogens is currently not easily manageable; but certain fungi, for example *Metarrhizium anisopliae,* have been mass produced in the laboratory for inoculative introduction for insect control.

**Bacteria**   The bacterium *Bacillus thuringiensis* (Bt) is the most widely used pathogen for biological control of insects. The bacterium produces an insecticidal crystal protein attached to the cell surface. This protein, known as an endotoxin, is toxic when ingested by larvae in the Lepidoptera (butterflies and moths), various Coleoptera (beetles), and certain Diptera (mosquitoes). The type of insect controlled depends on the Bt strain being used. The Bt endotoxin is nontoxic to warm-blooded animals. The use of Bt is basically an inundative tactic; a bacterial spore suspension is sprayed over the foliage to be protected, and the insect ingests the spores as it consumes the leaf. The ingested Bt spores release the endotoxin that causes the insect to stop feeding after a few hours, and to die within a day or two. Sprays of Bt are one of the few so-called insecticides that are acceptable for management of insects in organic farming systems.

Many of the genes in the bacterium that encode for production of the endotoxins have been identified, and have been inserted into several crops using genetic engineering technology (see Chapter 17 for more details).

There are two limitations to the use of Bt. First, ultraviolet light in sunlight kills the bacterial spores and thus Bt sprays are only effective for a few days. Repeated applications are often necessary, which is a desirable trait for a commercially viable product. Second, like synthetic pesticides, the target pests can develop resistance to Bt. Resistance has the potential to severely limit the utility of Bt in the future if sound IPM resistance management guidelines for its use are not followed.

Several other bacteria are used as insect biocontrol agents. Milky spore disease of Japanese beetle larvae, which is caused by *Bacillus popilliae,* is an example.

**Viruses**    Many viruses are known to attack insects. Polyhedrosis viruses, which are endemic, kill many species of caterpillars; and when an epizootic occurs, larval populations of cotton bollworm (Figure 13–4f) and beet armyworm, for example, can be decimated. Granulosis viruses are being evaluated for control of codling moth. The most successful use of a virus in insect pest management for biological control is that of the velvetbean caterpillar on soybean in Brazil with sprays of *Baculovirus anticarsiae* in water suspensions.

The use of viruses, however, has some limitations due to inability to maintain the viral agent in a stable state outside a living cell. Mass production of a virus on an artificial medium has not been achieved. Another problem is the relatively slow kill following virus infection. Unlike the Bt endotoxin, which causes the caterpillar to stop feeding in a few hours, a caterpillar infected with a virus may continue to feed for several days.

**Pathogens Versus Vertebrates**    The use of pathogens as a means to control vertebrates is not very popular. Such pathogens cause disease in the target animal, which often leads to a lingering death. Another serious difficulty lies with specificity. Any pathogen used for vertebrate control must not attack humans, pets, domesticated farm animals, or desirable wildlife.

Myxomatosis is an example of the use of a pathogen for vertebrate management. Myxomatosis is a benign virus disease of cottontail rabbits in North America, that is transmitted by fleas and ticks. This disease was introduced into Australia in 1950 as a biological control agent for rabbits that were causing devastating losses. Initial mortality was estimated to be as high as 99% in some regions of Australia, but populations recovered due to development of resistance in the rabbits. Rapid spread of the disease requires the presence of vectors; in some regions the lack of suitable vector insects limits the utility of the disease as a management tool. Myxomatosis was also introduced into several countries in Europe in attempts to reduce rabbit populations.

Myxomatosis serves as a further example that natural selection can reduce the utility of biological control due to the development of resistance. Virus strains with increased virulence against the resistant rabbits have also developed, and the disease continues to provide adequate control of the rabbit population in many regions of Australia. Due to the continued problem with rabbits in Australia, rabbit calicivirus is being evaluated as another possible biocontrol agent.

## Plants, Including Weeds, as Agents

Plants have a unique role as biocontrol agents that differs from other such agents because plants occupy the same trophic level as the crop.

**Plants versus Weeds**    Competitive exclusion of weeds is sometimes considered as biological control, but this is usually achieved by using crop, or non-weedy, plants. The concept is considered under the topic of cultural management tactics (see Chapter 16). Plants release chemicals that impact growth of other plants. This is called allelopathy (see chapter 14); and can be considered to be a form of biological control. The extent to which allelopathy might regulate weed growth is controversial, and to to-date there are no unequivocal examples in which allelopathy is used to manipulate weed growth.

**FIGURE 13–5**
Example of plants for biological control; African marigolds (*Tagetes* spp.) used in rotation to suppress nematodes.
Source: photograph by G. Caubel, with permission.

***Plants versus Nematodes*** Because plant-parasitic nematodes are obligate parasites, purposeful manipulation of available host plants may be used to reduce nematode densities. Several plants produce root exudates that are toxic to nematodes. Plants in the genus *Tagetes* (African marigolds) provide reduction in nematode numbers when planted in rotation with susceptible crops (Figure 13–5). Some plants, such as velvetbean, castor bean, and sword bean, reduce nematode populations in soil because they have rhizosphere bacteria that are antagonistic to plant-parasitic nematodes, such as the soybean cyst nematode and root-knot nematode. The bacteria associated with castor bean and sword bean roots, such as *Pseudomonas cepacia* and *Pseudomonas gladioli,* are different species than the bacteria commonly found on soybean roots.

***Plants versus Arthropods*** Plants indirectly support beneficial insects by providing alternate prey when the target crop pest is not present. Plants can also provide nectar and pollen to adult beneficial insects, which can improve both their longevity and fecundity. This role of plants, which can include weeds, forms the basis of conservation biological control. Plants can be sowed within fields or on field margins, or can be incorporated into noncultivated areas known as refugia to provide nectar, pollen, and other resources (see Figure 7–3). Plants also can be used to trap pest populations for destruction with an insecticide; these uses, however, are more appropriately discussed in Chapter 16.

## Nematodes as Agents

***Nematodes versus Weeds*** Nematodes can feed on weeds, and such feeding can reduce the weed fitness. The reduction in growth or reproduction is typically only partial and does not serve as adequate control by itself. Ring nematodes feeding on sedges in rice substantially reduce the competitive ability of the sedges, and gall nematodes on fiddleneck reduce seed production. There is currently no attempt to utilize nematodes for biological control of weeds.

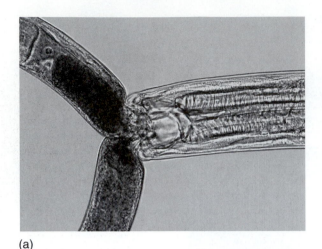

(a)

(b)

**FIGURE 13–6**   Examples of nematodes for biological control. (a) Nematode *Mononchus* (right) feeding on a nematode larva (left); (b) larva of fall armyworm infected by *Steinernema carpocapsae,* dissected to release nematodes.
Sources: photograph (a) by J. D. Eisenback, with permission; (b) by Arno Hara, with permission.

***Nematodes versus Nematodes***   Many nematodes prey on other nematodes. Such predatory nematodes have feeding structures, such as large teeth or oral spears, which allow them to gain access to the body contents of their prey (Figure 13–6a). In soil, the numbers of such predators are usually low relative to the numbers of their prey.

***Nematodes versus Mollusks***   Many nematodes (see lists in Godan, 1983) attack slugs and snails, and could be used for biological control. None are currently used, but several are being evaluated.

***Nematodes versus Insects***   Certain nematode species are insect parasites, and some are pathogenic to insects (called entomopathogenic nematodes). The entomopathogenic nematodes, *Steinernema* spp. and *Heterorhabditis* spp., carry internal symbiotic bacteria which are lethal to insects. The nematodes act as vectors and serve to move the bacteria from one insect host to another. The nematode enters the insect host through natural body openings and releases the bacteria into the insect hemocoel. The bacteria multiply within the insect, killing it, and the nematodes feed on the bacteria as they multiply within the insect cadaver (Figure 13–6b). Eventually the infective-stage of the nematodes emerge from the cadaver carrying their internal symbiotic bacteria. The entomopathogenic nematodes serve a beneficial role in pest management, as some are being successfully used in the biological control of soil-dwelling insect pests (e.g., Japanese beetle larvae, black-vine weevil, strawberry-root weevil, sod webworms, tent caterpillar, fungus gnats, billbugs, mole crickets). Mass production and commercialization of nematodes is being expanded for the biocontrol of insects.

## Mollusks as Agents
***Mollusks versus Weeds***   Although they have preferences regarding the plants they eat, slugs and snails are not sufficiently selective to be used for biological control of weeds.

**FIGURE 13–7** The decollate snail, used for biological control of the brown garden snail in citrus.
Source: photograph by J. Clark, University of California statewide IPM program.

***Mollusks versus Snails***   The decollate snail (Figure 13–7) is a herbivore and also a predator on the brown garden snail. The latter is a serious problem in citrus production, and the decollate snail is successfully employed as a biological agent against the brown garden snail. The decollate snail is also herbivorous and will eat seedling crop plants; thus, it should not be introduced into annual cropping systems.

## Arthropods as Agents

Feeding by arthropods damages or kills their lower trophic-level prey (for carnivores) or host plants (for herbivores). When the prey or host is a pest, the feeding can result in control of the pest. Examples of several arthropod biocontrol agents are shown in Figure 13–8; several additional examples are shown in Figure 2–12.

***Arthropods versus Pathogens***   There are no examples of managed biocontrol employing arthropods as agents against pathogens.

***Arthropods versus Weeds***   Herbivorous arthropods that feed on weeds typically reduce the vigor of the weed and make it less competitive, or they eat the weed seeds. To be used for biological control, a herbivorous arthropod must be selective and not attack nontarget plant species. Selectivity of the agent is proving to be a potentially serious problem for all classical biological control programs aimed at weeds. The seedhead weevil (*Rhinocyllus conicus*), which was introduced to control several weedy thistle species, is an example. The weevil has expanded its host range to include endangered thistle species in Nebraska.

There are several examples of classical biological control of perennial weeds using imported exotic arthropod species. Prickly pear was introduced into Australia in the nineteenth century where it rapidly invaded large areas of rangeland. As noted in Chapter 3, the beetle *Cactoblastis cactorum* was imported and released. The beetle damages the cactus by feeding on it, and wound-colonizing pathogens then kill the plant. The extent of the infestation was reduced by over 80% between 1929 and 1940, and the control is still satisfactory (see Figure 3–6).

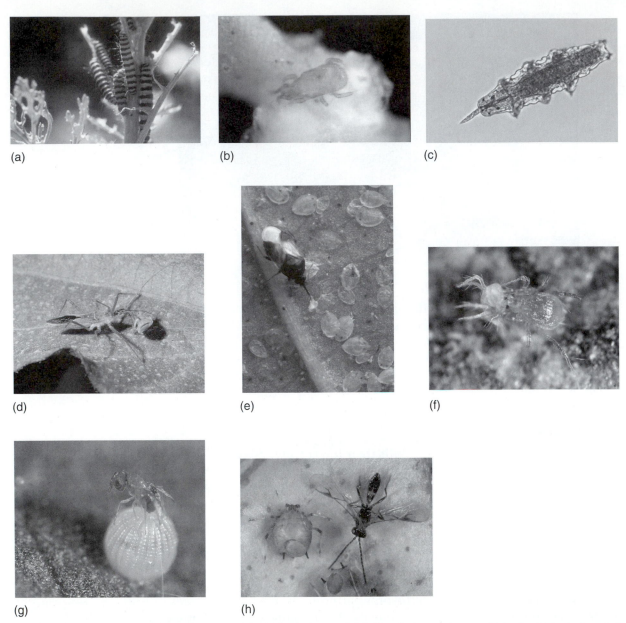

**FIGURE 13–8**  Examples of arthropods used for biological control. (a) Cinnabar moth larvae which give biological control of tansy ragwort; (b) soil-inhabiting mite eating *Meloidogyne chitwoodii* eggs; (c) soil-borne arthropod in the Tardigrada consuming a nematode; (d) generalist predator assassin bug feeding on a lygus bug; (e) generalist predator minute pirate bug feeding on whitefly nymphs; (f) the predatory mite *Phytoseiulus* feeding on a two-spotted mite; (g) parasitic wasp *Trichogramma* laying an egg in the egg of a tomato fruit worm; and (h) parasitic wasp of melon aphids, with aphid mummies (exit hole visible in aphid at bottom of photograph).

Sources: photograph (a) by California Department of Food and Agriculture; (b) by Renato Inserra, with permission; (c) by E. Bernard, with permission; (d), (f), (g), and (h) by Jack Clark, University of California statewide IPM program; and (e) by Jack Dykinga, USDA/ARS.

Tansy ragwort was introduced into rangelands in the western United States. The plant became dominant (Figure 13–9a), is poisonous to grazing livestock, and was targeted for biological control. The cinnabar moth, which has a foliar feeding larva (Figure 13–8a), was introduced, and has reduced the weed to acceptable levels (Figure 13–9b). Likewise, Klamath weed, also know as St. John's Wort was also introduced into the western United States and became dominant in many rangeland areas. This weed is now controlled biologically after introduction of the herbivorous beetle *Chrysolina quadrigemmina* and other insects. The control of *Lantana* in Hawaii serves as a further example of successful weed biocontrol.

**St. John's Wort is the basis of a now popular herbal remedy. Growers of the plant must now control the biological control agent.**

There are few examples of biological control of annual weeds. A seed weevil was introduced to control puncturevine in the southwestern United States. It is only a partial success because the seed weevil populations themselves are limited by a native parasite. Common purslane is attacked by two leaf miners, and despite the fact that the insects substantially damage the weed, they do not provide enough control to reduce the need for other control tactics. The nature of the disturbed ecosystem and leaf miner biology apparently limit the efficacy of the biological control.

Many arthropods that live on or in the soil, such as carabid beetles, are weed seedbank predators. The extent and actual long-term impact of this predation is not well understood, and no attempt is made to manage such predation for biological control purposes.

***Arthropods versus Nematodes***    Soil-dwelling microarthropods, such as mites (Figure 13–8b), collembola, and tardigrades (Figure 13–8c), are widespread and abundant in soils. Some mites appear to be nematode-specific predators, and they may have potential as biological control agents. These natural enemies are able to effectively access the larger pore spaces where nematodes live, although smaller pore spaces probably provide nematode refugia. Some mites that are nematophagous are also parthenogenetic; this trait is beneficial relative to biocontrol, because it allows their numbers to increase rapidly in soil. On the other hand, although obligate nematophagy might enhance the efficacy of these natural enemies, it is also a limitation because they are disadvantaged when nematode numbers are low. Many researchers consider these natural enemies as significant

(a)

(b)

**FIGURE 13–9**    Biological control of tansy ragwort, (a) before and (b) after introduction of the cinnabar moth.
Source: photographs from the collection of P. McEvoy, with permission.

nematode predators and that they have considerable influence on nematode numbers in natural ecosystems, particularly in sandy soils with larger pore spaces. The importance of mites and collembola in regulating nematode numbers in soil is poorly understood, and apparently no current attempts are made to manipulate these organisms for nematode biological control in agroecosystems.

***Arthropods versus Mollusks***    Carabid and other beetles attack slugs. Mites can also attack slugs and snails. Although such feeding is known to occur, there is no attempt to manipulate it for IPM purposes.

***Arthropods Versus Arthropods***    This interaction is the one that most people think of as biological control, supported by such images as spiders catching houseflies in their webs. Most arthropods are either beneficial or neutral and only a relative few are pests. Many beneficial insects are not very selective; they will eat their own species or another beneficial just as readily as they will a pest (e.g., preying mantis and lacewing larvae). Beneficial arthropods fall into two general categories—predators and parasitoids.

**Predators**    Generalist omnivorous predators eat many other insects. Examples of generalist predators include preying mantis (Figure 2–12a), certain species of lady beetles (Figure 2–12d), syrphid flies, assassin bugs (Figure 13–8d), damsel bugs (Figure 13–2b), big-eyed bugs (Figure 2–12c), minute pirate bugs (Figure 13–8e), various wasps, mites (Figures 2–12g and 13–8f), most spiders (e.g., see Figure 2–12f), and many others. Many programs aimed at conservation of biological control are targeted at maintaining populations of naturally occurring generalist predators.

> **Predator arthropods consume part or all of their prey, usually killing it.**

There are several difficulties associated with the effective use of generalist predators. Many of them are mobile and will move to a new location to find a food source if the current one is substantially reduced. Releases of lady beetles, for example, may be of little value for local insect control because they move away from the release point, although current research is suggesting that generalist predators may prefer hosts in the current vegetation rather than alternative hosts in another location. Another difficulty is that generalist predators will eat other beneficial insects if the opportunity arises. Despite these drawbacks, generalist predators are usually considered extremely useful in the integrated pest management context.

Predators that are limited in their prey species are termed specialists, and generally they will only prey upon the target pest. The vedalia beetle (Figure 3–7), introduced into California in the late nineteenth century to control the cottony cushion scale, is an example of a specialist predator. The larvae and the adults feed on all life stages of the scale insect. The biological control by the vedalia beetle is sufficiently high that the scale insect is currently not considered to be a problem. Predatory mites, such as *Metaseiulis,* feed mainly on pest mites or their eggs, providing high levels of biological control.

Numerous other examples of specialist predators have been introduced for biological control; 16% of biocontrol projects have been judged as completely successful in controlling the target pest.

**Parasites and Parasitoids**    Parasitoids attack different life-cycle stages of the host insect, and are classified according to the stage attacked. Most parasitoids are either wasps in the Hymenoptera or flies in the Diptera, but there are parasitic species in other orders of insects as well.

> **Parasites and parasitoids live inside their host and eventually kill it.**

Egg parasitoids, which are usually small wasps, lay an egg in each host egg. The parasitoid larva develops inside the host egg, eventually killing it. High levels of biological control can be achieved by egg parasitoids. Monitoring for percentage of eggs parasitized is part of IPM programs for several insect pests, such as the tomato fruit worm. Eggs of the fruit worm are counted and the degree of parasitization is determined (eggs that have a parasitoid inside are a different color). When a sufficiently high proportion of the eggs are parasitized, other management tactics (e.g., insecticides) are usually not necessary. Wasps in the genus *Trichogramma* (Figure 13–8g) can provide a high level of egg parasitism of pest insects. Inundative releases of *Trichogramma* are commonly and widely used to enhance the native populations, although the success of such releases is often equivocal.

Larval parasitoids attack the larval stage of the pest insect. The adults lay from one to several eggs in the host larva, which is killed as the parasitoid larva develops. Many wasps in the Ichneumonidae and Braconidae families provide substantial levels of biological control. The wasp *Apantales medicaginis,* for example, can provide sufficient control of the alfalfa butterfly such that no further treatment is necessary.

> Most wasps should not be thought of as pests, as they are extremely beneficial from an IPM perspective.

Adult parasitoids attack the adult stage of the pest insect. The adult lays an egg in the target host. The larva develops inside the host, eventually killing it and turning it into an empty skin called a mummy in which the larva then pupates. When the pupa hatches, the emerging adult cuts an exit hole in the mummy from which it escapes. Wasps such as *Aphidius smithii* can provide substantial control of aphids (Figure 13–8h).

It is actually the larval stage of many parasitoids that provide biological control. The adult stages either do not feed or utilize nectar or pollen as a food source. Certain specialized parasitoids of armored scales, however, feed in the adult stage on the same scales used by the larvae. In these cases, the effectiveness of the parasitoid is enhanced as both adults and larvae contribute to the destruction of the pest population. For some parasitoids that require supplemental feeding on nectar or pollen, provision of a suitable food source can improve the performance of the biocontrol agent. This is an underlying reason for increasing plant biodiversity surrounding agricultural fields.

**Hyperparasitoids**  "Parasitoids of the parasitoids," or higher trophic-order organisms, can use the biocontrol agents as a food source. Such secondary carnivore arthropods are called hyperparasitoids, and their activities may harm useful biological control. An example is provided by a hyperparasitoid of *Hypersoter exigua,* which is a parasitic wasp that provides effective control of beet armyworm in California until midsummer. After midsummer, the hyperparasitoid population increases and essentially eliminates useful biocontrol of the armyworm by *H. exigua.*

The case of the hyperparasite *Prochiloneurus* in citrus in Israel is a particularly complicated example. Whether the *Prochiloneurus* is beneficial or detrimental to an IPM program depends on the host that it uses (Figure 13–10). When *Prochiloneurus* is using *Anagyrus* as its host, it is a problem to an IPM program because *Anagyrus* is a biocontrol agent that attacks the citrus mealybug. On the other hand, when *Prochiloneurus* feeds on *Homalotylus,* it is beneficial to IPM because it is reducing a parasite of the beneficial *Chilocorus* lady beetle. This example highlights the value of a thorough understanding of the ecology of pests and their natural enemies, and the interactions between them. Detailed knowledge of the biology of all species is essential if the appropriate management decisions are to be made.

**FIGURE 13–10**
Interacting effects of higher trophic hyperparasitoid on functioning of biological control of citrus in Israel; organisms in ovals are pests; those in rectangles are beneficial (see text for further explanation).

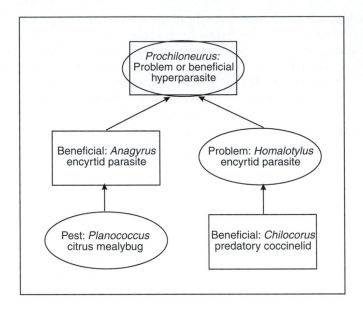

## Vertebrates as Agents

Vertebrates that feed on lower trophic organisms have the potential to provide control of such organisms in some situations (Figure 13–11). Human activity is not usually considered to represent biological control, yet such activities in pest management fit the typical definition of biological control when viewed from the ecological perspective of one organism influencing the dynamics of another. Humans are obviously important biotic components of agroecosystems. The difficulty is that human activity is rather generalist in that often the good and the bad are killed.

***Vertebrates versus Pathogens***    Vertebrates are not useful as biocontrol agents of pathogens.

***Vertebrates versus Weeds***    Vertebrate herbivores may eat weeds, and can thus provide weed control. However, selectivity is a problem; the vertebrate must eat only the weed. If a herbivorous vertebrate natural enemy of weeds is to be used, it must not find the crop palatable, the crop must be able to tolerate feeding damage, or the agent must be contained in some way so that it cannot access the crop.

The best known example of biological weed control by vertebrates is the use of weeder geese (Figure 13–11a). Geese prefer eating grass. They can selectively remove a weed such as barnyardgrass from cotton and tomatoes, because they do not like to eat these crops. Another example of vertebrates for biological weed control is the use of grazing sheep in alfalfa (Figure 13–11b). The sheep eat the alfalfa, but the alfalfa recovers because it is a perennial, while annual weeds are usually killed. Goats can be used for brush control as they eat anything; but they must be fenced away from crops. Selective grazing by cattle can be used to suppress seed production by yellow starthistle.

The use of live animals presents several problems in addition to the possibility that they will eat crops. They must be contained, even in large fields. An-

(a)

(b)

(c)

**FIGURE 13–11**  Examples of vertebrates for biological control. (a) Weeder geese eating grass in tomatoes; (b) sheep grazing weeds in alfalfa; and (c) hawk, a raptor that feeds on vertebrates such as mice and voles, sitting on a fence post.
Sources: photograph (a) by Tom Lanini, with permission; (b) by Robert Norris; and (c) by Terrel Salmon, with permission.

imals must be provided with water and protection from the weather and also from predators such as coyotes and dogs. For these reasons, the use of vertebrates for biological weed control is usually limited to systems where herbicides cannot be used.

Herbivorous fish have been used in some regions of the world to control weeds in aquatic systems. A fish, the grass carp, has been widely used and has provided effective control of weeds such as *Hydrilla.* The use of herbivorous fish poses serious problems however, because they are nonselective and may eat nontarget plants resulting in habitat degradation. They may displace native fish, thus affecting biodiversity.

Many birds and rodents eat weed seeds that have fallen on the soil surface. Such seed predation can result in 80% to 90% depletion of the weed seeds that are being deposited in the soil seedbank. This aspect of biological weed control is often overlooked but can be a significant component of the regulation of annual weeds because of the importance of the seedbank in maintaining weed populations. There is also some predation on the buried soil seedbank, but this is small in comparison with surface seed predation. Currently there is little attempt to manipulate levels of seedbank predation.

***Vertebrates versus Mollusks***    Some vertebrates will eat snails, providing biological control of the snails. Ducks, for example, are used by California citrus growers to control the common garden snail. Other birds, such as chickens and crows, will also eat snails.

***Vertebrates versus Arthropods***    Birds and other vertebrates eat insects, which can provide some measure of biological control. Control of the spruce sawfly by shrews in forest situations is probably the best example of this type of biological control; the shrews eat the pupae and are so effective that it is often not possible to locate any pupae where shrew activity occurs. Bats eat insects, and attempts are now being made to see if developing artificial bat roosts can increase the capability of bats to reduce insect populations in limited areas. The efficacy of this manipulation is yet to be determined.

Fish have been used for control of aquatic insects. The mosquito fish (*Gambusia* sp.) has been widely employed to control mosquitoes in several aquatic systems, such as in rice paddies.

***Vertebrates versus Vertebrates***    Higher trophic-level carnivores, such as owls and hawks (Figure 13–11c), feed on herbivorous pest vertebrates, such as ground squirrels, voles, and mice. This feeding activity is thought to provide biological control in some systems. There is, however, little scientific evidence either supporting or refuting those claims. The actual efficacy of predation by raptors in reducing numbers of these pests is questionable due to the population dynamics of the pest species, and accessibility of the animals to the raptors. For example, burrowing animals such as pocket gophers spend about 95% of their time underground, and raptors are therefore not effective against gophers. Roosting poles and perches, and nesting boxes, are often provided in an attempt to maintain raptor populations in areas where control of pest vertebrates is desired.

Target-prey specificity can also be a problem with raptors. Although they take pest species, they may also prey on nonpest native species and domestic species such as chicks and game-farm birds and snakes. In other words, they usually do not meet selectivity criteria used for other types of biological control agents. Vertebrate biocontrol provides an example of the complexity of attempting to manipulate populations of these vertebrate pests. Controlling coyotes has permitted the European red fox populations to increase, as the latter are prey for coyotes. Increased numbers of foxes have led to increased predation on duck nests. It is thus necessary to manage coyotes in such a way that duck nests are maintained.

## SUMMARY

The foregoing examples demonstrate the many ways in which biological control can help in managing pests. Natural enemies can provide a tremendous amount of pest control and it behooves pest managers to conserve as much biological control as possible.

Overall, biocontrol works well for many insect pests, and if not disrupted may be sufficient to keep many insect pests below the economic threshold. There are, however, exceptions. For a few perennial weeds in extensive stable ecosystems, biological control has been effective, but it has been of limited utility against annual weeds in arable cropping systems. Biological control of the weed seedbank probably provides a substantial reduction in weed populations, but this is something that cannot yet be manipulated. Managed biological control of soil-borne pathogens and nematodes is feasible in a few cases and endemic biological control is probably quite extensive; but manipulable, predictable biological control for these pests is not generally available. Biological control of aerial pathogens is limited and difficult, and will probably never be a major tactic for foliar disease management. For vertebrate pests, it is considered unlikely that biological control will be a major management method except in a few cases.

The proverbial bottom line regarding biological control is that management tactics should be used in ways to minimize the disruption of native biological control that is already occurring. Managers are getting better at this as research develops better information about pest biology and ecology.

## SOURCES AND RECOMMENDED READING

Good reference sources for biological control include *Biological Control* (Van Driesche and Bellows, 1996) and the *Natural Enemies Handbook* (Flint and Dreistadt, 1998), which provides excellent photographs of many beneficial organisms. Books edited by Andow et al. (1997) and Pickett and Bugg (1998) provide in-depth discussions of the ecology of biological control and its enhancement, mainly in relation to arthropod management. In-depth discussions of the different categories of pests are provided for weeds (Del Fosse and Scott, 1992), for nematodes (Stirling, 1991), for pathogens and nematodes (Mukerji and Garg, 1988), and for arthropods (Bellows and Fisher, 1999). Hokkanen and Lynch (1995), Follett and Duan (2000), Lockwood et al. (2001), and Wajnberg et al. (2001) provide discussions of the risks and benefits associated with biological control.

Andow, D. A., D. W. Ragsdale, and R. F. Nyvall, eds. 1997. *Ecological interactions and biological control.* Boulder, Colo.: Westview Press, xiv, 334.

Backman, P. A., M. Wilson, and J. F. Murphy. 1997. Bacteria for biological control of plant diseases. In N. A. Rechcigl and J. E. Rechcigl, eds., *Environmentally safe approaches to crop disease control.* Boca Raton, Fla.: CRC/Lewis Publishers, 95–109.

Bellows, T. S. and T. W. Fisher, eds. 1999. *Handbook of biological control: Principles and applications of biological control.* San Diego: Academic Press, xxiii, 1046.

Collins, W. W., and C. O. Qualset. 1999. *Biodiversity in agroecosystems.* Boca Raton, Fla.: CRC Press, 334.

Del Fosse, E. S., and R. R. Scott. 1992. *Biological control of weeds. VIII International Symposium on Biological Control of Weeds.* Lincoln University, Canterbury, New Zealand, 2–7 February 1992, Melbourne, Commonwealth Scientific and Industrial Research Organization Australia, xxiii, 735.

Flint, M. L., and S. H. Dreistadt. 1998. *Natural enemies handbook: The illustrated guide to biological pest control.* Oakland; Berkley, Calif.: UC Division of Agriculture and Natural Sciences, University of California Press, viii, 154.

Follett, P. A., and J. J. Duan. 2000. *Nontarget effects of biological control.* Boston: Kluwer Academic, xiii, 316.

Godan, D. 1983. *Pest slugs and snails: Biology and control.* Berlin; New York: Springer-Verlag, x, 445.

Hokkanen, H. M. T., and J. M. Lynch, eds. 1995. *Biological control: Benefits and risks. Plant and Microbial Biotechnology Research Series,* vol. 4. Paris; Cambridge; New York: Cambridge University Press, xxii, 304.

Khetan, S. K. 2001. *Microbial pest control.* New York: M. Dekker, xiv, 300.

Lockwood, J. A., F. G. Howarth, and M. F. Purcell. 2001. *Balancing nature: Assessing the impact of importing non-native biological control agents, an international perspective.* Lantham, Md.: Entomological Society of America, vi, 130.

Mehrotra, R. S., K. R. Aneja, and A. Aggerwal. 1997. Fungal control agents. In N. A. Rechcigl and J. E. Rechcigl, eds., *Environmentally safe approaches to crop disease control.* Boca Raton, Fla.: CRC/Lewis Publishers, 111–137.

Mukerji, K. G., and K. L. Garg. 1988. *Biocontrol of plant diseases.* Boca Raton, Fla.: CRC Press, 2 v.

Oatman, E. R., J. A. McMurtry, F. E. Gilstrap, and V. Voth. 1977. Effect of releases of *Ambylseius californicus, Phytoseiulus persimilis,* and *Typhlodromus occidentalis* on the twospotted spider mite on strawberry in Southern California. *J. Econ. Entomol.* 70:45–47.

Pickett, C. H., and R. L. Bugg, eds. 1998. *Enhancing biological control: Habitat management to promote natural enemies of agricultural pests.* Berkeley, Calif.: University of California Press, 422.

Rush, C. M., and J. L. Sherwood. 1997. Viral control agents. In N. A. Rechcigl and J. E. Rechcigl, eds., *Environmentally safe approaches to crop disease control.* Boca Raton, Fla.: CRC/Lewis Publishers, 139–159.

Simberloff, D. and P. Stiling. 1996. How risky is biological control? *Ecology* 77: 1965–1974.

Stirling, G. R. 1991. *Biological control of plant parasitic nematodes: Progress, problems and prospects.* Wallingford, Oxon, U.K.: CAB. International, x, 282.

Sutton, J. C., and G. Peng. 1993. Manipulation and vectoring of biocontrol organisms to manage foliage and fruit diseases in cropping systems. *Annual Review of Phytopathology* 31:473–493.

Van Driesche, R. G., and T. S. Bellows, Jr. 1996. *Biological control.* New York: Chapman & Hall, 539.

van Emden, H. F. 1990. Plant diversity and natural enemy efficiency in agroecosystems. In M. Mackauer, L. E. Ehler, and J. Roland, eds., *Critical issues in biological control.* Andover; Hants, England; New York: Intercept; VCH Publishers, 63–80.

Wajnberg, E., J. K. Scott, and P. C. Quimby, eds. 2001. *Evaluating indirect ecological effects of biological control.* Wallingford, Oxon, UK; New York: CABI Pub., xvii, 261.

# *14*

# BEHAVIORAL CONTROL

## CHAPTER OUTLINE

▶ Introduction
▶ Vision-based tactics
▶ Auditory-based tactics
▶ Olfaction-based tactics
▶ Food-based tactics

## INTRODUCTION

What animals do in nature to find food, escape from enemies, survive adverse weather, mate, reproduce, and raise offspring are aspects of the animal's behavior. Humans have observed the behavior of animals for thousands of years, particularly the animals that they hunted as food, or competed with for edible wild plants and, eventually, cultivated plants. As humans domesticated animals as livestock and as companions, the needs and uses of understanding animal behavior increased in scope. In modern days, the study of animal behavior helps scientists understand agricultural pests and their predators and parasites. The application of this knowledge in IPM is the realm of behavioral control. This chapter addresses the direct manipulation of animal behavior as a control tactic, and shows how modification of external signals that influence pest behavior may be used in pest management.

Nematodes, arthropods, and vertebrates possess nervous systems that modulate behavior, while pathogens and weeds do not. Accordingly, management tactics based on behavior are generally considered irrelevant for pathogens and weeds, because these pests cannot rapidly and obviously modify their behavior in relation to an external signal. This is not to say that plants and pathogens do not respond to external signals—they do—but their responses are generally modulated through modifications of development and growth that occur over longer time and shorter distances, respectively.

All animals perceive the external world by processing information detected by the peripheral sense organs of the nervous system, and include the sensory modalities of vision, smell, taste, touch, or hearing. The informational inputs from the environment elicit signals in the nervous system, and these signals are

**Behavior may be defined as the actions of organisms or responses of organisms to their environment.**

373

translated as shapes, colors, movement, odors, tastes, and sounds. The combined sensorial input to an animal's nervous system results in a range of actions and responses that represents the animal's behavioral repertoire. Of practical interest in IPM are the behaviors that are associated with the following:

- Mating and reproduction
- Foraging or host selection
- Defense and escape from predators
- Shelter selection
- General orientation

Both vertebrate and invertebrate pests are targets for behavioral control. The world of vertebrates is dominated by visual stimuli, and that of invertebrates is overwhelmingly defined by chemical stimuli. These differences are important in the selection and use of behavioral control techniques.

Techniques that exploit behavioral responses from animals are used in IPM not only for direct control, but also just as importantly for the purpose of monitoring pest populations (as discussed in Chapter 8). For example, attractant odor traps are used to detect the presence of fruit flies to determine the need to initiate chemical control actions.

## Advantages and Disadvantages of Behavioral Control

Behavioral control tactics usually are species specific. When properly applied, they affect only the target pest and, therefore, are compatible with other control tactics. Most behavioral controls use natural products and thus are environmentally benign. Because behavioral control methods interfere with behaviors that have developed over millions of years of evolution, it is generally assumed that resistance to the methods is not likely to develop; however, evolution can occur very rapidly in response to human manipulation of the environment. Vertebrates are capable of modifying a behavior by learning, and thus they can learn how to avoid or bypass a control technique. Interestingly, it turns out that insects too may alter their behavior in response to overuse of a given control method, so even behavioral control methods are not totally immune to problems of resistance in the pest (see Chapter 12).

## Animal Behavior

A detailed discussion of animal behavior is beyond the scope of this book. Knowledge of some basic concepts and a technical vocabulary, however, is necessary to understand how behavioral control methods operate and to access the technical literature.

Orientation is the most fundamental behavior of all animals, from protozoans to mammals. Primary orientation is the normal stance that the body takes at rest, for example, a butterfly on a leaf with the wings folded straight up. Secondary orientation is a directional response to a signal or stimulus. The signal represents information from the environment that is processed by the organism with behavioral responses resulting. Most responses are innate, or unlearned, behavior. These innate behaviors are commonly classified as follows:

1. Reflexes are the simplest forms of innate behavior. A puff of air blown behind a cockroach resting on the kitchen floor causes it to run forward on a straight line for a foot or so. A barrier placed in front of a moving

nematode will cause the nematode to stop moving and back up a short distance. These are instantaneous responses, and are escape or avoidance reactions.

2. Kineses are undirected responses, or increases in random movements, and the response intensity varies with the stimulus intensity. Kinetic responses result in increased movement without an influence on the direction. A response to a stimulus gradient that results in increased movement is called orthokinesis. A response that results in an increase in the rate of turning is called klinokinesis. Orthokinesis and klinokinesis result in animals moving to, or remaining in, areas where the environmental conditions are suitable, and are often responses to humidity, temperature, or light gradients.

3. Taxes are directional movements toward or away from the source of a stimulus. The directional orientation of movement is along a line that runs through the source of the stimulus and the long axis of the animal's body. Taxes are classified based on the type of stimulus that triggers the response. Following are some principal stimuli and the corresponding taxis.

   3.1. Light—phototaxis
   3.2. Darkness—skototaxis
   3.3. Gravity—geotaxis
   3.4. Wind—anemotaxis
   3.5. Chemicals or odors—chemotaxis
   3.6. Contact—thigmotaxis

4. Transverse orientation occurs when an animal aligns its body at a certain angle relative to the direction of the stimulus source. The most important kind of transverse orientation is the light compass reaction that allows animals to move over long distances following a distant (sun, moon, stars) light source. Ants, bees, butterflies, birds, and wasps use this type of orientation.

Reflexes, kineses, taxes, and transverse orientation are the elementary components of animal behavior. More often than not, animals display considerably more complex behaviors that are normally referred to as fixed action patterns. Insects and nematodes usually are assumed to be born with their behavioral patterns preprogrammed; they display what is called stereotyped behavior. Vertebrates, although born with stereotyped behaviors, expand their behavioral repertoire by learning. Only recently has it been demonstrated that insects and nematodes may also show learned behaviors.

Besides these basic behavioral patterns, animals have the capacity to signal or communicate with other members of the same species and to recognize signals emitted by other species. Intraspecies communication is critical for mating, for risk avoidance, or for aggregation. Interspecific signaling may facilitate predator tracking of prey or allow the prey to detect predators. The understanding and manipulation of signaling and communication has been the most promising area in applied animal behavioral control. Atkins (1980), Ellis (1985), and Matthews and Matthews (1978) provide further information on animal behavior.

## Modalities of Behavioral Control Methods

Control methods target specific behavioral systems in vertebrate and invertebrate animals. The modalities of behavioral control methods can be grouped by the sensory system involved and the nature of the signal. Most of these methods involve chemotaxis, anemotaxis, phototaxis, and often combinations of those and other behaviors.

# VISION-BASED TACTICS

The behavior of animals that have vision is modified by what they see. There is a wide range of acuity and wavelength sensitivity among the visual systems in different animals. Most vertebrate animals see the world within a range of the electromagnetic spectrum that is similar to that perceived by humans, although colors may be resolved differently depending on the specific light receptors in the eye. Many insects are able to detect the electromagnetic radiation in the ultraviolet wavelengths, and a few have special infrared receptors. Behavioral control tactics manipulate color, shape, and movement of various objects. It is important to determine the spectral range to which the pest is sensitive and responsive.

## Vision-Based Tactics for Arthropods

1. Various traps are colored or shaped so that they attract the target arthropod. The trap is typically covered with a sticky material in which the insect becomes trapped and dies. These traps are a combination of behavioral and physical control.

   1.1. Adult apple maggot flies are attracted to red spheres that mimic the color and shape of ripe apples (Figure 14–1a). These spheres are coated with sticky grease, and are strategically positioned along the periphery of apple orchards where the maggot occurs. The adult flies are attracted to the dummies and are caught in the grease. The efficiency of the spheres in attracting the flies is greatly increased if a food odor source (attractant) is placed in the proximity of the spheres.

   1.2. Yellow sticky traps are sometimes used to trap insects (Figures 14–1b, c, and d). Examples where yellow sticky traps are useful include whiteflies in greenhouses, leaf miners in spinach, and leafhoppers in several crops. White sticky traps are used for thrips. The sticky surface may be rendered ineffective by dust or high numbers of the target pest. These traps are also useful for detecting the presence and relative abundance of certain insect pests, and such information is useful for IPM decision making.

2. Reflective mulches inhibit some insects from landing on the target plants (Figure 14–1e). Flying aphids and whiteflies are thought to perceive the reflective mulch as sky, and do not land. The technique has had limited practical application, but has performed well in small research plots and offers considerable promise, especially for virus management, by deterring vectors.

3. Light traps attract and capture insects because of the ubiquitous positive phototactic response of insects. Ultraviolet (UV) light is particularly attractive to many insects, so insect light traps often use UV or blacklight. Blacklight traps are sometimes used to monitor incidence of certain pests (see Figure 8–7e), mostly moths, in agricultural systems, but they are not effective as a control method.

(a)

(b)

(c)

**FIGURE 14–1** Vision-based tactics for controlling arthropod pests. (a) An apple maggot fly trap consisting of red, sticky-coated spheres with removable plastic cover for easy recoating; (b) yellow sticky trap to trap leaf miners alongside organic spinach (yellow sticky tape is rolled onto top take-up from supply at bottom to expose new sticky area); (c) close-up of take-up spool in (b) to show trapped insects; (d) colored sticky card for control of insects such as whiteflies and thrips; and (e) reflective mulch in a cucurbit crop to deter flying insects such as aphids and whiteflies.

Sources: photograph (a) by Ron Prokopy with permission; (b) and (c) by Robert Norris; (d) by Larry Godfrey, with permission; and (e) by Jim Stapleton, with permission.

(d)

(e)

## Vision-Based Tactics for Vertebrates

Most animals have excellent vision and modify their activity in relation to what they see.

1. The technique of hanging dead animals in the crop is not widely used, and is of questionable efficacy, although it is sometimes used to scare birds.
2. Model figures of predators and raptors (e.g., owls, snakes) may be placed in areas of high visibility to discourage pests from visiting the area (Figure 14–2a, b). Although there is no evidence that these provide any useful control of the target pests, some birds may be discouraged to forage in certain areas for short periods where models of predators are displayed. On the University of California campus at Davis, some buildings have had ceramic owls placed where pigeons congregate. Pigeons could be seen perched, quite comfortably it appeared, on the owls heads.
3. The use of scarecrows (Figure 14–2c) has been a traditional tactic to frighten birds. Scarecrows are still recommended for discouraging sparrows.
4. Reflective materials that move and glitter in the wind are used to deter birds. The efficacy is better than that of predator models, but birds also get used to moving objects over time (see habituation).

**FIGURE 14–2**
Vision-based scare tactics for vertebrate pests. (a) Model of an owl raptor; (b) plastic kite in the form of a hawk; and (c) scarecrow in a pea field.

Sources: photograph (a) by Edward Caswell-Chen; (b) by Terrell Salmon, with permission; and (c) by Robert Norris.

(a)

(b)

(c)

## AUDITORY-BASED TACTICS

Animals perceive sounds, often over a range of the electromagnetic spectrum that is broader than humans can perceive. Both insects and vertebrates use sound for communication. Moths can detect the ultrasounds used by bats as sonar, and the moths dodge the bat attack by interrupting flight and suddenly dropping down. No control methods based on sound are presently used in insect IPM.

Sound and ultrasound, vibrations, and explosions are used to frighten vertebrate pests. Almost universally, scare tactics do not work very well. At best they provide a few days of control, and then the animal learns that the sound does not represent a real threat and ignores it. The animal becomes habituated to the sound. Examples of audible signal devices include:

**Habituation is the process by which an animal learns to accept an unusual stimulus.**

1. Ultrasonic devices for burrowing rodents. There is no evidence that these devices provide satisfactory control.
2. Cannons. These various devices create explosions that sound like a gun being discharged (Figure 14–3). Used alone, noise makers do not work well because the animals habituate to the sound. If used in conjunction with other scare tactics, sound makers can be reasonably effective. The sound generated by such devices can be extremely irritating to humans living nearby.
3. Recordings of distress calls. Many animals, especially birds such as crows, produce a distress call to warn other members of the group of

**FIGURE 14–3**
Explosive propane-fueled sound cannon alongside a vineyard as a sound-based scare tactic for birds.
Source: photograph by Terrell Salmon, with permission.

impending danger. The technique of electronically recording and then replaying distress calls has provided effective control when used correctly.

## OLFACTION-BASED TACTICS

The ability to detect chemical signals in the environment is highly developed in most vertebrates, arthropods, and nematodes. Because nematodes lack a visual system, much of the information they obtain from the environment is based on chemical signals, which are used for both short- and long-distance information transfer. Chemical signals that are detected by olfactory and taste organs are so important in animal interactions that their study gave rise to the branch of science known as chemical ecology. Tactics based on manipulation of pest chemosensory sensitivity provide the most effective behavioral control methods used for insect pests in IPM.

### Definitions and Principles

Semiochemicals induce behavioral changes, such as altered searching, aggregation, dispersal, or attraction for mating, or can result in physiological alteration in the recipient. Semiochemicals are divided into allelochemicals that impact organisms in different unrelated species, and pheromones that impact organisms within the same species (Figure 14–4).

> **Semiochemical is the generic name for chemicals that act as stimuli and mediate interactions among organisms.**

***Allelochemicals***  Allelochemicals are released semiochemicals that mediate interspecific interactions. The several categories are as follows:

**Kairomones**  These allelochemicals are produced by one species, called the source or transmitter, and are detected by another species, called the receiver. Kairomones act as feeding attractants, or oviposition excitants, depending on the behavioral response that they elicit on the receiver organism. Kairomones

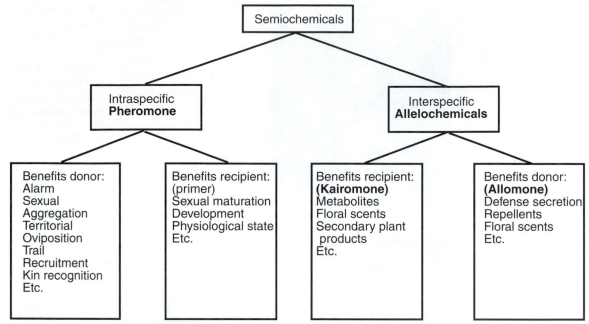

**FIGURE 14–4**   Diagrammatic representation of the relationships among different semiochemicals. Source: redrawn from Howse et al., 1998.

therefore provide an adaptive advantage to the receiver. For example, many plants in the family Brassicaceae (e.g., cabbage and broccoli, often called crucifers) produce a complex of chemicals commonly designated as mustard oil glucosides (mustard is a member of the Brassicaceae). The cabbage butterfly and many other crucifer-feeding insects use the smell and taste of mustard oil glucosides to find and forage on plants of the mustard family. The plant is the emitter of the signal that is used to the advantage of the receiver, the cabbage butterfly.

Plant-parasitic nematodes apparently detect and orient to chemicals, including carbon dioxide, exuded by host roots. Hatching in some cyst nematode species, such as the potato-cyst group, is dramatically stimulated by host-root diffusates. Hatching factors move readily in soil and stimulate hatch, even at low concentrations. Chemical fractionation of potato-root diffusate has allowed detection of multiple hatching factors, which stimulate hatch of potato-cyst nematode eggs. Inorganic ions, such as zinc, stimulate hatch of sugar beet and soybean cyst nematodes. Host-root diffusates also stimulate stylet thrusting, oriented searching behavior, and aggregation in infective juvenile stages of the sugar beet cyst nematode—all behaviors involved in finding food.

**Allomones**   These allelochemicals give an advantage to the producing organism, that is, the emitter. Allomones can act as repellents, feeding, or oviposition deterrents, and are thus detrimental to the receiver organism. Observations under laboratory conditions indicate that the mycelium of some nematode-trapping fungi is attractive to nematodes, but it is not clear whether attraction occurs under field conditions. Several allomones repellent to plant-parasitic nematodes have been detected in plant root exudates, although they have not been chemically characterized. African marigold roots produce a chemical, alpha-terthienyl, that may be toxic or repellent to nematodes; and African marigolds have been used to reduce numbers of some plant-parasitic nematodes in soils.

The same mustard oil glucosides that are attractive to crucifer insect-specialists are repellent to most other phytophagous insects. Consequently, the same compound in a plant may act as an allomone to many sympatric insect species and as a kairomone to a few feeding specialists. This dual role of many secondary plant compounds was the basis for the formulation of the theory of coevolution.

Repellents based on predator odors and other aversive smells are frequently used in vertebrate pest control.

**Sinomones**   These allelochemicals evoke a behavioral response in the receiver species that is beneficial to both the emitter and receiver. Sinomones mediate mutualistic relationships among organisms. Mutualisms are associations that have adaptive value to both species. For example, the smell of flowers that attracts bees and other pollinators is a sinomone.

**Pheromones**   Pheromones are chemicals that are released by an individual and affect other individuals within the same species (intraspecific effects). Pheromones are subdivided into different classes depending on the behavioral response that they elicit.

**Aggregation Pheromones**   These chemicals are released by individuals of one species that attract conspecific individuals to the source organisms. Certain stinkbugs emit aggregation pheromones that attract both male and female stinkbugs to their vicinity.

**Alarm Pheromones**   These chemicals elicit escape and defensive behavior. They are low molecular weight compounds that spread rapidly (e.g., through a colony), but are short lived. Certain species of aphids emit alarm pheromones in response to predators, and the alarm pheromone causes other individuals in the colony to drop from the plant where they were aggregated.

**Sex Pheromones**   These widely documented pheromone chemicals are involved with attraction between sexual partners to increase mating probability. They are emitted by one sex to attract the members of the opposite sex. Insects are extremely sensitive to minute amounts of sex pheromones. In the Lepidoptera (moths and butterflies), glands producing pheromones are located in the tip of the abdomen.

Male nematodes are attracted to females by sex pheromones. Although sex pheromones have been demonstrated in more than 30 nematode species, the only nematode sex pheromone that has thus far been chemically defined is vanillic acid, the sex pheromone of the soybean cyst nematode.

**Other Pheromones**   Epidiectic pheromones are chemicals that induce dispersal of individuals of a species, thus avoiding the overcrowding of resources.

Trail pheromones are semiochemicals involved in the foraging behavior and migration of ants.

Maturation pheromones are semiochemicals that alter the rates of certain physiological processes.

**Repellents**   Some simple inorganic ions, including $NH_4^+$, $K^+$, $Cs^+$, $Cl^-$, and $NO_3^-$, are highly repellent to root-knot nematode juveniles, and may repel infective juveniles at low concentrations (ca. 0.1 ppm). In greenhouse experiments, ion gradients in soil reduced root-knot infection in test plants. Successful manipulation of soil-ion gradients in the field to reduce nematode infection has not yet been achieved.

| TABLE 14–1 | Examples of Volatile Plant Kairomones Attractive to Insects (Based on Metcalf and Metcalf 1992). | |
|---|---|---|
| **Insect** | **Kairomone** | **Source** |
| **Lepidoptera** (moths and butterflies) | | |
| Corn earworm, European corn borer and other noctuid moths | phenylacetaldehyde | corn silk |
| Oriental fruit moth | terpineol acetate | peach |
| Codling moth | α-farnesene, ethyl-2,4-decadienoate | various |
| **Coleoptera** (beetles) | | |
| Striped cucumber beetle | indole | cucurbit blossoms |
| Root worms | eugenol, isoeugenol, cinnamyl alcohol, cinnammaldehyde, indole, estragole, β-ionone | cucurbit blossoms and synthetics |
| Colorado potato beetle | *trans*-2-hexen-1-ol, *cis*-3-hexen-1-ol | potato |
| Cabbage and striped flea beetles | allyl isothiocyanate | Brassicaceae |
| Sweet clover weevil | coumarin | sweet clover |
| **Diptera** (flies and midges) | | |
| Mediterranean fruit fly | terpineol acetate, α-copaine, α-ylangene | various |
| Onion maggot fly | dipropyl disulfide | onion |
| Cabbage maggot fly | allyl isothiocyanate | Brassicaceae |
| Carrot rust fly | *trans*-asarone, *trans*-2-hexenol, hexanal, heptanal | carrot |
| Apple maggot fly | butyl 2-methylbutanoate, propyl hexanoate, butyl hexanoate, hexyl propanoate, hexyl butanoate | apple |

## Chemistry and Synthesis of Semiochemicals

The first semiochemicals positively identified were kairomones from plants in the Brassicaceae. In 1910, the botanist Verschaffelt identified glucosinolates as the chemicals involved in the selection of Brassicaceae hosts by the cabbage butterfly. The glucosinolates act as attractants and feeding excitants to oligophagous Brassicaceae-associated insects. Examples of plant-derived kairomones are presented in Table 14–1.

The first sex pheromone for which the chemical structure was determined was from the common silk moth (Karlson and Butenandt, 1959). Since 1959, about 1,000 different chemicals have been identified that serve as sex pheromones in several insect orders. In the study of pheromones, the chemical is first isolated and concentrated from the source organism. This is not a simple task, as pheromones are often produced in very small quantities, especially in tiny organisms such as nematodes. Once sufficient chemical has been obtained, it is identified using combinations of techniques such as column or gas chromatography, coupled with mass, IR, or NMR spectrometry. The analytical techniques used may vary somewhat among pest organisms, as the chemical characteristics of volatile airborne insect pheromones may be quite different from combinations of volatile and nonvolatile pheromones produced in the aqueous phase of soil by nematodes. Following the isolation and identification of the chemical constituents in a pheromone, procedures are developed to synthesize the chemical.

At all stages of the isolation and identification process, the activity of the chemical must be evaluated using a bioassay. In the bioassays, live organisms are exposed to the odor and their behavior is observed. Certain male moths display a typical fluttering of the wings in response to the sex pheromone stimulus, and male nematodes will move toward an appropriate sex pheromone source. Electrophysiology is a common technique in the study of insect pheromones, wherein the male antenna is used as the detector in the gas chromatograph effluent. Electrodes are inserted at the base and tip of the antenna, and the depolarization of the antenna, in response to the chemical stimulus, is detected on an oscilloscope or chart recorder. Similarly, the depolarization of nerves in nematodes has been assessed relative to chemical stimuli. The nerve depolarization indicates the initiation of a neural response to a particular chemical fraction, and that fraction is collected for chemical characterization.

Pheromones include a range of chemical structures, from relatively simple long-chain alcohols, to aldehydes and ketones, to branched-chain hydrocarbons, to various cyclic compounds. Chapter 5 in Howse et al. (1998) provides an excellent review of pheromone chemistry.

The chemistry of allelochemicals is highly developed, particularly for allelochemicals produced by plants. Plants in different families produce distinct chemicals that are not essential for their primary metabolism. These are called secondary plant compounds, and it is generally agreed that for many of these compounds their main function is to defend against pathogens and herbivores. Secondary plant compounds are further discussed in Chapter 17. Allelochemicals are used in the behavioral control of insects and nematodes either as attractants and feeding excitants (kairomones) or repellents (allomones).

## Delivery Technology

For pheromones to be used for pest management, it was necessary to develop technology that would deliver the pheromone in such a manner that it could be manipulated. The requirements for an optimal release technology include the following:

1. A constant amount of pheromone should be released per unit time, independent of the crop or the weather.
2. The equipment should be able to release different pheromones.
3. The equipment also should be able to provide different release rates.
4. The pheromone must be protected from degradation.
5. Timing of the pheromone release must be appropriate for a particular pest and crop.
6. The device should release all of the pheromone.
7. The release technology must allow for easy pheromone application, including aircraft application.
8. The vehicle for pheromone delivery should be biodegradable, nontoxic, and inexpensive.

(a)                                                                    (b)

**FIGURE 14–5**   Examples of different types of pheromone dispensers. (a) A twist-rope dispenser in a peach tree; (b) a laminate membrane type dispenser, here for both oriental fruit moth and peach twig borer; note plastic clip at top for easy attachment to branches.
Source: photographs by Robert Norris.

Nothing to date fills all these requirements but delivery technology continues to evolve. The following vehicles have been used or are still in use for pheromone delivery, and all provide controlled release over an extended period. Hollow fibers that contain the pheromone and allow it to escape from the ends of thin impervious microtubes have been used for several pheromones. Laminated plastic flakes are used to sandwich the pheromone between two impervious outer layers, and microcapsules can be made with plastic to encapsulate the pheromone. These are intended to be used for widespread distribution of pheromone over large areas.

The following are used in more defined areas, such as a field or an orchard. Twist-tie ropes, which are similar to the hollow fibers noted previously but with a wire backbone, are used in orchards by twisting the ropes onto tree branches or twigs (Figure 14–5a). Pheromones can also be contained within plastic membranes through which the chemical slowly diffuses; the plastic laminates are sealed into a cardboard holder with a clip that can be attached to a tree branch (Figure 14–5b). Devices called puffers, which are essentially motorized aerosol cans on an automatic timer schedule that can spray puffs of pheromone, are currently being developed. Pheromone can be sealed inside rubber caps and widely dispensed in monitoring traps (see Figure 8–7d).

## Application of Semiochemicals in IPM

Semiochemicals are used in several ways as part of arthropod pest management.

***Monitoring***   Sex pheromones are widely used to monitor arthropod pest incidence and abundance (see also Chapter 8). The use of pheromones for monitoring arthropods began in the 1970s, and so predates their use for population reduction. Sex pheromone traps are designed to attract individuals, usually males, with a sex pheromone. Special traps have been developed in which the pheromone lures the insect into a tunnel-shaped container that is lined with a sticky paste. Traps based on kairomones are also used for monitoring. Traps containing feeding attractants are used, for example, to monitor egg-laying by the navel orange worm in almonds or to detect the presence of fruit flies in orchards. Trapping for monitoring purposes does not itself provide significant reductions in target pest numbers. Trapping does provide information that is essential in the timing of other control tactics, such as insecticide applications or decisions to initiate areawide eradication procedures, as in the case of the Mediterranean fruit fly in California.

**Insect numbers can be determined by using semiochemicals to attract individuals to a trap for counting.**

***Mass Trapping (Attract and Kill)***   Mass trapping attempts to attract and remove a sufficient number of target pests from the population so that the crop is protected from damage. In practice, successful use of this tactic faces several limitations. The primary one is the traps are not efficient enough, or they become saturated by high pest populations. Second, the traps need to attract and catch females, because they are the reproductive part of the populations; however, most sex pheromones attract males and sex pheromone–based traps usually cannot trap enough males to impact female fecundity. Certain kairomone traps, however, can be quite effective because they attract both sexes. Lastly, adequate control usually requires a high density of traps, and cost may become prohibitive.

Effective attract-and-kill methods have been developed that do not require traps. Beetles of the genus *Diabrotica* and other related genera have evolved a feeding relationship to plants in the family Cucurbitaceae (pumpkins, squashes, gourds). *Diabrotica* are strongly attracted to the flowers, and once they taste the bitter secondary compound cucurbitacin (of which there are many kinds), their feeding activity is stimulated, such that they will ingest even silica gel coated with the compound. In certain parts of South America, wild gourds are used to attract and kill pest *Diabrotica* spp. Gourds are cut in half, sprinkled with an insecticide, and left out to attract and poison the insects. Thousands of adult beetles can be found dead on the natural baits.

The discovery of the stimulative effect of these cucurbitacins on beetle feeding led to the development of a technique for the management of adult corn rootworms (three species of *Diabrotica*), which are some of the most serious corn pests in the midwestern United States. Wild buffalo gourd roots that are high in cucurbitacin content are ground to a powder and laced with a low dose of insecticide. The baited insecticide is sprayed on corn leaves, attracting the adult rootworm beetles to feed. Rates of 95% to 98% lower than the conventional spray rates can be used effectively in this way. Buffalo gourd roots with the bitter cucurbitacins are not used as food by most other insects, thus making the bait target specific and safe for beneficial insects. The successful management of rootworms by this type of adult management is that treatments must be applied at times when the gravid adults (ready to lay eggs) are actively feeding. Monitoring the presence of gravid females is an essential part of the program.

Some attract and kill methods have been developed that do not require traps. For example, one method to control boll weevils in cotton uses the "boll weevil attract and control tube" (BWACT). The BWACT is a 3-foot long cardboard tube that is painted yellow-green and contains the aggregation pheromone called Grandlure. The surface of the BWACT is treated with the insecticide malathion mixed with cotton oil, which adds a plant attractant factor.

The tubes are deployed vertically along the periphery of fields where they supposedly attract and kill weevils that are emerging from winter dormancy and are dispersing to find host plants. In this case, visual, olfactory, and gustatory stimuli are combined to maximize attraction and kill. Results with the tube in South America have been positive and they are used in extensive areas in Brazil and Paraguay. In the United States, however, experiments failed to confirm its effectiveness in reducing boll weevil populations so cotton IPM specialists do not recommend the BWACT for boll weevil control.

*The objective of mating disruption is to release sufficient sex pheromone over large enough areas so that males cannot find and fertilize females.*

***Mating Disruption***   Male moths and male nematodes find their prospective mates by following a chemical signal emitted by the female, a sex pheromone. Theoretically, once the male detects the pheromone odor it will fly or move up a concentration gradient of the pheromone (positive chemotaxis). The odor diffuses from the female into an odor plume that creates the phermone gradient. By following the gradient, the male will eventually come close enough to the female to find her (Figure 14–6a). For insects, the final orientation may involve vision as well as olfaction.

By saturating an area with the appropriate pheromone, the gradient that the searching male uses for orientation is effectively destroyed (Figure 14–6b). Disruption results in male confusion possibly through habituation to the pheromone. The local females remain virgin because males are unable to find them, and so no viable eggs are produced. Provided that no fertile females from outside the target area are likely to invade, pheromone confusion thus reduces or stops pest population increase in the target area.

As a behavioral control tactic the use of sex pheromones has achieved the greatest success in insect management. The use of nematode sex pheromones is not yet a practical management tactic, although greenhouse experiments have revealed that adding vanillic acid, the soybean-cyst nematode sex pheromone, to soil may reduce nematode populations.

In insect management, mating confusion is successfully used to control pink bollworm in cotton using the synthetic pheromone Gossyplure; sex pheromones are also used to control gypsy moth in hardwood forests, codling moth in pome fruit and walnuts, oriental fruit moth in stone fruits, and pinworm in tomatoes.

Following are limitations to mating disruption using sex pheromones.

1. Immigration of mated females into the target area can negate the impact of reducing reproduction by resident females. It is important that the area under mating disruption is of sufficient size to limit the influence of such immigration of mated females from outside the target area.
2. The success of the tactic requires the delivery and maintenance of adequate pheromone concentrations over large areas. This can be technologically difficult and expensive. The dissemination of nematode sex pheromones in soil is especially problematical.
3. Sex pheromones are usually species specific, and so they only control their target insect. If other pest species are present, pheromone confusion will not reduce their numbers, so other tactics may be necessary. However, successful pheromone use can result in a reduction of broad-spectrum insecticide use against the target key insect, which in turn can lead to improved control of the other pest insects by the natural enemies no longer killed by the pesticide.

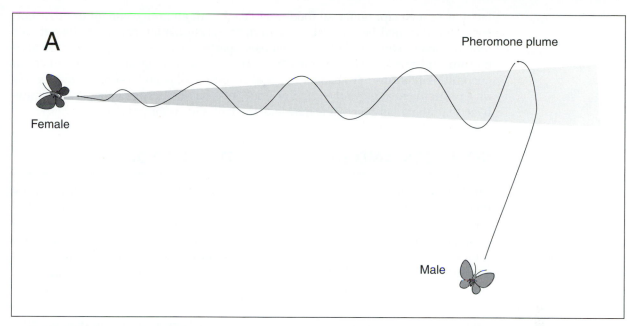

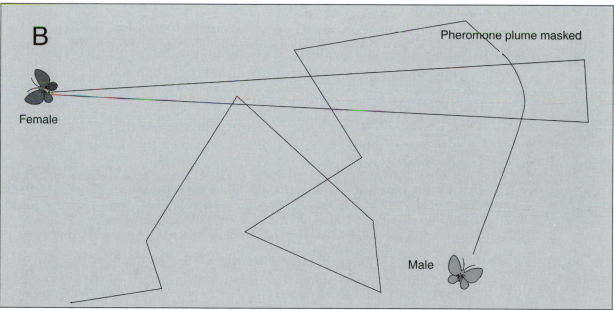

**FIGURE 14–6**  Diagram of sex pheromone mating disruption concept. (a) The male normally locates the female by orienting to the concentration gradient of the pheromone plume, and (b) the male cannot locate the female, because the gradient created by the actual pheromone plume is destroyed by inundation of synthetic pheromone released from dispensers.

# FOOD-BASED TACTICS

Perhaps the most fundamental behavioral patterns are those related to foraging and the attractiveness of food to pests. The combination of kairomones and pheromones has been mentioned relative to *Diabrotica*. A good example of the manipulation of foraging behavior in the control of a pest is from field rodent and rabbit control. The tactic uses a cheap, preferred food item laced with an anticoagulant or other poison. The poisoned bait is placed in an appropriate container where the animals tend to forage. This technique is often used, particularly in nonindustrialized agriculture, and examples include rat control in rice fields using baited rice grain, or in coconut plantations with baited pieces of coconut. All such food-based tactics rely on a pesticide to achieve the kill. Poisoned food bait tactics have the serious drawback that nontarget organisms may take the bait and suffer poisoning.

## SOURCES AND RECOMMENDED READING

The books by Howse et al. (1998) and Metcalf and Metcalf (1992) provide an up-to-date review of the topic of semiochemical use for arthropod management. The books by Ridgway et al. (1990) and Tan (2000) provide many examples of mating confusion. Perry and Wright (1998) provide a detailed overview of nematode physiology, behavior, and neurobiology. Several other books provide additional background material (Mitchell, 1981; Nordlund et al., 1981; Jutsum and Gordon, 1989).

Alcock, J. 1998. *Animal behavior: An evolutionary approach.* Sunderland, Mass.: Sinauer Associates, 625.

Atkins, M. D. 1980. *Introduction to insect behavior.* New York: Macmillan, vii, 237.

Brooks, J. E., E. Ahmad, I. Hussain, S. Munir, and A. A. Khan, eds. 1990. *A training manual on vertebrate pest management.* Islamabad, Pakistan: GOP/USAID/DWRC & Pakistan Agricultural Research Council, 206.

Cardé, R. T., and A. K. Minks, eds. 1997. *Insect pheromone research: New directions.* New York: Chapman & Hall, 684.

De Grazio, J. W., ed. 1984. *Progress of vertebrate pest management in agriculture, 1966–1982.* Denver, Colo.: U.S. Fish and Wildlife Service, Denver Wildlife Research Center and U.S. A.I.D., 60.

Dusenbery, D. B. 1992. *Sensory ecology.* New York: W. H. Freeman, 558.

Ellis, D. V. 1985. *Animal behavior and its applications.* Chelsea, Mich.: Lewis Publishers, xiv, 329.

Foster, S. P., and M. O. Harris. 1997. Behavioral manipulation methods for insect pest management. *Annu. Rev. Entomol.* 42:123–146.

Howse, P. E., I. D. R. Stevens, and O. T. Jones. 1998. *Insect pheromones and their use in pest management.* London; New York: Chapman & Hall, x, 369.

Jutsum, A. R., and R. F. S. Gordon, eds. 1989. *Insect pheromones in plant protection.* Chichester; New York: Wiley, xvi, 369.

Karlson, P., and A. Butenandt. 1959. Pheromones (ectohormones) in insects. *Annu. Rev. Entomol.* 4:39–58.

Matthews, R. W., and J. R. Matthews. 1978. *Insect behavior.* New York: Wiley, xiii, 507.

Metcalf, R. L., and E. R. Metcalf. 1992. *Plant kairomones in insect ecology and control.* New York: Chapman and Hall, x, 168.

Mitchell, E. R., ed. 1981. *Management of insect pests with semiochemicals: Concepts and practice.* New York: Plenum Press, xiv, 514.

Nordlund, D. A., R. L. Jones, and W. J. Lewis, eds. 1981. *Semiochemicals, their role in pest control.* New York: Wiley, xix, 306.

Perry, R. N., and D. J. Wright, eds. 1998. *The physiology and biochemistry of free-living and plant-parasitic nematodes.* London, UK: CABI Publishing, xviii, 438.

Ridgway, R. L., R. M. Silverstein, and M. N. Inscoe. 1990. *Behavior-modifying chemicals for insect management: Applications of pheromones and other attractants.* New York: M. Dekker, xvi, 761.

Tan, K.-H., ed. 2000. *Area-wide control of fruit flies and other insect pests.* Pulau Pinang, Malaysia: Penerbit Universiti Sains Malaysia, 782.

# C H A P T E R

# 15

# PHYSICAL AND MECHANICAL TACTICS

## CHAPTER OUTLINE

▶ Introduction
▶ Environmental modification
▶ Physical exclusion of pests
▶ Direct control of pests

## INTRODUCTION

Direct physical manipulation of the environment can be used as a pest control tactic. The mode of action is to kill or incapacitate pests by creating environmental stress that the pest cannot tolerate. The tactic is directed against the pest organisms and typically does not involve changes in the crop plant. Physical and mechanical manipulations are quite different from cultural techniques, because manipulations of cultural practices are intended to alter the development of the crop so that it becomes a less suitable host for pests, or is better able to tolerate their attack.

Physical pest control tactics are similar to pesticides in that their effect is typically rapid. Such rapid reductions in pest numbers may make physical control a better or more desirable control tactic than cultural manipulations, because cultural tactics take longer to yield results. Physical and mechanical manipulations are sometimes considered part of cultural management tactics. We argue that the ecological principles underlying the approaches are different, so they should be distinguished and considered separately. The primary ways that physical and mechanical tactics can be used for pest management include the following:

1. Environmental modification, such as altering temperature, water, and light, can change pest population development or actually kill pests.

**The physical environment can be modified to provide control of pests.**

2. Pests can be physically excluded by using barriers, or by trapping that can stop the pest from reaching the crop plants.
3. Pests can be destroyed by direct physical means, such as physical removal, tillage, drowning, burning, or shooting.

Pests have been controlled using physical and mechanical tactics since the origins of agriculture. This chapter explains various physical approaches to pest management and discusses their relevance to control of the different categories of pests.

## ENVIRONMENTAL MODIFICATION

Temperature is probably the most important environmental variable that can be modified, but manipulating the availability of water and light can also be used to control certain categories of pests.

### Temperature

Manipulation of temperature extremes can be used as a control tactic. Extremes of temperature can cause death of any category of pest organism, and temperature has a direct influence on rates of growth and development in all pest organisms except warm-blooded animals (see Figure 5–11).

***Heat***   High temperatures can desiccate organisms, and extreme temperatures denature proteins and destroy cell membranes. Excessive heat literally cooks cells and so they die. Different organisms have different tolerances to temperature extremes and some pests are killed by temperatures that are tolerated by other organisms. Raising the temperature of planting material can kill pests residing in the tissue without killing the plant and heating soil can kill pests residing there. Advantages to the use of heat include the transitory nature of the effect, and the fact that the heat itself leaves no residue in treated areas, although heat may stimulate decomposition of organic matter that may result in residues. Several of the more important ways that heat can be utilized to manage pests are outlined in the following sections.

**Fire**   This technique actually involves setting fire to dry vegetation, and may be used to destroy some pests or pest habitats. It has been used for disease management in several crops, such as stem rot in rice and several pathogens in grass seed crops. Partial control of weed seeds and insects is also obtained. Burning has been used for removing vegetation from ditches and canals, or as a forest management tool, mainly to remove excess vegetation or dead plants prior to planting new trees.

From an air pollution standpoint, it is often undesirable to burn plant residues, such as weeds, because of the smoke that is produced. Most uses of agricultural burning have been, or are being, limited or phased out because of the air pollution that is inevitably generated.

**Flaming**   Actually igniting or setting fire to organisms is not necessary to kill them. Flaming does not require that pest organisms be ignited, but rather relies on an external fuel source to elevate temperature sufficiently to kill the pest. The tactic is most often used for weed control. Flamers use propane, kerosene, or diesel to produce a hot flame that is briefly directed at the target weeds. An exposure of 1 to 2 sec is usually all that is required to kill the weeds.

Flaming is only effective when target weeds are small. Flamers can range from a single, handheld burner for spot applications to multiple burners mounted on a boom for use on large areas (Figure 15–1a). Flaming can be used in a nonselective manner to kill all aboveground vegetation. Examples include weed control in dormant alfalfa or in fallow beds prior to planting a crop. In the former situation, the deep-rooted perennial crop is able to sprout and resume growth, but annual seedlings are killed.

Flaming can be used to selectively control weeds without killing the crop. The flamer must be aimed so that it does not hit the crop but does hit the weeds. Selective flaming can be accomplished in several ways.

1. Weed seedlings that emerge before the crop has emerged can be killed with flaming. This works well when the crop germinates more slowly than the weeds. Examples include onions, carrots, and peppers.
2. If the emerged crop is taller than the weeds, such as cotton, the flame can be directed under the crop canopy.
3. In orchard crops, the flamer can be directed into the tree row to kill weeds (Figure 15–1b). The heat is not sufficient to penetrate and kill the tree bark, although caution is required when trees are young and have thin bark.

Flaming can kill insects, but is usually impractical for this purpose. Sufficient control of alfalfa weevil eggs can occur when alfalfa is flamed so that no other control tactic is required. Flaming has been used for grasshopper control when nymphs aggregate to swarming populations.

**Solarization**    This technique is also referred to as soil pasteurization, and elevates soil temperature using energy from the sun. Solarization is used to control some pathogens, weed seeds, and nematodes in relatively high-value cropping systems. The technique is most often successful in environments with long hours of sun and high temperatures, and is implemented by laying clear polyethylene tarps over moist soil for six to eight weeks (Figure 15–2a). The soil under the plastic is heated by sunlight, and the heat is trapped beneath the tarp

(a)

(b)

**FIGURE 15–1**    Examples of flaming devices used for pest control. (a) Broadcast flamer in alfalfa; (b) directed flamer in pears. Sources: photograph (a) by Robert Norris; (b) by Clyde Elmore, with permission.

(a)

(b)

**FIGURE 15–2**    Solarization for pest control. (a) Laying clear plastic on preformed beds for solarization; and (b) carrots sowed in a solarized bed (left) compared with a nonsolarized bed on right.
Source: photographs by Clyde Elmore, with permission.

resulting in elevated temperatures via a greenhouse effect. Solarization can result in temperatures between 50° and 60° C in the top few inches of soil, which is enough to kill many pests. Solarization can provide spectacular control of weeds (Figure 15–2b) and pathogens. A difficulty with the tactic, however, is solarization does not raise soil temperatures deeper in the soil profile, so pests such as nematodes and some pathogens survive in the subsurface soil. Thus, the efficacy of the tactic is limited.

Solarization has some serious drawbacks. Some pest species are heat tolerant, and are often not controlled. The plastic tarp is expensive, and eventually must be removed from the field, creating a disposal problem.

**Steam**    High-pressure superheated steam is used to elevate soil temperatures to kill all types of soil-borne pests. Steam sterilization of soil in large autoclaves is a standard greenhouse procedure that is used to ensure pest-free planting mix.

The use of steam for nematode and weed management in the field has been attempted, but heating a sufficient volume of the soil profile is difficult, and requires using a lot of energy. Using steam to heat field soils over large areas is thus prohibitively expensive.

**Heat Therapy**    Warm-water dips or exposure to elevated air temperature is used as a curative control for some important viruses and nematodes in plant material used for vegetative propagation. Heating is a tremendously important tactic, because there is no other method available to eliminate the viruses and nematodes from planting material. The success of this technique depends on the sensitivity of plant tissue to high temperatures, because the plant must be able to survive temperatures that are high enough to kill the pest. The heat therapy process requires accurate control of temperature, because the thermal death point for the pathogens is usually not much lower than that for the host plant. Accurate control of water temperature in large dip tanks can be difficult.

Vegetative structures, such as narcissus, tulip, and daffodil bulbs, are carefully subject to heat treatment to kill stem and bulb nematodes, and strawberry runners may be immersed in water at 46° C for 10 minutes to eliminate nematodes. Warm-water dips, or exposure to warm moist air, have been used widely to kill viruses from some clonally propagated crops such as strawberries and potatoes. Prolonged exposure to elevated temperatures (about 42° to 48° C) for several weeks can result in the shoot apex being pathogen free (mainly viruses). The virus-free apex is excised and grown in tissue culture to regenerate a new pathogen-free plant. This technique requires that the heat kills all pests in the tissues; otherwise, the pest populations will rebound as the plant grows.

**Other Uses of Heat**    Attempts have been made to use microwave energy to sterilize soil. Although this can work for small soil volumes, it is not effective for large soil volumes because the process is slow and energy intensive.

The use of Fresnel lenses to focus heat from the sun's rays to kill plant tissue has been attempted. No usable technology based on this technique has been developed due to difficulties of maintaining the lens orientation in relation to the sun, and keeping the lens clean.

High-voltage discharge has been attempted on numerous occasions as a means to "electrocute" weeds, and even nematodes. Problems with circuit completion and energy requirements have precluded development of commercially acceptable equipment.

**Problems with all Heat Approaches**    Although there are several ways heat can provide control of pests, the many difficulties limit its utility. Heating can be highly effective through applications such as solarization in the field, pasteurization of greenhouse soils, or hot-water dips for propagation stocks; but the use of heat or cold to kill pests generally requires large quantities of energy to modify the temperature of the target soil or plant material. If other tactics are available, they may be less expensive and thus, preferred. Heat therefore may not be cost effective in comparison with other tactics. Using heat is not pleasant for the equipment operator when open flames are employed, especially in the summer when heat-based tactics are often most effective. Any use of heat that produces smoke creates air pollution.

***Cold***    Low temperature, or cooling, is another form of physical control. Low temperatures kill many organisms that do not possess the ability to tolerate freezing. The low temperatures that occur in some climates limit the geographic range of many pests, because winter temperatures are lethal (Table 15–1). On the other hand, some insects, nematodes, pathogens, and weeds have stages that undergo diapause or become dormant, and so are adapted to withstand freezing winter temperatures.

The use of freezing or extreme cooling is not applicable to live plants so it has no use in preharvest settings. For postharvest protection of produce and grains, however, cooling is a powerful tactic. Grain in storage generates heat that

| TABLE 15–1 | Examples of Pests that are Limited in Their Geographic Range by Lack of Ability to Tolerate Freezing Conditions | | | |
|---|---|---|---|
| **Weeds** | **Arthropods** | **Pathogens** | **Nematodes** |
| Nutsedge species | Mexican bean beetle | Southern corn leaf blight | Root-knot nematodes (some) |
| Johnsongrass | Thrips | Blue mold of tobacco | Reniform nematodes |
| Itchgrass | Egyptian alfalfa weevil | | Citrus nematode |
| Water hyacinth | Velvetbean caterpillar | | |

often is favorable for the rapid development of insects and pathogens that attack stored products. It is common practice in grain elevators to circulate the grain along conveyers to reduce the grain temperature. In the midwestern United States, attempts are made to kill insects infesting grain by moving the grain at times when ambient temperatures are well below freezing. Cooling harvested fruit to near freezing is a routine practice to prolong shelf life, and to slow the development of pathogenic fungi.

The utility of freezing for pest control purposes is limited, except in the protection of postharvest produce. On a small scale it is possible to use freezing, for example, to kill eggs and larvae of codling moth in walnuts.

## Water

All living organisms require water. Excess water (flooding) or lack of water (desiccation) also can be used to control some categories of pests.

***Flooding***    Flooding can be used to kill some pests, because it creates conditions favorable for some beneficial water molds and also reduces oxygen concentrations in soils. Examples include ponding water on agricultural land to kill perennial weeds such as johnsongrass. The technique is also used to control soil-borne pathogens, such as black root of cotton, and some nematodes. The major drawbacks are the large amounts of water required, the water must be present for sufficient time, and the tactic can only be used in level fields. The practice is also only feasible on soils that do not drain quickly. Perhaps the best examples of the use of flooding in pest control are found in the extensive paddy rice plantations in Asia. Flooding is used mainly to control weeds in the rice paddy production system, but also drowns many insects such as mole crickets and armyworms.

Flooding can be effective for controlling ground-dwelling vertebrate pests such as gophers. These pests are normally kept under adequate control in alfalfa and in some orchard systems where flood irrigation is practiced, and changing to sprinkler irrigation from flood irrigation can result in increased vertebrate pest problems.

Water can also be used to physically dislodge some pests. Aphids can be knocked off plants by a water jet. This practice can provide temporary control in small-scale situations, such as roses in a garden setting or in small greenhouses, but is generally not practical for use in commercial field settings. Water may be used in other ways. For example, it has been observed in citrus groves in California that dust from dirt roads near the groves killed the tiny parasitic wasps that provide biological control of many scale insect pests. By periodically rinsing dust off the trees it has been possible to restore biological control of the pests.

In some situations it is possible to increase the relative humidity in plant canopies. Higher humidity discourages spider mites, which prefer dry and dusty conditions. A difficulty with the tactic is that increased humidity may lead to increased pathogen attack, demonstrating the requirement for knowledge of the pest complex ecology before implementing such a tactic. Conversely, by opening the crop canopy it is possible to improve airflow and decrease canopy humidity. This speeds drying and can decrease fungal infection. Grape leaves are removed from around developing fruit clusters as a physical means to decrease bunch rot incidence (Figure 15–3).

**FIGURE 15–3**
Grapevine with partial canopy removal as a tactic to reduce incidence of bunch rotting diseases.
Source: photograph by Doug Gubler, with permission.

***Desiccation*** For nonmobile pests that require free water to thrive, such as vegetative structures of perennial weeds and nematodes, it may be possible to withhold water, leading to desiccation and death. Thorough drying of soil can be used to kill rhizomes of perennial weeds such as johnsongrass. Drawdown and drying of irrigation ditches and ponds can be an effective way to control some aquatic weeds.

Drying soil may be useful as a nematode management tactic; however, the biology of the nematode must be understood because some nematodes can enter a state called anhydrobiosis, and thus survive total desiccation. As soil dries, a nematode enters anhydrobiosis by slowly eliminating water from its body, and replacing the water with sugars that stabilize its membranes and proteins. In the anhydrobiotic state, the nematode is dormant, has no detectable metabolism, and is resistant to standard nematode management tactics. When water returns to the environment, the nematodes rehydrate and resume activity. Some nematodes can survive for decades in anhydrobiosis. So, drying will not necessarily control all nematode species, and may actually aid the survival of some.

***Irrigation*** In irrigated agroecosystems, it is sometimes possible to modify the irrigation so that it favors the crop and not the pest. Factors such as amount of irrigation, timing, or method of water delivery can be used to modify pest abundance and damage. A disease known as blossom end rot on tomatoes is attributed to poor irrigation management.

Dry conditions can result in increased levels of dust, and as noted dusty conditions will exacerbate spider mite problems in cotton, tomatoes, and almonds, for example.

Excessive irrigation can result in waterlogged soil, which then makes soil-borne disease problems worse. *Phytophthora* produces mobile zoospores under wet conditions, and those zoospores infect the host. Poor irrigation practices can thus lead to increased pest problems.

The time of day when irrigation is carried out can impact pest severity. Sprinklers used in the middle of the day can make foliar disease problems worse, because such sprinkling extends the free water period. This can allow fungal spores to germinate and penetrate the plant when drier conditions would have prevented such infection from occurring. Sprinklers can also exacerbate

**Manipulation of irrigation can alter pest attack.**

many of the "rain splashed" bacterial diseases, such as fire blight in pears or bacterial speck in tomatoes. An apple orchard was designed with overhead sprinklers so that the sprinklers would cool the fruit; fire blight became so severe that the orchard was abandoned.

The timing of irrigation also may influence pest incidence. Under hot and dry conditions, if irrigation is spaced so that the crop is stressed between cycles, the conditions may become favorable for explosive spider mite outbreaks. Conversely, in potato, withholding irrigation shortly after planting has been tested as an aid to reduce the incidence of early dying syndrome, caused by the fungus *Verticillium dahliae* and the root-lesion nematode, because the pathogen cannot easily attack roots when moisture is limited.

The method of water delivery can alter pest management. Irrigation that keeps the soil surface dry can reduce weeds and will reduce the incidence of foliar diseases. For example, furrow irrigating so that the tops of the beds do not become wet reduces infection by *Sclerotinia minor,* which causes lettuce drop disease (Figure 2–2b). Recent advances in drip irrigation have allowed the delivery tube to be buried, which greatly aids in surface water management.

A practice referred to as pre-irrigation can be used in some crops, particularly for weed control. The irrigation is applied after soil preparation has been completed but prior to planting the crop. The irrigation germinates weeds, which can then be killed with either shallow cultivation or herbicides. The crop is then planted through the dry, cultivated soil. The tactic works well for large seeded crops such as corn, beans, and cotton. Pre-irrigation is not relevant for control of pests other than weeds, and is not very effective for control of perennial weeds. It can be used for weed management in rainfed agriculture, but has the hazard that too much rain can lead to difficulty planting the crop.

A variant of pre-irrigation is a practice referred to as stale seedbeds. In this situation, the seedbeds are prepared weeks or months in advance of sowing the crop. Weeds that emerge on the beds are controlled before the crop is planted to provide a substantial reduction in the number of weeds that emerge when the crop is sown. If such weeds are left too long, however, they may support the development of nematodes and pathogens.

**Limitations**　If the equipment and infrastructure required for irrigation is unavailable for general crop production, using it for pest management becomes an added cost. Likewise, if water is used as a pest management tactic in the absence of, or in addition to, the need for irrigation, it is an additional cost. Most of the irrigation tactics, with the exception of stale seedbeds, do not work well for small-seeded crops.

# Light

**Mulches that block light can provide excellent weed control.**

Light is a required resource for green plants, and barriers that block light may be useful in weed management. Such barriers have limited relevance to management of most other pests because the mode of action is to limit photosynthesis. Barriers that block light are usually referred to as mulches. Mulches may be used for purposes other than weed control, particularly water conservation, and reflective mulches may be used to deter insects from landing on crop plants. The several kinds of mulches serve slightly different purposes.

***Synthetic Mulches***　These types of mulches are made from such materials as solid black polyethylene plastic, spun-bonded dark-colored plastics, and various forms of woven black plastic. The latter type allows water penetration, and is usually made of UV-resistant polypropylene. In some situations clear plastic is used, but it does not block light and the pest management benefit is through heating (see solarization). Plastic mulches are often referred to as landscaping fabrics. They can be particularly effective for weed control in situations where close proximity of many different plant species makes the use of herbicides problematic (Figure 15–4a).

The disadvantage of synthetic mulches is that they are manufactured from petroleum products. They also degrade in sunlight and will eventually disintegrate into pieces that can be blown about by wind. In many situations the plastic must be further covered to block sunlight, to increase the longevity. There is also a serious problem with disposal at the end of the season. The cost of plastic mulches is relatively high, and they are usually only used in high-value crops and gardens.

***Natural Mulches***   Materials such as bark, wood chips, straw, lawn clippings, and various paper products can be used as mulches (Figure 15–4b). Except for paper products, such mulches must be several inches thick to achieve adequate weed control and even then perennial weeds will probably emerge through the mulch. Various kinds of rock can also be used, but they are primarily for decorative rather than pest management purposes. All mulches are effective for water conservation. Except for rocks these mulches are organic materials that eventually break down. This can be considered as an advantage, but it does mean that organic mulches must be replaced occasionally.

Organic mulches have the serious disadvantage that they create an ideal habitat for several pests, including slugs, snails, sowbugs, stinkbugs, and earwigs, and mulching vegetable crops has to be undertaken with extreme care to make sure that such pests are not increased. Organic mulches are useful in management of perennial tree and vine crops, except citrus where snails can become a severe problem.

(a)

(b)

**FIGURE 15–4**   Examples of mulches. (a) Synthetic black plastic mulch used for weed control; and
(b) chipped/shredded tree prunings used as a mulch in an ornamental landscape planting.
Source: photographs by Robert Norris.

***Living Mulches***   Living mulches are plants grown specifically to compete with weeds by intercepting the light needed by weeds. Although light exclusion is a major mechanism by which living mulches provide weed control, some may also produce allelopathic root exudates. Living mulches are discussed in Chapter 16 on cultural management, because living plants are involved rather than a purely physical process.

## PHYSICAL EXCLUSION OF PESTS

**Barriers can stop the pest from being able to reach the crop.**

The aim of physical exclusion is to stop the pest from reaching the crop. This tactic takes the form of either erecting barriers to the pest or trapping. Physical exclusion is practical on a small scale but is too expensive to apply to large-scale farming in most situations. Most of the following techniques are economically practical for pest management in gardens or in high-value vegetable and fruit production systems.

### Barriers

Barriers are used to keep the pest outside of the crop area. How barriers are used depends on the category of pest to be controlled.

***Barriers versus Slugs and Snails***   Barriers composed of materials that deter mollusks can be effective control tactics. For example, layers of diatomaceous earth or ashes may be sprinkled around plants. These materials are abrasive, and slugs and snails will not cross them when they are dry, thus excluding the mollusks. Water will reduce the efficacy of this type of barrier. Slugs and snails will not cross copper bands about 1 inch or more in width (Figure 15–5a). An electrolytic reaction occurs between the slime on their "foot" and the copper that apparently deters the animal. Copper barriers are not effective if there is an alternative route, or bridge, over the barrier copper bands. Low-hanging branches in citrus, for example, can eliminate the value of copper bands around the tree trunks.

***Barriers versus Arthropods***   Until recently, the exclusion of arthropod pests was practiced on a limited scale. The development of floating row covers made from lightweight polyester, or other fabric, provides the possibility of excluding flying insects from high-value crops (Figures 15–5b and c). The fabric is either laid directly on the crop plants, hence the name *floating row covers*, or is suspended above the crop on a frame. Row covers must extend to the soil surface, with the edges buried so that insects cannot crawl under the edges. Floating row covers provide excellent protection against aphids, whiteflies, moths, beetles, and other flying insects. By preventing insects from feeding, floating row covers also protect against virus diseases that are vectored by the insects. Additionally, floating row covers keep out cats and birds, provide protection from sunburn, and provide frost protection.

   The major drawback to the use of floating row covers is the cost of the fabric, so they are primarily used on high-value crops or in organic systems where the commodity demands a premium price. Another drawback to the row cover barrier is that the environment under the fabric is ideal for weed growth, and for slugs and snails. In IPM systems that use floating row covers, other pests must be monitored carefully.

(a)

(b)

(c)

(d)

(e)

**FIGURE 15–5** Various types of exclusion devices.
(a) Copper band for mollusk deterrent in citrus; (b) large-scale field use of floating row cover on celery in Denmark; (c) small-scale use of floating row cover supported on hoops in home garden; (d) netting over cherry tree to exclude birds; and (e) young trees with sleeves to protect the bark from chewing by rabbits.
Source: photographs by Robert Norris.

Another kind of barrier is to place a paper bag around individual fruit early in their development so that they are protected from insect feeding and oviposition. This approach is labor intensive, and is used for high-value fruit crops in the tropics, primarily against fruit flies.

***Barriers versus Vertebrates***    Exclusion is frequently the only option for managing some vertebrate pests.

**Birds**    Birds can severely damage many seedling crops and mature fruit crops, with ripe cherries being particularly vulnerable. Isolated fruit trees and small orchard blocks can be devastated by bird damage. For small, intensive operations, netting can provide excellent protection (Figure 15–5d). Physical protection from birds on a large scale is usually more difficult, but in areas where seed-eating birds are abundant, paddy rice fields are covered with netting that prevents birds from reaching the rice.

**Mammals**    Fences are routinely used to exclude animals, such as deer and rabbits, from backyard gardens and small truck farms. The fence height and depth in the soil must be appropriate for the animal to be excluded. Fences against rabbits, for example, do not have to be as tall as those used to exclude deer. With burrowing animals, such as rabbits and ground hogs, the fence must extend into the soil to a depth greater than the burrows. The use of fences is usually not effective in excluding large animals such as bears, elephants, and rhinoceros. Fences may also be the only acceptable protection from vandalism by humans.

Granaries have been built on mushroom-shaped posts to exclude rodents (Figure 3–3), and physical exclusion of rats and mice from storage facilities is still practiced.

In situations where rodents and rabbits might eat the bark of young trees, it is practical to encase the lower part of the trunk in a sheath to protect it from chewing damage (Figure 15–5e). The technique is widely used in establishing new orchards, and all forms of tree establishment in landscape restoration projects and forestry. Metal cages may have to be placed around young trees if deer are prevalent in an area.

## Physical Traps

**Traps can catch enough of the pest to lower the population to an acceptable level.**

Traps may be used to capture high numbers of pests, and reduce pest densities to acceptable levels. This use for traps is distinct from using traps to monitor population development. To be effective, traps must attract pests, and this is usually accomplished through the use of some type of bait or semiochemical (see also Chapter 14). None of the trapping techniques mentioned here provides more than partial pest control. Other limitations are that trapping provides only temporary control, is labor intensive, and relatively expensive. In some situations, however, the use of traps can be a significant component of an IPM program.

***Traps versus Weeds***    Weed seeds can float on irrigation water and be moved into fields. Filters placed in the irrigation flow can trap these floating seeds. This technique is often considered to be cultural management, but is in fact a purely physical exclusion. The technique is difficult to implement where high water flow rates are encountered, or where the water is heavily laden with debris.

***Traps versus Arthropods***    Of the various types of traps used to control arthropods, all are of limited utility. There are different attractants used to lure insects to traps. Insects are attracted to specific colors, and colored cards coated with sticky material lure and trap the insect. Sticky flypaper has been used for decades to kill houseflies, and likewise yellow sticky traps are used for flying insects such as leafhoppers, whiteflies, and adult leaf miners (see Figures 14–1b and c). Sticky traps are often used in greenhouses, but may be placed at field edges. Sticky trap efficacy for control of whiteflies in greenhouses has been well documented. In one study with greenhouse tomatoes—a favored host—the ratio of whiteflies caught on yellow sticky traps versus those caught on tomato leaves was 4.3:1. On greenhouse chrysanthemums, a less preferred host, the ratio was about 107:1. Used with other management tactics, the sticky traps are useful—if not for control, then as a reliable monitoring device.

Trapping insects that vector nematodes can decrease the incidence of some nematode diseases. The most effective trapping occurs when the traps are used in conjunction with a chemical attractant such as a pheromone or kairomone (as discussed in Chapter 14), and a pesticide (as discussed in Chapter 11), thus such physical traps may actually be a combination of several tactics. An example is management red-ring disease of coconut, a disease caused by a nematode that is vectored by the palm weevil. To attract, trap, and kill the weevil vector traps or guard baskets are filled with pesticide-treated chunks of fresh coconut tissue as an attractant, and placed on the ground in the plantation.

Boards, or other suitable habitat material such as corrugated cardboard, attract arthropods such as pill bugs and earwigs which take shelter in the habitat. The pest behavior is such that they orient to the close contact (thigmotaxis) with the sheltering substrate, and the trapped pests can then be killed. Slugs and snails can be controlled in the same way. Sticky bands and rolled cardboard bands on the trunk of fruit trees can trap crawling insects such as ants and codling moth larvae.

Effective mass trapping of crawling, swarming grasshoppers has been attempted with deep trenches dug along a line ahead of the advancing grasshopper swarm. The grasshoppers fall into the trench and are then killed by fire or some other means.

***Traps versus Vertebrates***   Trapping has been used to control vertebrate pests for thousands of years. There are two main types of traps for vertebrates. One type kills the animal, such as the Macabee trap for pocket gophers (Figure 15–6a). The

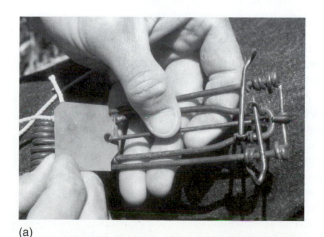

(a)                                                              (b)

**FIGURE 15–6**   Examples of devices to trap vertebrate pests. (a) Setting a Macabee trap; this is a kill trap with jaws on right that snap closed on an animal such as a gopher; and (b) rat caught in a live trap.
Source: photographs by Terrell Salmon, with permission.

trap is placed in the gopher burrow and catches the gopher as it crawls through the trap. The household mousetrap and rat trap are familiar kill traps. One problem with kill traps is that they do not always kill the animal. The other basic type of trap is the live trap, so named because they catch the animal without killing it (Figure 15–6b). Trapping can be an effective control tactic when vertebrate pest populations are low, but is usually not practical when populations are high, or where animals can move and recolonize from adjacent areas. The public attitude toward trapping agricultural pests depends on the pest; trapping rats is acceptable, but trapping rabbits or coyotes may be considered cruel.

## DIRECT CONTROL OF PESTS

**Pests can be killed by direct physical action.**

Direct control means that a physical process is used to kill the pest organisms. Cultivation for weed management is the most common form of direct control, but the tactic also has some application to other pest categories.

### Shooting

Shooting is relevant for vertebrate pests, and can sometimes be successful for protecting a high-value crop. Shooting may be locally successful for control of large animals, such as deer, coyotes, foxes, feral pigs, and goats, and is sometimes used for control of gophers, ground squirrels, and birds.

The major limitation to the efficacy of this tactic is the mobility of the pests. Even when shooting kills the offending animal(s), others can move into the area. Public acceptance and safety limit the use of this tactic on a routine basis. Shooting typically does not provide a long-term solution, because large vertebrates move from place to place.

### Hand Labor

Human labor can be used to physically remove pests. This usually takes the form of pest collection and destruction, or physical action that kills the pest in place. Direct control of pests by human labor can be highly effective, especially for weeds. In some cases, nematode-free planting material may be obtained by manually removing the nematode-infected surface layers from corms or bulbs. In some areas, the widespread use of labor for direct pest control is cost limited; however, the cost must be assessed relative to the production budget and the broader economic contributions to society, such as providing employment.

***Pest Collection and Destruction***   It is possible to handpick some pests, such as caterpillar larvae of moths (e.g., tomato hornworm) or slugs and snails. The collected pests are then killed. In Egypt, school children have been mobilized to go on field trips to handpick cotton leaf worms in outbreak years. This is done in large field situations and hundreds, if not thousands, of youngsters participate in the operation. Generally, pest collection techniques are used in gardens or very small farms, but are too labor intensive for use in most commercial agriculture in industrialized countries.

In some instances, the host plant may be pruned to remove infected tissue and the associated pest. Examples include pruning infected limbs of pear trees to remove fire blight infections (see Figure 4–4), or surgically removing crown gall from trees.

***Hand Pulling and Hoeing*** The technique of hand pulling and hoeing has been used for in-row weed control since crops were first cultivated. Only in the twentieth century, through the development of selective herbicides, has decreased reliance on this form of weed control become possible.

Pulling weeds by hand, coupled with hoeing, is probably still the best method for weed management in gardens, nonindustrialized agriculture, and intensive vegetable production systems (Figures 15–7a and b). It is the only completely effective method for in-row weed control in organic production systems. Systems of weed control that rely on hand labor must remove the weeds while they are still small, because the costs of control increase rapidly as weeds become larger. The high cost of weeding in organic production systems is one reason why produce from such systems must be sold at a premium.

In some production systems it may be cost effective to manually remove weeds remaining (sometimes called escapes) in row crops after the use of other control tactics. This is done to prevent the weed escapes from producing seed so that they do not add to the weed seedbank and create problems in future years. This concept is probably more important for organic farmers and systems that do not rely on herbicides, because such systems do not have a quick means to eliminate in-row weeds. Even in large-scale field crop production, hand removal is practiced under certain circumstances. In some areas of the midwestern United States, volunteer corn growing in soybean fields is often removed by hand, and occasionally even pockets of velvetleaf and cocklebur that escape herbicide control are manually removed by midseason.

In western societies, weed management in row crops is problematic because of cost and labor availability. Systems utilizing hand labor for weed control require a large workforce at critical times in the growth of the crop (see Figure 2–4). At current minimum wages in North America, the expense of hand weeding ranges from $300 to $800 per acre for crops such as sugar beets and tomatoes, to over $4,000 to $5,000 per acre for container-grown ornamentals. These costs are economically untenable. Agriculture in developed countries thus cannot rely on hand weeding as the sole means to achieve in-row weed control for both economic and sociological reasons.

(a)

(b)

**FIGURE 15–7** Human labor for weeding. (a) Hand crew hoeing weeds in processing tomatoes; and (b) hand weeding peppers in Morocco.
Source: photographs by Robert Norris.

Hand weeding provides the economic base to retain the labor force in major agricultural areas where labor is also needed for seasonal tasks of pruning and harvesting. This has been the case in some coffee and fruit-producing countries of South America.

## Mechanical Tillage/Cultivation

**Cultivation is the most widely used method of weed control for row-crop agriculture**

Tillage is intended to uproot plants or to modify the habitat so that pests do not survive. A primary function of all tillage is weed management. Weeds are hosts to other pests, such as nematodes, so conscientious weed management is particularly important, especially during crop rotations intended for nematode management. The widespread adoption of no-tillage agriculture, in which crop production is maintained without tillage, has resulted in reliance on herbicides for weed management.

*Tillage versus Pathogens*    For most pathogens there is no impact of tillage, but there are a few examples where tillage is an integral part of an IPM program. Tilling fallen apple leaves into the soil aids leaf decomposition which destroys the overwintering inoculum of the apple scab fungus, breaking the disease cycle by eliminating inoculum that is needed to start disease in the next year. For a few pathogens, such as the disease lettuce drop, deep burial in soil of survival structures such as sclerotia can reduce disease incidence. However, the lengthy survival of the buried sclerotia may actually limit the efficacy of tillage because surviving sclerotia can still be present when the next deep plowing occurs.

*Tillage versus Weeds*    The invention of the seed drill permitted accurate between-row tillage (see Chapter 3). Cultivation is the most widely used method for controlling weeds. Effective cultivation depends on the following factors.

1. Accuracy. Ideally the cultivation equipment is set close to the crop row (Figure 15–8), thus the uncultivated strip that must be hand hoed or sprayed with herbicide is narrow. It is critical that the crop is sowed accurately to allow subsequent close cultivation.

**"Get them when they are small" is the axiom of mechanical weed control.**

2. Weed size. Hoeing or cultivation is most easily performed when weeds are small. Large weeds are difficult to uproot, they clog equipment, and their removal may cause crop damage. Thus, correct timing is critical.

3. Integrated into other cultural practices. Cultivation must be integrated with other cultural practices. The soil moisture conditions must be correct, because if the soil is too wet or too dry, then poor weed control results. Ideally there should be no irrigation or rainfall following cultivation, because with moisture, the uprooted weeds can reroot and resume growth.

4. Annuals versus perennials. Mechanical weed control is much less effective for control of perennial weeds than it is for annual weeds. A single cultivation usually kills an annual weed, but it only removes the aboveground portion of the perennial weed, which then rapidly regenerates from roots, tubers, or rhizomes. Repeated cultivation is required to deplete the underground reserves of a perennial weed. It can take several years of diligent cultivation to substantially reduce the population of a perennial weed such as field bindweed.

5. Cultivation at night. Exposure to light triggers germination in weed seeds buried in the soil. Cultivating at night has been proposed as a way to stop

**FIGURE 15–8**
Sugar beet row between the top coulters of a cultivation rig showing close cultivation.
Source: photograph by Robert Norris.

light from stimulating such weed seed germination. The reduction in weed germination achieved by cultivating at night does not appear to justify the difficulty of carrying out the operation.

6. Deep burial. The use of special double-acting moldboard plows (see Tillage Equipment Selection) can bury perennial weed tubers and rhizomes under as much as 18 inches of soil. They cannot emerge from such depth and eventually die. The technique appears to hold promise for control of purple nutsedge. Deep burial has only limited utility against seeds, because they remain viable for long periods.

***Tillage versus Nematodes*** Nematodes are basically aquatic animals. Tillage turns the soil and exposes deeper soil strata and root fragments to rapid drying. If the soil dries rapidly, it can lead to desiccation and death of nematodes, including those within root fragments. This can effectively reduce the nematode inoculum present in the soil. Because nematodes occur throughout the soil profile, surface tillage may not reduce the population to below the damage threshold. As mentioned, some nematodes are able to survive drying by entering anhydrobiosis; and if they do, they will not be controlled by this tactic. The eggs of some nematode species survive slow drying relatively well, so this technique is usually of limited efficacy with those nematodes.

***Tillage versus Slugs and Snails*** Tillage destroys slug and snail habitat. It reduces cover and buries the surface organic debris into the soil, which in turn reduces their food supply, so slugs and snails are rarely problems in tilled agricultural fields. Adoption of no-till agriculture allows plant residue to remain on the soil surface, leading to increased mollusk problems.

***Tillage versus Arthropods*** Tillage has had considerable application in the management of soil-inhabiting arthropods, such as symphylans, and insects that spend a life stage in the soil, such as white grubs, wireworms, root worms, and the overwintering pupae of Lepidoptera and Coleoptera. Chopping and shredding the crop followed by moldboard plowing is recommended for control of

European corn borer, cotton bollworm and corn earworm, and pink bollworm of cotton. Tillage can also reduce survival of the grape root borer. Spring plowing is recommended for wheat stem sawfly management.

In some instances, tillage can lead to increases in the severity of an insect pest. In western Canada, tilling in late summer increases success of pale western cutworm because tilling loosens soil and reduces crusting, thus allowing improved oviposition (egg laying). Increased adoption of minimum or no-till systems has had some impact on some soil-borne arthropod pests, because they are allowed to complete development in the soil without disturbance.

***Tillage versus Mammals***    Tillage can be used to destroy the burrows of ground-dwelling vertebrate pests, such as pocket gophers and meadow mice and the warrens of cottontail rabbits. Pocket gophers are rarely a problem in tilled fields. Adoption of reduced- or no-till systems may lead to increased vertebrate pest problems in response to lack of burrow destruction.

***Tillage Equipment***    The types of implements used for tillage vary depending on the intended use and the status and type of the crop.

**No Crop Present**    When no crop is present, tillage is usually substantial and involves major broadcast disturbance of the soil from a few inches to 18 or more inches deep. This is often referred to as primary tillage when it takes place prior to sowing the crop. A range of tillage equipment may be used.

- Moldboard plows (Figure 15–9a) typically invert the soil to about 10 to 12 inches deep. These are used to bury large quantities of existing vegetation and require relatively slow movement through a field. Double depth moldboard plows bury residue to about 18 inches deep in the soil (Figure 15–9b).
- Discs are groups of sharpened concave circular steel blades that are pulled through the soil at an angle (Figure 15–9c). Large discs that are used for destroying vegetation often have fluted outer edges to aid in chopping plant debris. Discs can disturb soil from a few inches to about 8 to 10 inches deep, and are very effective for mixing existing vegetation and debris into the soil. Discing equipment can travel through fields at faster speeds than can plowing equipment.
- Chisels are vertical shanks of various sizes and types that are pulled through the soil to considerable depth. They are mainly used to loosen soil, and are less effective at killing weeds than plows or disks.
- Harrows are either spring-tooth (Figure 15–9d) or fixed spike-tooth (Figure 15–9e) devices that are pulled through the soil. Their action is relatively shallow, usually disturbing the soil to a few inches at most, and their primary use is to prepare the seedbed. They are effective in killing only small weeds.

**Between-row Cultivation**    This type of tillage is performed to remove weeds between the crop rows. It must be relatively shallow so that root damage to the crop is minimized. Examples include:

- Hand-pushed and animal-pulled wheel cultivators (Figure 15–10a) provide between-row weed removal.
- Vertical and angled knives, and horizontal sweeps, are pulled through the soil just under the surface uprooting weeds (Figures 15–10b and c).

(a)

(b)

(c)

(d)

(e)

**FIGURE 15–9**  *Examples of primary tillage implements.*
(a) Conventional single-depth moldboard plow;
(b) double-depth moldboard plow for deep burial of
residue; (c) large stubble disc (on field road to show the
discs and the double gang); (d) a spring-tooth harrow;
and (e) a spike-tooth harrow.
Source: photographs by Robert Norris.

• Ground-driven rolling cultivators are devices composed of groups of discs
made of curved teeth. The cultivator is pulled over the soil surface to
dislodge weeds (Figure 15–10d). Rolling cultivators have the advantage
that they are operated at relatively high speed and can pass through a
field quickly.

**Within-row Cultivation**   Mechanical weed control within the crop row is difficult. Special devices that work in the crop row have been developed, and
they are used to uproot small seedlings without killing the larger crop plants;

**FIGURE 15–10** Examples of between-row and within-row cultivation equipment. (a) Animal-pulled single-row cultivator in peppers in Morocco; (b) a typical between-row sled cultivator at work; (c) sled cultivator elevated to show knives and sweeps; (d) ground-driven rolling cultivator; (e) fingers of a Bezzerides cultivator for in-row mechanical weed removal; (f) a mechanical rotating brush hoe for removing in-row weeds; and (g) French plow being used for in-row weed control in a vineyard.

Sources: photographs (a), (b), (c), (d), and (e) by Robert Norris; (f) by Steve Fennimore, with permission; and (g) by Clyde Elmore, with permission.

however, in practice, considerable damage to the crop usually occurs. Examples of specialized equipment for weed control include:

- Rollers have been used in grass crops to kill broadleaf weeds but are only partially effective.
- Rod and finger weeders are mechanical devices that flick in and out of the crop row dislodging small weeds (Figure 15–10e).
- Rotary hoes are power-driven rotating devices that drive small knives through the crop row dislodging weeds. They can also remove crop plants, so they must be used carefully.
- Brush hoes are counterrotating brushes that pull out shallow-rooted weeds and do not remove the relatively deeper rooted crop (Figure 15–10f).
- French plow is a special plow developed for in-row weed control in tree and vine crops. The plow has a mechanism that allows it to swing out around the vine and then "drop" back into the row (Figure 15–10g).

**Specialized Systems**   These have been developed for specific situations.

- Renovation of alfalfa is an attempt to provide mechanical weed control in a broadcast perennial crop. Light discing or harrowing is used to dislodge weeds with minimal impact on the deep-rooted alfalfa.

- Capping and scalping is a technique used in direct seeded crops such as tomatoes and peppers that are relatively slow to germinate. After seeding, a 2- to 3-inch soil cap is placed over the top of the row. As the crop seeds start to germinate, the cap is physically removed with cultivation equipment so that germinated weed seeds in the cap soil are destroyed. The timing of removing the cap is critical.

***Disadvantages to Tillage***    Although tillage is widely used and can provide substantial control of weeds and several other types of pests, it is not without serious drawbacks that must be considered in an IPM program.

1. Soil erosion. Tilled soil is subject to serious erosion by water (Figure 15–11) and wind; soil that has not been tilled and is covered by plants is much less subject to erosion. Reduction in tillage for crop production can substantially reduce soil erosion. The use of no-till systems can, for example, reduce soil erosion from 2 tons/A in tilled systems to less than 2 lb soil/A in a no-till system during a single rainfall event. Soil erosion is probably the single greatest weakness of using tillage for pest management reasons. Erosion is especially serious with heavy rain (e.g., thunderstorms), or when high winds occur under dry conditions. The need to reduce soil erosion has been the primary driving force behind the adoption of no-till agricultural systems. The exacerbation of soil erosion through tillage may have more immediate, observable deleterious effects than result from herbicide use.

2. Energy use. Tillage requires moving large amounts of soil, which in turn requires using large amounts of energy. Animal energy was used for this purpose for hundreds of years. At the turn of the twentieth century, fossil

**FIGURE 15–11**  Soil erosion from a conventionally tilled wheat field.
Source: photograph by Jack Dykinga, USDA/ARS.

fuel used in internal combustion engines was substituted for animal energy to a large extent, and starting in the 1940s the use of chemical energy in pesticides was adopted.

3. **Soil habitat destruction.** Tillage causes destruction of habitat for nearly all soil-dwelling organisms. In addition to the intended target pests, wildlife from large animals such as burrowing owls and kit foxes, to earthworms and even microorganisms can be severely influenced. As much as 76% of earthworms can be destroyed by plowing and discing. Birds follow tillage operations to feed on organisms exposed by the tillage, and many of the organisms eaten are earthworms.

4. **Soil structure.** The physical structure of the soil is degraded by cultivation. The problem is considerably reduced by cultivation when soil is relatively dry, but tillage of wet soil causes considerable damage to soil structure. Power-driven rotary tillers are particularly damaging to soil structure.

5. **Soil compaction.** Driving heavy cultivation equipment across fields results in soil compaction and the formation of hardpan layers that cannot be penetrated by plant roots. Hardpans may exacerbate nematode problems because the roots are confined to upper soil strata where nematode numbers are highest. Breaking hardpans by deep tillage allows crop roots to penetrate through the compacted layer to access moisture and nutrients from deeper soil strata improving plant growth, and in some cases increasing tolerance to nematode parasitism.

6. **Rainfall.** Rain can make the soil too wet to cultivate, but weeds continue to grow. This can lead to severe crop loss, or even crop destruction, from lack of timely weed control.

7. **Dust.** Tilling dry soil creates dust. As tilled soil dries out, the loose topsoil can be blown, creating dust storms. Dust on plant leaves may exacerbate mite problems. Dust is a component of particulate air pollution and creates visibility and human health hazards. Laws to control dust pollution are being enacted in many regions of the world and have the potential to alter how tillage is used for pest management.

8. **Damage to crop.** Tillage may not be helpful, and may have a negative impact on the crop. Above- and belowground damage to the crop can occur. Equipment that has been improperly adjusted can kill the crop. Damage by cultivation equipment can cause wounds to leaves and stems that serve as entry points for bacterial diseases such as crown gall in orchard crops (e.g., walnuts) and soft rots in sugar beets. Damage to roots, called root pruning, can occur for many crops, such as where tillage is used for weed management in orchard crops.

9. **Broadcast crops.** Tillage is usually difficult or impossible to implement in broadcast crops.

10. **Transport of propagules within fields or from field to field.** Tillage equipment contributes to the spread of all pest organisms that either live in the soil or have a life stage that lives in the soil. This includes weed seeds, nematodes (see Figure 8–3c), many soil-borne pathogens, and some insects. The importance of this problem in the spread of pests is usually given insufficient consideration. Sanitation of equipment prior to moving it from one field to another is an important pest management concept. Sanitizing field equipment is considered in Chapter 16.

# Specialized Mechanical Tactics

***Weed Management***    The following are mechanical procedures that have been developed to control specific types of pests in limited situations.

**Deep Plowing**    This technique is used to bury pest propagules so that they cannot grow but are subject to death and decay. Examples include deep plowing to bury the sclerotia (Figure 15–9b) of lettuce drop fungus and yellow nutsedge tubers. The tactic requires the use of specially designed plows that can bury the surface soil to a depth of 18 to 24 inches. The efficacy of the technique is questionable because the next deep plowing returns surviving viable propagules to the surface. Additionally, some pathogens, such as nematodes, are able to survive in the deeper soil strata long enough for new plant roots to grow to them.

**Mowing**    This technique is used to limit weed growth without killing the plants. Mowing is used in orchards, noncrop areas, and roadsides where continuous vegetation cover is desired. Mowing very close to the soil surface can be used to control the parasitic weed dodder in alfalfa, as the weed attaches to the alfalfa stems about 2 to 3 inches above the soil. The equipment used for mowing includes rotary mowers, vertical flails, and sickle-bar mowers. Mowing is a temporary control, because the plants regrow, which necessitates repeated mowing.

**Dredging and Chaining**    Dredging, which is physical removal of soil and plant material from aquatic systems such as canals and reservoirs, is useful for aquatic weed control. It is roughly the aquatic equivalent of tillage, and is widely used in irrigation and drainage canals and ditches. Dredging is expensive and time consuming. The results are temporary, because the target plants are perennials and dredging does not remove all the propagules. Dredging fragments weeds and actually creates new propagules. Dredging must be repeated every few years, and with each use results in the need to dispose of large amounts of mud and plant material.

   A process called chaining is also used for aquatic weed control, and can also be employed in forestry as a preplanting weed management tool. Two large tractors are used to pull a heavy chain through the canal or vegetation to uproot unwanted plants. The limitations to chaining are much the same as those for dredging for aquatic situations. As the weeds are not removed from the canal by the chain, it is necessary to physically remove the uprooted weeds from the canal downstream from the chained area.

***Insect Management***    Few mechanical devices have been developed specifically for the control of insects. Two of these devices are worth mentioning.

**Hopperdozers**    In areas that are plagued by periodic outbreaks of migratory locusts, a device called a hopperdozer was used to mass collect and destroy them. Drawn through the field by horses, the dozer used shallow trays (of water, with a thin film of kerosene, crude oil, or tar) mounted on low wheels or runners to trap the disturbed nymphs and adults. Reports from the early 1920s claimed that farmers could collect as much as 14 bushels of grasshoppers per day using a hopperdozer.

**Suction Devices**    An extension of the suction sampling concept (see Chapter 8) has been the construction of suction-operated, mass collecting devices mounted on tractors or pickup trucks. These large vacuum devices have been occasionally used for the mass removal of insects from crops, with questionable efficacy and the removal of many beneficial insects. The devices are effective in strawberry production to control lygus bugs and other mobile insects, but have proved ineffective for removal of stationary insects such as aphids.

## SOURCES AND RECOMMENDED READING

There are no single reference sources that address the use of physical/mechanical tactics for pest management. Several texts included under the topic of cultural management (Chapter 16) also include discussion of physical/mechanical control tactics.

# *16*

# CULTURAL MANAGEMENT OF PESTS

## CHAPTER OUTLINE

## INTRODUCTION

The cultural management of pests involves changes to the way a crop is grown in order to make the crop less suitable for the pest, to make it more suitable for natural enemies, or to enhance the ability of the crop to withstand pest attack. The ultimate goal is to change the agroecosystem in such a way that the pest population remains below the economic injury level. The following are several general considerations about cultural pest management tactics.

A healthy, vigorous crop is one of the best forms of pest management.

1. Cultural tactics are fundamentally different from physical and mechanical tactics in that all effects of cultural tactics are mediated through the crop or the crop environment (for instance, changes in the microclimate within the crop canopy), rather than any direct effect on the pest organism. Because cultural management tactics only indirectly affect pests, they are usually relatively slow acting, and most cultural techniques are therefore of limited value in resolving pest problems that require rapid remedial action.

2. Cultural tactics for pest management exert their effect by changing the pest population dynamics so that the pest densities remain below the economic injury level (Figure 16–1). The employment of cultural tactics does not generally change the ecosystem carrying capacity for the pest, so that if the management tactic or tactics are relaxed, it means that the pest population can return to its original damaging level. Cultural management tactics therefore must be a continuous part of an IPM program; they cannot be stopped and started as needed to exert rapid pest control, as some other tactics can.

3. Cultural tactics generally reduce the magnitude of the pest problem with relatively low levels of inputs, and so have only minimal environmental impact. Cultural tactics can usually be implemented without specialized equipment and without use of large external inputs, relying mostly on equipment already available for normal cropping operations.

4. As single stand-alone practices, most cultural tactics do not completely control or eradicate pests. Cultural tactics must be part of a management program in which they are one of several tactics used to keep the pest population below damaging levels.

5. Most cultural management tactics require expertise and time on the part of managers, so that there are increased training costs for managers. Cultural control tactics must therefore be judged in relation to the increased management skills required and associated costs.

6. Cultural pest management tactics are often specific to a region; what works well in one region may not work elsewhere. Care should be exercised in transferring tactics from one region to another unless environmental conditions are appropriate.

Several topics presented as a part of cultural management tactics were discussed in Chapter 7, and will not be treated in depth again. Although cultural management tactics offer many advantages, they also have limitations that will be discussed.

**FIGURE 16–1**
Theoretical pest population dynamics in relation to cultural control tactics.

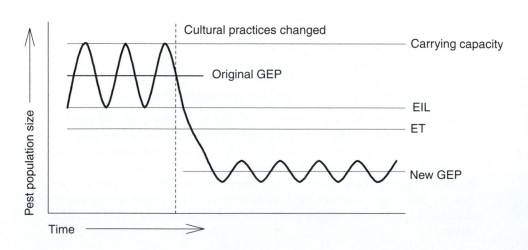

# PREVENTION

## Possibilities

If a pest is not present in a field or area, then preventing invasion is highly desirable. Prevention, as discussed in this section, is considered at the local field scale (see Figure 1–5) rather than the country and regional scale (as presented in Chapter 10). Prevention at the field scale is a tactic that can be used by the individual manager, and involves several components.

**It is always better to prevent a pest problem than to try and "cure" it.**

## Prevent Transport of Pests

Pests can invade fields by various means. Animals are capable of actively invading fields by walking or flying. Nematodes, pathogens, weeds, and many small insects and mites can invade fields by passive means. The chief means of passive invasion are via wind, water, and human activity. A major way that humans transport pests is in soil and on equipment, as well as either on or in crop seeds or transplants. Pest dissemination by means of seeds and transplants will be discussed in a separate section.

***Equipment***    Pests can be moved within fields or from one field to another by farm machinery. Virtually all weeds, and all soil-borne pathogens, nematodes, and arthropods can be dispersed in soil attached to farm machinery. To avoid spreading pests, the standard recommendation is that all equipment should be cleaned before it is moved between fields. Carefully cleaning all surfaces on which soil or plant debris can collect is especially important with primary tillage and harvesting equipment. Although the practice of cleaning equipment is recommended, it is often not followed. Figure 16–2 shows weeds growing from seeds in the soil contaminating the back of a landplane in a farmer's equipment yard. Although the photograph demonstrates weed contamination of soil, the concept applies to soil-borne nematodes and pathogens. Removal of the soil

**FIGURE 16–2**
Shepherds purse seedlings growing in soil carried on the back of a landplane scraper blade, in farm equipment yard, illustrating how soil-borne pests can be transported on equipment.
Source: photograph by Robert Norris.

from the equipment would have taken a few minutes at most, and would have stopped the spread of the weeds and other soil-borne pests to another field.

***Soil***    Soil can be contaminated with all types of soil-borne pests, including pathogens, weed seeds, and nematodes. A standard recommendation for nematode management is that soil should not be moved from one field to another. Control of sugar beet cyst nematode illustrates the problem of transporting pests in the soil. For many years in the Imperial Valley of California, the soil dislodged from sugar beets at the factory before they were processed (tare soil) was collected and returned to fields. The problem was that the soil was not returned to the same field it came from, so that the soil from a cyst nematode–infested field was returned to a different, possibly noninfested, field (see Figure 8–3c). This movement of soil virtually assured distribution of cyst nematodes to all fields being serviced by the factory. Once the problem was recognized, the process of returning tare soil to fields was halted.

Carefully cleaning all farm equipment should be a standard component of on-farm pest management. Some consider cleaning equipment a physical control tactic, but equipment is involved in most cultural control operations and therefore is discussed here in conjunction with cultural control tactics.

## Plant Crop Free of Pests

**Do not sow pests with planting material.**

The desirability of planting pest-free seed is obvious, and certified seed is free of pests. In many instances, however, there is no way of knowing if seed is contaminated with a pest, especially if that pest is a pathogen. In some instances, pests may be planted along with the crop, such as when growers plant their own seed that is contaminated with weed seed.

***Pathogens***    Many pathogens can be introduced in infected transplants, or can be seed borne. The pathogen can either be external or internal; the latter is much more insidious. For some pathogens (e.g., viruses), the use of seed certified as pathogen free is essential if an IPM program is to be successful.

Pathogen contamination is a major reason for planting certified seed or certified plants. Transplant material or seeds that are certified have been grown or produced so that they are free of pathogens. Certification usually entails growing the mother crop in areas that are pathogen free and inspecting the mother crop to ensure that it is pathogen free. Certification is administered by regional or national government agencies, often in conjunction with agricultural universities.

Planting certified stock is considered to be the main way to limit the introduction of virus problems in crops such as grapes. If virus-infected planting stock is used, the IPM program is compromised before it is started. The use of certified seed is an important part of IPM for limiting virus and other pathogens in potatoes.

Viruses can be seed borne in some crops. *Lettuce mosaic* is a good example. It is not possible to visually inspect a lettuce seed and see if it is infected. Commercial lettuce seed lots are therefore indexed to determine if mosaic is present. Seed lots found to contain mosaic are automatically rejected. The use of pathogen-free seeds is only effective when combined with sanitation and host-free periods (see appropriate sections later in this chapter).

| Pure Seed | Variety | Origin | Germination |
|-----------|---------|--------|-------------|
| 39.29% | AZTEC<br>TALL FESCUE | OR | 90% |
| 34.38% | ADOBE<br>TALL FESCUE | OR | 90% |
| 24.56% | TITAN 2<br>TALL FESCUE | OR | 90% |

Other Ingredients
0.09% Other Crop Seed
1.67% Inert Matter
0.01% Weed Seed
NOXIOUS WEED SEEDS: NONE
Net Weight 3 lbs. (1.36 kg)

Tested: OCTOBER 1997

(a)

(b)

**FIGURE 16–3**    Weed seed contamination of crop seed. (a) Notification on lawn grass seed package showing percentage of contamination with weed seeds; and (b) dodder seeds mixed into alfalfa seeds, which resemble each other and are very difficult to detect or to remove mechanically.

Source: photographs by Robert Norris.

***Weeds***    Weed seeds are sometimes planted with the crop seed. For some crops such as cereals, farmers often plant their own seed held back from the previous harvest. Weed seeds may be harvested with the crop seeds. Surveys in England and the United States (Utah) found that over 20% of the grain drills inspected had grain contaminated with wild oat seeds.

Weed seeds are frequently contaminants of commercial crop seed. A familiar example of this is seen on the label of lawn grass seed in any local garden store (Figure 16–3a). Seed of the parasitic flowering plant dodder is a very serious problem in alfalfa seed (Figure 16–3b). Weed seed contamination of crop seeds is an ongoing problem, and although modern seed-cleaning equipment has reduced the magnitude of the problem, it has not been eliminated.

Use of weed-free certified seed is a sound IPM practice. The use of such seed assures the manager that he or she will not be creating a weed problem by sowing weeds with the crop.

Paradoxically, although weed prevention is desirable, some weedy plants are sold commercially as useful or ornamentals plants. Examples include chufa (or *horchata* in Spain), which is yellow nutsedge and it is arguably the world's worst weed. Annual morningglories and water hyacinth are sold as ornamentals, yet the former is a serious weed in several crops, such as cotton and beans, while the latter is one of the most serious aquatic weeds worldwide. The aquatic weed *Hydrilla* is widely sold in the aquarium trade, yet millions of dollars have been spent removing this plant from irrigation and drainage canals, marinas, and docks. Finally, commercially sold bird seed often includes weed seeds.

***Nematodes***    Just as with fungal and viral pathogens, nematodes can contaminate seeds and vegetative planting stock, and can be readily moved into non-infested fields. This can be particularly problematic with bulbs, corms, and tubers. The use of certified seed or planting stock is again the preferred IPM tactic.

***Insects***    Contamination of seed by insects is less frequent than with other pests, and can often be avoided by treating seeds with insecticides or fumigants.

Insects in cuttings or roots used in the vegetative propagation of crops can initiate infestations. Sugar cane is planted as cane cuttings. In certain sugar cane regions, lepidopterous borer larvae can be living inside the cane pieces and start infestations as the plants grow.

## Limitations

Although the use of certified seed is a sound IPM practice, noncertified seed or planting stock may be slightly cheaper and may be used. The slight cost savings can be offset by a pest introduction.

# SANITATION

## Possibilities

**Sanitation removes debris and vegetation that can serve as hosts of pest inoculum.**

Sanitation is used to remove crop and other plant debris that may harbor pests. Carrying out sanitation tactics removes overwintering inoculum and potential food sources when the crop is no longer present. It thus decreases the likelihood of pest carryover from season to season. Effective sanitation requires that the life cycle of the target pest be understood.

There is a conflict between sanitation as an IPM tactic and no-till agricultural systems. A major objective of no-till systems is to maintain plant cover at all times, which can provide the resources for pests that sanitation removes. Implementation of no-till systems has required the development of new pest management tactics as alternatives to the previous pest control that relied on tillage and vegetation removal.

The options for sanitation vary with pest category and the crop involved. In the strict sense, sanitation has no application to weed management. Following are a few examples of IPM tactics based on sanitation.

***Pathogens***   Destruction of plant debris in which fungi, such as apple scab, overwinter is a key strategy for managing some pathogens. The apple scab fungus overwinters in dead leaves on the ground. In the spring the fungus in the dead leaves produces spores that infect the newly emerged leaves on the tree and the developing apples. The key IPM tactic is to till the old leaves into the soil, or add nitrogen fertilizer to speed leaf decomposition to decrease inoculum. Brown rot disease of peaches is handled in much the same way; the fungus overwinters on infected fruit mummies on the ground, and fungal inoculum develops from the mummies in the following year.

Infected plant parts can be pruned away so that they cannot provide inoculum. This procedure is only applicable to perennial crops and is used, for example, for fire blight suppression in pears, and to reduce powdery mildew of apples.

***Weeds***   Stopping weeds from growing and producing seed is sometimes termed a sanitation approach to weed management.

***Nematodes***   In some cases, infected plant material provides a source of nematode inoculum. Nematode-infected plant material may need to be destroyed as part of a management strategy. Palms infected with red ring of coconut are usually cut down and burned to ensure that the weevil vectors do not emerge from the diseased bole and infect new palms.

***Mollusks***    Removal of plant debris is one of the best methods for reducing slug and snail numbers, as they use debris for both cover and food. This tactic is particularly important in small farm and garden situations.

***Insects***    Sanitation is considered to be an essential component of several insect management programs.

1. The navel orange worm is a serious pest in almonds. The key management tactic is to remove mummy (or sticktight) nuts from the trees, because the nuts contain the overwintering pupae. Once the infested nuts are on the soil they must be disced in or physically removed. This sanitation tactic can keep the pest below economic threshold levels.
2. One of the main control tactics for cotton insect pests worldwide is the destruction of the stubble by cultivation or plowing. This practice destroys overwintering sites for pests such as cotton bollworm, boll weevil, and pink bollworm. In some regions the practice is mandatory and has become a part of area-wide efforts to manage those pests. For example, the Boll Weevil Eradication Program in the U.S Cotton Belt mandates the destruction of crop residues.
3. The overwintering stages of several corn borers, such as the European corn borer, the southwestern corn borer, and the southern cornstalk borer, are killed by shredding the corn stalks, a regular practice in much of the corn growing regions of the world.
4. The survival of overwintering populations of codling moths on pear can be greatly reduced if fruit left over in the field after harvest is collected and destroyed. This practice, if adopted over a large area, can substantially reduce the endemic populations, their rate of increase within the growing season, and, consequently, the overall damage caused by the pest.

***Vertebrates***    Destruction of vegetation removes habitat and food for ground squirrels, gophers, meadow mice, and other ground-dwelling vertebrate pests. Vegetation removal is one of the most important tactics for mitigation of these pests.

## Limitations

There really are no disadvantages to sanitation, except as those noted for no-till agricultural systems. Carrying out sanitation measures makes sense in most situations.

# CROP HOST-FREE PERIODS

## Possibilities

If host plants are not available between cropping periods, the pest cannot carry over from one crop to the next. The concept of host-free periods must encompass not only crop hosts, but also all possible alternative hosts in the region,

**If there is no host, the pest cannot survive.**

such as weeds. Host-free periods do not work in agricultural regions where the climate allows the crop to be grown continuously, because overlapping crops permits the pest to move to each succeeding crop. For host-free periods to work in regions where continuous culture is possible, it is necessary to change the cropping sequence to ensure that no overlap occurs between successive crops. Host-free periods have no relevance to management of weeds per se; lack of weed management can negate the utility of host-free periods for management of other pests.

**Pathogen Management**    The use of host-free periods is critical for management of many diseases, but is most important for control of the fastidious pathogens that are vectored by arthropods, including the phytoplasmas, some bacteria, and all viruses. Lettuce mosaic virus management in the Salinas Valley of California is an example. Lettuce cropping used to be continuous in the mild climate of the Salinas Valley, and lettuce mosaic was an ever-present problem. The disease is now managed by limiting the availability of virus-infected hosts through a combination of tactics; but mandating that no lettuce plants are growing in December is the key to the program. December is now designated as a host-free month. Management of the aphid-vectored yellows virus complex in sugar beets in regions where the crop can be overwintered also requires a complex scheme for limiting the occurrence of infected hosts so that the pathogen cannot be transferred from one crop to the next. Numerous other viruses can only be managed through the use of host-free periods; and in the absence of an effective host-free period, some viruses cannot be managed.

**Nematodes**    The host-free period is the underlying premise for nematode management using rotations to nonhost crops (see Rotations later in the chapter). The timescale for such host-free periods depends on the death rate of nematodes without a host, and usually has to be on the order of several years rather than the few weeks as noted for viruses. Factors that determine rotation length include the nematode species, temperature, spontaneous hatching, and the age of eggs. Some nematodes, such as the cyst nematodes, are capable of surviving as eggs in cysts for long periods (e.g., 10 years), so host-free periods need to be lengthy to allow nematode numbers to decline to levels below the damage threshold. For example, rotations as long as 6 years were ineffective at reducing populations of the pea-cyst nematode, a parasite of lentils, lupines, pea, and vetch. Long periods without hosts may lead to selection of nematode races that are longer lived and so can survive long rotations.

**Insect Management**    Host-free periods are critical for some insect management programs. An example is management of the southern green stinkbug in India. The species, although quite polyphagous, has a strong preference for grain legumes, feeding especially on developing pods. In certain regions of India, many different species of grain legumes were cultivated yearround except during the hottest months of August and September. Under those conditions, there was a host-free period that resulted in the crash of the stinkbug populations that had developed when hosts were available. When industrial soybean production was introduced to the region, pod maturation occurred in August and September, eliminating the host-free period, and damaging populations of the bug occurred throughout the year.

The pink bollworm of cotton is another example of the importance of a host-free period. The bollworm overwinters in cotton stalks and debris. There must be a cotton-free period from January to March in the Central Valley of California. In regions where stub cotton is grown, the pink bollworm is always present, because the host is always present. Stub cotton is produced by allowing the cotton plant, which is perennial, to regrow after the first harvest, thus producing a second, or stub, crop.

**Vertebrates** Host-free periods are not used for vertebrate management due to the mobility of the organisms, and because of their ability to use several different food sources.

## Limitations

There are some serious impediments to the use of host-free periods as an IPM tactic. Host-free periods are of limited value for management of many mobile polyphagous insects because of the abundance of alternative hosts. A major difficulty is that the host-free tactic may require regional cooperation among farmers, because implementation at the single-field scale is usually not effective with mobile pests. The use of host-free periods can require foregoing planting at a time when it is agronomically feasible and economically advantageous to grow the crop.

# CONTROLLING ALTERNATIVE HOSTS

## Possibilities

An extension of the concept of host-free periods is the control of alternative hosts. By controlling plants that can serve as alternative hosts when the crop is not present, it is possible to limit the overwintering population and reduce the growth rate of many pest populations. Although this concept is important for all pests, it is particularly important for mobile pests such as insects, pathogens, and some vertebrates. The vegetation serving as the alternative host does not necessarily have to be in the crop field, but may be in surrounding areas.

**Pathogens** Many plant pathogens have wide host ranges, and plants other than a particular crop serve as alternative hosts when the crop is not present. Examples of pathogens that occur in alternative hosts to the detriment of the primary crop include *lettuce mosaic virus* and the *sugar beet yellows virus* complex. Controlling alternative hosts is one means of reducing the inoculum sources for viruses, and this is particularly valuable as mobile insects vector viruses.

Some pathogens require an obligate alternate host to complete their life cycle, such as the pathogen causing cereal rust diseases. Control of alternate hosts is an important aspect of management of these diseases.

**Weeds** Weeds as primary producers require no hosts, with the exception of some parasitic weeds such as dodder, so control of alternative hosts to enhance weed management is not relevant. Weeds do serve as alternative hosts to many pests of all other categories; thus, control of weeds may have a significant influence on the dynamics and impact of all those other pests (see Chapters 6 and 7).

***Nematodes***   The presence of alternative host weeds in field borders and areas surrounding crops allows nematodes to survive and reproduce. Although nematodes do not move over distances by themselves, wind, water, and animals can transport nematodes from such vegetation into fields. In addition, control of alternative hosts is a serious problem in rotations intended to reduce nematode numbers (see Rotations).

***Insects***   The control of alternative hosts is a proverbial two-edged sword in relation to arthropod management, as both pests and beneficials may be living on alternative hosts. The role of alternative hosts to maintain beneficials was discussed in Chapters 7 and 13. Many pest insects can use weeds and other noncrop vegetation on which to overwinter, or on which populations can increase, including the following examples.

1. Johnsongrass is an excellent host of sorghum midge. Elimination of the wild hosts around fields is an important method for controlling the midge.
2. Beet leafhopper increases on many cool-season annual broadleaf weeds.
3. The false chinch bug increases on mustard and then can migrate to crops such as grapes.
4. Lygus bugs develop on many weeds. Monitoring lygus on weeds (see Figure 8–5) is used to estimate the potential for later impact on nearby crops.
5. Whiteflies use many broadleaf weeds as alternative hosts when suitable crops are not present; this can occur even under greenhouse conditions.
6. Over 80% of the leaf miners in vegetable crops in Florida move to the crops from weeds.
7. The beet leafhopper uses grasses for overwintering and spring survival in the desert and range areas of the Snake River basin in southern Idaho. Destruction of the grasses and their replacement with broadleaf annual plants has successfully reduced the impact of the leafhopper on many commercial crops in the region.

The concept of alternative host destruction to reduce buildup of insects is one of the rationales for area-wide IPM programs. In the previous situations, and in many others, it may be better to control the insects on the vegetation external to field rather than later having to control them in the crop (see Figure 8–5). Such control is a component of area-wide insect management. The area-wide IPM program for the cotton bollworm in the southeastern United States is a good example of this strategy.

Lettuce aphid provides an example of the importance of obligate alternate hosts. Similar to cereal rust and the barberry bush, the lettuce root aphid requires poplar trees as an obligate alternate host. If poplar trees are present in regions where lettuce is grown, the root aphid will be a problem. In the lettuce growing region of the Salinas Valley in California, the lettuce root aphid was a recurring problem until the importance of poplar trees in the aphid life cycle was understood. Removal of poplar trees solved the root aphid problem. This approach may seem rather draconian, but the alternatives may be even less acceptable, such as not growing the crop or substantially increased use of potent insecticides.

***Vertebrates***   Alternative hosts can serve as food for ground squirrels and meadow mice. Removal of noncrop host vegetation is one of the best methods for reducing populations.

## Limitations

The major limitation to alternative host removal is that effectiveness of this control tactic may depend on its implementation at the regional level. The alternative host vegetation may be on land that is not managed by the same individuals that are experiencing the pest problem. This can impede the use of a strategy, because different people consider the vegetation from different perspectives.

# ROTATIONS

## Possibilities

Rotation means that different crops are grown in sequence rather than growing the same crop repeatedly over time. Many pests feed on a limited number of host-plant species; changing crop species therefore changes the associated pest complex. Rotation often ensures that a particular problem pest in one crop does not continue to increase year after year. This is especially important for pests that have resting stages in the soil, including many pathogens, weed seeds, nematodes, soil-dwelling invertebrates, and vertebrates. The principles involved in rotation are to deprive the pests of their host plants and to utilize different management practices in different crops.

**Rotation of crops modifies the pest complex and the management practices, which can reduce buildup of crop-specific pests.**

*Pathogens*    Without a suitable host on which to reproduce, the propagules that serve as pathogen inoculum decline. The decrease in disease incidence over time of eyespot in wheat, as the years of rotation without the wheat host increase, is a particularly clear example of this tactic (Figure 16–4). Most soil-borne pathogens and nematodes exhibit this type of population decline over time in the absence of a host crop, including lettuce drop, *Verticillium* wilts, and some *Phytophthora* root rots. Rotation to nonhost crops is only effective if alternative host weeds are also not permitted to grow (see below for nematodes). If the pathogen is a facultative parasite that can also live as a saprophyte, then the use of rotation to nonhost crops is not so effective. Attempting to reduce the inoculum of oakroot fungus by not growing trees for several years is only marginally valuable as the fungus can live in dead and dying tree roots for several years.

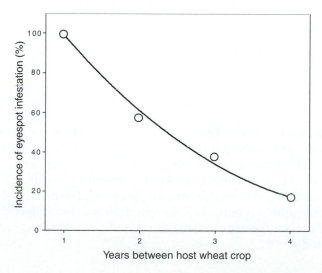

**FIGURE 16–4**
Reduction in cereal eyespot disease in wheat in relation to number of years in rotation to nonhost crops.
Source: drawn using data from Schulz (1961).

***Weed Management***    Rotation of crops permits changes in control technology, especially with herbicide use, where rotation permits weed control in the rotation crop by a herbicide that could not be used in the previous crop. Rotation between perennial forage crops and annual row crops results in changes in tillage practices or herbicide use, for instance, that keeps weeds in both systems at reduced levels. Similarly, rotation between a broadcast cereal crop and annual row crops can also be useful for weed management. Rotation to nonhost crops is a recommended tactic for controlling root-parasitic weeds in the genera *Striga* and *Orobanche*.

***Nematodes***    Rotation to nonhost crops is the standard recommendation for cultural control of many nematodes. This tactic is routinely employed to manage sugar beetcyst nematode; sugar beets should not be planted more often than every third or fourth year in soil infested with cyst nematodes, depending on the nematode population density. Rotation is not useful if the target nematode pest has a wide host range among agronomic plants, as is the case with lesion, and many root-knot nematodes.

Weeds in crops can negate the effect of crop rotation. Because some nematodes have wide host ranges, they can live and reproduce on many common weeds. The presence of the alternative host weeds in rotational crops can eliminate the benefits to nematode management derived from rotation to nonhost crops. Allowing pigweed, for example, to grow in corn can maintain some root-knot nematodes even though corn is not a host. Good weed management thus becomes an essential component of rotation for nematode management.

***Insects***    Rotation to nonhost crops can be effective for control of some insects that have a nonmobile soil-dwelling stage in the life cycle. A two-year rotation to soybeans almost eliminates the corn rootworm as a problem in corn in the Midwest. In recent years, however, the western corn rootworm evolved strains that spend two years as diapausing eggs in soil and other strains have adapted to ovipositing in soybean fields, thus greatly reducing the effectiveness of standard rotations for the management of the rootworms in the Midwest. Care should be taken in designing rotations, because the incorrect choice of crop sequence can result in an elevated insect problem; wireworms, for instance, are a more severe problem in potatoes following red clover or sweet clover.

***Vertebrates***    Rotation between a tilled cropping system and a system based on a perennial crop can provide short-term vertebrate control, but pest mobility limits the long-term utility.

## Limitations

A major limitation to the use of rotation as a pest management tactic is the fact that the best crop for pest management reasons may not be the best crop economically. If a farmer cannot make a profit on a crop, rotation to the crop for IPM reasons may not be economically tenable. A case in point is the continuous cultivation of corn in the Midwest United States despite the risk of elevated corn rootworm infestations. A corn–soybean rotation was recommended by research and extension, but for many years crop economics caused growers to ignore the recommendations.

## Fallow

Fallowing is a special type of rotation in which no crop is grown for an extended time. Fallow can function to reduce pest densities by starvation. For example, some nematodes cannot survive for extended periods without a host, and fallow can reduce their numbers; however, if weed hosts are present, the fallow will not work. Fallow is also a period when pest control tactics can be carried out that are not feasible when a crop is present, including large-scale nonselective approaches such as tillage and desiccation. Fallow is particularly important for reducing weed seedbanks, by using repeated light tillage to kill each flush of seedlings, or to deplete the underground root reserves of perennials. Because no commodity is produced during fallow loss of crop production and revenue is a serious limitation to the use of fallow for pest management purposes. Due to this financial cost, the use of fallowing has decreased as alternative control tactics have been developed.

# PLANTING DATES

## Possibilities

The planting date can be adjusted for some crops. By adjusting the planting date, the pest can be avoided or the impact of the pest on the crop can be reduced. The extent to which planting dates may be manipulated for pest management reasons varies depending on climatic region, type of crop, and the nature of the pest.

**Plant at the time of year that is good for crop germination and growth, but is inappropriate for pest establishment.**

*Pathogens*   By altering planting dates it is possible to (1) avoid seasons in which pathogen vector activity peaks (see following arthropod section), (2) maximize crop growth rates in relation to pathogen activity, and (3) avoid seasons when pathogen inoculum is at its greatest. Planting early or late can lead to reduced incidence of many diseases depending on the crop and the region where it is grown (Table 16–1). These examples emphasize that knowledge of crop and pathogen biology is essential to make the correct disease management decision.

*Weeds*   In some situations, it is possible to time the planting of a crop so that crop growth is favored and weed growth is not. Planting cereals in the spring can minimize the impact of downy brome, wild oats, and other weed grasses that grew during the winter, because they can be killed before the crop is planted. In some regions, the planting date for alfalfa can also be changed to favor the crop and disfavor the weeds. Alfalfa planted in late summer in Central Valley, California, escapes summer and winter weeds, whereas late fall planting coincides with winter weed emergence, and spring planting coincides with summer weed emerge (Figure 16–5).

*Nematodes*   Nematodes are poikilothermic and thus their activity is reduced when soil temperatures are cool. Planting susceptible crops when soil temperatures are low can result in reduced infection. Sugar beets planted in early spring when soil temperatures are low allows for beet establishment but discourages nematode attack until after the beets have grown somewhat. In mild

| TABLE 16–1 | Selected Examples of Impacts of Sowing Date on Incidence of Diseases[1] | | |
|---|---|---|---|
| **Crop** | **Disease** | **Country** | **Affect of sowing date** |
| Broad (fava) bean | Subterranean clover *red leaf virus* | Tasmania | September sowing coincides with peak vector flights; May, July, or November sowings less impacted. |
| Carrot | *Motley dwarf virus* | Australia | Delay sowing until vector's spring dispersal period has passed. |
| Corn (maize) | Stalk rot | Yugoslavia | Sow in May to avoid drought and high temperatures when crop is most susceptible. |
| | Smut | USA | Sow early in spring; rising soil temperatures later favor disease. |
| Oilseed rape | Blackleg | France | Sow in August; plants have time to become resistant before autumn weather favors disease. |
| Rice | *Pratylenchus* sp. nematode | USA | Sow late in autumn when soil temps. below 13° C. |
| Tobacco | Downy mildew | Greece | Sow in January or later to shorten period of mildew attack. |
| Wheat, spring | Root rots | Canada | Sow early May; rising soil temperatures. in later sowings favor disease. |
| Wheat, winter | Covered smut | USA and Germany | Sow early in fall so that seedlings are less susceptible by the time secondary sporidia develop. |

[1] Data obtained from Tables 2.24 and 2.25, pages 170 to 173 in Palti, 1981.

**FIGURE 16–5**
Generalized germination patterns for cool-season and warm-season weeds in relation to time of year in a Mediterranean climate; arrows indicate optimal planting dates for alfalfa that minimize impact of weeds during stand establishment.

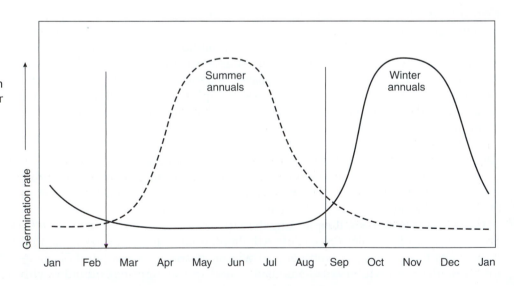

climates, beets can be planted in the fall and can escape major impact of cyst nematodes until the following spring, at which time the crop is more able to tolerate the pathogen. Planting carrots in late fall in southern California reduced galling and forking damage by root-knot nematodes (see Figure 2–1i). Approximately 50% of carrots were unmarketable due to galling in mid-October plantings, but only 11% of carrots were unmarketable in mid-December plantings.

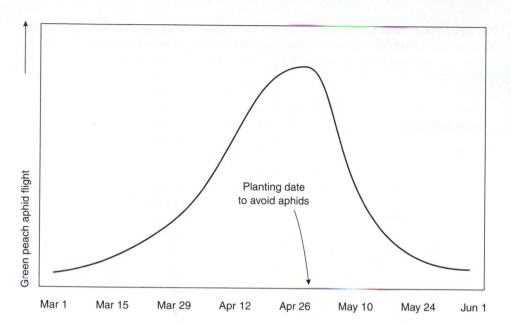

**FIGURE 16–6**
Typical generalized green peach aphid flight pattern in the spring in central California. The aphid is a major vector of sugar beet virus diseases; delaying crop planting to avoid the aphid flight minimizes disease infection.

***Arthropods*** There is a close relationship between the phenology of many insects and the phenology of the crop. Many host-specific insects require a near perfect synchronization between the life stage of the insect and the stage of development of the crop. By altering planting dates in some crops, it is possible to disrupt this synchrony to the benefit of the crop and the detriment of the pest. Insect activity is to a large extent also regulated by temperature. When the seasonal timing of flight patterns is known, it is feasible to plant crops to avoid such flights. Sugar beets are routinely planted after green peach aphid flights in Central Valley, California, to avoid transmission of *yellows viruses* to the crop (Figure 16–6). Planting too early results in sugar beets emerging while the aphids are flying, and thus high incidence of virus disease can result. This example demonstrates the complexity of developing true IPM programs, because in California sugar beets, the planting dates that would be selected for nematode management are in conflict with the dates that would be selected for yellows virus management. Monitoring pest status, understanding pest biology, the costs of tactics, and crop economics demand consideration to determine the best management given such conflicts.

Severe fall infestations of the Hessian fly in the central United States are usually caused by the existence of two favorable conditions: an earlier-than-normal planting date and a Hessian fly–susceptible variety of wheat. Simply delaying planting until the fly-free period would, in most years, reduce or prevent infestation, but still provide ample time for wheat growth and development.

In some regions planting early, growing short-season varieties, and defoliating early will accelerate maturation of the cotton crop before boll weevil and bollworm populations reach damaging levels. In other regions, however, a better strategy is late planting so that when the overwintered population emerges, all will die due to a lack of suitable hosts. This again emphasizes that detailed knowledge and understanding of pest biology relative to local environmental conditions is needed to design the best IPM program.

***Vertebrates***   Changing the crop planting date is generally not very effective due to the nature of vertebrate population dynamics. Because vertebrates are warm blooded, their activity is not as tightly linked to environmental temperature.

## Limitations

Manipulating the planting date may result in missing the best market price, particularly for fresh market crops. In many regions, weather dictates when land preparation is done. Early planting may not be feasible due to weather and soil conditions, such as spring rains, and delayed planting can be problematic if rain becomes more likely with a later harvest.

There is a fundamental ecological difficulty with using planting date for management purposes. Many pests associated with crops have evolved with the crop, and thus conditions that are unfavorable for the pest often are unfavorable for the crop. If an altered planting date were used widely as a management tactic on a regular basis, it would select for strains of the pest that are best adapted to that time, especially for weeds but possibly also for insects and nematodes. Planting date manipulation is an important tactic for some systems, and is mainly used in combination with other tactics.

> A crop is not attacked by a single pest, and altering the sowing date to avoid one pest has the potential to make attack by another pest more severe.

## CROP DENSITY/SPACING

### Possibilities

Increasing crop density can increase competition with weeds, using the density dependence of competition to IPM advantage. Increasing stand density can also allow some loss of plants to occur, without the crop density dropping below that required for optimum yield. Increased stand can be obtained by increasing the rate of seeding, by obtaining more uniform crop emergence, and by planting to minimize gaps, for example cross-drilling in high-density crops such as alfalfa. The greatest potential benefit of increased crop density is weed management. Decreased crop density can reduce the spread of insect, disease, and nematode posts, but increased crop density may offset losses due to pests. Crop density therefore presents a dilemma in relation to management of these pests.

***Pathogens***   Seedling diseases such as damping off can sometimes be offset by a higher seeding rate. This must be judged against the possibility that higher densities may increase the rate of disease spread. Local conditions will dictate whether this tactic is worth attempting.

***Weeds***   Crop plants compete with weeds. A weed management axiom states that a healthy, vigorous crop is the best form of weed control. Weak or low crop stands are notorious for encouraging weed invasion. Conversely, dense crop stands inhibit weed invasion and growth. For instance, elevating density of wheat sowing rate from 120 lb/A to 180 lb/A decreased general weed growth by about 50%. The impact of crop density on weeds applies to all crops, but is usually more important in broadcast crops such as cereals and alfalfa.

***Arthropods*** Phytophagous insects usually colonize row crops earlier than their natural enemies. The ratio of plant cover to bare ground, called the leaf area index, for newly emerged crops is small. Generalist predators that orient to the foliage where prey is likely to be found usually colonize the crop when the canopy is dense. Reducing row spacing causes the canopy to close early and improves the conditions for predator colonization of the crop. In soybean, it was observed that corn earworm moths prefer to oviposit in open canopy fields. Again, by reducing row spacing, the canopy closes earlier and conditions become less favorable for moth oviposition.

Similar to pathogens, the damage done by insects that kill seedlings, such as cutworm and darkling ground beetles, can sometimes be offset by higher seeding rate.

***Vertebrates*** Seedling losses to birds can also be partially offset by higher seeding rate in some situations.

## Limitations

The cost of purchasing extra seed is a disadvantage to increasing crop density. This increased cost must be balanced against the cost of other tactics that might be employed. Because increased seeding rate must be decided before the pest damage occurs it is prophylactic, with all the inherent disadvantages of such tactics. If increased crop density aims only at suppressing weeds, then other tactics must be employed for the management of other pests. If the latter includes the use of a herbicide that can remove the weeds regardless of crop density, then there is no reason to manipulate crop density for weed management. Often weeds and other pests are present only in some parts of a field, yet increased seeding rate would occur over the entire field, unnecessarily increasing cost. Higher seeding rate must be judged carefully for root crops (e.g., sugar beets and carrots), because crop density that is above optimum can reduce marketable yield.

# PRIMING OR PREGERMINATING THE CROP

## Possibilities

Any tactic that allows the crop to establish more quickly has the potential benefit that it reduces the time the crop is exposed to pests associated with stand establishment. Faster establishment can minimize seedling disease problems and decrease the length of time that the seedlings are most sensitive to insect attack. Rapid germination and emergence permit a crop to establish faster than weeds, and thus increase early competitive ability. Pregerminating is used for the water-seeding method for direct-seeding of rice. It provides a gain in competitive ability, but the main reason for using water seeding is not pest management.

Another approach to pregermination is to allow crop seed to imbibe water and become physiologically active (germinating) prior to sowing. This process is referred to as priming the seed. The tactic has seen the greatest use in intensive small-scale systems, although it is being evaluated for large-scale commercial use. The technique has more potential value with crops such as carrots, peppers, and parsley that germinate slowly than it does with crops such as corn and cereals that germinate rapidly.

## Limitations

The process of pregermination or priming makes the seed difficult to handle. If the tender radicle has emerged, the seed must be handled in such a way to avoid physical damage and not disrupt the physiological processes that have been initiated. Pregermination is expensive relative to conventional seeding. A process called gel seeding is being tested for use in row crops, where the imbibed germinated seeds are extruded in a gel matrix during sowing.

## DEEP PLANTING

By planting the crop seeds deeper than usual it is possible to reduce losses to some pests. This tactic is sometimes used to reduce losses due to bird predation. It has no relevance to the management of other pests, and even has the disadvantage that seedlings are weaker and emergence is slower. Weakened seedlings generally are more susceptible to other pests, such as damping off and cutworms, and are at a competitive disadvantage against early developing weeds.

## TRANSPLANTING

### Possibilities

**Transplants are large enough to better tolerate pest attack.**

Transplanting is the process in which actively growing plants are planted in the field rather than establishing plants by sowing seeds. Transplanting avoids many of the problems associated with seedling stand establishment. The plants are grown under controlled conditions, and when they are large enough they are transferred to the field. The small plants that are ready to be planted in the field are referred to as transplants. There are agronomic advantages to using transplants, such as generally more uniform crop stand and reduced seed cost. The latter can be particularly important for crops with high seed cost, such as hybrid tomatoes, or when seeds are in short supply. By using transplants it may be possible to achieve earlier harvest, a condition that often provides a premium price in the market or can help avoid late-season pests at harvest.

***Pathogens and Nematodes***    Transplants are usually beyond the stage of growth that is susceptible to diseases of germinating seeds, such as damping off. The handling required, however, can also result in mechanical spread of virus diseases in susceptible crops such as tobacco. The process of transplanting also damages the roots; such wounding can provide entry for some pathogens, such as crown gall of fruit trees. Transplanting delays nematode infection and, in general, larger plants are better able to tolerate nematode attack. As with most cultural control approaches, agronomic considerations usually prevail and pest management advantages or disadvantages are secondary, except, perhaps, in the management of weeds.

***Weeds***    The transplanted crop is much larger than germinating weeds (Figure 16–7a) and thus has a competitive advantage in comparison with crop seedlings that emerge from seeds at the same time as the weeds (Figure 16–7b). The larger plants also allow piling soil into the crop row for weed management,

(a)                                                          (b)

**FIGURE 16–7** Weed management in direct-seeded versus transplanted tomatoes. (a) Almost weed-free transplanted tomatoes about three weeks after transplanting; and (b) uncontrolled weeds emerging with direct-seeded tomatoes completely smothering the crop in the absence of a preplant herbicide application (left rows) about three weeks after sowing. Source: photographs by Robert Norris.

a cultural practice that is much more difficult with a direct-seeded crop. Crops that can be transplanted may often be grown without the use of herbicides, because of the advantages presented above. Transplanting can also permit use of herbicides that cannot be used for a direct-seeded crop; trifluralin, for example, can be used in transplanted tomatoes but is too phytotoxic to use in direct-seeded tomatoes. Transplanting has been used for millennia to combat weeds in rice production. The economic and social costs of this practice are increasingly questioned because of the time required to transplant a densely spaced crop.

***Arthropods/Mollusks*** The presence of the established root system and top growth leads to reduction, or even avoidance, of many of the stand establishment problems that can be caused by cutworms or slugs.

## Limitations

The expense of growing transplants and then planting them in the field is much higher than direct seeding for most crops. Due to the additional costs of transplanting, the tactic is usually only economically favorable for high-value fresh market crops, or where labor is cheap. Transplanting is generally not applicable for root crops such as carrots, beets, and turnips due to damage that inevitably occurs to the root during the transplanting process.

## SOIL CONDITIONS

## Possibilities

Two aspects of the soil environment can be manipulated to help in pest management. First, changing drainage patterns so that water does not collect in low areas can alter soil moisture. This can be achieved by improving water runoff or by physically changing the overall topography to fill and level low areas. Secondly, it is possible to adjust the soil pH, which can alter the ability of microorganisms to survive and attack hosts.

***Pathogens***    Wet soil can result in increased pathogen attack. Many soil-borne pathogens have a motile stage in the life cycle, such as zoospores for *Phytophthora* which require free water to swim to the host. Without excess water in the soil, the pathogen cannot infect plant roots. Management practices that reduce free water in the soil tend to reduce the incidence of soil-borne diseases.

Changing the soil pH (acidity/alkalinity) can change the suitability of the soil environment for many microbial organisms. The fungal pathogen that causes the devastating disease in cole crops, called clubroot (Figure 16–8), only infects the host when soil pH is below 7.0. Raising the soil pH to above 7.0 by liming provides excellent disease control. This single practice is the heart of a clubroot management program. Potato scab, caused by a species of *Streptomyces,* is exactly the opposite. Acidifying the soil to pH 5.3 or below is the management strategy for this pathogen, as it only infects potatoes when the soil pH is above about 6.0. This difference in response to pH by different pathogens emphasizes the need to correctly identify the pathogens and to understand their biology.

***Weeds***    There are many weeds that can tolerate wetter soil than most crops. Both canary grass and wild oats grow better than most wheat varieties under wet soil conditions. Attempting to grow crops in wet areas almost always leads to higher weed incidence. The roots of crop plants growing in wet soil are subject to low oxygen concentrations, and are often less vigorous or may even die, and are thus less able to compete with weeds. Improved soil drainage and elimination of wet areas can help reduce the severity of weed problems.

Soil compaction can also favor weeds; common knotweed and some spurges can thrive in compacted soil, but most crops cannot.

Manipulation of soil pH typically cannot provide control of weeds, but may alter the species composition of an area due to individual species preferences for acidic (calcifuge) or basic (calcicole) conditions.

***Nematodes***    Soil moisture that is sufficiently high to result in anoxic conditions can be lethal to some nematodes. As noted in Chapter 15, flooding soil can

**FIGURE 16–8**
Brussels sprouts plants showing severe example of clubroot disease. Note swollen base of stems (club) with almost no fibrous root development; at soil pH above about 7.0, root development would be normal without incidence of the disease.
Source: photograph by John Marcroft, with permission.

be used as a management tool to reduce nematode numbers in soil. Soil pH can also influence nematodes, and the optimal pH varies among nematodes. In the early 1900s, the root-knot nematodes were the primary nematode problem in Hawaiian pineapple production. As ammonium sulfate fertilizers were used in pineapple fields, the soils became acidic and by the 1960s the reniform nematode came to be the predominant nematode parasite. Soil pH can also influence microorganism activity and the level of natural nematode biological control that occurs in fields.

*Insects*  Sorghum at the whorl stage suffers the greatest injury from the fall armyworm because the whorl is the preferred feeding site for the worms. If sorghum is growing under acidic conditions (pH less than 5.4), instead of under optimal pH levels (greater than 6.0), the low pH retards plant development causing the whorl stage to last longer and so be more exposed to armyworm feeding, resulting in greater damage to the plants.

## Limitations

Such changes as noted should be made when economically profitable, as they are probably the easiest way to resolve a problem that is amenable to this type of manipulation.

# FERTILITY

## Possibilities

Soil fertility can also impact pest management because the crop nutrient status may determine its susceptibility to pest attack. It is universally recognized that an optimally fertilized crop is best able to withstand pest attack, and that both under- and overfertilization often lead to increased attack and losses. Nitrogen is the nutrient most often associated with altered pest attack. Literally thousands of studies have been conducted on the effect of mineral nutrients on pests associated with crops.

**Nutrient status can alter pest attack and the level of damage sustained.**

*Pathogens*  High fertility, particularly nitrogen, leads to lush vegetative growth, which typically increases the crop susceptibility to pathogen attack. The effect of nitrogen fertilizer on eyespot disease in wheat demonstrates the complexity of pathogen interactions (Figure 16–9). Nitrogen applied in the spring leads to an increasing yield in relation to increased fertilizer in the absence of any change in disease incidence. Nitrogen application in the fall, on the other hand, causes increased disease incidence in relation to nitrogen application which results in reduced crop yields. The response to nitrogen therefore varied in relation not only to amount applied but also time that it was applied. The book edited by Engelhard (1989) provides an in-depth coverage of the impacts of nutrients on disease incidence.

*Weeds*  Mineral nutrients serve as direct resources for weeds. The effect of altered nutrient supply on weeds is thus different from the effect on other types of pests. High nitrogen fertility increases the competitive ability of weeds. Higher levels of nitrogen fertilizer result in decreased wheat yields in the presence of

**FIGURE 16–9**

Impacts on timing of application and quantity of nitrogen fertilizer applied to wheat on severity of eyespot disease and yield.

Source: redrawn from Huber, 1989.

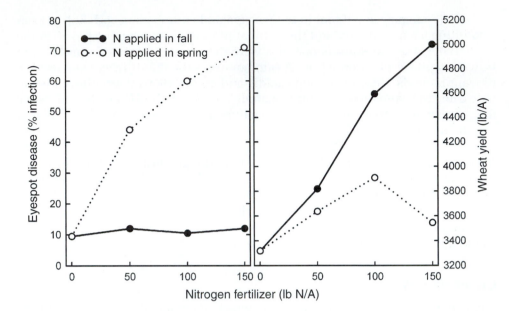

wild oats, because the higher fertilizer leads to differentially higher competition by the weed. A similar phenomenon has been reported for sedges in rice. This type of response is especially true when the fertilizer is applied broadcast; banding fertilizer along the crop row reduces the impact of high fertilizer on weed growth. High phosphorus fertilizer at seeding can increase the competitive ability of redroot pigweed in the seedling stages.

***Arthropods***    As a rule, insects develop faster and better on plants with high water and nitrogen contents. Experimental results are, however, highly variable particularly with reference to nitrogen (N). Differences in responses of insects to N nutrition of the host plant may be due to either the nutritional status of the plant or to the conversion of N to nitrogen-rich defense chemicals such as alkaloids. Low N or high potassium in the soil causes a reduction of available soluble nitrogen in the phloem of Brussels sprouts, a condition that depresses fecundity and reproductive rates of the aphids feeding on those plants.

## Limitations

The crop must usually be adequately fertilized to achieve optimum yield based on conditions when the pest is absent. Withholding fertilizer is typically not economically justified for pest management reasons, although pest attack provides another reason why excess fertilizer should not be applied.

# NURSE CROPS

## Possibilities

Stand establishment of some perennial crops is fairly slow; alfalfa and most orchard crops are examples. By sowing a fast-growing "nurse" crop at the same time as the primary crop, it is possible to obtain crop protection from environmental extremes and some types of pests, hence the term *nurse* crop. The nurse crop can provide increased competition with weeds during stand establishment at a time when the crop provides little competition. The sowing of oats in alfalfa, and cereals or beans in young orchards is practiced in some areas. A properly chosen nurse crop adds diversity to the system and can reduce pest attack on the crop. The nurse crop can also provide income before the main crop produces harvestable product.

## Limitations

Nurse crops can serve as hosts to pests, and can also compete with the primary crop resulting in yield loss if not carefully managed. The degree of yield loss is modified by weather conditions, making their use somewhat risky. Nurse crops invariably increase the complexity of management, such as irrigation and fertilizer use.

# TRAP CROPS

## Possibilities

Trap crops are crop or non-crop plants or even weeds, that can be used in several ways to manage pests. There are two different ecological processes involved with the use of trap crops. In some cases, pests are attracted to the trap crop; in others, the trap crop may disrupt the normal pest/crop interaction. Trap or catch crops are typically grown around or adjacent to the target crop, or may be grown during intercrop periods.

**Trap crops are plants grown to attract a pest away from the primary crop, or to lure a pest into the trap crop so that it can be destroyed.**

## Trap Crop as Preferred Hosts

The first ecological process involves the ability of the trap crop to attract the pest. The use of plants that are more attractive to the pests than the main crop can serve to divert mobile pests from the main crop to the trap crop. The pest population concentrates on the trap patch, allowing the main crop to escape damage, or the pest can be controlled on the trap crop before it does damage to the main crop. Depending on the nature of the trap plants used, the trap crop eventually may be destroyed.

***Arthropods***   A well-documented example of trap cropping involves *Lygus* management in cotton. *Lygus* bugs prefer alfalfa to cotton. By planting alfalfa in strips in cotton fields it is possible to attract the *Lygus* away from the cotton. This can provide effective control of *Lygus* for cotton production. Although the tactic works, it is almost never utilized. The limitation on the use of the tactic is that alfalfa and cotton have completely different agronomic requirements, particularly irrigation and harvesting schedules.

Planting of 8 to 10 rows of early maturing soybean on the borders of fields, planted adjacent to a later maturing variety, is an example of using the same crop plant as the trap crop. The early maturing plants set pod several weeks before the main crop reaches the same stage, and stinkbugs prefer to attack at the pod setting stage, and so they are attracted to the trap border strip (Figure 16–10). Stinkbug populations on those rows may represent 70% to 85% of the total

**FIGURE 16–10**

Early maturing soybean variety (lighter color in foreground) planted as a trap crop for stinkbugs to stop them from damaging the main late-variety planting (darker color in background).

Source: photograph by M. Kogan, Integrated Plant Protection Center at Oregon State University.

population in the field, and yet the trap rows represent less than 10% of the area. Spraying the border rows kills the stinkbugs and greatly reduces the level of the endemic population through the rest of the season, often eliminating the need for insecticide applications to the main crop.

A successful variant of the technique has been developed and deployed in Brazil. Instead of spraying the trap strip, the egg parasitoid *Trissolcus basalis* is released. The parasitoid oviposits on the egg masses that are concentrated on the trap strip and they multiply rapidly. When the main crop reaches the pod-set stage and becomes attractive to the stinkbugs, a large parasitoid population is present and ready to destroy most of the stinkbug eggs, again eliminating the need to use an insecticide. The technique is used because the parasitoid, under normal conditions, cannot build up effective populations in time to exert a significant reduction of stinkbug numbers to rescue the main crop.

Management of the rape blossom beetle in Finland is an example of the use of a variety of plants as trap crops. The beetles adapted to feed on cauliflower and became a limiting production factor. Trap crop strips were strategically distributed within the crop oriented transversely to the direction of beetle spread. The trap crop strips contained a mixture of Chinese cabbage, oilseed and turnip rape, sunflower, and marigold. At bloom, the trap crop strips were bright yellow and extremely attractive to the beetles. The strips were sprayed several times during the season when they were overwhelmed by the beetles. Marketable yields of cauliflower in fields with the trap crop were 20% higher than in other fields.

***Vertebrates***    With vertebrates, trap cropping involves planting a preferred alternative food source around the perimeter of the primary crop to which the pest is attracted. Crops planted alongside wildlife refuges or riparian systems can be decimated by animals moving into the crop from such areas. Poisoning or shooting is usually not a management option in such situations. The use of a sacrificial trap crop planted between the wildlife area and the crop can provide protection to the crop.

## Trap Crop as Hatch/Germination Stimulant

The second ecological process involved in trap cropping is to grow plants that will induce a root-parasitic organism to resume active growth, such as egg hatch or seed germination, in the absence of the host crop. The concept involved is that a chemical signal released by the host plant causes eggs to hatch, or weed seeds to germinate, only when a host is present. Some plant species, referred to as trap plants, produce the chemical signal but do not serve as hosts for the parasite, while others produce the signal and also can serve as a host for the pest. If the trap crop can act as a host, it must be destroyed before the parasite reaches the reproductive phase of its life cycle. The use of catch crops is applicable with parasitic weeds, but using the technique effectively with nematodes can be difficult because the nematodes can live on roots that survive even after the top has been killed.

**Weeds**   Trap cropping can be useful for weed management in the special case of the root parasitic weeds *Striga* spp. and *Orobanche* spp. As with egg hatch in some nematodes, these weeds require a chemical stimulus to trigger seed germination. Sowing plants that can induce germination but which are not hosts can lead to reduction in the seedbank of the parasitic weed. For example, cowpea and cotton can stimulate germination of some species of *Striga,* and sorghum and barley stimulate germination of *Orobanche crenata;* in both cases, the crops are not hosts of the respective weed.

**Nematodes**   Nematode egg hatch may be increased by chemicals released from host-plant roots. Some plants are either poor hosts or nonhosts, but do produce hatch-inducing chemicals. Such plants can result in decline of nematode populations, because they induce hatching, but the nematodes have very limited success in feeding and reproduction. Varieties of oilseed radish and mustards have been developed in Europe that stimulate hatch of the sugar beet cyst nematode but are poor hosts and the nematodes do not reproduce well on them. They are used as trap crops to reduce nematode numbers in sugar beet fields.

## Limitations

The acceptability of the control achieved is based on economics, and if the cost of implementing the control tactic exceeds the value of control, then the tactic is of limited utility. Implementation of the tactic often requires the use of potential cropland to grow the trap plants, which reduces potential income (termed an "opportunity cost" in economics). The tactic does not work for weeds and nonmobile pests, with the exception noted above, or those that have limited host specificity, such as many pathogens.

# ANTAGONISTIC PLANTS

## Possibilities

Antagonism occurs when chemicals released from a plant reduce the population of a pest organism. Antagonism based on chemicals released by plants is termed allelopathy. This topic may be considered as part of biological control, but we consider it here as part of cultural management.

*Weeds*    An example of allelopathy is the inhibition of tomatoes and other plants grown under walnut trees. Straw from some varieties of rye releases allelopathic compounds as it decomposes, and can provide control of some weed seedlings and some nematodes. Practical application of allelopathy for weed management is limited.

*Nematodes*    Many plants release allelopathic compounds that alter nematode behavior, and some produce chemicals that actually kill nematodes. Growing such plants in rotation can reduce nematode numbers in the soil. An example of this tactic is the growing of certain species of African marigolds in the genus *Tagetes* (see Figure 13–5). Some members of the Brassicaceae may also produce natural nematicidal compounds. Although the phenomenon of nematode-killing plants has been known for decades, it has not been widely exploited.

*Insects*    The effect of allelopathy on insects and other arthropods is not clearly defined.

## Limitations

Antagonistic plants have limited practical use in mainstream agriculture, although there is use in smaller organic operations. A major limitation is that the chemical-producing plants may have to be grown at a time when there are no crops, which can limit economic returns.

# MODIFYING HARVEST SCHEDULES

## Possibilities

**By modifying the harvest schedule it may be possible to avoid pest attack.**

The concept underlying this tactic is that it may be possible to harvest the crop early, before pest attack occurs or at a time when pest impact is minimized. Early harvest may require that planting is also done early, or that early maturing varieties must be selected when planted later. In either case, harvest must be planned in advance. Early harvest has essentially no relevance to weed management.

*Pathogens*    The severity of smut in corn, or black mold and fruit rots of tomatoes, can be reduced by harvesting early in season. The converse is also true: These diseases are often more serious on late harvests.

*Arthropods*    Arthropods, as mentioned earlier, have a phenology that must be closely synchronized with the phenology of the crop. Early harvest often produces phenological asynchronies capable of disrupting the arthropod's life cycle, allowing harvest of the crop before the damaging stage occurs. By moving the first alfalfa cutting date about two to three weeks earlier, much of the damage by the alfalfa weevil can be avoided. The corn earworm is a late-season insect in sweet corn, and can be almost completely avoided by early harvest in regions with a long growing season (e.g., Mediterranean climate zones).

Harvesting alfalfa on an alternating strip sequence maintains foliage in the field so that lygus bugs are not forced to move into adjacent crops, such as cot-

**FIRST CYCLE**                    **SECOND CYCLE**

Plan views of field

Cross-section views

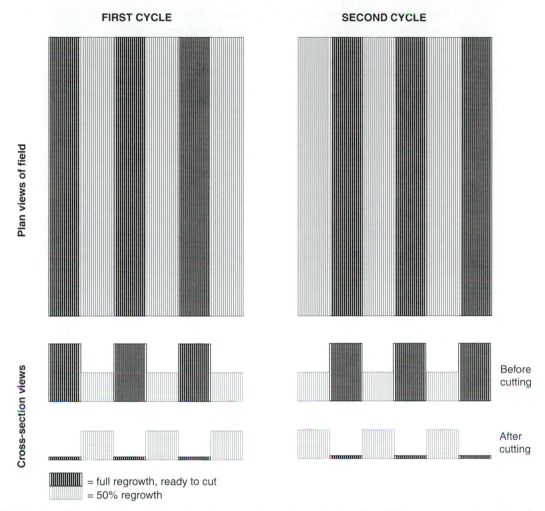

Before
cutting

After
cutting

= full regrowth, ready to cut

= 50% regrowth

**FIGURE 16–11**   Diagram demonstrating the alternate strip-cutting technique for alfalfa as a means of retaining lygus bugs and beneficial insects in alfalfa hay fields.

ton (Figure 16–11). This an example of an area-wide approach to management that relies on altered harvest schedules.

One of the most successful applications of the early maturing–early harvesting of a crop for insect management is the technique known as the "reproduction-diapause system of cotton boll weevil control." Successful onset of overwintering diapause in the boll weevil, and consequently survival of a viable adult population the following year, depends on adequate fat accumulation that results from adult feeding on squares and bolls. The reproduction-diapause system of boll weevil management is a combination of insecticide application, defoliation, and rapid harvest and stalk destruction with the objective of starving or killing the weevils so that they do not accumulate enough fat to overwinter. This system of reduced insecticide use, coupled with cultural control practices such as early crop maturation and rapid destruction of crop residues, greatly decreased outbreaks of secondary pests such as the bollworm. The system remains as an effective component of the boll weevil management programs of cotton-producing areas of the U.S.A. where the boll weevil occurs.

***Vertebrates***	Early harvest can sometimes reduce losses to migrating birds as the crop is removed before birds move into the area.

## Limitations

Crops that require processing, such as sugar beets, sugarcane, and processing tomatoes, have harvest schedules and quotas set by the processors. Typically the grower is under contract to deliver the produce at a specific date and in a specified quantity. Changing harvest schedules is not under the grower's control. Furthermore, changing harvest schedules may complicate general production practices. Although the alternate strip cutting of alfalfa as described is effective in keeping lygus in the alfalfa, it is not widely used because the cost of scheduling harvesting in the field twice and the difficulties in irrigation are more costly than other available alternatives for lygus management. Produce may be required throughout the growing season, such as sweet corn, and so concentrating harvest at a time when the pest is not present may not be an option. Cash or industrial crops, on the other hand, may not suffer such limitations and the system has been successfully implemented on an area-wide basis in cotton and soybeans, for instance.

# CROP VARIETIES

The use of crop varieties for managing pests may be conceived as a cultural tactic, although some have considered resistant varieties as a biological control tactic. The topic is of sufficient importance to IPM, however, that in most IPM texts it stands alone under the title "Host Plant Resistance," an approach adopted in this book (see Chapter 17).

# INTERCROPPING

## Possibilities

Intercropping is the practice of growing two or more crops simultaneously in a single field. Intercropping increases within-field diversity that can conceivably help to stop crop-specific pests from building up populations as rapidly as in monoculture. Whether such reductions result from greater effectiveness of biocontrol agents, or from changes in the main resource concentration for the pest, varies with the system; and in many instances benefits of intercropping have been difficult to demonstrate. Yields of each crop are often lower than those of the same crop grown without the companion crop, but total yield per unit area is often higher in the mixture. Intercropping has a margin of risk reduction built in, since if one crop is badly damaged by a pest the other crop can still be harvested. Intercropping now sees very little use in developed countries, because of the limitations noted below. Small-scale farmers in less-developed countries extensively use intercropping; intercrops of maize, beans, and a squash are commonly employed. It has been estimated that 98% of the cowpeas grown in Africa and 90% of the beans grown in Colombia are intercropped.

*Insects* Benefits of intercropping have been studied with arthropod pests more than with any other pest category. In many studies, numbers of natural enemies in polycultures are increased, but there is no evidence that these increases are reflected in a reduction of pest incidence. For example, populations of the flower thrips in Nigeria were reduced by 42% on cowpea intercropped with corn, but there was no effect on other important pests such as the bean pod borer, pod-sucking bugs, or blister beetles. Such ambiguous results are typical of many analyses of intercropping systems. The overall benefit in promoting natural enemies in itself justifies the adoption of such systems wherever they are economically sustainable, as long as they do not complicate management of other pests.

## Limitations

Intercropping poses serious weed management difficulties as the presence of two or more different crop species essentially eliminates the possibility of using selective herbicides, and can even limit mechanical cultivation. Constraints to weed management limit the application of intercropping as a pest management tactic to regions where cheap human labor is available. The overall limitations of diversity in relation to weed management were discussed in Chapter 7.

## HEDGEROWS, FIELD MARGINS, AND REFUGIA

Noncropped refugia can be used to harbor beneficial organisms, especially insects and spiders (see earlier discussion under diversity in Chapter 7). Hedgerows provide definite benefits in terms of soil conservation and enhancement of natural enemies. Benefits of hedgerows in the regulation of aphid populations have been observed in the United Kingdom where, for many years, hedgerows had been a landmark in rural areas. Mechanization and gradual increase in field sizes have led to a reduction in hedgerows and other refugia; the implications to pest management are still not clear (see Chapter 7).

## SUMMARY

Most cultural tactics for pest management are based on common sense, and should be incorporated into all IPM systems where appropriate. There are, however, several reasons why cultural tactics will almost never stand alone as management techniques. Most cultural tactics are only partially effective, which means that another tactic will also have to be employed. The cost of implementation may be more than the gain, such as rotation to a nonprofitable crop. Cultural tactics do not act quickly and thus cannot resolve an immediate problem. The need for management expertise and knowledge of pest and crop/ecosystem biology is greater than that needed by other methods if cultural management tactics are to be implemented with satisfactory results. Despite these possible limitations, cultural control methods are the backbone of ecologically based agricultural systems.

# SOURCES AND RECOMMENDED READING

The following texts provide general background on the topic of cultural management of pests; these texts also cover material included in Chapter 15. The book by Palti (1981) and the edited book by Rechcigl and Rechcigl (1997) provide thorough discussions of the cultural management of pathogens. Rechcigl and Rechcigl (1999) also edited a useful series of reviews on cultural methods for arthropod management. Van Vuren and Smallwood (1996) review ecological approaches to vertebrate management, and the book edited by Singleton (1999) covers ecological management of rodents.

Altieri, M. 1994. *Biodiversity and pest management in agroecosystems.* New York: Food Products Press, xvii, 185.

Engelhard, A. W., ed. 1989. *Soilborne plant pathogens: Management of diseases with macro- and microelements.* St. Paul, Minn: APS Press, vi, 217.

Huber, D. 1989. Introduction. In A. W. Engelhard, ed., *Soilborne plant pathogens: Management of diseases with macro- and microelements,* St. Paul, Minn.: APS Press, 1–8.

Palti, J. 1981. *Cultural practices and infectious crop diseases.* Berlin; New York: Springer-Verlag, xvi, 243.

Rechcigl, J. E., and N. A. Rechcigl, eds. 1999. *Insect pest management: Techniques for environmental protection.* Agriculture and Environment Series. Boca Raton, Fla.: CRC/Lewis Publishers, 422.

Rechcigl, N. A., and J. E. Rechcigl, eds. 1997. *Environmentally safe approaches to crop disease control.* Agriculture and Environment Series. Boca Raton, Fla.: CRC/Lewis Publishers, 386.

Schulz, H. 1961. Verhütung von Fusskrankenheiten beim Getreide. *Mitteilungen deutsches landwirtschafts gessellschaft* 76:1390–1392.

Singleton, G. R., ed. 1999. *Ecologically-based management of rodent pests.* ACIAR Monograph Series, no. 59. Canberra: Australian Centre for International Agricultural Research, 494.

Van Vuren, D., and K. S. Smallwood. 1996. Ecological management of vertebrate pests in agricultural systems. *Biol. Agric. Hortic.* 13:39–62.

# CHAPTER

# *17*

# HOST-PLANT RESISTANCE AND OTHER GENETIC MANIPULATIONS OF CROPS AND PESTS

## CHAPTER OUTLINE

▶ Introduction
▶ Host-Plant Resistance
▶ Conventional plant breeding
▶ Genetic engineering
▶ Applications of pest genetics in IPM
▶ Summary

## INTRODUCTION

This chapter discusses genetic alterations of crops and pests as they relate to IPM systems, and includes both classical breeding and genetic engineering approaches. Host-plant resistance is a fundamental approach to the management of pathogens, nematodes, and arthropod pests. New technologies offer opportunities to enhance even further the utility of this approach in IPM. Understanding the inheritance of biological traits and the development of techniques to manipulate genes that govern those traits constitutes a revolution in applied biology, including agriculture and pest management. The techniques of genetic engineering make it possible to modify organisms to suit human objectives in ways that could not be accomplished just a few decades ago. Recently developed techniques in molecular biology that are useful in integrated pest management include methods for detecting genes, and techniques that allow the transfer of genes between organisms and even between different species. In addition to the consideration of resistance in crop plants to animal pests and pathogens, changing crops to make them resistant to herbicides, and manipulating pest reproductive systems and genetics to induce sterility also are considered in this chapter because they are extensions of the broader field of genetic manipulation of organisms for the purpose of reducing pest impact.

**Most plants are resistant to most pathogens and herbivores.**

443

# HOST-PLANT RESISTANCE

Despite the problems with agricultural pests around the world, only a relatively few pest species attack any particular plant species in any given region. Plants have some resistance to most herbivores, nematodes, and pathogens; if they did not, there would be far fewer plants around. Plants are primary producers and are the base of the food chain. Plants cannot move and escape from attack; therefore, over time, there has been strong selection pressure for evolution of resistance to the organisms that would use them as food. Likewise, as plants evolved defenses to the various organisms that would use them as food, so did the organisms that use plants as food evolve to overcome those plant defenses. These are the organisms that we now call pests when they attack crop plants. The concept of coevolution, wherein traits of one organism serve as a selection pressure on another organism and vice versa, explain the genetic interplay between plant defenses that act as selection pressure causing evolution in herbivores and pathogens to overcome those defenses.

The many defenses that plants have evolved to ward off pests have been manipulated by humans through plant breeding to develop host-plant resistance. This use of plant breeding to produce varieties that are resistant to pests is often considered to be the ideal approach to management of many pests, especially pathogens, nematodes, and many insects.

The concept of host-plant resistance works well for pests that feed on plants, because the host (crop) interacts at the biochemical level with the organism feeding on it. Because there is no direct transfer of resources from crop plants to weeds, plant breeding has little application for management of weeds. The development of herbicide-resistant crops, discussed later in this chapter, represents a new way in which genetic manipulation of the crop contributes to weed management.

## Advantages of Host-Plant Resistance for Pest Management

Pest-resistant crops that have been produced through plant breeding have several advantages over other tactics used for IPM. Some advantages are real, but some are perceived, as follows:

1. Host-plant resistance is usually the only tactic required to manage a pest when a highly resistant variety is available.
2. The protection is season-long, as the protectant is always present.
3. The pest is always influenced at the most sensitive stage.
4. Protection is independent of weather.
5. Generally all plant parts are protected, including those that are difficult to treat using conventional pesticides.
6. Only pests that feed on the crop are exposed, eliminating most nontarget effects.
7. Only the target pests are controlled; most beneficials are not influenced; that is, plant resistance tends to be selective to the target pest. Certain resistance mechanisms, however, also may affect beneficials (see tritrophic level interactions in Chapter 13). The property of selectivity applies to pathogens, nematodes, and arthropods.

8. For the most part, plant resistance is compatible with all other management tactics, including pesticides. So it is ideally suited for multitactic IPM systems.

9. Plant resistance for the control of arthropod pests often has an additive effect. If resistance is sustained over the years and the resistant variety is grown regionally, the pest will be reduced to a residual population.

10. The protectant factor is usually active in the tissue in which it is expressed, not in the environment; therefore, it has minimal environmental impact. Host-plant resistance is environmentally friendly.

11. The protectant principles are biodegradable, and the choice of suitable genes ensures that those principles are not toxic to humans or domestic animals.

12. There are economic advantages because the only cost to the grower is the seed, which is already part of any farm budget. In addition, there is no need for repeat applications within a season. As long as host-plant resistance is reliable, and does not quickly select for resistance-breaking pests, it helps to stabilize an IPM program. Conversely, breakdown in resistance can destabilize an existing IPM program that uses host-plant resistance.

Although there are definite advantages to using host-plant resistance as a means to manage pests, the tactic is not without limitations. These are discussed later in the chapter.

Understanding the genetics of both pests and crops has permitted development of management strategies based on resistance to pests. A systematic effort to identify sources of resistance started at the beginning of the twentieth century, but selection of resistant varieties had probably been inadvertently practiced for centuries by farmers that kept the seeds from plants that performed the best. The governing ecological principle behind the use of resistance is disruption of the food web by rendering a suitable, concentrated food resource into an unsuitable resource through host modification (the food source) (Figure 17–1). Rendering a food source unsuitable can be a permanent change in the agroecosystem, unless the pest develops the ability to overcome the resistance.

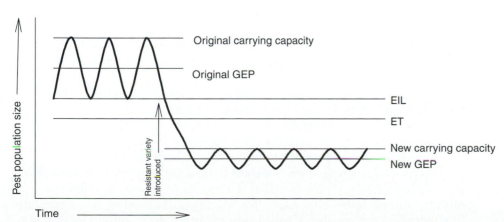

**FIGURE 17–1**

Theoretical pest population dynamics in relation to control tactics based on plant breeding that lower the carrying capacity for the pest.

# CONVENTIONAL PLANT BREEDING

## Principles

The Austrian monk, Gregor Mendel, laid the foundation for plant breeding, including breeding for resistance to pests, when in 1865 he published his findings on inheritance in peas. He discovered that traits were passed on from parent to offspring in a predictable fashion, and that when parents with different traits were crossed, the offspring possessed a combination of these traits. He also established that the genes for the traits were either recessive or dominant. Mendel's concepts comprise what is now called Mendelian genetics, and the principles form the basis of all conventional plant breeding.

**Crop varieties that have resistance to pests or have desirable agronomic characteristics, can be developed by cross-breeding crop varieties with resistant or desirable plants in the same species.**

Mendel's discoveries were essentially lost for 30 years, but were rediscovered early in the twentieth century at the time when researchers began to realize that resistance to rust disease in wheat was an inherited trait. They observed that the inheritance of rust resistance followed the mechanism that Mendel had discovered. That observation allowed agriculturists to develop the techniques necessary to transfer desirable traits from one variety to another. Cross-breeding crop varieties with plants that are resistant to particular pests is used to move resistance genes into the crop varieties. In this way, crop varieties that have resistance can be developed and used to modify the crop/pest interaction.

The basic approach underlying plant breeding for pest management is to develop a crop variety that is no longer susceptible to the pest (the variety is resistant), or is able to tolerate pest attack and is less susceptible to the damage caused by pests. The process used to develop a resistant variety usually starts with the screening of appropriate plant accessions of seeds, clones, cuttings, or even tissue culture for the crop of interest. Government research institutions and international crop research centers maintain extensive germplasm collections of most existing varieties, landraces, and wild relatives of the major crop species. The National Plant Germplasm System of the U.S. Department of Agriculture is an excellent source of information on collections for the major crops. Typically the following steps are followed in the development of resistant varieties.

1. Identify plants in the same species as the crop that seem to be less susceptible to the pest than are the standard varieties. Access to seeding materials from germplasm collections to establish screening nurseries is a common procedure under this first step.
2. Once a prospective source of resistance is identified, the next step is to move (or introgress) the gene(s) that confer the resistance into the crop cultivar (cultivated variety). The introgression is accomplished by making hybrids between the desirable susceptible crop variety and the resistant plants.
3. Following the first cross, only 50% of the genome is from the parent crop variety; the other 50% is from the resistant plant. The resulting hybrid will probably not be acceptable as a commercial variety because of undesirable characteristics that were transferred in the cross. It is therefore frequently necessary to perform a variable number of backcrosses between the new resistant hybrid and the original susceptible parent crop variety, which is called the recurrent parent, in order to eliminate undesirable characteristics. During backcrossing, it is necessary to ensure that the resistance trait is moved into the progeny. The number

of backcross generations required to obtain resistant plants that are carrying a minimum of undesirable characters depends on the following factors.

    3.1. Number of genes that control the resistance—the higher the number, the longer the process

    3.2. Degree of linkage between or among resistance genes—the more tightly linked, the better

    3.3. Degree of linkage between genes for the resistance and genes that control undesirable traits—the weaker the linkage the easier it is to introgress resistance genes without undesirable traits, such as shattering in grain legume crops.

    3.4. Dominance of the genes—work is easier with dominant genes

4. Eventually a resistant variety that is similar in agronomic characteristics to the original susceptible variety is developed and released for commercial use.

Because backcrossing may have to be carried out many times, resistance gene introgression into a crop variety can require many plant generations. This means that it takes years before a new resistant variety can be developed in annual crops using classical breeding techniques. The process can take many years or even decades for perennial tree crops, and hence there has generally been less effort invested in the breeding of resistant tree and vine varieties. The decision to attempt breeding a resistant variety must always include consideration of the time required for breeding against the likelihood that the pest will overcome the resistance.

**Attention to terminology.** In *plant breeding, resistant* is a crop qualifier. A wheat variety is resistant to the rust pathogen. In *toxicology, resistant* is a pest qualifier. The Colorado potato beetle is resistant to parathion.

## Mechanisms of Resistance

Several different mechanisms confer pest resistance in a crop plant. Not all mechanisms apply to all categories of pests and there are some differences in terminology used in nematology, plant pathology, and entomology as applied to resistance mechanisms.

1. Antixenosis. Pest behavior is altered through chemical or physical means to deter or reduce colonization of the host plant. It is also referred to as nonpreference by the pest. This means that a herbivore will choose not to eat the plant, or the pest will be repelled by the plant and will avoid it for oviposition.

    1.1. Physical antixenosis. The morphology of crop makes it less likely that the plant will be attacked. Examples are leaf characteristics such as hooks, hairs (Figure 17–2a and b), glands, and thicker cuticle.

    1.2. Chemical antixenosis. Many secondary plant compounds inhibit the pest's host selection behavior for feeding (feeding deterrents) or for oviposition (oviposition deterrents).

2. Germination inhibition. This mechanism is usually of a chemical nature and applies to spores of pathogens, or eggs of nematodes and arthropods.

3. Antibiosis. The plant produces a toxin, usually a secondary metabolite, which affects organisms that feed on the plant. Only those organisms that have developed the ability to overcome the toxins can become a pest on the plant. Antibiosis also may result from the activity of plant enzymes that reduce the ability of herbivores to digest the food (digestibility reducing factors).

(a)                                                                    (b)

**FIGURE 17–2**   Resistance in soybean to leafhopper attack. (a) Leafhopper on hairy variety cannot gain access to the plant; and (b) variation in resistance by different soybean varieties, shown as variation in height of the different accessions. Source: photographs by Marcos Kogan.

**Tolerance terminology.** In *arthropod management,* a plant is tolerant if it can sustain injury without yield or quality loss. In *nematode management,* a tolerant plant can support feeding and reproduction by nematodes without loss of yield or quality. In *weed science,* a plant is tolerant if it is inherently not affected by a herbicide.

4. Hypersensitive response. This is an active, inducible resistance and is a host defense response to the pest. Hypersensitivity is an incompatible reaction between the host and the pest, and is a defense response to certain pathogens and nematodes. When the plant is infected, specific pathogen proteins (elicitors) interact with host receptor proteins, eliciting an active defense response wherein the cells in contact with the pest immediately die. This essentially walls off the pathogen before it can spread into noninfected tissue. Resistance genes are thought to function as the host receptors for the eliciting proteins from the pest.

5. Tolerance. The resistant cultivar supports feeding and reproduction by the pest, and accordingly sustains some damage, but without loss in quantity or quality of product.

6. Immunity. This term expresses the highest level of resistance to a pathogen. Plant species that are nonhosts to particular pests are said to be immune. Resistant varieties are rarely immune to the respective pests.

## Physiological Bases of Resistance

Plant-derived chemicals form the bases of both antibiosis and antixenosis. These chemical compounds arise as products of metabolic processes and are designated as secondary plant compounds. Many secondary plant compounds are characteristic of specific plant families; for example, the Solanaceae are rich in alkaloids, the Apiaceae usually contain high levels of furanocoumarins, and the Cruciferae typically contain mustard oil glucosides. Depending on the processes of biosynthesis and accumulation of these compounds, resistance involving the compounds can be constitutive or induced.

***Constitutive Resistance***   If the compound accumulates in a plant as part of its normal metabolic processes regardless of the presence of pest injury, then the resistance is said to be constitutive. The mustard oil glucosides of the crucifers are constitutive compounds. The pungent flavor of cabbages and kale, or the even more pronounced flavor of mustard, results from these compounds. Most insects are repelled or are deterred from feeding on these plants; however, coevolution has allowed a few crucifer specialists to bypass this chemical defense and use the odor and flavor to locate and identify such plants as preferred hosts. This insect-plant relationship poses a dilemma in breeding for resistance, because some of these chemical factors reinforce resistance to many herbivores but increase susceptibility to a few of the specialist insects.

***Induced resistance*** When pests attack plants, or when plants are subject to environmental stresses, certain physiological responses occur in the plant. These responses are termed induced responses, and they may or may not alter pathogen infectivity or herbivore fitness. Induced responses that result in the accumulation of compounds or physical changes that reduce herbivory or pathogen virulence are called induced resistance. Elicitors stimulate the induced response. The primary plant response to an elicitor involves localized reactions such as hypersensitivity, and is followed by secondary responses in tissues immediately surrounding the area of primary response. Finally, after the localized secondary responses, systemic defense responses occur throughout the plant as elicitors trigger a physiological cascade of responses that spread within the plant. An insect chewing on a bottom leaf of a potato or soybean plant will cause the leaves several nodes above them to be less palatable to the same or other insects. Similarly, mechanical injury or infection of a leaf by a pathogen may cause other leaves on the same plant to be more resistant to the pathogen. Induced resistance responses vary among host-pest interactions, and all three types of response do not always occur.

The discovery of induced resistance has led to the suggestion that plants could be "immunized" against insects and pathogens by a controlled exposure of the plants to elicitors. For example, infesting wine grapevines with the Willamette mite very early in the season induces resistance in the vines. The vines are then resistant to later infestations by the more damaging Pacific mite. The concept is similar to the immunization of humans and domestic animals against infections with pathogen strains that have reduced virulence (hypovirulence) or are nonvirulent. Widespread practical application of the concept in plants has been limited.

In soybean, infection by the fungus *Phytophthora megasperma* f.sp. *glycinea,* the causal organism of Phytophthora rot, elicits a systemic response that results in increased concentrations of defense compounds called phytoalexins. Studies with Mexican bean beetle feeding on soybean have revealed that the same or similar phytoalexins were feeding deterrents for the beetles. A single gene named *Mi* confers resistance to root-knot nematodes in tomato by an induced, localized, hypersensitive response to nematode infection. The *Mi* gene has been widely used in processing tomatoes for root-knot nematode management. It has been observed that the *Mi* gene also confers resistance to aphid feeding. These are examples of convergence of resistance mechanisms to multiple pests.

# Genetics of Resistance

The genetic basis for host-plant resistance is explained using terms that describe the plant genes involved. Resistance that is determined by multiple genes is termed polygenic, or horizontal resistance. Resistance that is determined by a single gene or a few genes is termed monogenic or oligogenic, respectively, or is more generally termed vertical resistance. These concepts concerning the genetic determination of resistance were developed by plant pathologists and apply more readily to pathogen and nematode resistance, and they have found only modest use in arthropod resistance, although they are applicable in some instances.

1.  Horizontal resistance is usually incomplete resistance that extends across many pest races (Figure 17–3). Horizontal resistance is controlled by many genes, perhaps in excess of 100 in some cases. Polygenic resistance is usually stable and is not easily overcome by pests, but may vary with environmental conditions. It has also been termed *durable* resistance. Horizontal resistance shown in variety A (Figure 17–3) is only 20%, whereas that in variety B is 50%.
2.  Vertical resistance gives complete resistance to a race, or several races, of a pathogen, but may have no impact on another race. Vertical resistance is usually conferred by one or as many as three major genes. A new race of the pest (usually a pathogen) can completely overcome this resistance. Varieties A and B (Figure 17–3) show different combinations of vertical resistance to different races of the hypothetical pest.

Resistance genes may also be referred to by the magnitude of the phenotypic effect they exert on a pest, with major genes exerting a large effect and minor genes exerting smaller effects.

The gene-for-gene hypothesis states that for every gene for resistance in the host there is a corresponding and specific gene for virulence in the pathogen (Flor, 1942).

Flor (1942) was one of the first to assess the genetics of both host and pathogen. He conducted experiments on flax and the flax rust fungus, leading to the gene-for-gene hypothesis to explain the genetic basis for interaction of pathogens and hosts. The gene-for-gene hypothesis assumes that pathogen recognition (resistance) is conditioned by single-resistance genes (R genes) in the host, and by single avirulence (Avr) genes in the pathogen. Pathogens carrying Avr genes are unable to reproduce in the presence of the specific R gene(s). Virulence may refer to the ability of a pathogen to reproduce on a host, or to the tendency of a pest to cause damage to the host. The gene-for-gene recognition of the pest by the host results in resistance through a cascade of plant defense responses.

The gene-for-gene relationship has been demonstrated for many pathogen/host complexes, including many rusts, powdery mildew, apple scab, many bacteria, and many viruses, and is recorded for nematodes and some insects such as Hessian flies on wheat and brown planthopper on rice. The gene-for-gene relationship is particularly important in efforts to breed cultivars with vertical pest resistance. As noted previously, plants and their pests have evolved with each other. Generally speaking, changes in virulence of the pathogen seem to be continuously balanced by changes in host resistance. The end result is a dynamic equilibrium over evolutionary time between pathogen virulence and host resistance, and this equilibrium is maintained by the gene-for-gene relationship. When the pathogen evolves a new virulence gene, it acts as a selection pressure that leads to evolution of new resistance in the host specific to the new virulence gene. Plant species likely include many, perhaps thousands, of different R genes that constrain pathogens, and R genes may be involved in general aspects of plant defense responses, other than gene-for-gene responses, that limit pests.

Knowing the races of pest present in a field is essential for choosing the crop variety that has resistance against that race. Figure 17–3 offers a theoretical example. If race 6 was present, the best choice would be variety B, but the crop

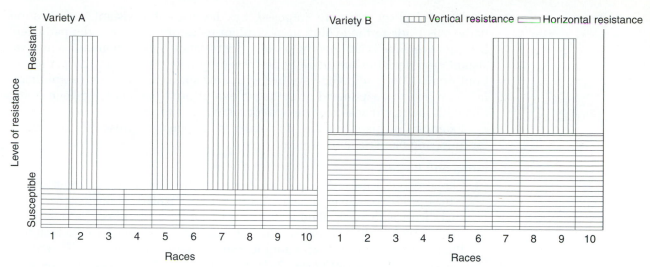

**FIGURE 17–3** Theoretical example of horizontal and vertical resistance in two crop varieties (A and B) in relation to 10 races of pest.

would still be partially damaged. If races 2 or 5 were present, then variety A would be the best choice; and if races 8 or 9 were present, then either variety could be used. In the latter situation, the variety selection would be based on agronomic qualities other than pest resistance.

For most resistance breeding based on Mendelian genetics, the sources of resistance usually come from genetic variation that exists among related species, including wild relatives of cultivated crop plants. A limitation with traditional breeding efforts is imposed by the difficulties encountered in obtaining hybrids between species. This limitation has now been reduced for some pests and crops through the use of genetic engineering.

The amount of genetic variability that exists within and among species is one aspect of biodiversity. The desirability of finding new resistance genes from the genetic variability that exists within a species has been a major reason for attempts to maintain collections of genetically diverse plants. The need for within-crop resistance genes was recognized by plant breeders, and collections of crop varieties and related plants are held at germplasm centers around the world, as mentioned. Such repositories of genetic variability are of great importance to humankind, because they provide the genetic resource for efforts in both conventional breeding and genetic engineering.

## Examples of Conventional Plant Breeding for Resistance

***Pathogens*** Resistant varieties are the mainstay of disease management for numerous annual crops. In many instances, using resistant crop varieties is the only viable way to manage the pathogen, especially soil-borne pathogens and viruses. The rust diseases of cereals are examples of pathogens that are managed by use of resistant varieties, because there are no other management tactics that adequately control these diseases. A recent example of successful breeding for resistance is provided by resistance to black sigatoka disease in bananas, where chemical control of this pathogen would be impractical and economically unrealistic.

The selection pressure imposed by the use of resistant varieties on pathogens leads to the evolution of virulent races or variants in the pathogen. Loss of resistance is an ongoing problem in resistance management, as new resistant crop varieties are literally destroyed by race changes in the pathogen. Continuous efforts to breed varieties with new sources of resistance are therefore required to keep ahead of pathogen evolution. Breeders have been doing this with the rust diseases of cereals for almost 100 years.

Examples of plant breeding to combat pathogens are presented separately for fungi and bacteria, and viruses.

### Fungi and Bacteria

1. Cereal rust diseases, as noted in Chapter 10, have been the scourges of wheat production ever since wheat was domesticated. Management of rust diseases through plant breeding is the oldest example of pathogen control by breeding-resistant crop varieties, and dates back to the turn of the twentieth century. Different races of rust pathogens attack different varieties of the crop (Figure 17–4). All cereal crop breeding programs still maintain a rust resistance component, as the pathogen typically takes about three to five years to develop a new strain, at which time the formerly resistant variety succumbs to attack.

**To use resistance against soil-borne pests, it is essential to know which species are present.**

2. *Fusarium* and *Verticillium* wilts are caused by soil-borne pathogens that essentially are uncontrollable by any tactic other than host resistance. Examples of crops that resist these pathogens include cotton and tomatoes. Tomatoes that carry the abbreviation VFNT as part of their variety name are resistant to *Verticillium, Fusarium,* root-knot Nematode, and *Tobacco mosaic virus.* The VFNT varieties can prevent losses to these pathogens in regions where these pathogens are endemic.

3. *Phytophthora* spp. cause root rot disease in many crops; and the problems are exacerbated in perennial crops such as alfalfa and fruit trees. Partial resistance to *Phytophthora* is routinely incorporated into new alfalfa varieties. Because alfalfa is a segregating genetic population, the variety is a mixture of resistant and nonresistant genotypes. By allowing for loss of the susceptible members of the population, by practices such as increased

**FIGURE 17–4**
Side-by-side comparison of a barley variety susceptible to stripe rust (left side) and a variety that is resistant to the particular race of the rust (right).
Source: photograph by Lee Jackson, with permission.

seeding rate, the manager can allow for the attrition caused by *Phytophthora* that will inevitably occur.

4. Rootstocks of some tree crops are resistant to root rot organisms and nematodes. Selection of the appropriate rootstock for a specific site is the foundation for long-term management of soil borne pest problems in orchard crops.

**Viruses**   The primary long-term control tactic for virus diseases is host-plant resistance. Control of the vector organisms does not necessarily provide adequate disease control, or is too costly in terms of economics and ecosystem disruption. In the absence of host-plant resistance, vector exclusion is the only other potentially effective solution.

Breeding for resistance to viruses can reduce application of insecticides used for vector control, and so can help to stabilize insect IPM programs that involve natural enemies because insecticide use can result in pest replacement (see Chapter 12). The following are examples of plant breeding for virus management.

1. Rhizomania, caused by the *beet necrotic yellow vein virus*, is a disease that attacks sugar beets, and which rapidly spread through all regions in the world where sugar beets were grown in the 1980s. The virus is vectored by a soil-borne pathogen. Development of varieties that have resistance to both the vector and the virus saved sugar beet production from severe yield losses.

2. Soybean mosaic or soybean crinkle, caused by the *soybean mosaic virus,* can limit soybean production in some parts of the world. The virus is seed transmitted, and can also be transmitted nonpersistently by some 30 species of aphids. In most of the world's soybean production areas, the common varieties have genes for resistance derived from two lines that originated in Asia, with resistance conditioned by a pair of alleles, one of them being dominant and conferring high levels of resistance.

**Nematodes**   Resistant varieties are one of the best ways to manage nematodes. Plants are considered resistant to nematodes when they limit nematode development and reproduction. Resistance can range from low to moderate, where partial or intermediate reproduction occurs, to high resistance where nematode development and reproduction is almost eliminated.

Many of the nematode-resistance genes that are known appear to have gene-for-gene relationships with particular nematode species or a small group of species. The use of crop resistance in nematode management is limited by the availability of suitable resistance genes. When resistance to particular nematode species is available, it is highly effective; however, resistance is a selection pressure on nematode populations, and the selection of resistance-breaking pathotypes and races is a problem, particularly in species such as the cyst nematodes that reproduce sexually. The parthenogenetic root-knot nematodes can also break resistance; for example, some root-knot nematode pathotypes are capable of circumventing the *Mi* gene in tomato.

Traditional breeding efforts have resulted in the creation of crop cultivars that are resistant to nematodes. Resistance is available in many crop cultivars, and examples of major crops in which nematode resistance is available include soybean, alfalfa, clover, wheat, barley, oats, potato, bean, cowpea, sweet potato, tomato, apricot, citrus, grape, walnut, and *Prunus* rootstocks.

Several nematode-resistance genes have been cloned, including a gene from a wild relative of beet that confers resistance to sugar beet cyst nematode, a tomato gene (*Mi*) that is effective against several important root-knot nematode species, and a gene that confers resistance against some isolates of potato-cyst nematode. Many resistance genes have some commonality of structure, although the timing, location, and mechanisms of defense responses conditioned by the genes seem to vary. The cloning of resistance genes allows for detailed analysis of gene and protein structures, investigation of the biochemical mechanisms of resistance, and potentially the engineering of new resistant varieties.

Most fruit and nut trees are grafted onto a rootstock. Because trees are perennial crops, nematode management tactics such as rotation and fumigation cannot be used regularly. Accordingly, nematode-resistant rootstocks, such as 'Nemaguard' rootstocks for *Prunus,* are useful in nematode management programs.

***Slugs and Snails***   Resistant crop varieties are normally not employed to manage mollusk pests, probably due to the lack of within-species variation in crop tolerance to mollusk feeding. The lack of mollusk-resistant varieties may also reflect the general severity of the problem, because pest mollusks are highly polyphagous. Their abundance in crops is the result of favorable ecological conditions, such as abundant straw left over after harvest of grass seed production fields. Mollusks seldom cause economic losses in most food and fiber crops.

***Insects***   Breeding for resistance is considered to be one of the best means to manage arthropod pests, and arthropod-resistant crop varieties have been used for over 150 years. In many cases, such as the following, the tactic literally saved the crop from total failure.

The use of broad-spectrum insecticides to manage insect pests can kill beneficials and release a nontarget insects from natural control agents so that they become pests (see upsurges of secondary pests and pest replacement in Chapter 12). In many instances the use of a resistant variety can alleviate the need for the insecticide, which then allows beneficials to build up and control other less important pests. This is a major reason why the use of host-plant resistance can have a stabilizing influence on IPM programs. Even low levels of resistance in a variety are useful because of interactions with biological control that increase the efficacy of both control tactics.

Following are examples of host-plant resistance for arthropod management.

1. Grape phylloxera. An aphidlike insect that attacks the roots of susceptible grapes, the grape phylloxera causes severe stunting and even death of the vines. Wine and table grape varieties are all susceptible, but native North American grapes are almost symptomless hosts. Phylloxera was transported in the mid-1860s from the United States to France, where it almost eliminated the wine grape industry. Researchers solved the problem by grafting the French wine grape varieties to North American grape rootstocks. Since that time, many resistant rootstocks have been developed using the resistance in the North American grapes and others. This was one of the earliest examples of using resistance to manage a pest. It seems that all mechanisms of resistance (antibiosis, antixenosis, and tolerance) are involved. In the late 1980s, a new biotype of the phylloxera appeared in California which overcame the resistance in the current

rootstock, causing destruction of many vineyards. This necessitated the development of rootstocks that were resistant to the new biotype, as phylloxera cannot be managed with insecticides.

2. European corn borer. The European corn borer is one of the most destructive corn pests in the midwestern United States. Eggs are laid on leaves, but the emerging larvae rapidly bore into the stems and out of reach of common contact insecticides. The pest typically has two generations per year, called first and second broods. In corn, resistance to the first brood is well understood and is mostly due to the presence of the compound DIMBOA. This compound in corn provides an example of using plant breeding to manipulate a secondary metabolite level in a crop as a means to control arthropods. Development of resistant varieties was the only practical long-term solution for control of the European corn borer, and now about 90% of the corn varieties grown in the region have the resistance.

3. Resistance in rice to insect pests. Rice, one of the most important food crops, is grown worldwide on over 50 million acres. About 1,000 species of arthropods have been recorded in association with rice, and about 30 can cause damage. One serious pest of rice in Asia is the brown plant hopper. The first brown plant hopper resistant variety was developed in 1973 by the International Rice Research Institute in the Philippines. Since then, resistant rice varieties have been planted in about 38 million acres. At least three biotypes of the brown plant hopper have been identified, requiring a continued effort to develop new varieties that have resistance to all biotypes.

4. Spotted alfalfa aphid. The spotted alfalfa aphid causes growth reduction in alfalfa by feeding on photosynthates, but the more serious damage is caused by injection of a toxin into the alfalfa. The toxin stops stem growth. The aphid was introduced into the United States in 1954, and caused devastation in all warmer regions. The alfalfa variety 'Lahonton' was found to contain a multigenic resistance that was about 90% resistant, and this resistance saved the crop. This discovery was serendipitous, as the resistance was noted in the field and did not require extensive screening of the alfalfa germplasm. The aphid cannot be managed with insecticides because of interfield movement of the aphids, short harvest intervals leading to insecticide residue problems in a feed crop, and the cost of multiple treatments in a relatively low-value crop. Resistance genes are incorporated into any variety that is to be used in regions where the spotted alfalfa aphid is established. The resistance is a striking example of a long-term durable resistance; there is still no change in aphid's ability to overcome resistance after nearly 50 years. The aphid remains present in alfalfa growing regions, and nonresistant varieties are attacked.

5. Resistance to the wheat curl mite. Resistance to the wheat curl mite was first observed in a hybrid x *Agropyron*. The resistant gene from *Agropyron* was identified as a dominant trait, but, other resistance genes have since been identified. In addition to direct damage by the mite, the mite transmits the *wheat streak mosaic virus*. Planting the mite-resistant cultivars reduces the costs of miticide treatments and significantly reduces virus incidence.

***Vertebrates***    The use of resistance to vertebrates is theoretically possible, but is unlikely to be used, particularly for mammalian pests, because the crop plant is produced to feed other mammals. Making the crop resistant to attack by a pest mammal is likely to make it unusable as a food source for humans or domesticated animals. There are some bird-resistant sorghum varieties, with the resistance due to higher levels of tannins in the seed; however, the tannins can make the variety less palatable to humans as well.

## Limitations

The major factor limiting conventional plant breeding to produce resistant varieties is that the process is time consuming. In addition, it requires that resistance genes be in a species that is the same as or closely related to the crop. When no genes are known that confer resistance to a particular pest, conventional plant breeding cannot be used to develop resistant cultivars. Due to limitations imposed by the availability of resistance genes in the same species, it is necessary to establish as wide a genetic base as possible for each crop species. Most germplasm collections include native or wild crop relatives and landraces that provide a potential reservoir of diverse resistance genes for many important crops and their pests. There is a great need to preserve genetic diversity for breeding purposes. Most modern crops are based on very limited genetic backgrounds, and are thus susceptible to evolving pests. Genetic diversity is lost as local varieties preserved within small family farms (landraces) are replaced by higher yielding, but often more pest susceptible, commercial varieties. The so-called green revolution was a major factor in the loss of genetic diversity in crops such as rice and wheat.

Breeding new varieties of perennial fruit and nut tree crops takes many years, requiring long times for the application of plant breeding to these crops. Except for the development of resistant rootstocks, there has been no serious attempt to breed resistance into tree crops, primarily because of the time required to develop new varieties of these perennial crops, and because of the ability of pests to overcome the resistance in a few years.

The selection of pest biotypes that circumvent resistance in released varieties is a major limitation to the use of plant resistance in pest management. The solution to this problem has been to maintain continuous monitoring and breeding programs so that new resistant varieties can be produced as quickly as new biotypes appear. Wheat breeders have been successful in keeping up with the new biotypes of Hessian fly, and 16 biotypes have been identified to date. The evolution of biotypes that overcome resistance in extensively grown cultivars denies the two advantages as listed earlier in the chapter, and actually exacerbate selection of pests that can overcome the host-plant resistance. For this reason, the tactic of host-plant resistance should be used as part of an integrated approach to pest management. Reliance on a single pest management tactic usually leads to development of resistance to that tactic in the target pest; host-plant resistance is not immune to this problem.

The lack of appropriate resistance genes to certain pests has meant that those pests could not, in the past, be managed with host-plant resistance. Genetic engineering may reduce the importance of this limitation.

***Weeds***    Conventional plant breeding is much less relevant to weed management than it is with the other pest categories, except in the area of herbicide resistance, due to the lack of any direct trophic connection between weeds and crop plants. However, weeds do compete with the crop for light and nutrients, and it is theoretically possible to breed more competitive crops, which is being attempted. It may, for example, be possible to increase the speed at which the crop canopy develops to provide greater competition against weeds. The inherent problem is that plant canopy architecture is not determined by single genes, and canopy architecture varies in relation to the crop species. It will be difficult to achieve more than a small increment of change in the competitive ability of crops such as sugar beets, processing tomatoes, lettuces, and most vegetables, because these crops are typically shorter than the weeds with which they must compete.

Variation in tolerance to herbicides exists within crop species. This variation has permitted selection of herbicide-resistant crop varieties through the use of conventional plant breeding, as follows:

| Herbicide group | Crop | Comments |
| --- | --- | --- |
| Triazines | Canola | Commercialized in 1980s |
| Imidazolinones | Maize, wheat | Trade name for crops is Imi® |
| Sulfonylureas | Soybean, cotton | Under development |
| Cyclohexanediones | Maize | Under development |

The limitation to development of herbicide-resistant crops using this method is the lack of adequate genes within a species that confer selective resistance to herbicides.

# GENETIC ENGINEERING

## Principles

Research in cell and molecular biology reached the point in the early 1980s where it became possible to identify genes, characterize their function, and copy them (called cloning). The techniques for identifying and mapping genes in organisms have progressed rapidly since that time. Projects have been established with the objective of sequencing the entire genome of particular organisms, including humans, several crops, fruit flies, *Arabidopsis thaliana* (a small plant), and a nematode. Once science developed the ability to identify genes and associate them with functions, it rapidly became apparent that one application of this new technology would be in the area of pest management.

Genetic engineering is a branch of biotechnology. Genetic engineering is defined as techniques that modify genetic structure of the crop or pest genome, which is accomplished either by gene deletion or blocking, or by transferring genes from one organism to another. Crops that have been modified by one of these gene-modifying techniques are referred to as genetically modified organisms (GMOs), or GMO crops. Another similar term that has also been used is genetically engineered organisms (GEOs), or GEO crops. This terminology is used to differentiate genetically engineered crop varieties from varieties that have been developed using the conventional Mendelian plant-breeding processes. The term *transgenic crop* is applied when the genetic modification involves transfer of gene(s) from one organism to another, and will be used here. The genes that have been moved using genetic engineering are often referred to as transgenes to differentiate them from genes introgressed into crop cultivars using conventional plant breeding techniques.

There is a fundamental difference between genetic engineering and conventional plant breeding. The technology of genetic engineering allows that the trait, and its associated gene(s), do not necessarily have to come from the same species as the target crop, because the genetic engineering technology permits gene transfer between different species. It even permits gene transfer between completely dissimilar organisms, such as from bacteria to plants, and even from animals to plants. Overcoming the same-species limitation of conventional plant breeding opened up enormous potential for pest management, but also opened up serious ecological concerns.

**Identify gene(s) conferring pest resistance, and then transfer these genes to crop plants using genetic engineering techniques.**

The concept underlying all current applications of genetic engineering to pest control is to identify, isolate, and clone the gene (or genes) conferring a particular characteristic or trait, such as the ability to resist a pest or a pesticide. Once identified and cloned, the genes conferring the trait are transferred to nonrelated desirable crop plants using transformation technology. Alternatively, genetic engineering can be used to create new types of resistance that will function in particular plant tissues against specific pests.

There are a few caveats regarding cloning and transformation. The task of identifying and characterizing the genes is not easy, and has only been possible since the early 1980s. Acceptance of the technology has been rapid at the agricultural industry level, but the consuming public in developed countries has been reluctant to accept products resulting from genetic engineering. This reluctance may limit application of the technology on a commercial scale (more on this later in the chapter and in Chapter 19).

The following is an abbreviated description of the three techniques used to transfer genes between organisms using genetic engineering technology. For a more complete introduction to molecular biology applied to pest management, see Marshall and Walters (1994) and Persley and Sidow (1996).

1. Natural vector system. The pathogenic bacterium *Agrobacterium tumefaciens* causes infected plants to produce abnormal growth in the form of tumors. The bacterium accomplishes this by inserting special DNA (genes) into the host-plant cell which causes the cell to start proliferating. The special genes reside on small circular DNAs called plasmids. The bacterium has the ability to insert plasmid DNA into the host cell where it becomes incorporated into the host DNA. Through the use of molecular biology, the tumor-inducing genes can be deleted from the plasmids and cloned resistance genes can be inserted in their place. The bacterium then inserts the plasmids carrying the resistance genes into the target host cells where they become incorporated into the host genome and are expressed by the host plant. The technique is limited to plants that can be attacked by the pathogen, mainly dicotyledon species. Other techniques are required to transform grasses (all the major cereal crops).

2. Direct plasmid fusion into protoplasts. Protoplasts are naked plant cells from which the cell walls have been enzymatically removed. This technique also utilizes small plasmids, containing the desired genes, but in this case they are developed in *Escherichia coli*. The protoplasts are grown in a suspension culture that contains the plasmids developed in *E. coli*. The plasmid DNA can be taken up by the cells and integrated into the plant genome. This technique has been widely employed to transform cereal crops.

3. Particle bombardment. This technique involves shooting microscopic DNA-coated metal pellets at the target plant cells. It is also referred to as

microinjection or biolistic injection. The tissue used as a target is typically from callus culture, but imbibed seeds and embryos are also used.

The first two techniques utilize single-cell cultures during the transformation process, and bombardment uses callus or embryo culture. Because the transfer of DNA to a specific cell is uncertain, techniques were developed that allow identification of the cells or embryos that have been successfully transformed. Techniques are also available to regenerate whole plants from the individual transformed cells in the culture. Following the successful regeneration of the transgenic plants it is still necessary to go through conventional plant breeding to ensure that other desirable agronomic traits have been maintained.

Two or more transgenic traits can be combined into a single variety. This is referred to as stacking or pyramiding, and is being developed in several ways.

1. More than one resistance mechanism can be stacked against a single pathogen.
2. Resistances to more than one pest can be stacked (e.g., resistance to two different viruses, or for a virus and an insect).
3. Resistances to more than one herbicide can be stacked (e.g., glyphosate and glufosinate).
4. Resistance to a herbicide and resistance to a pest can be stacked (e.g., glyphosate and corn root worm).

The trend for stacking will probably accelerate and many transgenic crop varieties will ultimately have more than one inserted trait.

Transgenic crops are being developed for reasons other than pest management, including for agronomic and quality reasons. Examples include tomatoes engineered to ripen more slowly (first GMO approval in 1992), and alteration of the oil profiles of canola (oilseed rape) and soybeans. Such uses are not discussed here.

## Examples of Genetic Engineering for Pest Management

Advances in the area of molecular biology are occurring rapidly and many applications have relevance for integrated pest management. Some developments are still experimental and others are now in commercial use.

**Pathogens**   Several different approaches are being investigated for application of genetic engineering to pathogen management.

**Pathogen-derived Resistance**   Due to the difficulty of controlling virus diseases with other IPM tactics, the development of transgenic varieties to control viruses has been at the forefront of this technology. The basic approach used here is to inoculate the variety with a small part of the pathogen genome that does not cause disease, but which does elicit resistance to the pathogen (hence the term pathogen derived). Genetic engineering techniques were used to transfer genes that control production of the coat proteins from the virus to the crop. When inserted into crops, these genes provide protection from the virus from which the genes were derived. The mechanism by which the protection is achieved is not well understood. Examples of approved virus-resistant varieties include the following:

1. *Cucumber mosaic virus, watermelon mosaic virus 2,* and *zucchini yellow mosaic virus* resistance has been available in various squash varieties

since 1994. All references to approval dates for transgenic crop varieties are obtained from information posted at the USDA/APHIS website (Anonymous, 2000). These viruses are difficult to manage in many squash crops, and transgenic varieties offer much improved control.

2. *Papaya ringspot virus* resistance in papaya is one of the few examples of transgenic varieties developed by public institutions (Cornell University and the University of Hawaii), and has been of great value to the papaya industry.

3. *Potato leaf roll virus* resistance has been available since 1998 and only in combination with insect resistance (stacked with Bt endotoxin gene; see later in the chapter).

Plant pathologists believe that combining resistance genes may delay the selection of resistance-breaking pathogen strains because two mutations must occur in the pathogen to overcome two resistance genes. Since several different viral genes have been effective as sources of transgene resistance, the stacking of two or more viral-derived transgenes is a strategy that may be developed in the future. Transgenes may also be combined in a variety that has been bred conventionally for resistance. The combination of a transgene with a resistance gene may allow the use of the latter gene which, when used singly, did not provide sufficient control.

**Antifungal and Antibacterial (Host-Derived) Strategies**   The underlying strategy behind this concept is to identify the resistance genes and then transfer them to nonresistant crops using molecular genetics. Rather recently a few resistance genes have been identified and cloned. From a purely pest management point of view, this approach, if it can be achieved, appears to be a very good way to manage pathogens. There are, however, serious ecological implications that will need to be addressed (see later in chapter).

**Engineer Pathogen that has Reduced Virulence**   Attempts are being made to genetically engineer pathogens with decreased virulence. These nonpathogenic strains would then be used in a competitive exclusion tactic. This approach is currently experimental. It is being evaluated for fire blight control in pears, but there is no approval for this type of transgenic organism.

***Nematodes***   Efforts are underway to engineer crop varieties that are resistant to nematodes. For example, research is underway to move the *Mi* gene, that confers resistance to root-knot nematode in tomatoes, into other crop plants where no resistance to nematodes is available. Genetically engineered resistances are also being researched such as antifeeding genes that will destroy nematode feeding sites, or antinematode genes that will produce substances that are inhibitory or toxic to nematodes. Molecular techniques are being applied in concerted efforts to develop new, durable nematode resistances. Although there are no commercially available transgenic nematode-resistant varieties, it would seem that major breakthroughs are imminent.

***Arthropods***   Control of certain arthropod pests through the transfer of resistance genes is another area in which there has been considerable development since the mid-1980s. Two genetic engineering approaches are being employed to enhance arthropod pest control.

**Genetically engineer crops so that they produce toxins that kill arthropods attacking them.**

**Arthropod Toxicants Produced by Crop**   The concept involved is to identify genes that confer resistance to arthropods, and to transfer such genes to crops.

Several different plant-derived toxins are being evaluated, including inhibitors of proteolytic enzymes such as proteases and lectins (e.g., cowpea protease inhibitors, CpTI, are the most widely tested to date); secondary plant metabolites such as DIMBOA, nicotine, and rotenone; and production of chitinases. By summer, 2001, no varieties using plant-derived toxins had been approved.

The major activity has been in the area of toxins derived from the bacterium *Bacillus thuringiensis*. The genes encoding the production of the endotoxins have been isolated, cloned, and transferred to crops. The gene produces the toxin in the crop cells so that the plant is essentially producing its own insecticide.

The advantages of transgenic crop varieties to pest management are those listed at the start of this chapter. An additional advantage is the potential for reduction in the use of synthetic insecticides and improved control of the target pest over that achieved by other means, as graphically shown in relation to defoliation of potatoes by the Colorado potato beetle (Figure 17–5). In the absence of any control, the potatoes were totally defoliated by late July. A foliar insecticide protected the plants from defoliation early in the season, but later in the season the defoliation increased to a high level. This loss of control is attributed to breakdown of the insecticide in the environment, leading to loss of activity. The transgenic variety did not experience defoliation throughout the season because the toxin is continuously expressed. Until resistance develops in the pest population (see later), this type of effective control is observed in most Bt transformed crops.

A second major advantage of transgenic insect-resistant crops is manifested if control of the target pest leads to reduction in use of broad-spectrum insecticides. Such reduction has been demonstrated with a Colorado potato

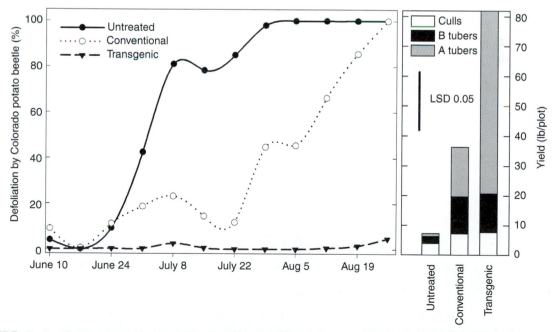

**FIGURE 17–5** Graph comparing defoliation of potato by Colorado potato beetle on Russet Burbank potatoes and impact on yield. Potatoes were nontransgenic that were untreated, treated with conventional insecticides, or were transgenic (gene expressing Bt CryIA endotoxin inserted).

Source: data from Casey Hoy, Ohio Agricultural Research and Development Center, Wooster, Ohio, with permission.

beetle resistant, Bt endotoxin transgenic potato variety (Figure 17–6). The use of a foliar insecticide to control beetle larvae resulted in an "explosion" of aphids which was attributed to reduced predator activity, because predators were killed by the insecticide. When a systemic insecticide was used, which had less effect on predators, the aphid populations did not reach high levels.

**FIGURE 17–6**

Graph comparing aphid and aphid-predator populations on nontransgenic potatoes treated with topical foliar applied insecticides for control of Colorado potato beetle or with systemic insecticide, versus Bt endotoxin transformed potatoes.

Source: data from Gary Reed, Hermiston, Oregon, with permission.

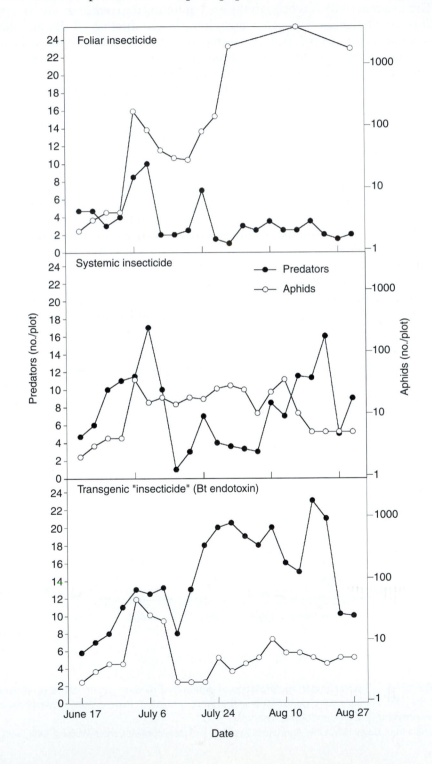

On the transgenic potato variety, the predator densities were higher than those on systemic insecticide treated plants, and the aphid population was kept at very low levels. The use of the transgenic variety demonstrates the importance of pest upsurges or replacement, as discussed in Chapter 12, that can occur when an insecticide suppresses natural enemies.

Several strains of *B. thuringiensis* produce slightly different toxins, which are specific to different target insects. The toxins are proteins that form crystals; CryIIIA, for example, is the toxin that kills many Lepidopterans (moths and butterflies) whereas CryIA kills beetle larvae and is used for control of Colorado potato beetles. The genes encoding production of the toxins are now available in the United States in approved, transgenic varieties of the following:

1. Cotton—for control of bollworms and other Lepidoptera
2. Corn—for control of European corn borer and other Lepidoptera
3. Potatoes—for control of Colorado potato beetle (Figure 17–7). Control of the tuber moth is expected in future.
4. Other crops—no approved transgenic varieties yet

Researchers are currently contemplating stacking of herbicide and insect resistance, and pathogen and insect resistance.

Resistance management will be a major problem for the long-term sustainability of transgenic crops for insect management (see later discussion). The current recommendation is that growers leave refugia of nontransgenic crops and plants. The implementation of this tactic is not completely resolved yet and guidelines change each season. The use of refugia requires that a certain percentage of the crop be planted to a nontransgenic variety and that areas of native vegetation that serve as alternate hosts not be treated. There are no experimental precedents on which to base the strategy, and, thus, considerable skepticism exists within the pest-biology research community regarding the success of the refugia strategy. The concept is based on the assumption that a residual susceptible population will be preserved and will interbreed with the population that is under the heavy selection pressure of the Bt crops. Interbreeding will avoid, or at least delay, the onset of resistance.

**Refugia maintain susceptible individuals in the pest population.**

**FIGURE 17–7**

Comparison of Colorado potato beetle damage to potatoes; defoliated conventional non-Bt potatoes in foreground and healthy transgenic Bt potatoes in background.

Source: photograph by Marcos Kogan, International Plant Protection Center at Oregon State University.

There is a regulatory dilemma surrounding the use of transgenic crops for insect control. As the transgenic crop plants are essentially producing an insecticide that they do not naturally produce, some argue that the transgenic variety should be registered as a pesticide. Under strict interpretation of FIFRA this would appear to be correct. Others argue that production of the Bt endotoxin is toxicologically no different than many other compounds that plants produce naturally that have insecticidal properties, and which are not subject to pesticide regulations.

For more information on the application of genetic engineering to arthropod management, we suggest the book edited by Carozzi and Koziel (1997).

**Genetically Modified Beneficial Insects**   The second approach has similarity with increasing crop resistance to herbicides. If beneficial arthropods can be made to resist pesticides, then the use of pesticides would not decimate that beneficial population. The concept is strictly at the research stage, with no commercial approvals for insecticide-resistant beneficials. Attempts to make predatory mites resistant to insecticides are currently being undertaken.

*Weeds*   Management of weeds by using crops that have been genetically engineered to tolerate a herbicide is one area that has seen rapid commercial development.

> **Genes conferring resistance to herbicide(s) have been moved into crops that do not normally have such resistance.**

The term *host-plant resistance* has taken on new meaning in relation to weed management with the advent of genetically engineered crops. In the past, resistance in crops referred strictly to the ability of the crop to withstand attack by a herbivore or a pathogen. Through the use of genetic engineering it now also means the ability of the crop to resist the effects of a herbicide. The concept is different from that used for changing host-plant resistance to pests. Genes that confer the ability to tolerate a herbicide are identified, regardless of source, and transferred to crop plants. In reality, the process changes the crop selectivity of the herbicide from nonselective to selective. For this reason, transgenic crops to which herbicide resistance has been transferred are referred to as herbicide-resistant crops (HRCs). Three different approaches are used.

1. Modification of the action site so that the herbicide cannot exert its effect. This is usually effected through a change in the target binding site.
2. Enhancement of the rate of herbicide degradation so that it is broken down before it has any major effect.
3. Overproduction of enzymes that are affected by the herbicide, so that the crop plant can tolerate the reduction caused by the herbicide.

Resistance to three different herbicides is being developed:

1. Glyphosate. This broad-spectrum translocated herbicide kills most annual and perennial plants. Nontransformed crops have little or no inherent natural tolerance to the herbicide. Glyphosate is also relatively environmentally benign, as it has low mammalian toxicity and is rapidly broken down in the soil. Genes conferring both type 2 and 3 selectivity (above) have been identified in microorganisms and transferred to crops. Soybean, in 1994, was the first glyphosate-tolerant transgenic crop to be approved for commercial use. Additional crops approved for commercial use by 1999 include cotton, corn, canola, and sugar beets. The use of glyphosate in these crops does not alter overall herbicide use because they are typically grown using other herbicides, but glyphosate-based weed control is more effective (Figure 17–8), cheaper, and is applied after

**FIGURE 17–8**
Winter annual weed control using glyphosate (foreground) in transgenic glyphosate-tolerant sugar beet variety compared with untreated (rear).
Source: photograph by Robert Norris.

weeds have emerged. The latter is a preferred IPM tactic over prophylactic treatments made in anticipation of a problem.

2. Glufosinate. This broad-spectrum contact herbicide kills most annual plants in the seedling growth stages. Nontransformed crops have little tolerance to this herbicide. Genes conferring resistance to this herbicide have been transferred to approved varieties of cotton, soybeans, sugar beets, corn, and canola.

3. Bromoxynil. This herbicide kills many dicotyledon plants, but has little activity on grasses and has been used widely for weed control in cereal crops. Bromoxynil tolerance in most nontransformed dicotyledon crops is low. Transgenic cotton varieties with resistance to bromoxynil were approved in 1994.

A transgenic canola variety with enhanced tolerance for sulfonylurea herbicide residues has been developed and was approved in 1999. The transgenic variety can be planted as a rotational crop where herbicide residues persist in the soil following use in a preceding cereal crop. This can permit changes in the herbicide used in the preceding cereal crops.

As mentioned under conventional plant breeding, the possibility that crops can be made more competitive with weeds is being investigated. It is conceivable that genetic engineering could be used for this purpose. Any potential in this area is in the future.

## Adoption of Transgenic Crops

Between 1996, when the first commercial planting of transgenic crops occurred, and 1999 there was a rapid increase in commercial adoption. The majority of this use was in the area of pest management. In 1998, 50 million acres of crops were planted to herbicide-resistant transgenic varieties, and 19 million acres were planted to insect-resistant varieties. Of these, the United States had 74% of all transgenic crops that were planted worldwide, Argentina had 13%, and Canada had 10%. The remaining 3% were scattered in various countries. Table 17–1 provides a breakdown of the major uses of transgenic crops in the United States in 1998.

| TABLE 17–1 | Adoption of Transgenic Crops in the United States in 1998 | | |
|---|---|---|---|
| Crop | Acres planted (millions) | % transgenic | Main transgenic trait(s) |
| Soybean | 72 | 32 | Herbicide resistance (glyphosate and glufosinate) |
| Corn | 80 | 25 | Insect resistance (European corn borer) |
| Cotton | 13 | 45 | Insect (bollworm) and herbicide resistance |
| Potato | 1.4 | 3.5 | Virus (leaf roll) and insect (Colorado potato beetle) resistance |

It is unclear how the future adoption of transgenic crops for pest management will proceed. Public unwillingness to accept GMO products in many developed countries has slowed adoption of the technology. This is discussed again in Chapter 19.

## Limitations

On the surface, genetic engineering technology seems to offer extremely attractive means to increase options for managing pests. There are, however, serious questions that need to be addressed before the technology is widely adopted.

***Ecological/Environmental Implications*** Genetically engineered or transgenic crops cannot necessarily be completely removed from the environment if a problem arises. Once transgenes are released in the environment it may be impossible to contain them. The concerns and uncertainties regarding release of genes into the environment in ways that do not happen naturally will be considered further under societal implications.

***Control of Other Pests*** The use of transgenic crops, and the traditional plant breeding approaches for pest management, necessarily targets a single pest or group of related pests. Pests other than those targeted can still attack the transgenic crop and require the application of other control tactics. The apparent advantage of reduced pesticide use may be diminished, as the noncontrolled species may require additional action.

***Resistance in Pest Populations*** Pest resistance to transgenes is potentially the single most important factor that may limit the long-term use of transgenic crops. There are several cases of insects resistant to Bt based on spray application technology, yet genes encoding production of the toxin are being introduced into many crops, substantially increasing the selection pressure. Resistance to glyphosate has been found in ryegrass and horseweed, suggesting that the increased use of the herbicide as a result of glyphosate-resistant crops will result in more rapid development of herbicide resistance in weeds. The resistance question has not been adequately addressed, and it now appears that it is being answered with an enormous, unplanned field "experiment" from which it will be difficult to recover if widespread selection of resistance-breaking pests occurs.

**Herbicide Drift**   Drift of herbicides from the target crop onto a nontarget crop is a serious problem (see earlier discussion in Chapter 11). Herbicide drift from transgenic herbicide-resistant crops onto nonherbicide-resistant crops may be an even more serious problem because large areas of herbicide-resistant crops will be treated with the same herbicide, increasing the possibility of drift.

**Super Weeds**   There is concern that resistance genes will transfer from the crop to weeds. Such transfer will only occur between interbreeding plants, and in most cases can only occur between the crop and weeds in the same species. There are situations in which such crop to weed compatibility exists, for instance in some *Brassica* crops such as canola and wild mustard, and between wheat and some species of *Aegilops*. Introgression of transgenes to such weeds would only be a problem if they increased the fitness of the resulting offspring. Transfer of insect or pathogen resistance to weeds would be much more likely than transfer of herbicide resistance to create superweeds, as this could provide substantial increase in fitness through reduction in herbivory or diseases that currently help to regulate weed populations.

**Volunteer Transgenic Crops**   Related to the idea of super weeds is the problem caused by transgenic crop seed that persists from one season to the next. For those that are HRCs, it will be essential that an alternative weed management program be available that does not rely on the same HRC technology. Volunteer crops are a serious weed problem in some agricultural systems; the use of transgenic crops will exacerbate this problem.

**Allergic Reaction to Foreign Proteins**   The implementation of transgenic crops for pest management requires the insertion of novel proteins (mostly in the form of enzymes) into plants. There is concern that such proteins could cause allergic reactions in humans (or animals) eating the transgenic crop. Feeding, nutritional, and allergic reaction tests are conducted on all transgenic crops prior to release. It is thus unlikely that a transgenic crop could be released that produces a new allergic reaction. In at least one case, an allergic reaction was noted in testing which resulted in halting development of that particular transgenic variety.

**Fears about Nutritional Quality**   Lack of public acceptance of transgenic crops in developed countries is a serious impediment to the implementation of this tactic for pest management. Less-developed countries, however, could probably benefit from the technology, because risks from transgenic crops are probably less than the risks posed by some pesticides, or by malnutrition. The potential benefit may be enormous, in the form of alleviating hunger and malnutrition, and improvement of the general standard of living.

**Lack of Perceived Benefit**   The general public in the industrialized countries does not seem to appreciate the potential benefits from the current use of transgenic crops for pest management. When the only people perceived to benefit from the technology are the agrochemical and farming industries, there is no reason for the general public to embrace the technology.

**Intellectual Property Rights**   Who owns genes? Does a company have the right to patent a gene because they identified and cloned it? Do genes belong to the general human population? How does society pay for the costs involved in identifying and cloning genes? These serious issues have no clear-cut answers.

Due to the real and perceived problems noted, most concerned scientists would agree that the deployment of transgenic crops for pest management purposes, and probably other purposes too, should proceed with extreme caution.

# APPLICATIONS OF PEST GENETICS IN IPM

A special form of genetic pest management is based on the release of live male insects that have been sterilized by radiation or other means. This pest management tactic is known as the sterile insect release (SIR) technique. The concept is to take advantage of insects' mating behavior to lead them to self-destruction, what was called by Edward Knipling, a pioneer of the technique, the "autocidal method of insect control." The most common procedure in SIR is:

a. Mass rear the target insects in large production laboratories.
b. Sterilize the insects, usually by submitting them to a source of radiation.
c. Release the sterile insects in inundative quantities to mate with wild females and thus stop reproduction.

An essential requirement for the success of the technique is that the sterile males must be competent to compete with the wild males for mating with the wild females. A female mating with a sterile male produces eggs that are not viable. The SIR insect management strategy has no equivalent in other disciplines, although the concept has been proposed for rodent control but has not been implemented. Application of the SIR technique has many limitations and according to Knipling should be considered under four possible situations.

1. For the suppression of well-established pest populations when they naturally exist at low levels and are restricted in distribution during some period in their life cycle
2. For eliminating incipient populations in new areas
3. For preventing establishment of damaging populations in new areas
4. To eradicate well-established pest populations that can first be reduced by other methods

A requisite of the SIR technique is its application over a large area, and to be successful, the SIR tactic must be employed using an area-wide approach.

The first and still the most spectacular demonstration of the SIR approach is the eradication of the screwworm flies from cattle in the southeastern United States. The SIR technique has been used as part of the program to eliminate the Mediterranean fruit fly from California invasion locations. Flies seem to invade California from northern Mexico, so an area-wide eradication program was established to contain the northward spread of the pest. Another large program of mass releases of sterile insect is aimed at the containment of the spread of the pink bollworm, a devastating cotton pest. The SIR technique has been used in the huge San Joaquin Valley of California since 1970. Pink bollworm suppression was achieved and cotton was protected by an integrated approach. Since 1994, the program was extended to the Imperial Valley in southern California, where the pest was well established. With the advent of insect-resistant transgenic cotton, the combination of resistant cotton and the SIR technique opened new opportunities to achieve eradication in areas where the pest had been established for a long time. The use of SIR in combination with other tactics has been applied in the control of the codling moth on apples and pears in northern Washington State, and the Okanagan Valley of British Columbia, Canada. In a large area-wide program, mating disruption by sex pheromones and SIR are combined to achieve reduction of the pest to subeconomic levels. Serious questions remain about the effectiveness of the technique within the context of IPM and the ecological impact of large eradication programs.

## SUMMARY

Host-plant resistance achieved through conventional plant breeding is probably the best IPM solution to pest problems for which the approach is suitable. As host-plant resistance removes the crop as food source for the target pest, the tactic provides a permanent solution to the pest problem with essentially no economic or ecological disadvantages (see Figure 17–1). The major downfall of the host-plant resistance tactic is the ability of pests to develop new strains that are capable of overcoming the resistance. A major limitation to the universal application of the host-plant-resistance tactic has been lack of suitable resistance genes within the germplasm of the crop species. Genetic engineering permits incorporation of resistance genes from unrelated species into crops, and is providing spectacular improvements in control of some pests not amenable to conventional plant breeding. The use of transgenic crops (those into which genes have been moved using molecular techniques) raises serious environmental and ethical questions.

## SOURCES AND RECOMMENDED READING

For good general discussion of many aspects of pest resistance, see Young (1998), Van der plank (1984), Maxwell and Jennings (1980), Clement and Quisenberry (1999), and Panda and Khush (1995). Inducible host-plant defenses are discussed by Tallamy and Raupp (1991) and Karban and Baldwin (1997). Readshaw (1986) and Carey (1996) are suggested for expansion on the techniques of sterilization and the advantages and criticisms of the approach.

Anonymous. 2000. Current status of petitions (biotechnology permits), http://www.aphis.usda.gov/biotech/petday.html

Carey, J. R. 1996. The incipient Mediterranean fruit fly population in California: Implications for invasion biology. *Ecology* 77:1690–1697.

Carozzi, N., and M. Koziel, eds. 1997. *Advances in insect control: The role of transgenic plants.* London; Bristol, Pa.: Taylor & Francis, xvi, 301.

Clement, S. L., and S. S. Quisenberry, eds. 1999. *Global plant genetic resources for insect-resistant crops.* Boca Raton, Fla.: CRC Press, 295.

Flor, H. H. 1942. Inheritance of pathogenicity of *Melampsora lini. Phytopathology* 32:653–669.

Karban, R., and I. T. Baldwin. 1997. *Induced responses to herbivory.* Chicago: University of Chicago Press, ix, 319.

Marshall, G., and D. Walters, eds. 1994. *Molecular biology in crop protection.* London, UK: Chapman & Hall, xii, 283.

Maxwell, F. G., and P. R. Jennings, eds. 1980. *Breeding plants resistant to insects. environmental science and technology.* New York: Wiley, xvii, 683.

Panda, N., and G. S. Khush. 1995. *Host-plant resistance to insects.* Wallingford; Oxon, UK: CAB International, in association with the International Rice Research Institute, xiii, 431.

Persley, G. J., and J. N. Sidow. 1996. *Biotechnology and integrated pest management.* Oxon, UK: CAB International, xvi, 475.

Readshaw, J. L. 1986. Screwworm eradication a grand illusion? *Nature* 320:407–410.

Tallamy, D. W., and M. J. Raupp, eds. 1991. *Phytochemical induction by herbivores.* New York: Wiley, xx, 431.

Van der Plank, J. E. 1984. *Disease resistance in plants.* Orlando, Fla.: Academic Press, xiv, 194.

Young, L. D. 1998. Breeding for nematode resistance and tolerance. In K. R. Barker, G. A. Pederson, and G. L. Windham, eds., *Plant and nematode interactions.* Madison, Wis.: American Society of Agronomy, 187–207.

# CHAPTER

# *18*

# IPM Programs: Development and Implementation

## CHAPTER OUTLINE

▶ IPM revisited
▶ IPM program development
▶ Examples of IPM programs
▶ IPM program implementation
▶ Summary

The first 17 chapters of this text have examined pest biology and ecology, and the various tactics used to manage pests. The simple application of individual tactics to manage specific pests without consideration of the other pests in the agroecosystem does not constitute an IPM program. Individual tactics are the tools that serve as building blocks to create an integrated management program. This chapter examines how tactical components are integrated into strategies used to implement an IPM program.

## IPM REVISITED

### Definitions

A major difficulty arises when discussing what constitutes an IPM program because of the wide variation in definitions of IPM, and the perceptions and expectations of individuals as to how true IPM should be practiced.

The IPM concept originated in the 1950s, but the acronym was not actually coined until after 1972. The term IPM is now more or less universally understood (Leslie and Cuperus, 1993; Morse and Buhler, 1997; Kogan, 1998), but what actually comprises an IPM program is still open to discussion. Following are three variations of IPM.

1. A systems approach to pest control that utilizes all appropriate strategies to minimize pest impact while protecting the environment and providing acceptable economic return.
2. A comprehensive approach to pest control that uses combinations of tactics to reduce pest numbers to tolerable levels while maintaining a quality environment.
3. A coordinated, compatible combination of suitable tactics that, in the context of the associated environment and pest population dynamics, results in pest numbers that do not cause economic injury.

The definition of IPM used throughout this text is presented in the margin note on page 11; two of the preceding definitions fit within that concept; however, many definitions (e.g., variation 3) stress economic thresholds and are thus biased toward insect management. Such definitions usually provide only token recognition of the need to manage other categories of pests. It is imperative that integration of management of all pest categories be implicit in the definition, because the farmer, manager, or IPM practitioner has to integrate all aspects of pest management and production practices since pests do not occur in crops in isolation from one another. Most pest management programs, to date, have focused on single pest categories; they are at level I integration (to be discussed later). At more advanced stages of development, a pest management program must integrate management of all pest categories.

## Goals of IPM

It is useful to translate the basic elements of the various IPM definitions into a series of operational goals.

1. The IPM program *must* maintain economic reliability in managing pests. If IPM practices are suggested that are not economically sustainable, producers will not adopt them.
2. IPM practices should reduce the risk of crop loss. Any practice that leads to increased risk will probably not be employed.
3. Due to the importance of pest resistance to pest management tactics, particularly with pesticides, an IPM program must be designed to minimize selection pressure on pests to maintain the utility of the tactics in the future. Any program that does not address this issue is likely to experience problems with pest resistance.
4. An IPM program must strive to maintain environmental quality, and must avoid use of tactics that are unnecessarily disruptive or damaging to ecosystems, especially those ecosystems that are not the target of the management.

Reducing pesticide use is not stated as an explicit goal of most IPM programs, although reductions in pesticide use often result from properly implemented IPM. Most insect pest management systems, however, require the replacement of broad-spectrum insecticides with selective insecticides, so insecticide use per se does not necessarily decrease. Rather, in the case of insect pest management systems, reduction of broad-spectrum insecticide use is an explicit goal.

Another common misconception is to associate IPM with the principles of organic farming. Organic farming is not a direct goal of IPM, although organic farming certainly uses many IPM principles.

# IPM Strategies

A pest management strategy can be considered the optimum mix of pest management tactics for a specific agroecosystem and environment. Optimal strategies cannot be generalized across all pest categories. The mix of control tactics that is optimum for one category of pest will most likely not be optimum for another. Even within pest categories, the optimum mix of tactics will vary, depending on the following:

1. Managed ecosystem, such as short-season row crop versus perennial tree crop
2. Ecosystem and environmental constraints, such as irrigated versus rain-fed system
3. Production philosophy of the agroecosystem manager, such as organic versus conventional farming
4. Category(ies) of pests, and in particular the number of key pests, that must be managed, such as soil-borne versus aerial, or stationary versus mobile
5. Economics of the agroecosystem, such as $300/A wheat versus $10,000/A strawberries

Hoy (1994) used a series of diagrams (Figures 18–6 to 18–8) which illustrate the relative importance of the various arthropod management tactics in different management strategies. The diagrams provide useful paradigms by which to compare various management strategies between pest categories. Expanding the diagrams (Figures 18–1 to 18–5) depicts the relative importance of the different control tactics within an IPM system for each of the major pest categories. As with all generalizations, certain exceptions deviate from typical circumstances, and so the diagrams must be viewed with recognition of such limitation.

In the following diagrammatic generalizations, the relative size of the circle represents the magnitude of the contribution of that tactic to the overall IPM, and the weight of the arrow represents the importance of the tactic on the various impacted organisms. It is important to note that the relative importance of tactics used for the management of pests differs among agroecosystems, and is defined by many factors, including the intensity of management, the acceptability of particular tactics, and the elements of the pest tetrahedron (see Figure 1–6) that are most manipulable. Some agroecosystem and pest combinations have characteristics that dictate the use of tactics that would rarely be applied in other systems. The rationale for each paradigm is discussed separately.

***Pathogens***    The pathogen management paradigm reflects the need for major emphasis on the crop for management of many pathogens (Figure 18–1). Resistant crop varieties are the mainstay of many pathogen management programs. Soil and cultural management are important to many programs, and so is the use of pesticides (e.g., fungicides, antibiotics). Managed biological control currently plays a relatively minor role in the management of pathogens, and the use of biorational pesticides is relatively low. There is more emphasis on monitoring climatic conditions for the management of pathogens, particularly the polycyclic diseases, than is typical for management of other pest categories. Organic management of pathogens does not change the relative importance of most tactics, but does eliminate the use of most pesticides.

**Tactic means "a device to accomplish an end." Tactics are the control methods in IPM.**

**Strategy means "the art of devising or employing plans toward a goal." Strategies are combinations of tatics in an IPM system.**

**FIGURE 18–1**
Diagram depicting the
IPM paradigm for
management of
pathogens. The size of
the circles represents
the relative
contribution of the
tactic, and the weight
or thickness of the
arrows represents the
relative impact of the
tactic.

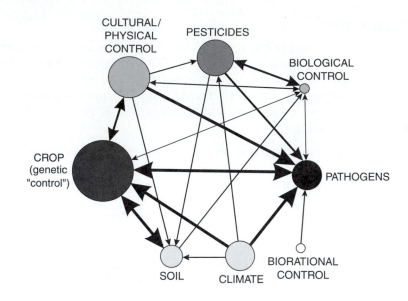

**FIGURE 18–2**
Diagram depicting the
IPM paradigm for
conventional weed
management in arable
crops. The size of the
circles represents the
relative contribution of
the tactic, and the
weight or thickness of
the arrows represents
the relative impact of
the tactic.

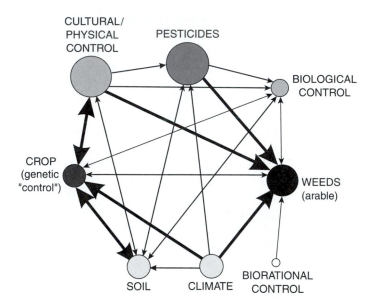

***Weeds***   The paradigm reflects the changes in importance of various tactics in relation to management of producers (plants), rather than primary consumers.

**Weeds in Arable Crops**   Directly manipulating or modifying the crop is less important for weeds than for pathogens and other pests (Figure 18–2), because there are no direct trophic connections between weeds and the crop. The most important tactics are cultural practices, physical controls, and pesticides (herbicides). The importance of managed biological control and biorational controls is low. Soil and climate play modest roles in weed management programs.

**Organic Weed Management**   Organic weed management relies heavily on cultural and physical control tactics, with a little more emphasis in biological control and crop traits than are used for conventional weed management (Figure 18–3). The relative importance of the different tactics for organic management

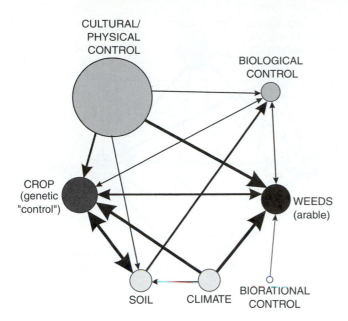

**FIGURE 18–3**
Diagram depicting the IPM paradigm for organic management of arable crop weeds without herbicides. The size of the circles represents the relative contribution of the tactic, and the weight or thickness of the arrows represents the relative impact of the tactic.

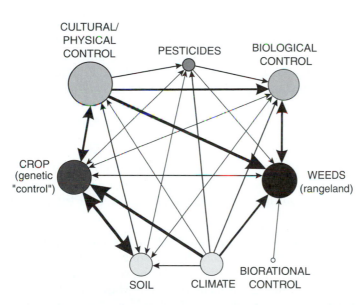

**FIGURE 18–4**
Diagram depicting the IPM paradigm for conventional management of weeds in rangeland/stable ecosystems. The size of the circles represents the relative contribution of the tactic, and the weight or thickness of the arrows represents the relative impact of the tactic.

of weeds should be contrasted with those for organic arthropod management in which biological control dominates (Figure 18–3; cf. Figure 18–6).

**Rangeland Weed Management** This paradigm reflects the reduced importance of pesticides (herbicides) and increased importance of biological control for weed management in the relatively stable rangeland system from which there are no high cash returns (Figure 18–4). It is probably also representative of weed management in most forest systems.

**_Nematodes_** In annual cropping systems, nematode management tactics are generally defined and applied before planting. If appropriate nematode-resistant cultivars are available, they are used. Nematicides are often the tactic of choice. In perennial cropping systems, tactics can be employed at replant, but fewer

**FIGURE 18–5**
Diagram depicting the IPM paradigm for management of nematodes. The size of the circles represents the relative contribution of the tactic, and the weight or thickness of the arrows represents the relative impact of the tactic.

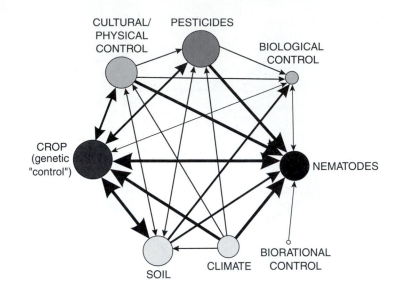

tactics are available for established plants. In general, the nematode paradigm is similar to that for arable weed management (Figure 18–5), but with much more emphasis on crop resistance and somewhat less on cultural and physical control tactics, although in some systems cultural and physical tactics work very well. Maintenance of general plant health and vigor relative to other stresses is important. Soil type and soil conditions are more important in nematode management than for other pests. Biocontrol and the use of biorational chemicals is generally low, although the use of such is increasing as new biorational chemicals are developed.

***Arthropods***   Hoy (1994) presented three paradigms for arthropod management that varied in relation to degree of pesticide use. These are conventional pesticides, organic arthropod management, and integrated arthropod management.

**Conventional Pesticides**   A paradigm that places major emphasis on use of pesticides, with relatively little use of biological control tactics or physical/cultural approaches (Figure 18–6), falls outside the scope of IPM and is therefore a pest control approach that remains below the IPM threshold as defined later in this chapter. Use of biorational tactics is also low. The conventional pesticide paradigm for arthropod control is not analogous to the pesticide-based approach to managing other categories of pests. For example, weed management places much greater emphasis on cultural/physical controls (Figure 18–2) and pathogen management utilizes crop-based management strategies to a much greater extent (Figure 18–1).

**Organic Arthropod Management**   The pesticide control tactic is greatly reduced, and "natural" insecticides that are derived from plants and oils, and insecticidal soaps, usually replace organosynthetic insecticides (Figure 18–7). There is a significant increase in reliance on biological control and the use of biorational tactics. This paradigm is not consistent with tactics for either organic weed or pathogen management (compare Figure 18–7 with Figures 18–1 and 18–2).

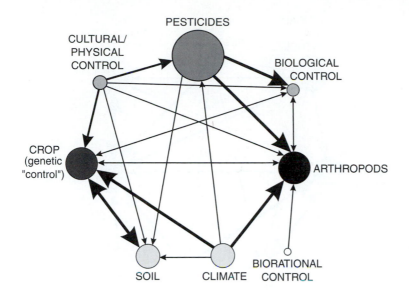

**FIGURE 18–6**
Diagram depicting the
IPM paradigm for
pesticide-based
management of
arthropods (redrawn
from Hoy, 1994). The
size of the circles
represents the relative
contribution of the
tactic, and the weight
or thickness of the
arrows represents the
relative impact of the
tactic.

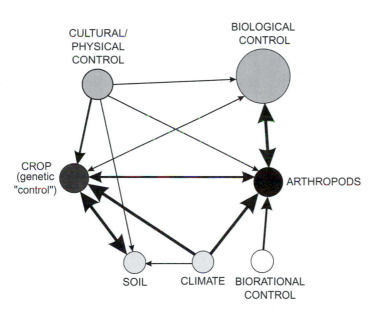

**FIGURE 18–7**
Diagram depicting the
IPM paradigm for
organic management
of arthropods
(redrawn from Hoy,
1994). The size of the
circles represents the
relative contribution of
the tactic, and the
weight or thickness of
the arrows represents
the relative impact of
the tactic.

**Integrated Arthropod Management** This paradigm probably reflects current conventional arthropod management for major field crops and most perennial crops (such as orchard and vines) (Figure 18–8). The major control tactics are based on cultural/mechanical, biological control, and biorational tactics, coupled with moderate reliance on pesticides and host-plant resistance. This paradigm probably does not represent arthropod management of short-season vegetable crops, which is probably much closer to Figure 18–6.

Comparisons of the relative importance of each pest management tactic to each category of pest are summarized in Table 18–1. IPM specialists must take these differences into account when designing systems for pest management for a crop, so that the overall management program is effective for all pest categories.

**FIGURE 18–8**
Diagram depicting the IPM paradigm for integrated management of arthropods (redrawn from Hoy, 1994). The size of the circles represents the relative contribution of the tactic, and the weight or thickness of the arrows represents the relative impact of the tactic.

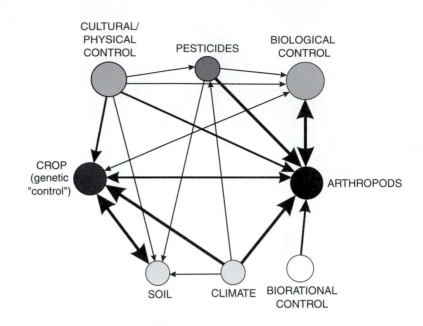

| TABLE 18–1 | Relative Importance of Different Tactics Used in IPM Programs to Manage Different Categories of Pests |
|---|---|

| Pest | IPM system | Crop genetics | Cultural/ physical | Chemical | Biological | Behavioral | Biorational | Climate | Soil |
|---|---|---|---|---|---|---|---|---|---|
| Pathogens | Conventional | +++ | +++ | ++ | + | NA | + | ++ | ++ |
|  | Organic | +++ | +++ |  | + | NA | + | +++ | +++ |
| Weeds | Arable, conventional | − | +++ | +++ | − | NA | − | + | + |
|  | Organic | + | +++ |  | + | NA | − | + | ++ |
|  | Rangeland | + | ++ | + | ++ | NA | − | + | + |
| Nematodes | Conventional | +++ | ++ | ++ | + | NA | − | − | ++ |
| Arthropods | Conventional | ++ | ++ | +++ | + | − | + | − | − |
|  | (Insect) IPM | ++ | ++ | + | ++ | + | ++ | − | + |
|  | Organic | ++ | +++ |  | +++ | ++ | ++ | − | + |
| Vertebrates | Conventional | − | +++ | + | ++ | ++ | NA | − | − |

Key to symbols indicating level of importance
+++       = major
++        = important
+         = minor
−         = minimal
NA        = tactic not applicable
Blank cell = tactic not used

## Levels of IPM and Integration

Pest management can be thought of as a continuum, ranging from doing nothing and risking substantial crop loss, to total reliance on calendar-based pesticide spraying, to extremely complicated regional programs that allow for complex interactions between pests, ecosystems, and society (Figures 18–9 and 18–10).

Likewise, integrated pest management can be considered as a continuum, ranging from cases where pests are managed individually by species without

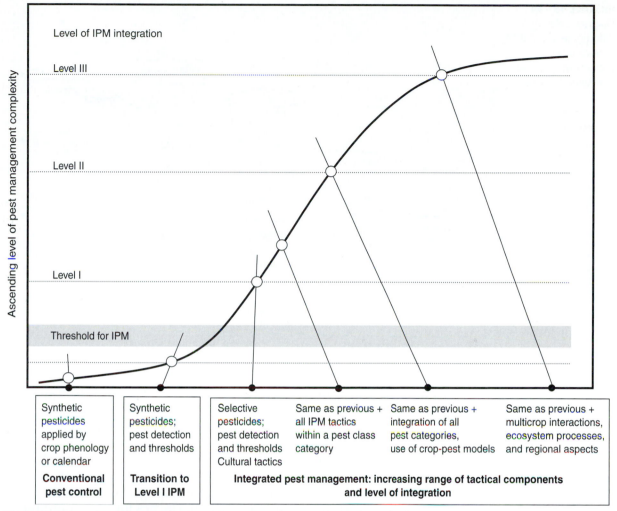

**FIGURE 18–9**   Diagram depicting the IPM continuum, and showing the relative complexity of different levels of integration. Source: redrawn from Kogan and Bajwa, 1999.

consideration of the other pests that are present, to the ideal of totally integrated management that includes consideration of all pest categories and beneficials simultaneously and uses complicated decision rule–based algorithms to optimize all tactics relative to multipest economic thresholds. The first extreme on the management continuum does not integrate tactics even within a class of pests. It is not really IPM, and such an approach resulted in many of the problems associated with pest control that relied solely on chemical pesticide tactics. The latter extreme on the continuum of management is an ideal that does not exist, because the mentioned decision rules and multipest economic thresholds have not been developed (and may not even be feasible except in extremely site-specific form).

IPM can be considered at different levels of integration, with each successive level incorporating the components of the preceding level. The most elementary form of integration in pest management occurs when tactics are integrated for control of a pest or a pest complex within a pest category. This integration of tactics is called level I IPM (Figures 18–9 and 18–10). Level I IPM

**FIGURE 18–10**
Diagram relating
different levels of IPM
to ecological,
socioeconomic, and
agricultural scales.

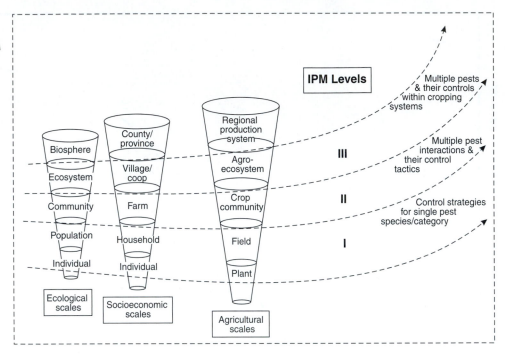

is exemplified by the paradigms described in the preceding section. This level of pest management involves the use of scouting and thresholds for decision making, and typically employs a mix of control tactics. Most current IPM programs are at level I integration.

As discussed in Chapter 6, pests do not occur in isolation, but rather are present in a mixture of categories that is typical of each ecosystem. To develop a truly integrated management system, it is necessary to understand the interactions that occur between pest categories and the tactics for their control (Figure 18–11). This is called level II IPM, and is only achieved in a few current IPM programs. Level II IPM is functional at the whole-farm scale.

Ultimately it is necessary to place the entire management system for the pest complex into a broad ecological and socioeconomic framework at a regional scale (Figures 18–10 and 18–11), termed level III IPM. The overall goal of an IPM program is thus to integrate the needs of all the different pest categories in a single comprehensive program. It is probably not achieved in any current IPM program.

## IPM PROGRAM DEVELOPMENT

Individuals typically do not develop IPM programs. Due to the complicated nature of IPM programs, teams of specialists in the various pest disciplines typically cooperate to develop them with agronomists, meteorologists, ecosystem ecologists, and economists. Once the program has been developed and tested, it is implemented by IPM practitioners. The advent of microcomputers has allowed the use of effective, predictive phenological and disease forecasting models. Similarly, microcomputers combined with modern electronic com-

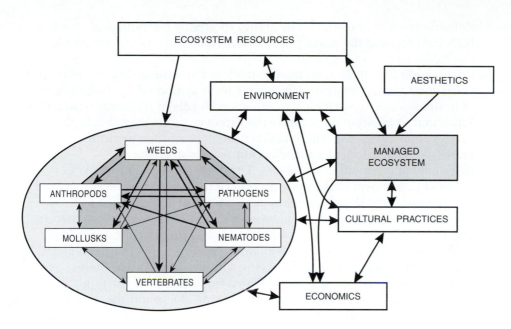

**FIGURE 18–11**
Diagram showing pest hexagon (Figure 6–1) placed into ecological and societal framework.

munications systems, particularly the World Wide Web (WWW), have greatly facilitated diffusion of IPM information and the use of decision-making tools (see later section in this chapter).

Certain types of information are necessary to develop an IPM program. Following are some of the most important aspects.

## Key Aspects of IPM Programs

IPM systems are knowledge intensive. An IPM program is based on the understanding of the biology and ecology of the system, the economics, and the sociological implications. The following types of information are essential.

1. Identification of key pests. Pests that are likely to have significant economic influence on the productivity of the system must be identified. Secondary pests that are likely to become key pests also should be identified. Where biological control is a key component of the ecosystem it is also critical to identify the natural enemies.
2. Pest biology and ecology. The life cycle, life history, reproductive habits, behaviors, feeding habits, host preference, activity patterns, dispersal mechanisms, sensitivity to environmental conditions, virulence, prey, predators, parasites, competitors, and aspects of density-dependent determinants of the preceding elements must be well understood for all pests in the system.
3. Characteristics of the regional crop-production system. An effective IPM program requires that all aspects of the crop-production system, biology and ecology of all pests present in the system, advantages and disadvantages of different pest control strategies, environmental constraints, societal demands, and local, state, and federal regulations be known and understood. Accessing and interpreting such complex sets of information is difficult and represents one of the impediments to adoption of IPM.

4. Reliable predictive models. The ability to predict phenological events of both the crop and the pests, plus predict yield impacts and economics, are desirable.

5. Cost-benefit information on control tactics. The availability, benefits, and costs of all management strategies have to be evaluated, including the role of judicious use of pesticides. At advanced levels of IPM integration, it is important to incorporate not only the direct costs and benefits to the producer, but also costs and benefits to society and the environment. However, such information is often not readily available.

6. Regional management components. The degree to which there is a regional component to the management of the pests must be known and incorporated into the program. Incorporation includes factors such as exclusion and detection (e.g., pheromone traps to detect pink bollworm and mass releases) and role of regional control programs, such as beet leafhopper in California to reduce *curly top virus* spread or the regional boll weevil eradication program in the southeastern United States.

7. Scouting and monitoring systems. The requirements for the types and extent of population scouting and monitoring to support the decision-making process must be determined, as well as how monitoring is to be carried out and who will pay for it.

8. Record keeping. The type of record keeping that will be required must be determined; requirements may be quite different among pest categories. Data on the following components of the system should be recorded.

   8.1. Pest identification, population size, and phenology of all types of pests (see Chapter 8 for more detail of pest records)

   8.2. Pest mitigation tactics, especially pesticide applications, including tactic(s) used and dates, kinds of chemical used, rates of application (including, if appropriate, biological control releases or installation of pheromone traps and dispensers), and methods of application of all tactics

   8.3. Results of pest control tactics: How well did control tactics work? These records are particularly important for monitoring rise in resistance levels to pesticides or occurrence of biotypes virulent to resistant varieties of plants.

   8.4. Crop yields. Yield records are becoming much more routine and more site specific in some crops with utilization of global positioning yield monitoring systems; crop yields can be related to the success or failure of IPM tactics.

   8.5. Nutrients used: records of types, quantities, and times of application are important for association with possible impacts on pests but also to allow correct budget analyses.

   8.6. Weather data on at least temperature and rainfall is important not only to interpret yield results but also for use in phenology models. Timing of irrigation where appropriate is another necessary element for record keeping.

   8.7. Other important records are crop variety, labor costs, and other circumstances.

9. Resistance management. Development and implementation of resistance management strategies must be incorporated into the program.

10. Environmental and social constraints. All relevant ecological and sociological constraints must be considered (see Chapter 19 for more details).

Planning ahead is a key to developing an IPM program that works. An IPM program will anticipate pest problems, rather than be a simple set of reactions that occur once the pest is present. Many mitigation tactics must be implemented prior to pest attack. Long-term planning, as outlined in the next section, is useful for both short-season and perennial crops.

## General Considerations

In developing an IPM program, the following specific types of information must be considered.

***Before-planting in Annual Crops***   It is important to determine the key pest species in all categories that typically attack or invade the crop, whether they are present, and their numbers and damage potential. With this information the manager does not have to check every organism in the field, but rather can target those reasonably expected to cause unacceptable injury. Knowledge of the key pests permits development of plans to contend with each pest. Prior knowledge about the major pests may be essential if the only appropriate control tactics must be implemented before planting the crop, such as the case for many nematode pests. Good records from past cropping seasons can help planning by allowing anticipation of pest problems. Planning requires, if possible, preplant sampling to define inoculum densities for pathogens and nematodes.

Regulations in some regions require that a written plan for the anticipated use of certain pesticides during the cropping season be developed and filed with appropriate authorities before the crop is planted (as noted in Chapter 11).

The need to consider the regional aspects of the program where appropriate must be made at this stage. It is futile to develop an IPM program only to have it negated because of actions occurring in adjacent areas. It is essential to recognize scale differences during the planning process.

***During the Crop Cycle***   The decision staircase approach described in Chapter 8 provides an appropriate guide for pest management activities required during the crop cycle. Crop growth and pest occurrence must be monitored, and appropriate mitigation actions taken. Specifics are region, crop, and pest dependent.

***After Crop Harvest***   Management of some pests requires that actions be taken after the crop is harvested. These types of actions are more likely as part of IPM for perennial crops, but they apply to annual crops in which crop residues can harbor inoculum of pathogens or dormant stages of arthropods that might be damaging during storage or in the following cropping season. Examples of actions following harvest include weed control action to stop seed production, crop destruction to remove inoculum and source of overwintering sites, area-wide controls also to reduce pest carryover or to enhance beneficials.

## EXAMPLES OF IPM PROGRAMS

Pest management exists for all crops, but the level of integration of control tactics is often minimal across pest categories. The specific details vary by region. There is a vast literature on IPM programs for a large number of crops. Most are

| TABLE 18–2 | Comparison of Cultural Characteristics of Selected Crops | | |
|---|---|---|---|
| **Cultural operation** | **Lettuce** | **Cotton** | **Apples and pears** |
| Sowing to harvest | 45 to 70 days | 7 or 8 months | Many years |
| Direct market | Yes | No | Yes |
| Harvesting | Hand | Machine | Hand |
| Started by | Seed or transplants | Seed | Transplants |
| Cosmetic standards | Yes | No | Yes |
| System stability | Highly disturbed | Disturbed | Relatively stable |
| Market Value/A | $2000 | $350 | $3500 |

focused on single pest categories, but in the aggregate they form a valuable resource for anyone who is interested in IPM. A comprehensive listing of major reviews of these programs, handbooks, and proceedings of workshops and symposia can be found in the website www.ippc.orst.edu/ipmreviews.

In this book three examples are selected to illustrate how IPM programs are developed for crops with different pest management requirements and constraints. Some of the characteristics important to developing an IPM program for these crops are summarized in Table 18–2.

1. Lettuce represents a specialty, nonstaple, short-season vegetable crop that is sold directly to the consumer without processing.
2. Cotton is a full-season, large-scale row crop that is processed before being sold to the consumer.
3. Pome fruits are one example of a perennial, tree-crop system with a harvested product that is sold directly to the consumer, although a portion of the crop can be directed to processing.

## IPM for Lettuce

Lettuce, and some high-value flower crops, exemplify one extreme in the spectrum of crop/consumer conditions that impose severe constraints to the development and adoption of IPM.

***Key Pests in Lettuce***    Lettuce is a cool-season crop, and the expected pests are associated with cool weather. This must be considered as the IPM program is developed.

**Pathogens**    Several virus diseases attack lettuce, including *lettuce mosaic, big vein,* and *corky root.* It is essential that the IPM program addresses how to manage these virus diseases. The virus problems can be so severe that their management dictates how several components of the program are designed. Other key pathogens include lettuce drop and powdery mildew.

**Weeds**    Cool-season annuals include mainly dicots, especially those in the Asteraceae, plus burning nettle, shepherds-purse, mallow, and common purslane. Lettuce is a poor competitor, which means that high levels of weed control are needed to obtain commercially acceptable yields. Burning nettle can be a serious problem at harvest because of injury to workers.

**Nematodes**   Root-knot nematode can be a problem, but cool temperatures for lettuce production typically limit the importance of nematodes.

**Arthropods**   Aphids, leaf miners, and various lepidopteran larvae are the most serious arthropod problems.

**Vertebrates and Mollusks**   Not normally a problem in commercial lettuce production, these can be very serious in small farming and backyard situations.

***Cultural Considerations***   Several factors dictate some of the IPM practices that can and cannot be used. Lettuce is a high-value crop, and therefore can justify relatively high management inputs. Lettuce is a fresh market crop, meaning that the crop is generally marketed directly to consumers without processing, although this is changing in industrialized countries where prepackaged, premixed salad greens are becoming more common. Lettuce grows rapidly and can be harvested in as little as 45 days. When lettuce is growing slowly, the crop is harvested in about 60 to 75 days. At harvest, lettuce is a low-growing crop, is relatively widely spaced, and often does not achieve complete coverage of the ground. The produce is picked by hand, so workers must be able to enter the field at harvest.

The market value of lettuce is variable: It can yield $2000 net per acre one week and not pay for harvesting two weeks later because of market volatility in relation to supply and demand. Market variability makes cost-benefit analysis of IPM decisions difficult.

***Constraints***   The cultural and marketing facts impose some limitations on the IPM tactics that can be used in lettuce. Due to the short duration of the crop there may be insufficient time between treatment and harvest for pesticide residues to decline to acceptable levels. Residue restrictions can preclude use of pesticides that do not break down rapidly, a limitation that can apply to a decline of residues in both the product and the soil. The latter can limit the choice of rotation crops or the choice of pesticide used in lettuce. Plant-back restrictions for herbicides and other pesticides become especially important in all short-season crops.

There is little tolerance for cosmetic damage in lettuce (see Chapter 19). The crop must be free of all insects and blemishes at harvest, otherwise the consumer in industrialized countries will not accept it. This poses severe difficulty for an IPM program as it essentially eliminates the use of biological control. Cosmetic standards are real; most consumers will search over the lettuce at the supermarket and select the head with no signs of pests or pest damage, such as aphids or holes in the outer leaves, or some mildew.

Generally speaking, biological control often does not work well in short-season crop situations when pests increase to damaging levels, because beneficials cannot build up sufficiently to provide adequate control (lag time, see Figure 13–1). Cosmetic standards apply to beneficials, as consumers object equally strongly to the presence of beneficial insects as they do to pest insects.

Price fluctuations for fresh market produce make the use of economic thresholds difficult. When this difficulty is coupled with the problem of cosmetic standards, most managers operate on the principle that if an insect or pathogen is present near harvest it cannot be tolerated.

## IPM for Cotton

Comprehensive IPM programs have been developed for cotton, and have been implemented to varying degrees in most of the regions of the world where the crop is grown. Adoption of IPM is high in Australia, much of the United States, Israel, and some countries in South America, particularly Brazil and Peru. In some regions the programs are approaching level III IPM.

***Key Pests in Cotton***    Cotton is a warm-season crop with warm-season pests.

**Pathogens**    *Verticillium* and *Fusarium*-wilt fungi are the most important pathogens. Black root (*Thielaviopsis*) can be a serious problem during seedling establishment.

**Weeds**    Warm-season annual weeds such as several pigweed species, night-shades, morning glories, and barnyardgrass are a major problem. Several perennial weeds can also be serious problems in cotton production, including john-songrass, yellow and purple nutsedge, and field bindweed.

**Nematodes**    Root-knot nematodes are a significant problem in cotton. The species and race of root-knot nematode must be determined, because *Meloidogyne acronea* and host races 3 and 4 of *M. incognita* are able to parasitize cotton. The reniform nematode is becoming more of a problem in many cotton-producing areas in the United States, and the Columbia lance nematode is also a problem in certain circumstances.

**Arthropods**    The key arthropod pests include the boll weevil, the pink and other bollworms, cotton leafworms, whiteflies, the cotton aphid, lygus bugs, other plant bugs, thrips, and spider mites. Not all these insects are problems in all regions where cotton is grown. Generally, areas that are not infested by either the pink bollworm or the boll weevil, which are Malvaceae specialists, have less severe arthropod pest problems.

**Vertebrates and Mollusks**    These are not normally a problem in commercial cotton production.

***Cultural Considerations***    Cotton is a perennial plant, but the crop is usually grown as an annual. Even as an annual crop it has a long growing season that typically exceeds six months. As all pests are not present in all cotton growing regions, this means that exclusion is a management strategy in regions where such pest is not present. With ease of international trade and travel, exclusion has been rather ineffective, as exemplified by the relatively recent invasion of the boll weevil into the cotton-producing regions of Brazil.

Cotton is machine harvested, it is mechanically ginned, and the cotton fiber is mechanically spun into thread. Several pest species impact the ability to process the cotton, which must be considered when designing an IPM program. All harvested material that is not seed or lint (cotton fibers) is removed during the ginning process. This material is known as gin trash, and potentially contains weed seeds, pathogen inoculum, and insect eggs. Disposal of gin trash is an IPM problem because the trash may spread pests if returned to the cotton fields.

***Constraints***    Because cotton is a perennial plant, it has the potential to survive from one season to the next, and thus support pest inoculum from year to year. Crop destruction is thus more significant for IPM in cotton than it is in most annual crops. Humans do not eat cotton, and it is processed before it can be sold, so there are some IPM tactics that are perfectly acceptable for cotton IPM that cannot be used for a fresh market crop.

***Timeline***   Cotton exemplifies the way that an IPM program for a single-season irrigated row crop comes together under semiarid conditions, for example, in the San Joaquin Valley of California (Figure 18–12). The cotton IPM program for the valley is used to describe the basic system components; it has been expanded to include components that are relevant to rain-fed cotton production regions, and to regions where the pink bollworm and the boll weevil are prevalent key pests. The IPM program has to be modified to fit local conditions, such as pests present and climate, for any other cotton growing region.

**Prior to Planting**   The following pest management considerations must be made in relation to choice of crop variety.

1. Disease resistance. The only effective way to manage *Verticillium* wilt is to plant a tolerant variety, as no true resistance is available. To a considerable degree, the same is true for *Fusarium* wilt. If these diseases cannot be adequately controlled through the use of tolerant varieties, then it may be necessary, for a few years, to plant a crop other than cotton that is not a host of the pathogens to allow the inoculum level in the soil to decline.
2. Root-knot nematode resistance. Preplant soil samples should be taken to identify potential nematode problems. The use of resistant varieties is an effective tactic for root-knot nematode management, and the use of a nematode-resistant variety will also help reduce *Fusarium* wilt, because root damage by the nematode leads to higher levels of infection by the pathogen (see Figure 2–7i). Rotation to a nonhost crop may be necessary if nematode populations are too high.
3. Insect resistance. Low levels of insect resistance are present in many current varieties, but none are adequate to preclude use of other management tactics and so insect resistance is not a major consideration in selecting a variety. This is changing with the introduction of Bt transgenic varieties that are resistant to some of the most serious lepidopterous pests of cotton, such as the cotton bollworm, budworm, and pink bollworm. Use of these varieties, however, has to follow current requirements for management of resistance in the pest population, which requires that part of the total acreage be planted to nontransgenic, susceptible varieties.
4. Herbicide resistance. If the use of nonselective herbicide is anticipated, then it is necessary to select an appropriate transgenic-resistant variety prior to planting.
5. Quality constraints. Fiber length (staple) and thickness (micronaire) impose special demands on variety selection and cultural methods.

If seedling diseases are expected to be a problem, then it may be necessary to ensure that fungicide-treated seed are used to control black root (*Thielaviopsis*). Flooding in the summer prior to the cotton crop can provide adequate control of the inoculum of *Thielaviopsis*.

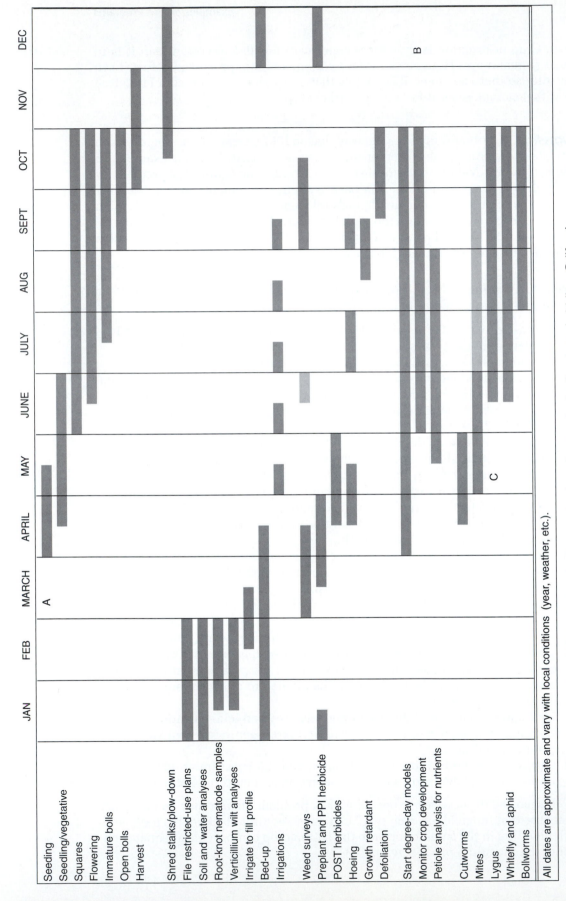

**Figure 18–12** Timeline for cultural and IPM activities for the production of cotton in the San Joaquin Valley, California.

All dates are approximate and vary with local conditions (year, weather, etc.).

In certain production regions, state regulations may require filing a plan, prior to the cropping season, with the appropriate authorities if any restricted-use pesticides are to be employed in the IPM program.

In many situations, weed management requires that herbicides must be applied prior to planting the crop, as there are no satisfactory postemergence herbicides available. A sound decision regarding use of preplant herbicides can only be made with the detailed knowledge of the species present in the weed flora of the field. Some preplant herbicides are sufficiently long-lived in the soil that they can be applied several months before planting. Such early application has the advantage that there is one operation less to be performed during the busy planting schedule.

**During Crop Growth**  During the period when the crop is growing and maturing in the field, pests must be monitored on a routine basis and the crop status carefully assessed. It is essential to know the phenology and status of both the crop and the pests in order to make the best management decisions. The following information is necessary.

1. Cotton plant mapping. Essential data to be recorded include the number of main stem nodes, positions of branches, positions on branches that have flowers, and bolls set.
2. Calculation of percent boll retention in relation crop growth stage. Boll retention is calculated based on the percentage of flowers that actually formed bolls; for example, feeding activity by lygus causes flowers and bolls to abort.
3. Assessment of foliage injury level (entomologists refer to defoliation level). Decisions regarding management of foliar-feeding arthropods usually are based on a combination of population density level and current level of injury. Defoliation is estimated visually, a method that requires training and is subject to wide variations.
4. Assessment of relative abundance of natural enemies, especially generalist predators, such as minute pirate bugs, lacewing larvae, and damsel bugs present in the crop field. Certain models provide a no-treatment decision if enough natural enemies are present to maintain the pest population below the economic injury level.
5. Access to tables and decision-making models. These determine if major arthropod pests can be tolerated or if pest populations need to be treated, that is, if they approach the economic threshold.

It is important to monitor surrounding areas (alfalfa fields, safflower fields, native vegetation and weeds) for the presence of lygus bugs (note C on Figure 18–12; also see Figure 8–5) that may move to the cotton as the other crops are harvested or the plants senesce.

In certain regions, an area-wide approach to managing key insect pests has been successfully used. These programs contain a regulatory component and involve the active participation of both state and federal public agencies.

In the southeastern United States, an extensive boll weevil suppression program has been implemented on an area-wide basis since 1978. The program comprises pheromone trapping, chemical treatments, and cultural practices employed over a 2.5-year period. In sensitive areas, such as near schools, hospitals, and housing developments, these techniques are supplemented with alternative control technology intended to minimize the use of pesticides. Spray operations begin about a month after trapping indicates the presence of weevil populations in the region. Continued trapping pinpoints hot spots of infestation and triggers necessary treatments until all weevils are gone. Full implementation of the program and virtual eradication of the weevils takes about 2.5 years. Malathion, an insecticide with very low mammalian toxicity, is the primary pesticide used in the program. It is applied at an ultra-low-volume rate from aircraft or in a ready-to-use formula from ground sprayers. Dimilin, an insect growth regulator, may be used around certain sensitive areas as a means of reducing the number of more toxic pesticide applications. The program dictates that insecticides are applied only on fields where infestations are detected and spray criteria are met. During late summer, traps are placed and checked to assess the extent of infestation in new program areas. Fields that are infested generally receive an average of seven applications in the fall.

Another area-wide suppression program has been deployed in California against the pink bollworm. In this case, the suppression is achieved with the release of sterile males (see Chapter 17).

**Following Harvest**    As a component of the pink bollworm and boll weevil exclusion programs, all crop residues must be shredded and disked under (see B on Figure 18–12). All weeds that may serve as overwintering habitat for the silverleaf whitefly should be removed.

## IPM for Pome Fruits

Pome fruits are perennial plants and include pears and apples. IPM programs for these crops, mostly at level I integration, are aimed at insect management and have been developed in North America, Europe, and New Zealand. Level II IPM that includes consideration of insect, weed, nematode, and pathogen management have been successfully deployed in the United States in Massachusetts.

### Key Pests in Pome Fruits

**Pathogens**    The major diseases are apple and pear scab. On pears, fire blight and pear decline are also devastating diseases. Mildew may be a problem on apples. Root-rot pathogens, such as *Phytophthora* and oakroot fungus, are problems for all types of fruit tree crops.

**Weeds**    Almost any weed species present in a region can occur in pome fruit orchards. Perennials are frequently more important than annuals.

**Nematodes**    Root-knot, root-lesion, dagger, and ring nematodes are all problems.

**Arthropods**    The key arthropod pest of pome fruits on a worldwide basis is the codling moth; other Lepidoptera include Pandemis leaf roller, orange tortrix, and other leaf rollers. Other insects include rosy apple aphid, the wooly apple aphid, and the apple maggot on apples; the plum curculio, mites, plant and stinkbugs, and pear psylla on pears.

**Vertebrates**    Due to the relative stability of the orchard ecosystem, vertebrates can cause significant damage, especially in young orchards. Field mice (voles) and various squirrels can be serious problems because they chew on trees and irrigation equipment, and tunnel in the soil. Vertebrates are especially important for young trees when girdling by rodents can easily kill the tree. In certain areas, deer may also be a problem.

***Considerations***    The crop is perennial and is maintained for many years; therefore, decisions made when the trees are planted are very important, because incorrect decisions can be costly. Furthermore, if a soil-borne pest becomes established, it may be very difficult to take useful remedial actions, because a perennial crop cannot be easily rotated to a nonhost crop for the pest. If the crop is transplanted, it is essential to ascertain that no pests are being carried in the transplant soil or the seedling. If soil-borne pests are suspected, pest identification is critical and the choice of appropriate resistant rootstocks is essential.

As the cropping system is not removed each year, it is amenable to long-term, sustainable IPM to preserve natural enemies. Because of the relative stability of the system, the pests may also come back each year, with more regularity. It is often possible to design control tactics that are implemented in the off season from which control benefits are expressed during the season. For example, the application of dormant oils in the winter kills arthropod eggs and overwintering forms of insect pests.

***Constraints***    Perennial crops that grow for many years in the same place pose different challenges than do annual crops relative to implementing IPM. One of the most obvious is the increased importance of vertebrate pests. It also becomes necessary to consider that misdiagnosing a vertebrate problem can result in loss of the trees, and the long-term consequences of management decisions are important because they may persist for many seasons. Managers typically have several years before there is a crop, which means no cash flow to support pest management practices that must be carried out in order to ensure that the crop establishes and grows well. Once established, the large, tall canopy makes weed management much easier. Many modern orchards keep a herbicide-treated weed-free area along the tree row, but manage the between-row understory vegetation by mechanical means. Appropriate orchard floor vegetation management conserves moisture and provides habitat for natural enemies of the fruit pests. Under all circumstances it is essential to plant disease-free stock, especially for viruses. Because of the perennial nature of the crop, certain cultural control options such as crop rotation are not applicable.

***Timeline***    We present a timeline for pear pest management in southern Oregon (Figure 18–13) to represent an IPM system for pome fruit; adjustments would be needed for other regions.

The most advanced IPM systems for pome fruit crops are being developed based on the concept of integrated fruit production (IFP). IFP has been adopted in several European countries, and represents an approach to level III IPM. IFP has been successfully tested in a few restricted fruit production regions in the United States, foremost in the Hood River area of Oregon.

# IPM PROGRAM IMPLEMENTATION

Who actually carries out IPM programs? There is no uniform, absolute answer to this question. All farmers carry out some form of control, as they try to protect their crop from losses caused by pests. By default they must manage all

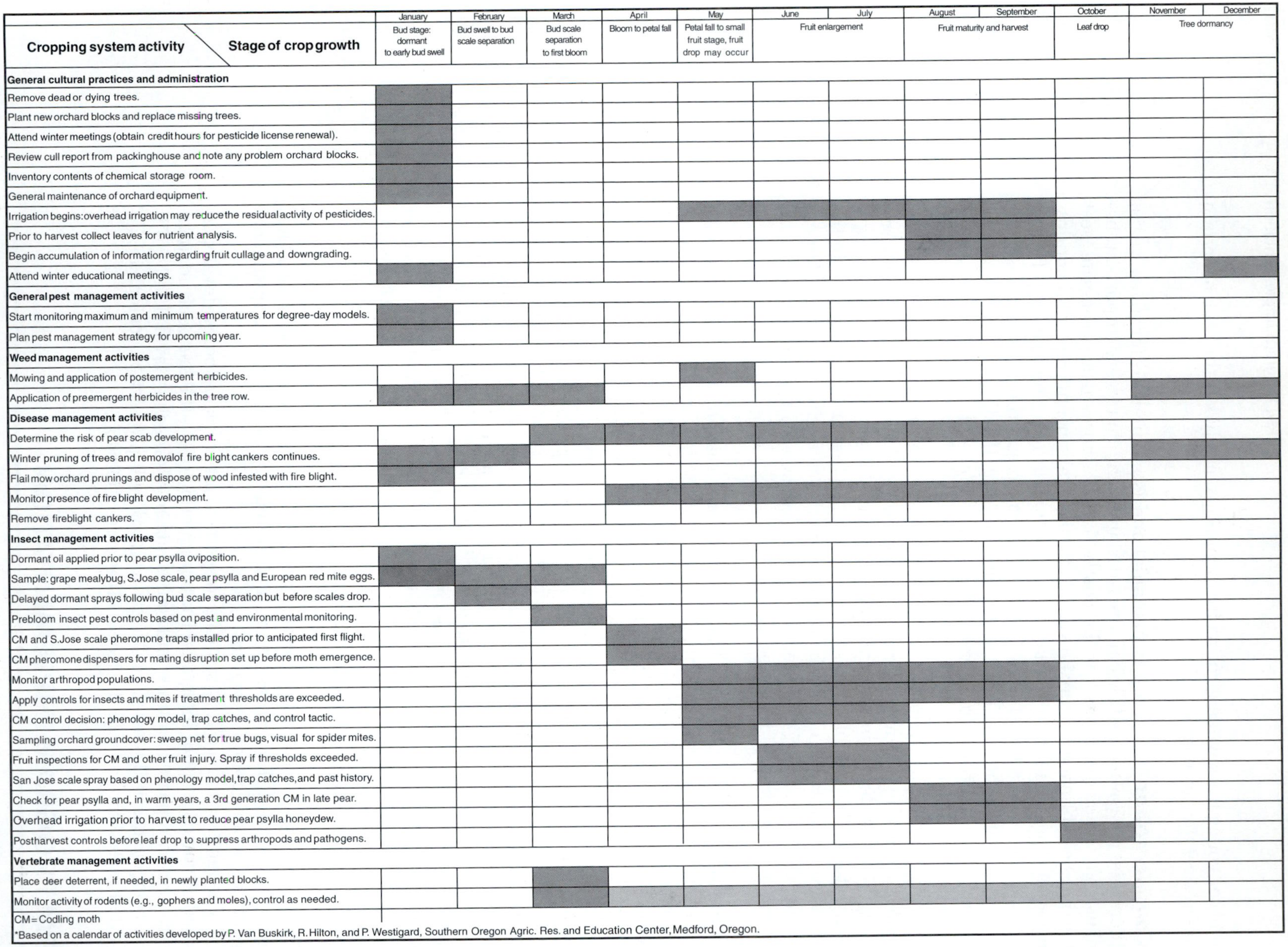

**FIGURE 18–13** Timeline for cultural and IPM activities for the production of pears in southern Oregon.

pests that attack their crops. Differences occur in how much integration of tactics is achieved, and the overall success of the program.

Several important issues must be addressed in relation to IPM decision making: What are the critical issues that should be considered when determining who is responsible for decision making in IPM? Which groups of people are actually involved in IPM decision making?

## Critical Issues

1. Knowledge. With increasing sophistication of IPM comes a greater requirement for biological and technical knowledge. All IPM systems, especially those that do not rely solely on pesticides, are said to be knowledge intensive. The question of who will have the required knowledge is not trivial. Computer-based knowledge systems are helpful in this regard as they can provide expert assistance rapidly assessing information from stored databases.

2. Conflict of interest. From a societal viewpoint, there is a fundamental question regarding who should make the decisions. Should it be those who are perceived to have a vested interest in the results, or should it be persons who have no biases or personal stake in the decisions? The topic of conflict of interest is addressed further in Chapter 19.

3. Personnel. Another fundamental question concerns who should develop and provide pest management information. Should it rest solely with the landowner/manager? What is the role of the pesticide industry, the biotechnology industry, growers' cooperatives and similar organizations, institutions of higher learning, or government? In reality, implementation of IPM programs probably involves a combination of all of these groups, but the agricultural community has experienced difficulty in coordinating such an activity.

4. Cost of IPM. As the use of IPM is more people intensive than scheduled pesticide application, certain questions exist about who pays for the extra costs that are incurred.

## IPM Practitioners

The people who make pest management decisions or are involved with any other phase of IPM are referred to as IPM practitioners. Any person who works in the area of IPM decision making for hire (i.e., is paid for services) is called a pest control advisor (PCA). PCAs normally perform a subset of the duties of a crop consultant, who also advises on cultural matters such as crop variety and fertilizer use. Others who participate in the IPM decision-making process are as follows:

1. Farm owner/manager. When pest control was simply a process of spraying a pesticide, many farmers carried out their own decision-making duties, often obtaining their information from pesticide sales representatives or suppliers of agricultural chemicals. Due to the complexities of carrying out even level I IPM, it has become increasingly difficult for farmers to have the time or expertise to implement IPM programs. Some very large farms hire pest management experts (PCAs) to do the specialized work required for an IPM program; however, most farmers who use IPM either hire or contract with PCAs from one of the next five categories to assist in their IPM program.

2. Chemical industry. The use of pesticides is a component of many IPM programs, but it is not likely that the pesticide industry can implement the total concept of level I, or higher, IPM.

3. Pesticide sales. It is unlikely that persons involved in selling pesticides can participate fully in applying IPM concepts that often go counter to their own personal interests (see conflict of interests).

4. Full-service companies. Such companies typically sell pesticides and fertilizers, can provide specialized equipment and application, and have trained PCAs who make decisions regarding IPM programs. These PCAs visit the fields of their farmer clients and assess the need for pest control. Full-service companies typically also assist their clients with record keeping and paperwork involved with pest management. Because such companies sell pesticides, the PCAs who work for the companies may have a conflict of interest in making pest management decisions.

5. Cooperatives. Farmers operating small farms form a cooperative that hires a PCA to do their IPM advising. In some situations, the cooperative may also handle general farm supplies and pesticides. In São Paulo, Brazil, farmers of Japanese descent formed a cooperative in 1921 that provided full services to members, even conducting research in certain areas. The cooperative became one of the most powerful economic forces in the agriculture of the state, and served as a model to many others in South America.

6. Independent consultants. This type of advisor charges a per-acre fee for providing IPM expertise, such as designing pest management strategies, scouting and monitoring pests, pest management decision making and recommendations, and assisting in paperwork involved with IPM, such as filing yearly plans and filing use reports. Independent consultants usually do not perform the actual pest management operations and do not sell pesticides. Independent consultants are considered to have less conflict of interest than PCAs in full-service companies. Their importance in the implementation of IPM is increasing and could become the major way that IPM is implemented in the future.

7. Public agencies. Various public agencies typically develop the information that supports IPM programs. The initial development and demonstration of the program is often carried out by the agency, but the agency seldom carries out the day-to-day IPM programs once they are commercially successful. Consultants, PCAs, and farmers implement the program once it has been shown to operate satisfactorily. The public agencies involved in IPM are as follows:

   7.1 Agricultural universities. Research departments and their network of experiment stations usually are the source of the basic research that provides and supports the informational database essential for implementation of IPM programs.

   7.2 Cooperative extension services. The U.S. Agricultural Extension Service is an integral part of the land grant university system. In some other countries, extension service functions are under ministries of agriculture or state departments of agriculture. Extension agents and specialists carry out applied research, test new technologies, and transfer the results of research to IPM practitioners.

   7.3 Federal departments or ministries of agriculture. The U.S. Department of Agriculture (USDA) has an independent research branch, the Agricultural Research Service (ARS); and through several other agencies, performs a coordinating function among cooperating state universities, establishes regulations regarding pests, and quarantine functions to prevent biological invasions.

7.4  State departments of agriculture. These departments promote agricultural practices, including IPM, that are relevant to the local agricultural systems. They assist in developing regionally adapted IPM practices, but are not usually involved in day-to-day practical farm-level IPM.

## Sources of IPM Information

The following section outlines where information on IPM can be obtained.

1. Printed sources. All pest disciplines have textbooks covering the respective pests and their management (see General IPM references at the end of this book). The nature of these books neither stresses nor leads to integration of management, but emphasizes control of the particular pest category, usually without any consideration for impacts on other pest management disciplines. There are also many regional publications prepared by universities and government agencies. These cover specifics of pest management in the region, but are also typically sectionalized by pest discipline. Industries involved in pest management products provide marketing brochures that explain their product, but these rarely involve integration of tactics; they usually extol the single tactic approach to pest control.

2. Professional organizations.

    2.1  Pest discipline professional organizations. Each pest discipline has an associated professional society at the international level, national level (see accompanying text box), and regional level (e.g., California Weed Science Society). These provide information and research for their associated discipline, but provide only token information on IPM that crosses discipline boundaries.

    **Professional Societies in the United States**

    | Pest type | Professional society | Acronym |
    |-----------|---------------------|---------|
    | Pathogens | American Phytopathological Society | APS |
    | Weeds | Weed Science Society of America | WSSA |
    | Nematodes | Society of Nematologists | SON |
    | Arthropods | Entomological Society of America | ESA |
    | Vertebrates | The Wildlife Society | TWS |

    2.2  Organizations representing integrated pest management. The Consortium for Integrated Crop Protection and the International Plant Protection Congress represent world-level organizations that provide information on IPM. Although initially oriented toward insect management, they are now expanding the focus to all plant protection disciplines. The Food and Agriculture Organization (FAO) also provides considerable information and expertise on IPM, especially for less developed countries. The United States has a national IPM program that fosters interaction between disciplines in pest management without providing routine information directly to practitioners.

3. Public meetings. IPM information is routinely presented to interested persons at public meetings organized by both private industry and government agencies (e.g., the cooperative extension service in the United States).

4. Field days and demonstrations. Many public institutions conduct field days and demonstrations of pest management practices; in most part, these are discipline oriented and contain little information about integration.

5. Mass media. Magazines and television are used to market pesticides.

6. Electronic. The availability of microcomputers has provided several improvements to the way that pest management information can be stored, retrieved, delivered, and utilized. The internet allows for widespread availability of pest management information via the WWW. Electronic IPM information delivery takes several different forms:

6.1 Diagnostic. Information such as pest photographs, symptoms, and simple keys to aid in diagnosis are available in CD-ROM format or directly from websites for many crops and pests. Quality is variable and most do not provide much information on integration of control tactics. The ability to access photographs in the field with a laptop computer is increasingly used as an aid to pest identification.

6.2 Modeling and prediction. The development of inexpensive, powerful computers and small, inexpensive environmental sensors has contributed to much activity in the general area of modeling since the early 1980s. Much effort remains at the research level, but several model-based systems, particularly for crop phenology and disease forecasting, are being used commercially.

6.2.1 Crop growth models. Many crop growth models have been developed, for example, alfalfa (ALFSIM), soybean (SOYGRO), wheat (CERES-WHEAT), rice (INTERCOMP), and cotton (GOSSYM). These models form the basis of crop loss forecasting and are used in expert systems (see 6.2.4). Crop growth models are used primarily at the research level. INTERCOMP also has the ability to simulate competition between crops and weeds.

6.2.2 Prediction of events. Models based on real-time weather data are used to predict phenological events in pest life cycles. Some effective models for use in IPM have been developed for fruit crop insects, including the codling moth and several leaf roller species. Effective disease forecasters, or disease warning systems, have been developed, and examples include a computerized forecast for potato late blight (BLITECAST), a forecaster of onion downy mildew (DOWNCAST), a forecast model of grapevine downy mildew (PLASMO), and the disease warning system called Apple Scab Predictor.

6.2.3 Herbicide decision aids. Models relate crop yield loss prediction based on weed density, herbicide selectivity, and cost, to arrive at improved herbicide-use decision making. An example is a model known as HERB.

6.2.4 Expert systems. These large models attempt to incorporate all aspects of crop growth modeling with all aspects of cultural management, pest management, and economics. Current examples include systems such as CALEX Cotton, GOSSYM/COMAX, TEXCIM for cotton, and others. They are called expert systems as they attempt to mimic the decision-making process of an experienced crop consultant. The use of expert systems in IPM is currently limited to a few effective examples.

6.3 Internet and WWW. The utility and use of electronic information dissemination tools increased rapidly in the 1990s and the use in pest management is expected to increase. The internet provides access to information relevant to many aspects of IPM (see text box and Anonymous, 2000). As mobile internet connections become easier to use, field-level access to information via the internet is becoming routine. It must be stressed that there is no oversight or control of the IPM information disseminated over the WWW, and the validity of WWW-based information should always be verified. Users should recognize that the so-called facts as presented on a website may not have been verified, so ascertaining the source of the information and comparing different sites is a prudent way of avoiding inaccurate information.

## IPM data available on the WWW:

1. Pest identification
2. Pest biology and ecology
3. Field records
4. Pesticide label information
5. Forecasting models
6. Phenological models
7. Weather data (historic and current)
8. Control recommendations

## Adoption of IPM Programs

A recently stated goal of the U.S. federal government was to have 75% of the agricultural acreage under IPM by 2000. Without a unanimously accepted definition (see earlier in this chapter and in Chapter 1) there is considerable difficulty in determining the extent to which IPM has been adopted. Depending on how IPM is defined, it can be argued that the 75% goal was or was not attained. In view of the small volume of available data, and in the absence of objective measurements of program performance, the extent to which adoption of IPM is a reality is unclear.

*A manager wants to minimize risk while maintaining adequate economic return. If an IPM program improves economic return or is perceived as less risky, then it will be adopted.*

Based on the definition of IPM adopted for this book, the use of truly integrated pest management is still relatively low. A worldwide review of the IPM literature suggests that there is little adoption of IPM on a global level (Kogan and Bajwa, 1999). Level II IPM was possibly used on about 5% of crops. A survey of farmers in Illinois (Czapar et al., 1995) indicated only modest use of IPM. Given the efforts that have been put into defining information for IPM, the low adoption of IPM might seem to be a concern. If IPM is a continuum, some elements have been more widely adopted than others. Certainly pest management in many agroecosystems has changed dramatically since the 1960s, and the IPM philosophy has made major contributions in that regard. IPM should probably be considered as a work in progress; and because the adoption of IPM is contingent on many unpredictable factors, including commodity prices and societal attitudes, adoption might change considerably at any time, and over a relatively short time.

Why has there been such slow adoption of IPM when programs labeled as IPM have been developed for nearly all our major crops? Crops for which IPM programs have been developed include wheat, rice, cotton, corn, soybean, sorghum, cowpea, cassava, potato, alfalfa, pepper mint, many vegetable crops, ornamentals, citrus, almonds, apple, pear, and many other smaller crops (see Anonymous, 2000). Reasons given by farmers and pest control practitioners in surveys for relatively low adoption of IPM programs include the following:

1. Adoption of IPM is difficult in low-value crops because of cost of population monitoring and deployment of alternative tactics to pesticides (see reason 4 below).
2. Growers perceive that IPM programs are more risky than conventional pest management. To be adopted, an IPM program must decrease the risk of economic loss to the farmer.
3. Many farmers note that most IPM programs to date have focused on arthropod management. Farmers emphasize that they cannot use an IPM program that does not consider weed management, as certain practices suggested for insect management may conflict with the long-term weed management. An example is the recommendation of habitat management to enhance natural enemies using potential weeds as alternative hosts for the natural enemies. IPM programs must address the integrated management of all major pest categories if the programs are to be widely adopted.
4. Adoption of IPM programs is often more expensive than conventional pesticide-based management, due to the increased need for population assessment and record keeping. Unless there is a clear economic advantage, an IPM program is not likely to be adopted. Clearly, in this

| TABLE 18–3 | Contrasting Features of Pesticide Technology and IPM as Possible Reasons for Rapid Adoption of the Former and Slow Adoption of the Latter |
|---|---|
| **Pesticides** | **IPM** |
| Compact technology from acquisition to application. Easily incorporated into regular farming operations. | Diffuse technology with multiple components. At times difficult to reconcile with current farming operations. |
| Promoted by the private sector. | Promoted by the public sector. |
| Strong economic interests; large budgets for research and development. | No economic incentive; limited budgets for research and development. |
| Aggressive sales promotion supported by professionally developed advertising campaigns. | Promoted by government and extension services personnel who are trained as educators not as salespersons. |
| Skillful use of mass communications media. | Limited support for use of mass media or for hiring communications media personnel. |
| Ability to provide incentives for adoption (free advice, glossy publications, bonuses, and small gifts). | No material incentives. Technical support provided by limited or inadequate staffing. |
| Results of treatments immediately apparent. | Benefits often not apparent in the short run; may be difficult to demonstrate (e.g., biological control). |
| **Consequently, pesticide technology was rapidly adopted.** | **Consequently, adoption of IPM has been slow.** |

regard, societal standards and feelings play a role in the adoption of IPM. If the nonfarming consumer public is willing to pay for the additional cost of IPM, perhaps through government subsidized programs, then adoption of IPM will increase. Crops for which IPM adoption provides an economic advantage do not require such subsidies.

5. Certain IPM programs can only be effective if adopted regionally. These have been known as area-wide programs. Most current IPM programs have been developed for implementation at the farm level; however, control of some pests—particularly mobile pests such as viruses, codling moth on apple and pear, lygus on alfalfa and cotton, *Helicoverpa* spp. on cotton, and some aphids—require management at the regional level. IPM is difficult to implement at a regional level because it requires cooperation of different segments of society.

Kogan and Bajwa (1999) assessed the reasons why conventional pesticide-based management was adopted over IPM. Their results are reproduced in Table 18–3, with some modification.

## SUMMARY

Adoption of IPM would seem to be logical, and indeed, no manager wants to destroy the ecological basis for an operation, or degrade the resources that are the very basis for their livelihood. Although there are a few unscrupulous entrepreneurs, most farmers are cognizant of the responsibilities associated with their stewardship of agricultural lands. They will attempt to use the IPM approach that fits harmoniously within the farm operation and will keep that operation profitable. Experience shows that adoption of IPM occurs when it meets the economic interests of the growers. However, it is becoming increasingly apparent that successful producers will need to factor in their budget analyses not only the costs and benefits at the farm gate, but also the long-term costs and benefits to the environment.

Consumers must recognize their significant role in the adoption of IPM programs, not only through desires for reductions in pesticide use, but also through the demands of the marketplace for quality, quantity, and price of produce. As an example, the attitude of consumers in the European Union toward genetically modified organisms has had consequences for agricultural production practices there. Consumer attitudes and demands for quality are evident to farmers through marketing orders and the details of commodity contracts. Consumers determine the acceptability of cosmetic damage and minor flaws in produce, and those demands are passed on to farmers. The success of organic farming in the United States and abroad has demonstrated the willingness of some consumers to pay a premium for produce that is perceived to be safer for human health and for the environment.

## SOURCES AND RECOMMENDED READING

Anonymous. 2000. Electronic list of outstanding IMP resources,
　　http://www.ippc.orst.edu/cicp/outstanding-resources/index.html

Czapar, G. F., M. P. Curry, and M. E. Gray. 1995. Survey of integrated pest management practices in central Illinois. *J. Prod. Agric.* 8:483–486.

Hoy, M. A. 1994. Parasitoids and predators in management of arthropod pests. In R. L. Metcalf and W. H. Luckmann, eds., *Introduction to insect pest management.* New York: Wiley, 129–198.

Hoy, M. A., and D. C. Herzog. 1985. *Biological control in agricultural IPM systems.* Orlando, Fla: Academic Press, xv, 589.

Kogan, M. 1998. Integrated pest management: Historical perspectives and contemporary developments. *Annu. Rev. Entomol.* 43:243–277.

Kogan, M., and W. I. Bajwa. 1999. Integrated pest management: A global reality? *An. Soc. Entomol. Brasil* 28:1–25.

Leslie, A. R., and G. W. Cuperus. 1993. *Successful implementation of integrated pest management for agricultural crops.* Boca Raton, Fla: Lewis Publishers, 193.

Morse, S., and W. Buhler. 1997. *Integrated pest management: Ideals and realities in developing countries.* Boulder, Colo.: Lynne Rienner Publishers, ix, 171.

# *19*

# SOCIETAL AND ENVIRONMENTAL LIMITATIONS TO IPM TACTICS

## CHAPTER OUTLINE

▶ Societal constraints and public attitudes
▶ Environmental issues
▶ Summary

The rapid pace of technological change during the past 30 years has opened new and exciting opportunities for improving pest management tactics and strategies; however, the pace at which the applications of new technologies can be assessed for long-term safety to humans and the environment contributes to uncertainties concerning the desirability of deploying those technologies. In addition, the needs and concerns of different societies vary over time, and those concerns place constraints on how innovative pest management tactics can be used. These are the limitations within which IPM must operate and be implemented. This chapter provides brief consideration of some problems that arise relative to the adoption of specific IPM tactics.

## SOCIETAL CONSTRAINTS AND PUBLIC ATTITUDES

Public perceptions and expectations often influence the tactics used and the rate and extent at which they are adopted within an IPM program. Public policies, both federal and state or regional, that affect IPM may be developed under the pressure of public expectations or perceptions. There is a risk that such policies may be unduly influenced by emotion or limited information rather than science, and so perhaps can be detrimental to the advancement of IPM (Figure 19–1).

**FIGURE 19–1**

Factors that influence public perception of risk.

Source: redrawn from Peterson and Higley, 1993.

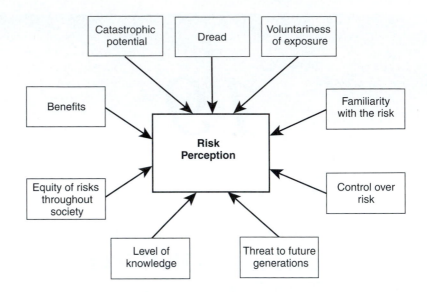

A historical perspective on how new ideas in science are accepted may help to reveal how the rate and direction of scientific discovery and its application are influenced by societal beliefs. Until the middle of the nineteenth century, religious belief dictated that diseases were an act of God, and that the fungi and bacteria associated with disease were the result of the disease, rather than the cause. Such beliefs delayed the understanding of the cause of disease. Similarly, the idea that organisms were produced by spontaneous generation misled scientists; for example, the idea that fly maggots arose from filth misled biologists for centuries regarding insect life cycles.

## Involvement in Agricultural Production

In agrarian societies, the majority of the people are directly involved with food production and most people realize that a reliable food supply is determined by their ability to control pests, especially weeds, or the extent to which they must grow extra crops to allow for the inevitable loss from pathogens and insects.

In industrialized nations, for example in the United States, only a small fraction of the population, about 2%, is directly involved with agricultural production (see Figure 1–1). Accordingly, it is easy for most people in these societies to underestimate the importance of pest management to the maintenance of a reliable food supply. Because most people are not involved in agriculture, the general public in industrialized countries has little appreciation for the significance of:

1. Crop losses caused by pests (see Chapters 1 and 2)
2. Costs of food production
3. Role that pesticides play in IPM programs in maintaining the quality and quantity of food produced
4. The complexity of managing pests, especially the enormous difficulty of implementing true IPM because of the detailed information required

The nonagricultural population in industrialized countries may have perceptions that differ greatly from the reality that exists on the farm. This nonagricultural public, about 98% of the population in the United States, may exert influence on public policy. Educating the public on the scientific basis of IPM is essential to avoid misconceptions that may result in misdirected public policy. In addition to the four points that relate to IPM implementation, the following are other societal issues that exacerbate crop pest problems.

## Pest Invasions

Many people do not consider the introduction of a nonnative organism a very serious problem (see Chapter 10 for examples). The general public needs to be aware of the concept of exotic species, and the consequences of their introductions (e.g., *Miconia* into Hawaii and potato cyst nematode into the Long Island area). Perhaps an even more serious problem for society is the intentional smuggling of plant materials and soil that can harbor pests. Smuggling activities range from large, illegal shipments of improperly handled produce, to seeds or plant material carried in a suitcase (dubbed "suitcase spread"). Illegal shipment of produce and undeclared importation of seeds and plant parts, such as cuttings of grape varieties carrying viruses or *Phylloxera,* have the potential to create serious problems for agricultural production.

World travel has increased over the past century. In the modern globetrotting society, intentional or unintentional movement of pests has increased (see Figure 10–2). People should abide by the laws (see Chapter 10) regarding the transportation of organisms in soil or plants, plant parts, and animals between regions and countries. Introduction of such organisms should occur through approved phytosanitary programs or established quarantine facilities.

Legal, or illegal, importation that results in the introduction of an exotic pest creates enormous problems for regulatory authorities. It often becomes necessary to treat large urban areas when they become infested. Such large-scale eradication efforts often face severe opposition from the public, even though regulatory officials are acting on behalf of the public good.

## Cosmetic Standards

Most consumers in industrialized countries expect blemish-free produce and produce free of pests (especially insects or insect parts). This expectation places major limitations on how IPM can be implemented in most fresh market crops.

When the pests or pest damage does not reduce yield per se, but does reduce the crop value, it is referred to as cosmetic damage (see Figure 8–9). A few scars on the outside of an orange or a few aphids on a head of lettuce do not alter the yield or food value of the produce. To a great extent, cosmetic standards necessitate the use of insecticides and fungicides on fresh produce as the crop approaches harvest. The standards are set by the U.S. Department of Agriculture in the USA, and are referred to as defect action levels (or DALs). Supermarket buyers, produce packers, and shippers enforce the DALs. In most situations, the only way cosmetic standards can be met is through the use of pesticides.

Why do cosmetic standards exist? In affluent societies, consumer expectations are the reason. Consumers in such societies select what appear to be the nicest fruits and vegetables. Few people willingly choose the damaged, scarred, or insect-contaminated produce. Describing a few insect parts in produce as "filth" does not help this situation. If a producer or produce manager does not keep produce clean and blemish free, then it probably cannot be sold. At best, the damaged produce goes to processing at a much lower economic return.

If consumers were more willing to accept blemishes on produce, and a low level contamination with insects and pathogens, then the pesticide use required to achieve high cosmetic standards could probably be reduced. Society has presented the IPM practitioner with an impossible conundrum to reduce pesticide use but keep the produce cosmetically perfect. There is no easy scientific solution to this puzzle.

## Preferences, Food Quality, and Pesticide Use in IPM

Pesticides are an important tactic in many IPM programs, but there are valid concerns relative to the use of pesticides in IPM.

***Risk Perception***    Humans have a limited capacity to accurately assess the magnitude of risks. As an example, consider the nervousness that people feel before boarding a commercial aircraft relative to the ease with which people venture forth in their automobiles. Which of the two forms of transportation actually presents the greater risk of personal injury? On average, it is clear that travel by personal automobile is less safe than travel by commercial aircraft. Risk perception plays a major role in how people react to pesticides and possible pesticide residues in their food. The factors that are involved in assessing risk are presented in Figure 19–1. An inability to accurately assess risk may increase fear of that risk.

**Fear and Familiarity with Risk**    Fear can arise for many reasons, including unfamiliarity, lack of understanding, misinterpretation or distrust of information, and comparison with known risks. Lack of understanding of pesticide chemistry and ecotoxicology may magnify the perception of risk and hence the fear of pesticide residues in food. Because of this lack of knowledge, it is difficult for most people to make accurate estimations of the risks associated with pesticide residues.

**Threat to Future Generations**    A threat to future generations is perceived when risks are considered over the long term. Possible effects of pesticides, such as the longevity of residues in the environment, the sequestering of toxins in the food chain, and unrecognized genetic damage caused by chronic exposure to pesticides are among the long-term concerns.

**Benefits**    The benefits derived from pesticides, such as their contribution to low food prices, may not be obvious to the general consumer in industrialized countries. A common misperception is that the only segment of society that benefits from pesticides is the chemical industry and possibly farmers.

**Catastrophic Potential**    The fear that risk implies the imminence of a major disaster often leads to mistrust of pesticide use. Based on the perception of risk, some people have adopted an antipesticide attitude. Organic farming is increasingly popular in meeting the demands of consumers who prefer to purchase commodities that have not been subject to pesticide applications.

***Pesticides and Public Health***    Pesticides are poisons and should be used in accordance with the proper safety protocols. Carelessness in handling pesticides may result in severe injury or even death. Due to enactment and enforcement of laws regarding pesticide use in industrialized nations, there are relatively few deaths caused by pesticides in those countries. Acute toxicity from a single dose is generally not an issue for most of the public, because they are normally not exposed to the concentrate; however, workers involved in pesticide application can be exposed to pesticide concentrate, and those involved in general field work can be exposed to high levels of pesticide residues. Laws have been established that, when followed (see Chapter 11), limit workers' exposure to injurious levels of pesticides. Between 1975 and 1986, eleven occupational deaths were related to pesticides in California (see Wilkinson, 1990). Conversely, during that same period, 130 farm workers died from tractor accidents and over 1,000 people died from work-related truck and automobile accidents.

There is always a need for awareness of pesticide-related illness, particularly in less developed countries. The World Health Organization reports many cases of poisoning and death from pesticides in less-developed countries, mainly of farm workers. This is attributed to lack of education and training on the hazards involved, lack of proper protective equipment, and lack of regulations or compliance with regulations. Problems can occur when pesticides are sold to people who do not know how to use them correctly or do not have the necessary equipment to apply them safely, resulting in injury to the personnel involved. Although local governments may have adopted pesticide regulations to protect the health of their agricultural workers and consumers, in many instances there has been no enforcement or supervision.

***Food Safety***    Pesticides are toxic compounds (see Chapter 11) and there are risks associated with their use. Chronic long-term exposure to traces of correctly used agricultural pesticides is a concern, but the risk from such exposure is generally considered to be low. Recent regulations in the USA mandate a reassessment of many pesticide registrations to include consideration of long-term chronic and aggregate exposure to pesticides (see FQPA in Chapter 11). The scientific assessment of such long-term risk is very difficult.

The Institute of Food Technologists (IFT), which represents 14 scientific societies involved in food and nutrition science, reached the following conclusions in 1989, subject to caveats raised by the 1996 Food Quality Protection Act

- The perception that the food supply is unsafe is not supported by scientific data; the American food supply is among the safest in the world.
- The scientific community needs to communicate the concept of risk more effectively to the public.
- Current allocation of resources devoted to food safety is not commensurate with the actual risks but more with risks perceived by the public.

Source: Institute of Food Technologists, 1989.

The IFT concluded that the risks from pesticide residues in food are orders of magnitude lower that those for several other aspects of food supply and nutrition. The following food safety issues were ranked in order of importance.

1. Food-borne diseases (e.g., salmonella and other enteric disease organisms)
2. Malnutrition (either through inadequate food supply or not eating a balanced diet)
3. Environmental contamination (natural and human made, but not including pesticides)
4. Naturally occurring toxicants (e.g, mushroom toxins and many mycotoxins in food, such as aflatoxin in peanuts)
5. Pesticides, food additives, and other agrochemicals

The toxicity and potential dangers of pesticides should not be ignored, but the hazard of pesticide residues should be placed in perspective relative to the other hazards that are associated with food.

A major public concern surrounds the possibility that pesticide residues in food cause cancer. Cancer is not a single disease and there is no single cause for all cancers. The occurrence of some cancer clusters and other illnesses have been attributed to pesticide use. Determining the true cause of such regional health problems is difficult, because many factors may be involved. Bruce Ames, of the University of California, Berkeley, in a series of papers with L.S. Gold, addressed the ability of pesticides to cause cancer (carcinogenicity) (Ames et al., 1990a, b; Gold et al., 1992; Ames and Gold, 1993). They concluded, "The major preventable risk factors for cancer that have been identified thus far are tobacco, dietary imbalance, hormones, infections, and high dose exposures in an occupational setting. . . . What is chiefly needed is to take seriously the control of the major hazards that have been reliably identified." Pesticides, in their view, were not a significant cancer-causing hazard for the general public when pesticides were properly used under an IPM program.

## Pest Management Alternatives to the Use of Pesticides

IPM strives to promote ecologically sound and economically feasible approaches to pest population regulation. Given the nature of modern agricultural production, there are instances when the rapid, remedial power of pesticides is needed to control pests. This fact is often true for weed control, and also for many arthropod, nematode, and pathogen infestations.

## Conflict of Interest

There is a possible conflict of interest when the person who recommends the use of pesticides is the same person who sells them. The pesticide industry argues that such conflict of interest is self-regulating, in that a person who recommends the use of pesticides when none are needed will not remain in business for long due to competitive pressures. The potential conflict of interest may make it necessary in the future to separate pest control advising from the selling of pest control technology.

## ENVIRONMENTAL ISSUES

Concerns about environmental quality or function can impact utilization of pest control tactics employed in IPM programs in several ways. They are outlined here, but are not prioritized.

## Soil Erosion

Cultivation, which is the most widely used weed management tactic, has the environmental drawback that it exacerbates soil erosion (see Chapter 15). Reducing tillage has been proved to result in much lower levels of soil erosion, especially following heavy rains when as much as 1,000-fold reductions in soil loss have been recorded. However, reducing tillage places greater reliance on herbicides for weed management.

# Air Pollution

***Dust Hazards***   Soil cultivation in dry climates can lead to wind erosion and airborne dust. Urban societies may deem such dust to be unacceptable. Regulations are being developed to control emission of particulate matter under $10\mu m$ in size, the so-called $PM_{10}$ standards. Implementation of regulations to reduce $PM_{10}$ particulates has the potential to alter the use of tillage for weed management, and could also lead to increased reliance on herbicides.

***Smoke***   Fire has been used for pest management for many centuries, and burning crop residue has been a standard practice to reduce overwintering inoculum of pathogens in several crops, such as grass seed crops and rice. However, burning crop residue creates air pollution. Many societies now deem such air pollution unacceptable. IPM systems that use alternative tactics to manage the pests have been developed or are being developed.

***Volatile Organic Compounds***   Many pesticides are formulated in volatile organic solvents. After the pesticide has been applied, the solvent evaporates, causing air pollution. The volatile organic compounds (VOCs) then combine with nitrous oxides and result in production of ozone. VOC pollution is not a problem for pesticides that are either formulated dry or in water. Due to concerns about VOC pollution, several pesticides have been, or are being, reformulated without such solvents. This reformulation may compromise efficacy.

***Drift***   Virtually all spray applications of pesticides are susceptible to drift (see Chapter 11). Use of a pesticide can be restricted if it is found to pose unacceptable drift hazard on adjacent crops or ecosystems. Restriction can take the form of specifying the type of application equipment, the time of year when applications cannot be made, or bans on application. For example, many pesticides cannot be applied by aircraft because of the drift hazard. Federally mandated use restrictions are stated on pesticide labels (see Chapter 11). In the United States, implementation of regional use restrictions vary by state.

# Endangered Species

An IPM program should not harm rare nontarget organisms. In the United States, the Endangered Species Act is used to designate species as endangered, which means that any farming practice, including IPM, must not endanger such a species. Presence of an endangered species on a farm, or in the local area, can restrict choice of pest management tactics such as limiting pesticide use, stopping deployment of biological control, and even decreasing the use of tillage. Protection of endangered species is important; thus, IPM programs must resolve the pest problems without impacting any endangered species in the region.

# Food Chain Considerations

If a chemical or chemical compound has characteristics that make it poorly metabolized and poorly excreted by animals, and if the chemical accumulates in fat, then the potential exists for biological magnification, or biomagnification, to occur. Biomagnification can result in the concentration of pesticides and heavy metals in organisms such as fish and raptors.

Biomagnification is diagrammed in Figure 19–2. The pesticide is present in the plants at relatively low concentration following application. When a herbivore eats the plants, it ingests some of the pesticide; but because the pesticide is not broken down and accumulates in fat, it is retained in the body of the herbivore. Of course the herbivore does not eat just a single plant, but rather continues to consume plants that contain low levels of pesticide, and over the life of the herbivore, the pesticide slowly accumulates and eventually reaches a higher concentration than originally present in the plants (shown in Figure 19–2). When a primary carnivore eats the herbivore, the process is repeated, and the pesticide reaches an even higher concentration in the fat of the carnivore. With each successively higher trophic level, the relative pesticide concentration continues to increase. It may happen that levels of pesticide can be accumulated in higher-order consumers that cause physiological problems or even death.

The eggs of certain birds became weakened as a part of the physiological consequences of biomagnification of DDT, with resultant death of bird chicks. Rachel Carson brought this to the attention of the world in her book *Silent Spring* (1962). Most current pesticides do not have the characteristics that lead to biomagnification, although a few (e.g., biodifacoum, a rodenticide) have a weak tendency to accumulate so their use is carefully monitored.

DDT, and most chlorinated hydrocarbon insecticides, had the chemical characteristics of environmental persistence and accumulation in fat that resulted in their biomagnification in the food chain. As the significance of biomagnification was recognized, the use of such chemicals was either restricted or banned. Concern for biomagnification in the food chain is sufficiently great that all new pesticides are tested for biomagnification.

**FIGURE 19–2**
Diagrammatic representation of pesticide residue biomagnification in the food chain. Each dot represents pesticide residue, and each box represents the decreasing biomass at each higher trophic level. The number of residue dots is the same in each box to demonstrate how the pesticide becomes more concentrated at each higher trophic level.

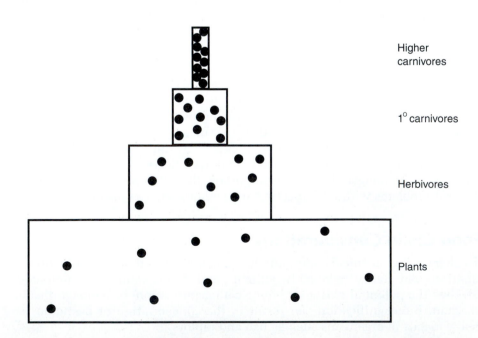

# Wetlands

Many areas are being set aside as wetland preserves. Potential pesticide runoff, or drift, into wetlands is of greatest concern. Pesticide use in the vicinity of an area designated as wetlands may be restricted to avoid contamination.

# Groundwater Contamination

Probably one of the most dangerous effects of pesticides is their unintended contamination of groundwater. Pesticide movement in water is an important mechanism by which pesticides are transported from the site of application to other areas. Under certain conditions, some pesticides have moved into groundwater. Groundwater is a source for drinking water, and decontaminating groundwater to make it suitable for human consumption is expensive. Accordingly, keeping pesticides out of groundwater is a high priority.

Many regions have instituted routine monitoring for pesticide contamination of water in wells. Data for the United States are available from the U.S. Geological Survey. Over the past two decades, pesticides or their by-products have been detected in groundwater in 43 states. High-use pesticides are frequently detected, such as the triazine and acetanilide herbicides (examples include atrazine, simazine, alachlor, and metolachlor). Pesticides for which sampling has been more intensive because of contamination problems also were detected more frequently. The insecticide-nematicide aldicarb and its transformation products, and the fumigants DBCP and ethylene dibromide, are among them. Over 143 other pesticides, plus 21 transformation products, have been detected in a limited number of wells.

Most surveys of pesticides in groundwater have reported that over 98% of the detections did not exceed 1 microgram per liter (1 part per billion). Maximum contaminant levels (MCLs) established by the U.S. EPA for drinking water were exceeded in less than 0.1% of the sites sampled. The extent of hazard to health is thought to be limited, but there is legitimate concern that multiple pesticides present in groundwater, where such contamination occurs, and long-term exposure might pose a greater hazard than when present singly.

Contamination of groundwater is most likely to occur in the following situations.

1. In locations where the water table is close to the surface
2. For pesticides that leach readily, degrade relatively slowly, and are applied multiple times
3. Soil profile characteristics that influence the movement and breakdown of pesticides as they pass through the profile, for example, depth, organic matter content, texture, hydraulic conductivity, permeability, impervious layers, and microbial activity
4. Major runoff events, such as flooding
5. Damaged or poorly constructed wells
6. Accidents, carelessness, and disregard for proper pesticide application

Reduction in pesticide contamination of groundwater requires recognition and modification of pesticide availability or of use patterns. Where contamination has occurred, changes must be instituted to reduce or eliminate the chance of the pesticide reaching the groundwater. Best management practice (BMP) standards have been introduced to mitigate the likelihood that pesticides will move to groundwater. In some regions (e.g., the Netherlands), pesticides that do not meet persistence and leaching criteria are used under restricted conditions, or are no longer used. All uses of bentazon herbicide in California rice, for example, were canceled because of groundwater contamination problems.

## Significance of Release of Genes

Society has not traditionally been concerned about the development of pest-resistant crop plants through conventional plant breeding. However, the use of genetic engineering to develop crops that either resist pests or pesticides has generated concerns. These concerns have placed constraints on the adoption of such crop varieties for pest management purposes in some parts of the world.

Genetic engineering allows genes to be moved in ways that were not possible using the methods of traditional breeding. Genes can be transferred within a species, between different species, between different genera, and even between organisms in different kingdoms using genetic engineering technologies. In addition, new forms of genes can be designed, and organisms transformed with those genes. The widespread deployment of genes in the genetic backgrounds of organisms that have not previously contained those genes present new concerns. The concerns arise for numerous reasons, and several are presented in the following points.

1. Development of resistant pests to engineered resistance. Development of resistance in the target pest may occur. Widespread adoption of a tactic based on genetic engineering will place a very high selection pressure on pests, with the attendant potential for evolution of resistance-breaking pest lines. Relative to this concern are efforts to develop resistance mechanisms with a mode of action that targets basic physiological processes in the pest so that selection will be unlikely to circumvent the mechanism.
2. Gene flow from engineered plants to wild plants. The term "super weeds" has been used to label plants that might acquire herbicide resistance from crop plants that carry engineered herbicide resistance. The possession of such resistance genes would be a benefit to the weed only under selection pressure from the particular herbicide. If insect or pathogen resistance were to be accidentally transferred to weeds from a crop, weed fitness might increase because of decreased endemic biological control of the weed by insects or pathogens.
3. Ecosystem disruption. Ecologists consider that herbivores and pathogens are involved in many important ecosystem processes. If resistance genes escape from crop plants and move into other plant species, it could conceivably affect ecosystem stability due to altered fitness of the component plant species. Changes in ecosystem features could happen subtly, and major effects might not be detected until too late.
   3.1. Changes in insect fauna. If insect-resistance genes, such as those conferring Bt endotoxin production in plants, escaped into wild plants, insect guilds and all the higher-order organisms that feed on them might be altered.
   3.2. Changes in pathogens. If genes conferring resistance to pathogens are inserted into crop plants and they were to escape into wild plants, it

would possibly alter the ability of the wild plants to resist diseases, which might in turn alter the competitive relationships in the community.

## Significance of Release of Biological Control Agents

Ecologists have questioned the long-term effects of releasing non-indigenous organisms into an ecosystem. There is potential for unexpected and undesirable ecological interactions among organisms. The introduction of biological control agents that are pathogens, predators, or parasitoids has potential to influence the population dynamics of nontarget organisms. For this reason, extensive testing is conducted to assess potential undesirable trophic interactions prior to the release of any biological control agent into the environment (see Chapter 13).

There are many examples of problems caused by intentional release of nonindigenous species for biological control purposes. A seedhead weevil was introduced as a biological control agent for weedy thistles, but has now extended its range to native nonweedy thistles, some of which were considered endangered even before the weevil started to attack them. Another example is the discovery that the tachinid fly released repeatedly for gypsy moth control in the eastern United States has probably eliminated some large native moths. Other examples noted in Chapters 10 and 13 include the rosy wolfsnail in Hawaii and the mongoose in several regions. These examples show that release of nonindigenous species can potentially have serious repercussions.

The possibility of such unforeseen interactions exists for biocontrol agents, because it is impossible to determine with certainty all possible trophic interactions for an organism prior to its release into a new environment. It is also impossible to anticipate potential host shifts that may occur after an introduced organism adjusts to a new environment. The implications of insect and pathogen release, for which effects cannot be observed until too late, are significant. Reevaluation of how release of non-indigenous species should be conducted, as mandated by the Plant Protection Act of 2000, may preclude use of such organisms in some biological control programs.

## Impacts on Biodiversity

Many human activities impact biological diversity. There is growing awareness about the benefits of preserving the biological diversity of ecosystems. The application of principles of biological conservation in agriculture, however, is complex and still tentative for most cropping systems. The following is based on the conclusions of a paper regarding the role of insect conservation in IPM (Kogan and Lattin, 1992). Although the focus is entomological, some concepts and assumptions apply to all components of the crop community.

Even the most casual examination of crop communities reveals highly simplified systems with only a few species involved, many of which are exotic, including the crops themselves, weeds, animal pests, and beneficial organisms. This is a biological legacy from historical developments in agricultural production systems and past practices based in large part on pesticide use. The concepts of IPM, and more recently of sustainable agriculture, have shifted the emphasis toward programs that promote a more diverse fauna and flora in the cropping system. The success of these programs rests upon a melding of biological, cultural, and chemical control tactics and an understanding of multiple crop/pest/natural enemy interactions. The success of the biological and cultural controls depends largely on the skillful manipulation of a number of species of plants and animals in time and space. This shift in the control paradigm immediately results in increased biological diversity of both plants and animals. In theory, the objective of conservation biology and enhancement of biodiversity has been and is being achieved if numbers of species only are the objective. A further shift toward true conservation of species requires at least some reduction in nonindigenous species and the enhancement of indigenous species.

Most crop plants are exotic species, and this fact is not likely to change. Most weeds and many of the major insect pests are exotic as well. At least some beneficial insect species are exotic too. By and large, there are more exotic parasites in U.S. cropping systems than exotic predators. The predator complex is more likely to be composed of indigenous species. If these generalities are true, there are several portions of the cropping system that present areas most likely to reduce exotics and enhance indigenous species. The functional group most likely to benefit from an IPM approach to conservation probably is the predator complex. It would appear that the greatest opportunity to enhance native insect conservation is to put effort into increasing predator diversity and numbers. Since many predators feed on both native and exotic prey species, sustaining adequate population levels of predators usually involves maintaining alternate prey species as well—further increasing insect diversity.

Another option to promote conservation of native species involves the reduction of the exotic noncrop plants through the use of biological control agents. Biocontrol of weeds, however, creates the dilemma of introducing additional exotic organisms to the system (i.e., insects or diseases) with the attendant danger that those agents might switch to attacking nontarget plants. Effective cultural control methods might be a more desirable action. One might also examine the use of multiple cropping systems, using one crop as a competitor to weeds rather than the reverse. Cover crops and living mulches are gaining increased acceptance; if varied native rather than exotic plants are selected, enhancement of biodiversity will result. These efforts will loop back to the diversification of the native plants in and around the cropping system, further enhancing biological diversity.

Certain practical aspects limit the extent to which native plants can be incorporated into the broader cropping system. The diversity of native insects can be increased with greater use of native plants as part of an effort to enhance the diversity and species richness of the predator complex. A diverse predator fauna, involving a broad spectrum of species, is likely to be more effective than relying on only a few exotic species often selected against a single pest species. Use of tactics to increase diversity must, however, be analyzed from an ecosystem perspective before adoption into an IPM program.

## SUMMARY

Actions of society place many constraints on the way that tactics can be deployed in the implementation of an IPM program. Many of these constraints have forced pest management researchers and practitioners to develop improved tactics to manage pests. Others have severely limited the ability of control pests, and in a few cases crop production has been forced to change.

# SOURCES AND RECOMMENDED READING

The following topics and related textbooks are suggested for further reading: risk concepts (Richardson, 1988; Huber et al., 1993; Peterson and Higley, 1993); food safety (Baker and Wilkinson, 1990; Tweedy et al., 1991; Pimentel and Lehman, 1993); and pesticide contamination of groundwater (Gustafson, 1993; Vighi and Funari, 1995; Barbash and Resek, 1996; Barbash et al., 1999).

Ames, B. N., and L. S. Gold. 1993. Environmental pollution and cancer: Some misconceptions. In K. R. Foster, D. E. Bernstein, and P. W. Huber, eds., *Phantom risk.* Cambridge, Mass.: The MIT Press, 153–181.

Ames, B. N., M. Profet, and L. S. Gold. 1990a. Dietary pesticides (99.99-percent all natural). *Proc. Nat. Acad. Sci. USA* 87:7777–7781.

Ames, B. N., M. Profet, and L. S. Gold. 1990b. Nature's chemicals and synthetic chemicals—Comparative toxicology 3. *Proc. Nat Acad. Sci. USA* 87:7782–7786.

Baker, S. R., and C. F. Wilkinson, eds. 1990. The effects of pesticides on human health: Proceedings of a workshop, May 9–11, 1988, Keystone, Colo. *Advances in modern environmental toxicology,* vol. 18. Princeton, N.J.: Princeton Scientific Pub. Co., xxi, 438.

Barbash, J. E., and E. A. Resek. 1996. *Pesticides in groundwater: Distribution, trends, and governing factors.* Chelsea, Mich.: Ann Arbor Press, xxvii, 588.

Barbash, J. E., G. P. Thelin, D. W. Kolpin, and R. J. Gilliom. 1999. *Distribution of major herbicides in groundwater of the United States.* Sacramento, Calif.: U.S. Geological Survey, 58.

Carson, R. 1962. *Silent spring.* New York: Fawcett Crest, 304.

Gold, L. S., T. H. Slone, B. R. Stern, N. B. Manley, and B. N. Ames. 1992. Rodent carcinogens; setting priorities. *Science* 258:261–265.

Gustafson, D. I. 1993. *Pesticides in drinking water.* New York: Van Nostrand Reinhold, xii, 241.

Huber, P. W., K. R. Foster, and D. E. Bernstein. 1993. *Phantom risk: Scientific inference and the law.* Cambridge, Mass.: MIT Press, x, 457.

Institute of Food Technologists. 1989. *Assessing the optimal system for ensuring food safety: A scientific consensus.* IFT office of scientific public affairs, IFT Toxicology and Safety Evaluation Division, released April 5, 1989. Chicago, Ill.: Institute of Food Technologists, 25.

Kogan, M., and J. D. Lattin. 1993. Insect conservation and pest management. *Biodiversity and Conservation* 2:242–257.

Peterson, R. K. D., and L. G. Higley. 1993. Communicating pesticide risks. *Amer. Entomol.* 39:206–211.

Pimentel, D., and H. Lehman, eds. 1993. *The pesticide question environment, economics, and ethics.* New York: Chapman and Hall, xiv, 441.

Richardson, M., ed. 1988. *Risk assessment of chemicals in the environment.* London, UK: Royal Society of Chemistry, xx, 579.

Tweedy, B. G., H. J. Dishburger, L. G. Ballantine, and J. McCarthy, eds. 1991. *Pesticide residues and food safety: A harvest of viewpoints,* ACS Symposium Series 446. Washington, D.C.: American Chemical Society, xv, 360.

Van den Bosch, R. 1978. *The pesticide conspiracy.* Los Angeles, Calif.: The University of California Press, xiv, 226.

Vighi, M., and E. Funari, eds. 1995. *Pesticide risk in groundwater.* Boca Raton, Fla.: Lewis Publishers, 275.

Wilkinson, C. F. 1990. Introduction and overview. In S. R. Baker and C. F. Wilkinson, eds., The effects of pesticides on human health: Proceedings of a workshop, May 9–11, 1988, Keystone, Colo. *Advances in modern environmental toxicology,* vol. 18. Princeton, N.J.: Princeton Scientific Pub. Co., 5–33.

# 20

# IPM in the Future

> *You got to be careful if you don't know where you're going, because you might not get there.*
> *The future ain't what it used to be.*

—Yogi Berra

## CHAPTER OUTLINE

▶ Introduction
▶ How to measure progress in IPM
▶ Directions of possible changes
▶ Strategic developments
▶ Summary

## INTRODUCTION

Since the expression "integrated pest management" was explicitly presented in the late 1960s, the concept has been under close scrutiny. It has been criticized, alternative names have been proposed, and qualifiers added, but IPM has remained as one of the most robust concepts to appear within the agricultural sciences during the second half of the twentieth century. It is now generally agreed that adoption of IPM, although lagging in most crops and in most parts of the world, is not an option; it is an imperative if agricultural production is to keep up with the demands of a growing human population. IPM is a work in progress, and although ideal IPM has probably not been achieved in any particular cropping system, tremendous strides have been made toward IPM in many systems.

In real life, managers must integrate all aspects of pest management within the framework of an agricultural production system. All components of the production system—from the choice of crop, variety selection, soil preparation, preplant pest management tactics, and on through harvest and storage—are closely interconnected and all have an impact on the nature and severity of pest problems. The challenge for the future will be the extent to which agricultural scientists, together with producers and pest management practitioners, can develop and use the ever-increasing knowledge of pest biology and ecosystem function to sustain the productivity of agricultural ecosystems, while at the same time achieving the lowest possible collateral impact on the environment and the highest possible return to producers and to society as a whole.

Looking into the future is fraught with uncertainty, but a few things may be predicted with some confidence. Humans will continue to grow food and fiber to satisfy their basic needs, as well as some that are not so basic but become requisites of increasingly more affluent societies. Of equal certainty is that other organisms—the pests—will attempt to share those commodities with humans, so competition, or confrontation, between humans and pests will continue, requiring that humans do their best to manage pests. How will this be done in the future? As with all human endeavors, it is difficult to predict, as scientific breakthroughs cannot be foreseen. As the physicist Niels Bohr said, "It is difficult to make predictions, particularly about the future!" The following ideas are therefore presented with the caveat applicable to all long-term predictions that they stand a good probability of being wrong.

Pest management is intended to alter the ecosystem to benefit humans' interests. The goal of IPM should be to achieve the desired level of interference with ecosystem processes needed to regulate pest impacts with minimum undesired impacts on other ecosystem functions. It may be necessary in the future, as has happened in the past, to cease practices that are found to have too large a negative impact. The cessation of most uses of the insecticide DDT and the nematicide DBCP are examples of such decisions.

## HOW TO MEASURE PROGRESS IN IPM

A stated goal established by the United States government was to promote adoption of IPM practices so that by the year 2000, 75% of agriculture would be under some level of IPM. The problem with this goal was the difficulty in measuring IPM adoption due to lack of a common definition of IPM and a well-established baseline of the level of adoption in most crops (see Chapters 1 and 19).

The multiplicity of current definitions for IPM and the questions that are now being raised about the extent to which truly *integrated* pest management is practiced, suggests that a first consideration for the future is to develop a universally accepted definition. Until this is accomplished, it will be difficult to assess IPM adoption from either the research or practical level.

As noted previously, pest management will be required for the foreseeable future, and managers will necessarily carry out some form of IPM within the bounds of their knowledge. The key factor is the extent to which IPM will be integrated within and across disciplinary lines, that is, whether pests are considered individually or as part of the multiple interactions in the agroecosystem. Progress in adoption of IPM has been made, and more is anticipated in the future.

## DIRECTIONS OF POSSIBLE CHANGES

Chapters 10 through 17 introduced the control tactics that form the basic arsenal of IPM. There are many ways that pest management may change in the future. Most of the changes will certainly occur in development and improvement of existing tactics; but some essential developments will occur relative to the acquisition, organization, and availability of the information that is required for improved decision support systems.

### Pest Biology and Ecology

Knowledge is the underpinning for IPM. Although this statement seems obvious, it is still underappreciated. The continued improvement in understanding the biology and ecology of pest organisms is necessary to allow higher levels of IPM integration. Such understanding often seems to be ivory-tower basic biology and ecology far removed from application, but the contributions of such information to applied pest management cannot be overstated. Understanding multiple pest interactions and those of biological control agents with their environment will lead to improved management strategies and increasingly complex decision making. The role and impact of ecosystem diversity in relation to IPM will gradually be clarified, which will lead to improvement of regional-level IPM programs. IPM at this level will require significant strategic shifts, and area-wide IPM is an example of such a shift.

### Pest Monitoring and Decision Making

Technological developments have changed the approaches to arthropod pest monitoring and decision making starting in the 1980s. Advances in the identification, purification, and use of pheromones and kairomones as attractants in traps have been of great benefit in increasing the accuracy of monitoring arthropod pests.

Other important developments are based on the rapidly expanding computer and global communications technologies since the early 1980s. Increases in computer power, and reduction in size and weight, should contribute to greater use of computers for field diagnosis and real-time decision making. Data retrieval and recording, running models, and using expert systems in the field will increase through the use of powerful portable computers and data loggers. With these tools available in the field, where the data are collected, improvements in speed and accuracy in the decision-making process will occur. Conceivably, the use of computers to evaluate data will be the only possible way to achieve integration of pest management because of the enormous amount of information that must be evaluated. Expected future changes include the following:

1. Pest identification. Access to expert system pest identification will become routine, including picture catalogs of the pests. It is conceivable that computers will, through digital imaging technology, be able to rapidly identify pest organisms. Identification of pests will be expedited by genetic profiles and the use of genetic markers for identification will become more of a standard practice, particularly for cryptic organisms such as pathogens and nematodes.

2. Forecasting. Improvements in the ability to run models that forecast events will be an important area for disease and insect pest management. Improvements in forecasting for weed emergence in some regions also may provide a valuable tool in weed management.

3. Improved thresholds and monitoring. Development of user-friendly expert systems operating on handheld field computers will greatly improve the ability to use economic thresholds where such decision-making parameters are useful. Another computer application is the use of degree-day models to predict pest incidence and possible abundance. Degree-day models are an essential component of decision support systems for IPM and are becoming readily usable in the field. Such models are currently available through the WWW with accurate weather data downloaded in near real time for use in connection with the models.

4. Data storage and retrieval for recommendations. The ability to quickly and easily access large databases from the field, such as field historical records, pest biology, control recommendations, and pesticide label information, will greatly enhance the ability of PCAs to make IPM decisions on the spot.

5. Computer digital imaging technology. Such advancements will be developed to permit real-time crop and pest (especially weed) identification by machine. This will in turn permit programming of robotic equipment to control sprayer and other mechanical field equipment (see later), thus decreasing pesticide use and risk to human operators.

***Precision Farming***   The use of computers for precision farming (also called site-specific farming or prescription farming) through a global positioning system (GPS) and a geographic information system (GIS) will result in site-specific application of control tactics in real time. Farmers will apply pest management tactics over less than field-scale areas, and to vary the tactic depending on small-scale variations in pest distributions within spatial scales as small as a few feet. This will allow the use of tactics in those areas where actually needed. Yield mapping is already done, but the following also may be available in the future.

1. Mapping pest distributions within fields especially for less mobile pests such as weeds and soil-borne pathogens and nematodes

2. Monitoring the spread of mobile disease epidemics and arthropod invasion

3. Monitoring the movement of biological control agents released at points in fields

This information will then be used to better predict pest development and spread. For example, applying herbicide to control weeds may not be the best tactic when the weed presence was due to lack of competition from the crop, which had died because of a disease resulting from poor soil drainage. Such problems may be better diagnosed with more historically complete and spatially accurate information.

## Legislative Controls

Laws and regulations aimed at restricting invasion of pests into areas that are not already infested will almost certainly be increased on a worldwide basis.

From a historical perspective it is likely that these will not be very successful unless public attitude toward the spread of nonnative species shifts dramatically. Attaining any meaningful changes in this area will require public education. The regulations governing pesticide use, and the use of GMOs are, for the most part, always under consideration relative to current regulatory developments and availability of new information.

## Pesticides

The use of pesticides will continue, but it is anticipated that the efficiency of the products, safety of use, and the nature and mode of action (particularly of insecticides) will change in the future. Development of novel synthetic pesticides will be at a slower pace than in the second half of the twentieth century. The use of more selective insecticides and biorational pesticides derived from living organisms will increase. Of all the pesticides, herbicides for weed management will continue to be used in the largest amounts and on the greatest area.

Pesticide-delivery technology is changing, and "smart sprayers" that can detect the presence of weeds will be used for postemergence weed management. These weed-seeking sprayers will turn on and off depending on weed presence, which will greatly reduce herbicide use. Orchard sprayers will be able to sense tree size and automatically adjust spray patterns. Nozzles and spray technologies will be improved to reduce the risk of drift and nontarget effects.

## Managing Resistance, Resurgence, and Replacement

Long-term IPM sustainability requires increased awareness that control tactics exert selection pressure on pests, and that increased diligence will be needed in managing the three Rs: resistance, resurgence, and replacement of pest organisms. In the past, single tactics were considered expendable because of the seemingly ready availability of new tactics to replace those lost to pest resistance. Both technological and economic barriers prevent rapid development of new tactics, particularly pesticides, in the future. Thus, replacement of lost tactics cannot be assured. The multitactic IPM approach will become the norm for managing the three Rs of pest management.

## Biological Control

Biological control will continue to be the underpinning of most arthropod management programs. There will be further developments for biocontrol of pathogens and nematodes, and, perhaps more modestly, for weeds. There will also be much closer scrutiny for unintentional effects of introduced organisms, and there will be more regulation of importation of natural enemies. Such regulation may stifle efforts to control exotic invasive species in a timely fashion.

## Behavioral Controls

The use of semiochemicals, especially sex pheromones, to manage arthropod pests will increase in the future. Application of modern chemical tools to quickly identify semiochemicals, including pheromones, and development of novel ways of delivering them in the context of IPM systems will open new avenues for the management of arthropod pests.

# Cultural Tactics

Two areas of crop culture have been changing pest management since the mid-1970s. Both irrigation management and no-till or reduced tillage systems may further impact pest management.

*Irrigation Management*    In regions of the world where irrigation is part of the agricultural system, changing the method of water application may alter several aspects of IPM. The use of localized delivery systems, such as drip irrigation, will improve the ability to manage pests. Compared with sprinkler or flood irrigation, drip systems will reduce problems caused by pathogens, nematodes, and weeds that need water to infect or germinate, and through reduction in the amount of wetted soil and reduction in waterlogged conditions.

*No-till and Other Reduced Tillage Systems*    No-till or reduced tillage systems have been widely adopted in several regions of the world, but particularly in parts of the midwestern United States. Increased utilization of no-till agricultural systems to reduce soil erosion is anticipated. Changes in tillage systems will impact population dynamics and control of soil-borne pests and those that utilize plant cover. Further development of IPM systems for no-till agriculture will be necessary.

# Physical and Mechanical Controls

Restrictions on tillage may occur to reduce soil erosion and particulate air contamination. Such restriction could impact many IPM programs. Further restriction of tactics based on burning and flame are likely to occur because of energy use (fuel costs) and air pollution.

Computer imaging technology, coupled with global positioning systems, will be developed permitting equipment to more accurately follow the crop row and to adjust position automatically. This technology will allow cultivation closer to the crop and at much higher speeds than is possible with human tractor drivers.

# Host-plant Resistance/Plant Breeding

Classical plant breeding using Mendelian genetic principles will continue to be important to the development of pest resistance in crops. Such breeding efforts will be complemented by molecular genetic approaches. Molecular genetics has revealed that genes themselves, and gene functions in genomes, are complicated. The expanding knowledge about the mechanisms of gene function in both the crop and the pests may allow crop plant modifications tailored for particular pests species, and to ameliorate selection for resistance-breaking types.

Genetic engineering offers great possibilities relative to host-plant resistance to pests. The genetic engineering of crop plants by moving genes for desired characteristics into crop cultivars without transferring linked undesirable traits represents a significant advance over classical breeding for the development of new crop cultivars. This technology has the potential to revolutionize disease, insect, and nematode management completely. The development of herbicide-resistant crops may alter the whole concept of selective weed control using herbicides.

Developments in the area of transgenic crops may, however, have to be tempered relative to understanding the long-term consequences of such transgenes on consumers (in the ecological sense) and the environment, and the public acceptance of the technology. The technological innovations need to be introduced with adequate monitoring and safeguards to prevent problems similar to those that were caused by some pesticides. At the time of their introduction, pesticides were also viewed as a revolutionary but safe innovation, and many potential problems were simply unanticipated. Unlike pesticides, if a widely deployed transgenic crop is later found to pose an unacceptable risk, it may not be possible to recall novel genes that have invaded the environment.

## IPM Systems

The level of sophistication and true integration in IPM programs will slowly increase, and will eventually include integration of tactics across all pest categories.

## Pest Control Advising

People who make pest control recommendations will almost certainly be better trained, especially in less developed countries. Several universities already offer degrees in pest management. Due to the increasing complexity of decision making for IPM, it is likely that farmers and homeowners will not be permitted to make decisions concerning the use of pesticides because of their lack of expertise and training. It is conceivable that a professional board of conduct will be established to accredit doctors in plant health (our name), similar to the procedures used by the medical and legal professions. Some professional societies have already established such boards; for example, the Entomological Society of America sponsors a board-certified entomologist program that requires applicants to pass a board-supervised examination.

## STRATEGIC DEVELOPMENTS

If future developments can be predicted for individual tactics by extrapolating historical events in each of the various fields, it is much more difficult to anticipate major strategic shifts. IPM was originally conceived and developed for implementation at the single-field level. Most of the decision-making tools, such as economic injury levels and monitoring systems, were designed for single-field use. Many pests, however, are highly mobile and occurrence in one field is likely to influence events in neighboring fields. The advancement of IPM to higher levels of integration will require planning and implementation over larger spatial scales. A recent expansion of the concept of area-wide IPM represents a significant move in this direction. In some instances, area-wide IPM places the focus of the program at the landscape level. Future strategic shifts will encompass multiple ecosystems within the context of entire ecological regions. The level of organization and complexity of these programs will far exceed any that currently exists.

# SUMMARY

In addition to the purely technological advances that are intrinsic to IPM, future directions of IPM will be, perhaps even more profoundly, affected by extrinsic factors. IPM systems are impacted by dynamic natural processes of a global nature, as well as by processes determined by societies and their governments. Ever-changing natural processes result in the need to constantly monitor the conditions of crops and the pests that attack them. Changing regulations and legislation, or consumer demands that affect certain practices, also have a powerful effect on which tactics can be incorporated into an IPM strategy. Examples of societal-generated factors are the Food Quality Protection Act in the United States (see Chapter 11) and local, regional, and global market restrictions. Examples of factors resulting from natural events are biologically invasive species and climate change. These factors will greatly influence directions in IPM in the near future.

Although people around the world differ in religious beliefs, political philosophy, cultural standards, experiences, and individual financial situations, they are united in their need to eat high-quality, nutritious food. The need to increase food and fiber production to satisfy a world population growing at about 1.8% per year challenges pest managers to constantly improve IPM systems to reduce losses due to pests. Such losses are still estimated at 30% annually, despite new technological developments.

Experience with pesticides—early euphoria and optimism for miraculous pest control followed by a more tempered reality—has taught that no technological silver bullet will solve all pest problems. The experience with IPM in the past 30 years, however, has demonstrated that the integration of control tactics is the only ecologically sound approach to sustained pest management. The IPM paradigm has become a model for other components of sustainable agriculture.

# SOURCES AND REFERENCES

The following is a list of general reference material relating to the topics in this book. It is listed alphabetically within each general subject area.

## GENERAL IPM

Anonymous. Integrated pest management. University of California Statewide Integrated Pest Management Project, DANR publications. Year of publication varies. Available for the following crops: alfalfa (4104), rice (3280), tomatoes (3274), cotton (3305), lettuce and cole crops (3307), apples and pears (3340), almonds (3308), walnuts (3270), citrus (3303), small grains (3333), potatoes (3316), strawberries (3351), and pests of landscape trees and shrubs (3359). Some now in second editions.

Anonymous. 1992. *Beyond pesticides. Biological approaches to pest management in California.* Oakland, Calif.: DANR Publications, University of California, 183.

Beirne, B. P. 1967. *Pest management.* London: L. Hill, 123.

Bellows, T. S., and T. W. Fisher, eds. 1999. *Handbook of biological control: Principles and applications of biological control.* San Diego, Calif.: Academic Press, xxiii, 1046.

Burn, A. J., T. H. Coaker, and P. C. Jepson. 1987. *Integrated pest management.* London; San Diego, Calif.: Academic Press, xi, 474.

Cook, R. J., and R. Veseth. 1991. *Wheat health management.* St. Paul, Minn.: APS Press, x, 152.

Delucchi, V. L., ed. 1987. *Integrated pest management protection intégrée: Quo vadis? An international perspective.* Geneva, Switzerland: Parasitis, 411.

Dent, D., and N. C. Elliott. 1995. *Integrated pest management.* London; New York: Chapman & Hall, xii, 356.

Ennis, W. B., ed. 1979. *Introduction to crop protection,* Foundations for Modern Crop Science Series. Madison, Wisc.: American Society of Agronomy, xv, 524.

Fletcher, W. W. 1974. *The pest war.* New York: Wiley, x, 218.

Flint, M. L. 1998. *Pests of the Garden and Small Farm: A grower's guide to using less pesticide.* Oakland, Calif.: Statewide Integrated Pest Management Project, University of California Division of Agriculture and Natural Resources, x, 276.

Flint, M. L., and S. H. Dreistadt. 1998. *Natural enemies handbook: The illustrated guide to biological pest control.* Oakland; Berkley, Calif.: UC Division of Agriculture and Natural Sciences, University of California Press, viii, 154.

Flint, M. L., and P. Gouveia. 2001. *IPM in practice; Principles and methods of integrated pest management.* Oakland, Calif.: University of California, Division of Agriculture and Natural Resources, Publication 3418, xiii, 296.

Flint, M. L., and R. Van den Bosch. 1981. *Introduction to integrated pest management.* New York: Plenum Press, xv, 240.

Gaston, K. J. 1996. *Biodiversity: A biology of numbers and difference.* Oxford; Cambridge, Mass.: Blackwell Science, x, 396.

Glass, E. H. 1992. Constraints to the implementation and adoption of IPM. In F. G. Zalom and W. E. Fry, eds., *Food, crop pests, and the environment: The need and potential for biologically intensive integrated pest management.* St. Paul, Minn.: APS Press, 167–174.

Hawksworth, D. L. 1994. *The identification and characterization of pest organisms. Third Workshop on the Ecological Foundations of Sustainable Agriculture (WEFSA III).* Wallingford,UK: CAB International in association with the Systematics Association, xvii, 501.

Heitefuss, R. 1989. *Crop and plant protection: The practical foundations.* Chichester, England: Ellis Horwood; New York: Halsted, 261.

Henkens, R., C. Bonaventura, V. Kanzantseva, M. Moreno, J. O'Daly, R. Sundseth, S. Wegner, and M. Wojciechowski. 2000. Use of DNA technologies in diagnostics. In G. C. Kennedy and T. B. Sutton, eds., *Emerging technologies for integrated pest management.* St. Paul, Minn.: APS Press, American Phytopathological Society, 52–66.

Heywood, V. H., and R. T. Watson. 1995. *Global biodiversity assessment.* Cambridge, Mass.; New York: Cambridge University Press, x, 1140.

Hoy, M. A., and D. C. Herzog. 1985. *Biological control in agricultural IPM systems.* Orlando, Fla.: Academic Press, xv, 589.

Kennedy, G. C., and T. B. Sutton, eds. 2000. *Emerging technologies for integrated pest management.* St. Paul, Minn.: APS Press, American Phytopathological Society, xiv, 526.

Kogan, M. 1986. *Ecological theory and integrated pest management practice.* New York: Wiley, xvii, 362.

Kranz, J., H. Schmutterer, and W. Koch. 1978. *Diseases, pests, and weeds in tropical crops.* Chichester, England; New York: Wiley, xiv, 666, [32] leaf plates.

Landis, A. D., F. D. Menalled, J. C. Lee, D. M. Carmona, and A. Pérez-Valdéz. 2000. Habitat management to enhance biological control in IPM. In G. C. Kennedy and T. B. Sutton, eds., *Emerging technologies for integrated pest management.* St. Paul, Minn.: APS Press, American Phytopathological Society, 226–239.

Leslie, A. R., and G. W. Cuperus. 1993. *Successful implementation of integrated pest management for agricultural crops.* Boca Raton, Fla.: Lewis Publishers, 193.

Mengech, A. N., K. N. Kailash, and H. N. B. Gopalan, eds. 1995. *Integrated pest management in the tropics: Current status and future prospects.* Chichester, England; New York: Published on behalf of United Nations Environment Programme (UNEP) by Wiley, xiv, 171.

Morse, S., and W. Buhler. 1997. *Integrated pest management: Ideals and realities in developing countries.* Boulder, Colo.: Lynne Rienner Publishers, ix, 171.

Olkowski, W., S. Daar, and H. Olkowski. 1991. *Common-sense pest control.* Newtown, Conn.: Taunton Press, xix, 715.

Pimentel, D. 1991. *CRC handbook of pest management in agriculture.* Boca Raton, Fla.: CRC Press, 3 v.

Reuveni, R., ed. 1995. *Novel approaches to integrated pest management.* Boca Raton, Fla.: Lewis Publishers, xiv, 369.

Roberts, D. A. 1978. *Fundamentals of plant-pest control.* San Francisco: W. H. Freeman and Company, 242.

Ruberson, J. R., ed. 1999. *Handbook of pest management.* New York: Marcel Dekker, xvii, 842.

Sill, W. H. 1982. *Plant protection: An integrated interdisciplinary approach.* Ames, Ia: Iowa State University Press, xiii, 297.

Thresh, J. M., ed. 1981. *Pests, pathogens and vegetation.* London, UK: Pitman Books Limited, 517.

United States Congress. Office of Technology Assessment. 1987. *Technologies to maintain biological diversity.* Washington, D.C.: Congress of the U.S. Office of Technology Assessment. For sale by the Superintendent of Documents. U.S. Government Printing Office, vi, 334.

Van Driesche, R. G., and T. S. Bellows, Jr. 1996. *Biological control.* New York: Chapman & Hall, 539.

Ware, G. W. 1996. *Complete guide to pest control: With and without chemicals.* Fresno, Calif.: Thomson Publications, xii, 388.

Weinberg, H. 1983. *Glossary of integrated pest management.* Sacramento, Calif.: State of California, Department of Food and Agriculture, 84.

Zalom, F. G., R. E. Ford, R. E. Frisbie, C. R. Edwards, and J. P. Tette. 1992. Integrating pest management: Addressing the economic and environmental issues of contemporary agriculture. In F. G. Zalom and W. E. Fry, eds., *Food, crop pests, and the environment: The need and potential for biologically intensive integrated pest management.* St. Paul, Minn.: APS Press, 1–12.

Zalom, F. G., and W. E. Fry, eds. 1992. *Food, crop pests, and the environment: The need and potential for biologically intensive integrated pest management.* St. Paul, Minn.: APS Press, vi, 179.

# PATHOGENS

Agrios, G. N. 1997. *Plant pathology.* San Diego, Calif.: Academic Press, xvi, 635.

Ainsworth, G. C. 1981. *Introduction to the history of plant pathology.* Cambridge, England; New York: Cambridge University Press, xii, 315, [1] leaf plate.

Blakeman, J. P., and B. Williamson, eds. 1994. *Ecology of plant pathogens.* Wallingford, England: CAB International, xv, 362.

Brunt, A. A. 1996. *Viruses of plants: Descriptions and lists from the VIDE database.* Wallingford; Oxon, UK: CAB International, 1484.

Butler, E. J., and S. G. Jones. 1949. *Plant pathology.* London: Macmillan, xii, 979.

Ebbels, D. L., and J. E. King. 1979. *Plant health: The scientific basis for administrative control of plant diseases and pests.* Oxford, UK: Blackwell Scientific Publications, 217–235.

Fox, R. T. V. 1993. *Principles of diagnostic techniques in plant pathology.* Wallingford, England: CAB International, viii, 213.

FRAC. 2000. Fungicide Resistance Action Committee homepage, http://PlantProtection.org/FRAC/

Fry, W. E. 1982. *Principles of plant disease management.* New York: Academic Press, x, 378.

Hadidi, A., R. K. Khetarpal, and H. Koganezawa. 1998. *Plant virus disease control.* St. Paul, Minn: APS Press, xix, 684.

Hall, R., ed. 1996. *Principles and practice of managing soilborne plant pathogens.* St. Paul, Minn: APS Press, xii, 330.

Holliday, P. 1998. *A dictionary of plant pathology.* Cambridge; New York: Cambridge University Press, xxiv, 536.

Horsfall, J. G., and E. B. Cowling, eds. 1977. *Plant disease: An advanced treatise.* New York: Academic Press, 5 v.

Maloy, O. C. 1993. *Plant disease control: Principles and practice.* New York: J. Wiley, x, 346.

Maloy, O. C., and T. D. Murray, eds. 2000. *Encyclopedia of plant pathology.* New York: John Wiley and Sons Inc., 2200.

Manners, J. G. 1993. *Principles of plant pathology.* Cambridge; New York: Cambridge University Press, xii, 343.

Maramorosch, K., and K. F. Harris, eds. 1981. *Plant diseases and vectors: Ecology and epidemiology.* New York: Academic Press, xii, 368.

Matteson, P. C., M. A. Altieri, and W. C. Gagne. 1984. Modification of small farmer practices for better pest management. *Annu. Rev. Entomol.* 29:383–402.

Matthews, R. E. F. 1992. *Fundamentals of plant virology.* San Diego, Calif.: Academic Press, xii, 403.

Narayanasamy, P. 1997. *Plant pathogen detection and disease diagnosis.* New York: Marcel Dekker, vi, 331.

Nyvall, R. F. 1989. *Field crop diseases handbook.* New York: Van Nostrand Reinhold, 817.

Parker, C. A., and J. F. Kollmorgen, eds. 1985. *Ecology and management of soilborne plant pathogens.* St. Paul, Minn.: American Phytopathological Society, ix, 358.

Rechcigl, N. A., and J. E. Rechcigl, eds. 1997. *Environmentally safe approaches to crop disease control,* Agriculture and Environment Series. Boca Raton, Fla.: CRC/Lewis Publishers, 386.

Roberts, D. A., and C. W. Boothroyd. 1984. *Fundamentals of plant pathology.* New York: W. H. Freeman and Co., xvi, 432.

Schots, A., F. M. Dewey, and R. P. Oliver, eds. 1994. *Modern assays for plant pathogenic fungi: Identification, detection and quantification.* Wallingford; Oxford, UK: CAB International, xii, 267.

Schumann, G. L. 1991. *Plant diseases: Their biology and social impact.* St. Paul, Minn.: APS Press American Phytopathological Society, viii, 397.

Sutic, D. D., R. E. Ford, and M. T. Tosic. 1999. *Handbook of plant virus diseases.* Boca Raton, Fla.: CRC Press, xxiii, 553.

Vidhyasekaran, P., ed. 1997. *Fungal pathogenesis in plants and crops: Molecular biology and host defense mechanisms,* Books in Soils, Plants, and the Environment. New York: Marcel Dekker, viii, 553.

# WEEDS

Aldrich, R. J., and R. J. Kremer. 1997. *Principles in weed science.* Ames, Ia: Iowa State University Press, 472.

Anderson, W. P. 1996. *Weed Science: Principles and applications.* Minneapolis/ St. Paul, Minn.: West Pub. Co., xx, 388.

Ashton, F. M., T. J. Monaco, and M. Barrett. 1991. *Weed science: Principles and practices.* New York: Wiley, vii, 466.

Buhler, D. D., ed. 1999. *Expanding the context of weed management.* Binghampton, N.Y.: The Haworth Press, Inc., 289.

California Weed Conference 1989. *Principles of weed control in California.* Fresno, Calif.: Thomson Publications, 511.

Cousens, R., and M. Mortimer. 1995. *Dynamics of weed populations.* Cambridge; New York: Cambridge University Press, xiii, 332.

Holm, L. G. 1997. *World weeds: Natural histories and distribution.* New York: Wiley, xv, 1129.

Holm, L. G., D. L. Plucknett, J. V. Pancho, and J. P. Herberger. 1977. *The world's worst weeds: Distribution and biology.* Honolulu, Hawaii: Published for the East-West Center by the University Press of Hawaii, xii, 609.

HRAC. 2000. Herbicide Resistance Action Committee homepage, http://PlantProtection.org/HRAC/

Inderjit, K. M. M. Dakshini, and C. L. Foy. 1999. *Principles and practices in plant ecology: Allelochemical interactions.* Boca Raton, Fla.: CRC Press, 589.

Leck, M. A., V. T. Parker, and R. L. Simpson, eds. 1989. *Ecology of soil seed banks.* San Diego, Calif.: Academic Press, xxii, 462.

Parker, C., and C. R. Riches. 1993. *Parasitic weeds of the world: Biology and control.* Wallingford; Oxon, UK: CAB International, xx, 332, [16] plates.

Radosevich, S. R., J. S. Holt, and C. Ghersa. 1997. *Weed ecology: Implications for management.* New York: J. Wiley, xvi, 589.

Rees, N. E., P. C. J. Quimby, G. L. Piper, E. M. Coombs, C. E. Turner, N. R. Spencer, and L. V. Knutson, eds. 1996. *Biological control of weeds in the west.* Bozeman, Mont.: Western Society of Weed Science in cooperation with USDA Agricultural Research Service, Montana Dept. of Agriculture, Montana State University, 1 v. (unpaged).

Ross, M. A., and C. A. Lembi. 1999. *Applied weed science.* Upper Saddle River, N.J.: Prentice Hall, viii, 452.

TeBeest, D. O. 1991. *Microbial control of weeds.* New York: Chapman and Hall, viii, 284.

Weed Science Society of America. 1994. *Herbicide handbook.* Champaign, Ill.: Weed Science Society of America, x, 352.

Weed Science Society of America. 1998. *Herbicide handbook—Supplement to the seventh edition.* Lawrence, Kans.: Weed Science Society of America, vi, 104.

Zimdahl, R. L. 1989. *Weeds and words: The etymology of the scientific names of weeds and crops.* Ames, Ia: Iowa State University Press, xix, 125.

Zimdahl, R. L. 1999. *Fundamentals of weed science.* San Diego, Calif.: Academic Press, xx, 556.

# NEMATODES

Barker, K. R., G. A. Pederson, and G. L. Windham, eds. 1998. *Plant and nematode interactions,* Agronomy 36. Madison, Wisc.: American Society of Agronomy, xvii, 771.

Dropkin, V. H. 1989. *Introduction to plant nematology.* New York: Wiley, ix, 304.

Evans, K., D. L. Trudgill, and J. M. Webster. 1993. *Plant parasitic nematodes in temperate agriculture.* Wallingford, UK: CAB International, xi, 648.

Khan, M. W., ed. 1993. *Nematode interactions.* London; New York: Chapman & Hall, xi, 377.

Nickle, W. R., ed. 1991. *Manual of agricultural nematology.* New York: Marcel Dekker, Inc., 1035.

Weischer, B., and D. J. F. Brown. 2000. *An introduction to nematodes: General nematology; A student's textbook.* Sofia, Bulgaria: Pensoft, xiv, 187.

Whitehead, A. G. 1998. *Plant nematode control.* Oxon, UK; New York: CAB International, viii, 384.

# MOLLUSKS

Barker, G., ed. 2001. *The biology of terrestrial molluscs.* Wallingford, UK: CABI Publishing, xiv, 558.

Barker, G., ed. 2002. *Natural enemies of terrestrial molluscs.* Wallingford, UK: CABI Publishing, 320.

Godan, D. 1983. *Pest slugs and snails: Biology and control.* Berlin; New York: Springer-Verlag, x, 445.

Henderson, I. 1989. *Slugs and snails in world agriculture.* Thornton Heath, UK: British Crop Protection Council, 422.

Henderson, I. 1996. *Slug and snail pests in agriculture.* Farnham; Surrey, England: British Crop Protection Council, 450.

# ARTHROPODS

Arnett, R. H. 1993. *American insects. A handbook of the insects of America and Northern Mexico.* Gainesville, Fla.: The Sandhill Crane Press, Inc., 850.

Borror, D. J., C. A. Triplehorn, and N. F. Johnson. 1992. *An introduction to the study of insects.* Philadelphia; Fort Worth, Tex.: Saunders College Pub.; Harcourt Brace College Pub., xiv, 875.

Dent, D. 2000. *Insect pest management.* Ascot, UK: CABI Bioscience, xiii, 432.

Elzinga, R. J. 1997. *Fundamentals of entomology.* Upper Saddle River, N.J.: Prentice Hall, xiv, 475.

Evans, H. E., and J. W. Brewer. 1984. *Insect biology: A textbook of entomology.* Reading, Mass.: Addison-Wesley Pub. Co., x, 436.

Gordh, G., and D. H. Headrick. 2001. *A dictionary of entomology.* New York: Oxford University Press, 900.

Hill, D. S., and J. D. Hill. 1994. *Agricultural entomology.* Portland, Ore.: Timber Press, 635.

Horn, D. J. 1988. *Ecological approach to pest management.* New York: Guilford Press, xiii, 285.

Huffaker, C. B., and A. P. Gutierrez. 1999. *Ecological entomology.* New York: John Wiley & Sons, xix, 756.

IRAC. 2000. Insecticide Resistance Action Committee homepage, http://PlantProtection.org/IRAC/

Knipling, E. F. 1979. *The basic principles of insect population suppression and management.* Washington, D.C.: U.S. Dept. of Agriculture. For sale by U.S. Superintendent of Documents, U.S. Government Printing Office, ix, 659.

Metcalf, R. L., and W. H. Luckmann, eds. 1994. *Introduction to insect pest management,* Environmental Science and Technology. New York: Wiley, xiii, 650.

Metcalf, R. L., R. A. Metcalf, and C. L. Metcalf. 1993. *Destructive and useful insects: Their habits and control.* New York: McGraw-Hill, 1 v. (various pagings).

Pedigo, L. P. 1999. *Entomology and pest management.* Upper Saddle River, N.J.: Prentice Hall, xxii, 691.

Rechcigl, J. E., and N. A. Rechcigl, eds. 1999a. *Biological and biotechnological control of insect pests,* Agriculture and Environment Series. Boca Raton, Fla.: CRC/Lewis Publishers, 386.

Rechcigl, J. E., and N. A. Rechcigl, eds. 1999b. *Insect pest management: Techniques for environmental protection,* Agriculture and Environment Series. Boca Raton, Fla.: CRC/Lewis Publishers, 422.

Schoonhoven, L. M., T. Jermy, and J. J. A. van Loon. 1998. *Insect-plant biology: From physiology to evolution.* London; New York: Chapman & Hall, 409.

van Emden, H. F. 1989. *Pest Control.* London: Edward Arnold, x, 117.

# VERTEBRATES

Anonymous. 1970. *Vertebrate pests: Problems and control.* Washington, D.C.: National Academy of Sciences, National, Research Council (U.S.), Committee on Plant and Animal Pests, Subcommittee on Vertebrate Pests, 153.

Bäumler, W., J. Godinho, C. Grilo, M. Maduriera, I. Moreira, C. Naumann-Etienne, M. Ramalhino, T. Sezinando, and A. Vinhas. 1989. *Field rodents and their control.* Eschborn, Germany: Deutsche Gesellschaft für Technische Zusammenarbeit, 151.

Buckle, A. P., and R. H. Smith. 1994. *Rodent pests and their control.* Oxford, UK: CAB International, x, 405.

Dolbeer, R. A. 1999. Overview and management of vertebrate pests. In J. R. Ruberson, ed., *Handbook of pest management.* New York: Marcel Dekker, Inc., 663–691.

Prakash, I. 1988. *Rodent pest management.* Boca Raton, Fla.: CRC Press, 480.

Putman, R. J., ed. 1989. *Mammals as pests.* London; New York: Chapman and Hall, xi, 271.

Richards, C. G. J., and T.-y. Ku. 1987. *Control of mammal pests.* London; New York: Taylor & Francis, x, 406.

Singleton, G. R., ed. 1999. *Ecologically-based management of rodent pests.* ACIAR Monograph Series No. 59. Canberra: Australian Centre for International Agricultural Research, 494.

Van Vuren, D., and K. S. Smallwood. 1996. Ecological management of vertebrate pests in agricultural systems. *Biol. Agric. Hortic.* 13:39–62.

Wright, E. N., I. R. Inglis, and C. Feare, eds. 1980. *Bird problems in agriculture.* Croydon, UK: BCPC Publications, 210.

# PESTICIDES

Altman, J., ed. 1993. *Pesticide interactions in crop production: Beneficial and deleterious effects.* Boca Raton, Fla.: CRC Press, Inc., 592.

Anonymous. 1998. *Crop protection chemicals reference: CPCR.* New York: Chemical and Pharmaceutical Press, viii, 2101, S63.

Anonymous. 2000. *The future role of pesticides in US agriculture.* Washington, D.C.: National Academy Press, xx; 301.

Atkin, J., K. M. Leisinger, and B. Angehrn, eds. 2000. *Safe and effective use of crop protection products in developing countries.* Wallingford, UK: CABI and Novartis Foundation for Sustainable Development, xvii, 163.

Bohmont, B. L. 2000. *The standard pesticide user's guide.* Upper Saddle River, N.J.: Prentice Hall Inc., 544.

ExToxNet. 2000. Pesticide information profiles, http://ace.ace.orst.edu/info/extoxnet/pips/

Gustafson, D. I. 1993. *Pesticides in drinking water.* New York: Van Nostrand Reinhold, xii, 241.

Harris, J. 2000. *Chemical pesticide markets, health risks and residues.* Wallingford; Oxon, UK; New York: CABI Pub., vii, 54.

Hewitt, H. G. 1998. *Fungicides in crop protection.* Wallingford; Oxon, UK; New York: CAB International, vii, 221.

Kamrin, M. A., ed. 1997. *Pesticide profiles: Toxicity, environmental impact, and fate.* Boca Raton, Fla.: CRC/Lewis Publishers, 676.

Linn, D. M., T. H. Carski, M. L. Brasseau, and F. H. Chang, eds. 1993. *Sorption and degradation of pesticides and organic chemicals in soil,* SSSA Special Publication 32. Madison, Wisc.: Soil Science Society of America, American Society of Agronomy, xix, 260.

Marer, P. J., M. L. Flint, and M. W. Stimmann. 1988. *The safe and effective use of pesticides.* Oakland, Calif.: University of California Statewide Integrated Pest Management Project, Division of Agriculture and Natural Resources, x, 387.

Matthews, G. A. 1999. *Application of pesticides to crops.* London: Imperial College Press, xiii, 325.

Matthews, G. A., and E. C. Hislop, eds. 1993. *Application technology for crop protection.* Wallingford, UK: CAB International, viii, 359.

Perkins, J. H. 1982. *Insects, experts, and the insecticide crisis: The quest for new pest management strategies.* New York: Plenum Press, xviii, 304.

Prakash, A., and J. Rao. 1997. *Botanical pesticides in agriculture.* Boca Raton, Fla.: Lewis Publishers, 480.

Tomlin, C., ed. 2000. *The pesticide manual: A world compendium.* Farnham; Surrey, UK: British Crop Protection Council, xxvi, 1250.

Turnbull, G. J., D. M. Sanderson, and J. L. Bonsall, eds. 1985. *Occupational hazards of pesticide use.* London; Philadelphia: Taylor & Francis, xii, 184.

Tweedy, B. G., H. J. Dishburger, L. G. Ballantine, and J. McCarthy, eds. 1991. *Pesticide residues and food safety: A harvest of viewpoints,* ACS Symposium Series 446. Washington, D.C.: American Chemical Society, xv, 360.

Vighi, M., and E. Funari, eds. 1995. *Pesticide risk in groundwater.* Boca Raton, Fla.: Lewis Publishers, 275.

Ware, G. W. 1994. *The pesticide book.* Fresno, Calif.: Thompson Publications, 386.

Waxman, M. F. 1998. *Agrochemical and pesticide safety handbook.* Boca Raton, Fla.: Lewis Publishers, 616.

# PESTICIDE RESISTANCE

Alstad, D. N., and D. A. Andow. 1995. Managing the evolution of insect resistance to transgenic plants. *Science* 268:1894–1896.

Anonymous, ed. 1986. *Pesticide resistance: Strategies and tactics for management.* Washington, D.C.: National Research Council, National Academy Press, xi, 471.

Delp, C. J. 1988. *Fungicide resistance in North America.* St. Paul, Minn.: APS Press, American Phytopathological Society, v, 133.

Denholm, I., J. A. Pickett, and A. L. Devonshire, eds. 1999. *Insecticide resistance: From mechanisms to management.* Wallingford; Oxon, UK; New York: CABI Pub., vi, 123.

Feldman, J., and T. Stone. 1997. The development of a comprehensive resistance management plan for potatoes expressing the Cry3A endotoxin. In N. Carozzi and M. Koziel, eds., *Advances in insect control: The role of transgenic plants.* London; Bristol, Penn.: Taylor & Francis, 49–61.

FRAC. 2000. Fungicide Resistance Action Committee homepage, http://PlantProtection.org/FRAC/

Georghiou, G. P. 1972. The evolution of resistance to pesticides. *Annu. Rev. Ecol. Syst.* 3:144–168.

Georghiou, G. P., and A. Lagunes-Tejeda. 1991. *The occurrence of resistance to pesticides in arthropods.* Rome: Food and Agriculture Organization of the United Nations, xxii, 318.

Green, M. B., H. M. LeBaron, and W. K. Moberg. 1990. *Managing resistance to agrochemicals: From fundamental research to practical strategies.* Washington, D.C.: American Chemical Society, xiii, 496.

Heaney, S., D. Slawson, D. W. Holloman, M. Smith, P. E. Russel, and D. W. Parry, eds. 1994. *Fungicide resistance,* BCPC Monograph No. 60. Farnham; Surrey, UK: British Crop Protection Council, xii, 418.

Hoy, M. A. 1999. Myths, models and mitigation of resistance to pesticides. In I. Denholm, J. A. Pickett, and A. L. Devonshire, eds., *Insecticide resistance: From mechanisms to management.* Wallingford; Oxon, UK; New York: CABI Pub., 111–119.

HRAC. 2000. Herbicide Resistance Action Committee homepage, http://PlantProtection.org/HRAC/

IRAC. 2000. Insecticide Resistance Action Committee homepage, http://PlantProtection.org/IRAC/

McKenzie, J. A. 1996. *Ecological and evolutionary aspects of insecticide resistance.* Austin, Tex.: R. G. Landes, 185.

Powles, S. B., and J. A. M. Holtum, eds. 1994. *Herbicide resistance in plants: Biology and biochemistry.* Boca Raton, Fla.: Lewis Publishers, 353.

Roe, R. M., W. D. Bailey, F. Gould, C. E. Sorensen, G. G. Kennedy, J. S. Bacheler, R. L. Rose, E. Hodgson, and C. L. Sutula. 2000. Detection of resistant insects and IPM. In G. C. Kennedy and T. B. Sutton, eds., *Emerging technologies for integrated pest management.* St. Paul, Minn.: APS Press, American Phytopathological Society, 67–84.

Roush, R. 1997. Managing resistance to transgenic crops. In N. Carozzi and M. Koziel, eds., *Advances in insect control: The role of transgenic plants.* London; Bristol, Penn.: Taylor & Francis, 271–294.

Roush, R. T., and B. E. Tabashnik, eds. 1990. *Pesticide resistance in arthropods.* New York: Chapman and Hall, ix, 303.

Tabashnik, B. E., Y.-B. Liu, T. Malvar, D. G. Heckel, L. Masson, and J. Ferre. 1999. Insect resistance to *Bacillus thuringiensis:* Uniform of diverse. In I. Denholm, J. A. Pickett, and A. L. Devonshire, eds., *Insecticide resistance: From mechanisms to management.* Wallingford; Oxon, UK; New York: CABI Pub., 75–80.

# REGULATORY

Bright, C. 1998. *Life out of bounds: Bioinvasion in a borderless world.* New York: Norton, 287.

Brock, J. H. 1997. *Plant invasions: Studies from North America and Europe.* Leiden, Netherlands: Backhuys, vi, 223.

Chapman, S. R. 1998. *Environmental law and policy.* Upper Saddle River, N.J.: Prentice Hall, xviii, 258.

Ebbels, D. L., and J. E. King. 1979. *Plant health: The scientific basis for administrative control of plant diseases and pests.* Oxford, UK: Blackwell Scientific Publications, 217–235.

Foster, J. A. 1982. Plant quarantine problems in preventing the entry into the United States of vector-borne plant pathogens. In K. F. Harris and K. Maramorosch, eds., *Pathogens, vectors, and plant diseases: Approaches to control.* New York: Academic Press, 151.

Garner, W. Y., P. Royal, and F. Liem. 1999. *International pesticide product registration requirements: The road to harmonization.* Washington, D.C.: American Chemical Society, xi, 322.

Kahn, R. P. 1989. *Plant protection and quarantine.* Boca Raton, Fla.: CRC Press, 3 v.

Pyšek, P. 1995. *Plant invasions: General aspects and special problems.* Amsterdam: SPB Academic Pub., xi, 263.

Singh, K. G., ed. 1983. *Exotic plant quarantine pests and procedures for introduction of plant materials.* Serdang, Malaysia: ASEAN Plant Quarantine Centre and Training Institute, viii, 333.

Stout, O. O., H. L. Roth, J. F. Karpati, C. Y. Schotman, and K. Zammarano. 1983. *International plant quarantine treatment manual.* Rome: Food and Agriculture Organization of the United Nations, x, 220.

# SOCIETAL/ENVIRONMENTAL CONCERNS

Avery, D. T. 1995. *Saving the planet with pesticides and plastic: The environmental triumph of high-yield farming.* Indianapolis, Ind.: Hudson Institute, x, 432.

Baker, S. R., and C. F. Wilkinson, eds. 1990. The effects of pesticides on human health: proceedings of a workshop, May 9–11, 1988, Keystone, Colo. *Advances in modern environmental toxicology,* v. 18. Princeton, N.J.: Princeton Scientific Pub. Co., xxi, 438.

Barbash, J. E., and E. A. Resek. 1996. *Pesticides in ground water: Distribution, trends, and governing factors.* Chelsea, Mich.: Ann Arbor Press, xxvii, 588.

Beatty, R. G. 1973. *The DDT myth: Triumph of the amateurs.* New York: John Day Co., xxii, 201.

Benbrook, C. 1996. *Pest management at the crossroads.* Yonkers, N.Y.: Consumers Union, xii, 272.

Carson, R. 1962. *Silent spring.* New York: Fawcett Crest, 304.

Curtis, C. R. 1995. *The public & pesticides: Exploring the interface.* Columbus, Ohio; Washington, D.C.: Ohio State University and the National Agricultural Pesticide Impact Assessment Program, U.S. Dept. of Agriculture, viii, 96.

Dinham, B. 1993. *The pesticide hazard: A global health and environmental audit.* London; Atlantic Highlands, N.J.: Zed Books, 228.

Evans, L. T. 1998. *Feeding the ten billion: Plants and population growth.* Cambridge, UK; New York: Cambridge University Press, xiv, 247.

Foster, K., D. Bernstein, and P. Huber. 1993. Phantom risk: Scientific inference and the law. *Risk management* 40:46–56.

McHughen, A. 2000. *Pandora's picnic basket: The potential and hazards of genetically modified foods.* New York: Oxford University Press, viii, 277.

Pimentel, D., and H. Lehman, eds. 1993. *The pesticide question environment, economics, and ethics.* New York: Chapman and Hall, xiv, 441.

Taylor, S. L., and R. A. Scanlan, eds. 1989. *Food toxicology: A perspective on the relative risks,* IFT Basic Symposium Series. New York: Marcel Dekker, xiii, 466.

Van den Bosch, R. 1978. *The pesticide conspiracy.* Los Angeles, Calif.: The University of California Press, xiv, 226.

van Emden, H. F., and D. B. Peakall. 1996. *Beyond silent spring: Integrated pest management and chemical safety.* London; New York: Chapman & Hall, xviii, 322.

Van Ravenswaay, E. 1995. *Public perceptions of agrichemicals.* Ames, Ia: Council for Agriculture Science and Technology, vi, 35.

# PEST ORGANISMS

| Common name | Latin binomial | Chapters |
|---|---|---|
| **Pathogens** | | |
| Anthracnose | *Elsinoe ampelina* | |
|   Grape | | 10 |
| Bacterial speck (in tomatoes) | *Pseudomonas syringae* pv *tomato* | 2, 5 |
| *Barlety yellow dwarf virus* | BYDV | 6 |
| *Beet western yellows virus* | BWYV | 6 |
| Black root of cotton | *Thielaviopsis bassicola* | 15, 18 |
| Black sigatoka disease | *Mycospaerella musicola* | 17 |
| Brown rot (of fruits) | *Monilinia fructicola* | 5 |
| Canker, citrus | *Xanthomonas axonopodis* | 10 |
| Chestnut blight | *Cryphonectria parasitica* | 2, 10 |
| Clubroot, of Brassica spp. | *Plasmodiophora brassicae* | 10, 16 |
| Corn stunt | Corn stunt spiroplasma | 2 |
| Crown gall (in trees) | *Agrobacterium tumefaciens* | 2, 3, 13, 17 |
| *Cucumber mosaic virus* | CMV | 2, 6, 17 |
| *Beet curly top geminivirus* | BCTV | 5, 6 |
| Damping off | *Rhizoctonia solani* | 6, 13 |
| | *Pythium* | |
| Dutch elm disease | *Ophiostoma ulmi* | 2, 5, 6, 10 |
| Ergot of cereals | *Claviceps purpurea* | 2, 3, 6 |
| Fanleaf virus, grape | GFLV | 2, 6 |
| Fire blight (in pears and apples) | *Erwinia amylovora* | 2, 3, 4, 5, 6, 13, 17, 18 |
| Foolish seedling disease of rice | *Gibberella fujikuroi* | 2 |
| Grey mold (also bunch rot) | *Botrytis cinerea* | 5, 6, 12, 13, 15 |
| Late blight of potatoes | *Phytophthora infestans* | 1, 3, 12 |
| Leafspot | *Pseudomonas syringae* | 5 |
| Lettuce drop | *Sclerotinia minor* | 2, 5, 18 |
| *Lettuce mosaic virus* | LMV | 2, 6, 18 |
| Mildew, general | Several genera | 2, 12, 18 |
|   Downy of grapes | *Plasmopara viticola* | 3, 6, 10 |
|   Powdery | *Erisyphe polygoni* | 5 |
| Oakroot fungus | *Armillaria mellea* | 2, 5, 8, 16, 18 |
| *Papaya ringspot virus* | PRSV-p | 17 |
| Peach leaf curl | *Taphrina deformans* | 2 |

| Common name | Latin binomial | Chapters |
|---|---|---|
| **Pathogens—*continued*** | | |
| Peach yellows | Peach yellows phytoplasma | 2 |
| Pear decline | Pear decline phytoplasma | 2, 18 |
| Pierce's disease of grapes | *Xylella fastidiosa* | 2, 6 |
| *Potato leafroll virus* | PLRV | 17 |
| Rhizomania | *Beat necrotic yellow vein virus* (BNYV) | 2, 6, 10, 17 |
| *Rice tungro spherical waikavirus* | RTSV | 6 |
| Root rot, black, of cabbages | | |
|   Fusarium | *Fusarium* spp. | 6 |
|   Phytophthora | *Phytophthora* spp. | 6, 8, 17, 18 |
| Rust, cereal | Several genera | 2, 3, 5, 17 |
|   Barley | *Puccinia graminis hordei* | 5, 17 |
|   Coffee | *Hemileia vastatrix* | 2, 3, 10 |
|   Wheat | *Puccinia graminis tritici* | 3, 5, 10 |
|   White pine blister | *Cronartium ribicola* | 10 |
| Scab, apple | *Ventruria inaequalis* | 5, 12, 18 |
|   Pear | *V. pyrina* | 18 |
|   Potato | *Streptomyces* sp. | 16 |
| Sclerotium rots | *Sclerotium rolfsii* | 8 |
| Smut, general | Several genera | 2, 5 |
|   Corn | *Ustilago maydis* | 2, 3 |
|   Wheat (covered, or bunt) | *Tilletia* spp. | 3 |
| Soft rots | *Erwinia* spp. | 2, 6 |
| Southern corn leaf blight | *Cochliobolus heterostrophus* | 2, 5 |
| *Soybean mosaic virus* | SoyMV | 17 |
| Stem rot, rice | *Sclerotium oryzae* | 5 |
| *Beet yellows virus* | BYV | 2 |
| *Tobacco mosaic virus* | TMV | 2, 3, 5, 6 |
| *Tobacco rattle virus* | TRV | 2 |
| Wildfire of tobacco | *Pseudomonas syringae* pv *tabaci* | 5 |
| Wilt | | |
|   Verticillium | *Verticillium albo-atrum* and *V. dahliae* | 2, 5, 6, 8, 17, 18 |
|   Fusarium | *Fusarium oxysporum* and other spp. | 2, 6, 13, 17, 18 |
| **Weeds** | | |
| Barberry | *Berberis vulgaris* | 3, 5, 10 |
| Barnyardgrass | *Echinochloa crus-galli* | 4, 5, 6, 10, 18 |
| Bermudagrass | *Cynodon dactylon* | 5, 6, 10 |
| Blackberry | *Rubus* spp. | 6, 13 |
| Blackgrass | *Alopecurus myosuroides* | 6, 12 |
| Bluegrass, annual | *Poa annua* | 6 |
| Bromegrasses | *Bromus* spp. | 6 |
| Broom, Scotch | *Cytisus scoparius* | 2, 5, 10 |
| Broomrapes | *Orobanche* spp. | 2, 5 |

| Common name | Latin binomial | Chapters |
|---|---|---|

**Weeds—*continued***

| Common name | Latin binomial | Chapters |
|---|---|---|
| Cat's ear | *Hypochoeris* spp. | 6 |
| Chickweed, common | *Stellaria media* | 4, 6 |
| Cocklebur | *Xanthium trumarium* | 6 |
| Crabgrass | *Digitaria sanbuinalis* | 6 |
| Cutgrass | *Leersia* spp. | 6 |
| Dallisgrass | *Paspalum dilatatum* | 6 |
| Dandelion | *Taraxacum officinale* | 5, 6 |
| Docks | *Rumex* spp. | 5 |
| Dodder | *Cuscuta* spp. | 2, 15 |
| False flax | *Camellina* | 5 |
| False hellebore | *Veratrum californicum* | 2 |
| Ferns | | 2 |
| Water | *Salvinia molesta* | 10 |
| Fiddleneck, coast | *Amsinkia intermedia* | 2, 13 |
| Field bindweed | *Convolvulvu arvensis* | 5, 6, 18 |
| Foxtail | *Setaria* spp. | 6, 10, 12 |
| Giant | *S. faberi* | 4 |
| Green | *S. viridis* | 6 |
| Yellow | *S. glauca* | 5, 6, 12 |
| Goosefoot, nettleleaf | *Chenopodium murale* | 6 |
| Goosegrass | *Eleucine indica* | 5, 6 |
| Gorse | *Ulex europea* | 10 |
| Ground-cherry | *Phyusalis* spp. | 6 |
| Groundsel, common | *Senecio vulgaris* | 2, 4, 6, 12 |
| Halogeton | *Halogeton glomeratus* | 2 |
| Henbit | *Lamium amplexicaule* | 6 |
| Horsenettle | *Solanum carolinense* | 2 |
| Horsetails | *Equisetum* spp. | 2 |
| Horseweed | *Conyza canadensis* | 6 |
| Hydrilla | *Hydrilla verticillata* | 10, 13 |
| Itchgrass | *Rottboellia exaltata* | 6 |
| Johnsongrass | *Sorghum halapense* | 2, 5, 6, 10, 15, 18 |
| Jointvetch, Northern | *Aeschynomene virginica* | 13 |
| Jungle rice | *Echinochloa colona* | 6 |
| Klamath weed (Saint-John's-wort) | *Hypericum perforatum* | 2, 10, 13 |
| Knapweed | *Centaurea* spp. | |
| Russian | *Acroptylon* or *C. repens* | 10 |
| Spotted | *C. maculosa* | 10 |
| Knotweed | *Polygonum* spp. | 5 |
| Kudzu | *Pueraria lobata* | 10 |
| Lady's thumb | *Polygonum persicaria* | 6 |
| Lambsquarters | *Chenopodium album* | 5, 6, 10 |
| Loosestrife, purple | *Lythrum salicaria* | 10 |
| Lupins | *Lupinus* spp. | 5 |
| Mallow (cheeseweed) | *Malva* spp. | 18 |
| Mayweed | *Matricaria* | 6 |
| Medusahead | *Taeniatherum caput-medusae* | 10 |

| Common name | Latin binomial | Chapters |
|---|---|---|
| **Weeds—*continued*** | | |
| Melaleuca | *Melaleuca* | 10 |
| Miconia | *Miconia calvescens* | 2, 10 |
| Milk thistle | *Sylibum marianum* | 5 |
| Mistletoe | | 2, 3 |
| Dwarf | *Arceuthobium* | 2 |
| Large leaved | *Viscum* spp. and *Phoradendron* spp. | 6 |
| Morningglory, annual | *Ipomoea* spp. | 10 |
| Mustards, wild | *Brassica* spp., *Sinapis arvensis* | 2, 4, 5, 6, 17 |
| Nettles | *Urtica* spp. | 2, 18 |
| Nightshades | *Solanum* spp. | 5, 6, 10, 12, 18 |
| Nutsedges | *Cyperus* spp. | 2, 4, 5, 10, 18 |
| Yellow | *C. esculentus* | 10 |
| Purple | *C. purpurea* | 2 |
| Pampasgrass | *Cortaderia* spp. | 10 |
| Pigweed | *Amaranthus* spp. | 4, 6, 10, 18 |
| Redroot | *A. retroflexus* | 5, 6 |
| Plantain, buckhorn (narrowleaf) | *Plantago lanceolata* | 6 |
| Poison hemlock | *Conium maculatum* | 2 |
| Poison ivy | *Toxicodendron radicans* | 2 |
| Poison oak | *T. diversiloba* | 2 |
| Poplar, Lombardy | *Populus nigra* | 5 |
| Prickly lettuce | *Lactuca serriola* | 5, 6 |
| Prickly pear | *Opuntia* spp. | 2, 10, 13 |
| Puncturevine | *Tribulus terrestris* | 13 |
| Purslane, common | *Portulaca oleracea* | 4, 6, 13, 18 |
| Ragweed, giant | *Ambrosia artemisifolia* | 2 |
| Parthenium | *Parthenium hysterophorus* | 10 |
| Ragwort, tansy | *Senecio jacobaea* | 13 |
| Reed, giant | *Arundo donax* | 10 |
| Ryegrass | *Lolium* spp. | 6, 10, 12, 17 |
| Salvinia | *Salvinia* spp. | 13 |
| Shepherdspurse | *Capsella bursa-pastoris* | 6, 18 |
| Skeletonweed, rush | *Chondrilla juncea* | 13 |
| Snakeweed, broom | *Gutierrezia sarothrae* | 8 |
| Sowthistle, common | *Sonchus oleraceus* | 6, 13 |
| Sprangletop | *Leptochloa* spp. | 6 |
| Spurge, leafy | *Euphorbia esula* | 10 |
| Starthistle, yellow | *Centaurea solstitialis* | 2, 10, 13 |
| Strangler vine | *Morrenia odorata* | 13 |
| Tamarisk (also called salt cedar) | *Tamarix ramosissima* | 10 |
| Teasel | *Dipsacus* spp. | 3 |
| Thistles | Various | 13 |
| Canada (creeping) | *Cirsium arvense* | 6, 10 |
| Russian (tumbleweed) | *Salsola* spp. | 5, 6 |
| Scotch | *Onopordum acanthium* | 10 |

| Common name | Latin binomial | Chapters |
|---|---|---|
| **Weeds—*continued*** | | |
| Tumbleweed | See Thistle, Russian | |
| Velvetleaf | *Abutilon theophrasti* | 2, 5, 10 |
| Water hyacinth | *Eichornia crassipes* | 1, 2, 10 |
| Water lettuce | *Pisita statiotes* | 10 |
| Watergrass, late | *Echinochloa oryzoides* | 5 |
| Wild oat | *Avena fatua* and other spp. | 2, 5, 6, 10, 12 |
| Witchweed | *Striga* spp. | 2, 9, 10 |
| **Nematodes** | | |
| Burrowing | *Radopholus similis* | 5, 6 |
| Needle | *Longidorus* | 6 |
| Columbia-lance | *Hoplolaimus columbus* | 18 |
| Cyst | *Heterodera* spp. | 2, 5 |
|   Cereal | *H. avenae* | 13 |
|   Potato (golden) | *Globodera* | 5, 10, 17 |
|   Soybean | *H. glycines* | 6, 10, 13 |
|   Sugar beet | *H. schachtii* | 2, 3, 4, 5, 6, 10, 17 |
| Dagger | *Xiphenema index* | 2, 3, 6, 18 |
| Pinewood | *Bursaphelenchus xylophilus* | 5, 6, 10 |
| Red ring | *Bursaphelenchus cocophilus* | 6 |
| Reniform | *Rotylenchulus reniformis* | 12, 18 |
| Rice root | *Hirschmanniella* | 6 |
| Ring | *Criconemoides* spp. | 13, 18 |
| Root lesion | *Pratylenchus* spp. | 2, 6, 18 |
| Root knot | *Meloidogyne* spp. | 2, 3, 6, 12, 13, 17, 18 |
| Stem and bulb | *Ditylenchus* spp. | 6 |
| Stubby root | *Trichodorus* spp. | 2, 6 |
| | *Paratrichodorus* spp. | |
| **Mollusks** | | |
| Slugs | Various | 2, 6, 13 |
|   Garden | *Limax flavis* | 2 |
|   Gray garden | *Agriolimax reticulatus* | |
| Snails | Various | 2, 13 |
|   Brown garden | *Helix aspersa* | 13 |
|   Decollate | *Rumina decollata* | 13 |
|   Giant African | *Achatina fulica* | 10 |
|   Golden apple | *Pomacea canaliculata* | 10 |
|   Rosy wolfsnail | *Euglandina rosea* | 10, 13 |
| **Arthropods** | | |
| Adelgid, woolly balsam | *Adelges piceae* | 10 |
| Ants, general | Hymenoptera: Formicidae | 2, 3, |
|   Fire | *Solenopsis* | 10 |
| Aphid, general | Homoptera: Aphididae | 1, 2, 4, 5, 6, 8, 13, 15, 18 |
|   Cabbage | *Brevivoryne brassicae* | 13 |

| Common name | Latin binomial | Chapters |
|---|---|---|
| **Arthropods—*continued*** | | |
| Corn leaf | *Rhopalosiphum maidis* | 8 |
| Cotton | *Aphis gossypii* | 18 |
| Green peach | *Myzus persicae* | 5, 6, 12, 18 |
| Pea | *Acyrthosiphon pisum* | 5 |
| Potato | *Macrosiphum euphorbiae* | 5 |
| Rosy apple | *Dysaphis plantaginea* | 2, 13, 18 |
| Russian wheat | *Diuraphis noxia* | 10 |
| Spotted alfalfa | *Therioaphis maculata* | 10, 17 |
| Wooly apple aphid | *Eriosoma lanigerum* | 3, 18 |
| Armyworms, general | Lepidoptera: Noctuidae | 6, 13 |
| Beet | *Spodoptera exigua* | 13 |
| Bees | Hymenoptera: Apidae | 3, 13 |
| Beetle | Coleoptera | |
| Asian longhorned | *Anoplophora glabripennis* | 10 |
| Bean leaf | *Cerotoma trifurcata* | 1 |
| Carabid | Carabidae | 6, 8, 13 |
| Cereal leaf | *Oulema melanopus* | 10 |
| Colorado potato | *Leptinotarsa* | 3, 5, 10, 12, 17 |
| Elm bark | *Decemlineata scolytidae* | 2, 5, 6, 10 |
| Flea | Chrysomelidae | 5 |
| Ground (see carabid) | Carabidae | 4 |
| Japanese | *Popillia japonica* | 13 |
| Landbird | Coccinellidae | 2, 4, 6, 13 |
| Longhorn | Cerambycidae | 6 |
| Mexican bean | *Epilachna varivestis* | 2, 13, 17 |
| Rape blossom | *Meligethes aeneus* | 16 |
| Spotted cucumber | *Diabrotica undecimpunctata* | 2 |
| Striped cucumber | *Acalymma vittatum* | 14 |
| Vedalia | *Rodolia cardinalis* | 3, 13 |
| Bollworm | Lepidoptera | 17, 18 |
| Cotton | *Heliothis virescens* | 2, 5, 12, 13 |
| Pink | *Pectinophora gossypiella* | 8, 10, 16, 17, 18 |
| Bug | Heteroptera | |
| Assassin | *Zelus* | 13 |
| Bigeyed | *Geocoris puntipes* | 2, 13 |
| Damsel | *Nabis* | 13 |
| Minute pirate | *Orius tristicolor* | 6, 13 |
| Bulbfly, wheat | Diptera: Anthomyidae | 6 |
| Butterfly | Lepidoptera | |
| Alfalfa | *Colias eurytheme* | 13 |
| Monarch | *Danaus plexippus* | 5 |
| Cabbage worm, imported | *Pieris rapae* | 5 |
| Chinch bug, false | *Nysius raphanus* | 6 |
| Cicada, periodical | *Magicicada septendecim* | 5 |
| Codling moth | *Cidya pomonella* | 1, 3, 8, 17, 18 |
| Corn borer, European | *Ostrinia nubilalis* | 1, 10, 17 |
| Corn earworm | *Helicoverpa zea* | 1, 2, 5, 12, 16 |
| Cotton bollworm | *Helicoverpa zea* | 2, 12, 13 |
| Crayfish | Astacoidea | 2 |

| Common name | Latin binomial | Chapters |
|---|---|---|
| **Arthropods—*continued*** | | |
| Cricket, general | Orthoptera: Gryllidae | 2 |
|   Mole | Orthoptera: Gryllotalpidae | 3, 15 |
| Cutworm, general | Noctuidae | 2, 3, 4, 6, 8 |
|   Black | *Agrotis ypsilon* | 5 |
|   Variegated | *Peridroma saucia* | 6 |
| Cynipid wasps | Hymenoptera: Cynipidae | 2 |
| Earwig | Dermaptera: Forficulidae | 2 |
| Fly | Diptera | |
|   Apple maggot | *Rhagoletis pomonella* | 14 |
|   Cabbage maggot | *Delia brassicae* | 14 |
|   Carrot rust | *Psila rosae* | 14 |
|   Hessian | *Mayetiola destructor* | 5, 17 |
|   Onion maggot | *Delia antiqua* | 14 |
|   Syrphid | Syrphidae | 6, 13 |
|   Tachinid | Tachinidae | 6, 10, 13 |
|   Tsetse | Glossinidae | 5 |
|   Walnut husk | *Rhagoletis completa* | 8 |
|   Warble | *Hypoderma* | 4 |
| Fruit fly | Tephritidae | 13 |
|   Mediterranean (medfly) | *Ceratitis capitata* | 9, 10, 17 |
| Grasshoppers | Orthoptera: Acrididae | 2, 5 |
| Gypsy moth | *Lymantria dispar* | 2 |
|   Asian | *Lymantria dispar* | 10 |
| Lacewings | Neuroptera | 2, 4, 6, 13 |
| Leafhopper, general | Homoptera | 2, 6, 8, 12, 17 |
|   Blackberry | *Dikrella californica* | 6, 13 |
|   Grape | *Erythroneura elegantula* | 2, 6, 13 |
|   Sugar beet | *Eutettix tenellus* | 5, 6 |
| Leaf miners, general | Diptera: Agromyzidae | 8, 18 |
|   Pea | *Liriomyza trifolii* | 10 |
| Leaf roller | Lipidoptera: Tortricidae | 13, 18 |
|   Apple | Several spp. | 5, 18 |
|   Omnivorous | *Platynota stultana* | 6 |
| Locusts | Orthoptera: Acrididae | 2, 3 |
| Looper | Lepidoptera: Noctuidae | |
|   Soybean | *Pseudoplusia includens* | 5 |
| Lygus bugs | *Lygus* spp. | 1, 2, 6, 18 |
| Maggot, apple | *Rhagoletis pomonella* | 5, 8 |
| Mantis, praying | Orthoptera: Mantidae | 2, 5, 13 |
| Mealybug, coffee root | *Geococcus coffeae* | 13 |
| Mites | Acarina | 1, 5, 6, 12, 13 |
|   Spider | Tetranychidae | 2, 5, 6, 13 |
|   Predaceous | Phytoseiidae and Stigmaidae | 2, 13 |
|   Two spotted | *Tetranychus urticae* | 13 |
|   Wheat curl | | 17 |
| Moth, cactus | *Cactoblastis cactorum* | 3, 10, 13 |
|   Diamondback | *Plutella xylostella* | 12 |
|   Potato tuber | | 17 |

| Common name | Latin binomial | Chapters |
|---|---|---|
| **Arthropods—*continued*** | | |
| Naval orangeworm | *Amyelois transitella* | 5, 8 |
| Oriental fruit moth | *Grapholita molesta* | 8 |
| Phylloxera, grape | *Daktulosphaira vitifoliae* | 3, 5, 10, 17 |
| Planthopper, brown | *Nilapavata lugens* | 5, 17 |
| Psyllids | Homoptera: Psyllidae | 5, 18 |
| Rootworms | *Diabrotica* spp. | 12, 14 |
| Western corn | *Diabrotica virgifera* | 5 |
| Sawfly | Hymenoptera: Tenthredinidae | |
| Larch | *Pristiphora erichsonii* | 13 |
| Spruce | *Pikonema alaskensis* | 13 |
| Scale, general | Homoptera: Diaspididae | 2, 3, 8 |
| Cottony cushion | *Icerya purchasi* | 3, 13 |
| San Jose | *Quadraspidiotus perniciosus* | 3, 5, 12 |
| Red | *Aonidiella aurantii* | 13, 14 |
| Sharpshooters (glassy winged) | *Homalodisca coagulata* | 6 |
| Shrimp, tadpole | *Triops longicaudatus* | 5 |
| Silkworm | *Bombyx mori* | 2, 3 |
| Sowbug | Isopoda: Asellidae | 2 |
| Spider, general | Araneae | 2, 13 |
| Stem borers | Lepidoptera: Pyralidae | 2, 13 |
| Stinkbug | Heteroptera: Pentatomidae | 6 |
| Southern green | *Nezara viridula* | 13 |
| Termites | Isoptera | 2, 5 |
| Thrips | Thysanoptera: Thripidae | |
| Flower | *Megalurothrips sjostedti* | 16 |
| Six spotted | *Scolothrips sexmaculatus* | 13 |
| Tomato fruit worm | See Corn earworm | |
| Twig borer, peach | *Anarsia lineatella* | 5 |
| Velvetbeat caterpillar | *Anticarsia gemmatalis* | 5, 13 |
| Wasps, various | Hymenoptera: Vespidae | 2, 5, 6 |
| Cynipid gall | Hymenoptera: Cynipidae | 2 |
| Trichogramma | *Trichogramma* spp. | 13 |
| Weevil | Coleoptera: Curculionidae | |
| Alfalfa | *Hypera postica* | 5, 6 |
| Black vine | *Otiorhynchus sulcatus* | 13 |
| Cotton boll | *Anthonomus grandis grandis* | 2, 3, 5, 10, 14, 18 |
| Egyptian alfalfa | *Hypera bruneipennis* | 5, 6, 10 |
| Imported crucifer | *Baris lepidii* | 10 |
| South American palm | *Rhynchophorus palmarum* | 6 |
| Whiteflies, general | *Bemisia* spp., *Trialeurodes* spp. | 2, 8, 12, 18 |
| Wireworms | Coleoptera: Elateridae | 8 |
| **Vertebrates** | | |
| Bat | Chiroptera | 2, 13 |
| Bears | Ursidae: *Ursus* spp., *Euarctos* spp. | 2 |
| Beavers | Rodentia: Castoridae | 2 |
| Blackbirds | Icteridae | 2 |
| Carp | Cyprinidae: *Cyprinus carpio* | 10, 13 |

| Common name | Latin binomial | Chapters |
|---|---|---|
| **Vertebrates—*continued*** | | |
| Cattle | *Bos tarus, Box indicus* | 13 |
| Coyote | Canidae: *Canis latrans* | 2, 13 |
| Crow | Corvidae: *Corvus* spp. | 2 |
| Deer | Cervidae | 2 |
| Ducks | Anatidae: *Anas* spp. | 13 |
| Elephant | *Loxodonta africana* | 2, 3, 8 |
| Ferret | Mustelidae: *Mustela* spp. | 13 |
| Fox | Canidae: *Urocyon cinareoargenteus* | 13 |
| Geese | Anatidae: *Branta canadensis, Anser anser* | 13 |
| Goats, | Bovidae | 13 |
| Feral | | 10 |
| Golden eagle | *Aquila chrysaetos* | 2 |
| Gopher, pocket | Hetromyidac | 2, 13, 15 |
| Hawks | various | 13 |
| Hippopotamus | *Hippopotamus amphibius* | 2 |
| Horned lark | *Eremophila alpestris* | 2 |
| Human | *Homo sapiens* | 2, 13 |
| Kangaroo | *Macropus* sp. | 2 |
| Mongoose | *Herpestes* spp. | 2, 10, 13 |
| Mosquito fish | *Gambusia* spp. | 13 |
| Mouse | *Mus musculus* | 2, 3, 6, 12, 13 |
| Opossums | *Didelphis marsupialis* | 2 |
| Owl | Strigiformes | 13 |
| Barn | *Tyto alba* | |
| Pigs, feral | Suidae: *Sus scrofa* | 10 |
| Possum, brushtail | *Trichosurus vulpecula* | 2 |
| Prairie dogs | *Cynomys ludovicianus* | 2 |
| Rabbit | *Oryctolagus cuniculus* | 2, 6, 10, 13 |
| Cottontail | *Sylvilagus* spp. | 2, 13 |
| Jack (hare) | *Lepus* spp. | 2 |
| Rat | *Rattus* spp. | 2, 3, 10, 12, 13 |
| Norway | *Rattus norvegicus* | 2 |
| Roof | *Rattus rattus* | 2 |
| Sheep | *Ovis* spp. | 13 |
| Shrews | Soricidae | 13 |
| Skunk | *Mephitis* spp. | 2 |
| Snake | | |
| Brown tree | *Boiga irregularis* | 10 |
| Squirrel | *Spermophilus* spp. | 2, 18 |
| Ground | | 2, 5, 13 |
| Starling | *Sturnus vulgaris* | 2, 10 |
| Vole (field mouse) | *Microtus* spp. | 2, 5, 6, 13, 18 |
| **Crops and Other Plants** | | |
| Alfalfa (lucerne) | *Medicago sativa* | 1, 2, 4, 6, 13, 17 |
| Apples | *Malus sylvestris* | 1, 2, 3, 18 |

| | | |
|---|---|---|
| Asparagus | *Asparagus officinalis* | 6, 11 |
| Banana | *Musa* spp. | 7, 17 |
| Barley | *Hordeum vulgare* | 5, 17 |
| Beans, snap, string, dry | *Phaseolus vulgaris* | 2, 6 |
| Broccoli | *Brassica oleracea* var *botrytis* | 13 |
| Brussels sprouts | *Brassica oleracea* var *gemmifera* | 13 |
| Cabbage | *Brassica oleracea* var *capitata* | 2, 12, 17 |
| Canola (oilseed rape) | *Brassica napus* | 17 |
| Carrots | *Daucus carota* | 2 |
| Castor bean | *Ricinus communis* | 6 |
| Cauliflower | *Brassica oleracea* var *botrytis* | 13 |
| Cherry | *Prunus cerasus* | 2 |
| Clover | *Trifolium* spp. | 17 |
| Coca | *Erythroxylum coca* | 10 |
| | *Erythroxylum novogranatense* | |
| Coffee | *Coffea* spp. | 2, 10 |
| Cole crops | *Brassica oleracea* derived | 2 |
| Corn (maize) | *Zea mays* | 1, 2, 3, 4, 10, 11, 17 |
| Cotton | *Gossypium hirsutum* | 2, 3, 6, 12, 13, 15, 17, 18 |
| Cucumber | *Cucumis sativus* | 17 |
| Grape | *Vitis vinifera* | 3, 5, 6, 10, 13, 17 |
| Horseradish | *Armoracia rusticana* | 10 |
| Lettuce | *Lactuca sativa* | 1, 2, 4, 13, 16, 18 |
| Marigolds, African | *Tagetes* spp. | 13 |
| Melon | *Cucurbita melo* | 3 |
| Millets | *Pennisetum* and *Setaria* spp. | 2 |
| Oat | *Avena sativa* | 17 |
| Onions | *Allium cepa* | 13 |
| Oranges and other citrus | *Citrus* spp. | 2, 3, 10, 13 |
| Palm | | 6 |
|   Coconut | *Cocos nucifera* | |
|   Oil | *Elaeis guineensis* | |
| Peach | *Prunus persica* | 2 |
| Pear | *Pyrus communis* and other spp. | 2, 3, 4, 5, 17, 18 |
| Peas | *Pisum sativum* | 2, 3, 15, 17 |
| Potatoes | | |
|   Irish or white | *Solanum tuberosum* | 1, 2, 3, 6, 10, 17 |
|   Sweet | *Ipomoea batatas* | 13 |
| Rice | *Oryza sativa* | 1, 2, 3, 11, 15, 17 |
| Rubber | *Hevea brasiliensis* | 2, 4, 13 |
| Rye | *Secale cereale* | 2 |
| Sorghum, grain | *Sorghum bicolor* | 2, 11 |
| Soybean | *Glycine max* | 1, 3, 4, 6, 11, 13, 17 |
| Strawberry | *Fragaria vesca* | 2, 13, 18 |
| Sugar beet | *Beta vulgaris* | 1, 2, 4, 5, 10, 13, 17 |
| Tobacco | *Nicotiana tabacum* | 3 |
| Tomato | *Lycopersicon esculentum* | 2, 5, 17 |
| Wheat | *Triticum aestivum* | 1, 2, 3, 4, 5, 6, 10, 17 |
| Zucchini | *Cucurbita pepo* var *medullosa* | 2, 17 |

# GLOSSARY

This glossary is based on usage as published by Weinberg (1983), with additional reference to Agrios (1997), Caveness (1964), Flint and Gouveia (2001), Herren and Donahue (1991), Lincoln, Boxshall and Clark (1998), Pedigo (1999), Ricklefs and Miller (2000), Shurtleff and Averre (1997), and the University of California IPM project (Anonymous, 2000).

**abdomen**  Posterior part of the insect body containing reproductive organs in adults and sometimes locomotor appendages on immatures.

**abiotic**  The nonliving components of a system, such as temperature, mineral nutrients, water, soil type, sunlight, and air pollutants.

**abiotic disease**  A disease caused by nonliving factors, such as nutrient deficiency. Sometimes referred to as an abiotic condition.

**abscission**  The natural process by which leaves, fruits, or other organs become detached from plants.

**absorption**  The assimilation of molecules into cells, e.g., the process by which plants take up nutrients through their roots.

**acaricide (miticide)**  A pesticide toxic to mites and ticks (pest arthropods that belong to the class Acarina).

**accumulative pesticide**  See *biomagnification*.

**acetylcholine (Ach)**  A chemical that functions as a synaptic neurotransmitter in the nervous system of animals.

**acid equivalent**  The theoretical yield of parent acid from the active ingredient of a weak acid pesticide formulation.

**acre**  A unit of measurement for area that is equal to 43,560 square feet or 0.414 hectares, 640 A is equal to 1 square mile or a section of land.

**activator**  A compound or material added to a pesticide that either directly or indirectly increases the potency or activity of the pesticide. See *synergist*.

**active ingredient (ai)**  Chemicals in a pesticide formulation that are biologically active as toxins. A given pesticide formulation may have more than one active ingredient.

**acute toxicity**  An expression of the degree to which a toxicant is poisonous to an organism after a single exposure, contact, inhalation, or ingestion (cf., chronic toxicity).

**adaptation**  (i) Any genetically determined characteristic of an organism that enhances the fitness (survival and reproduction) of that organism. (ii) The result of natural selection and evolution that leads to organisms becoming better suited to their environments.

**additive effect**   The efficacy of a pesticide mixture that is equal to the sum of the toxicities of the individual pesticides (cf., antagonism and synergism).

**adjuvant**   Any nonpesticidal substance in a pesticide formulation that improves the physical, chemical, or biological properties of the active ingredient, or improves application efficacy.

**adsorption**   The adhesion of molecules to the surface of a solid. Pesticide binding to soil particles is an example.

**adult**   The reproductive, sexually mature, final stage in the life cycle of an organism.

**adulticide**   A pesticide that targets adult insect pests.

**adventitious**   Tissues or organs that arise in other than the usual location on a plant. Usually with reference to the production of lateral roots or buds.

**aerobic**   A process or an organism that requires oxygen.

**aerosol**   Very fine droplets (0.1 to 5 μm in diameter) suspended in air; generated by a container pressurized with a gas propellant, or by aerosol generators such as fogging machines or ultra-low-volume (ULV) equipment.

**aestivation**   Dormant or inactive during the summer (cf., hibernation).

**aflatoxin**   Potent mycotoxins that are carcinogenic and are naturally produced by the fungus *Aspergillus flavus* when it is growing on peanuts, corn, cereals, and other crops.

**agitate**   Stirring or shaking a pesticide mixture so that the components will not separate or settle in the application tank.

**agroecosystem**   An ecosystem manipulated by humans that typically has few common or major species (crops) and numerous rare or minor species (mainly pests).

**air-blast sprayer**   Spray equipment that uses a fan to create a high-speed air stream to transport and deposit pesticide solutions. Primarily used for perennial orchard and vine crops.

**air-carrier**   Use of air to transport pesticide to target. See *air-blast sprayer*.

**algicide**   Pesticide used to kill or control algae.

**alkaloids**   Nitrogen-containing chemical compounds produced by some plants that are toxic or inhibitory to herbivores.

**allele**   One of two or more alternative forms of a gene (or genes) at corresponding sites on a particular chromosome.

**allelopathy**   Direct inhibition of one organism by another organism, mediated through the release of toxic or noxious chemical compounds. Originally applied to the inhibition of one plant by another plant, as mediated by toxic compounds released by plant roots.

**allomone**   A chemical messenger produced by one species that has an effect on another species, with a result that benefits the donor but not the recipient (e.g., repellent).

**alpha diversity**   The variety of organisms occurring in a particular location.

**alternate host**   A plant host that is different from the primary host, on which a pathogen must develop to complete its life cycle (e.g., a heteroecious rust such as wheat rust). Often used in the entomological literature as synonymous with alternative host.

**alternative host**   A host used by a pest or pathogen when the primary or preferred host (crop) is not present. Alternative hosts are not required for the completion of the pest or parasite life cycle.

**ametabolous**   Insects that go through three life stages: egg, young, and adult. The young stages look the same as adults but do not have reproductive organs. Examples include silverfish and firebrats.

**anaerobic**   Refers to chemical processes or organisms that occur or live, respectively, without oxygen.

**angiosperm**   Flowering plants that produce seed in a closed ovary.

**anhydrobiosis**   Life without water. Some nematodes and other organisms can survive desiccation by eliminating water from their bodies and becoming quiescent with no detectable metabolism; when water again becomes available, the organisms rehydrate and resume normal activities.

**annual**   A plant that completes its life cycle in less than one year and then dies. Sometimes classified as summer and winter annuals, depending on the time of year that the plant is growing and reproducing.

**anoxybiosis**   Life without oxygen. Some organisms can survive oxygen-depleted conditions by lowering their metabolism and becoming quiescent until oxygen again becomes available.

**antagonism**   When two or more chemicals have opposing actions, such that the action of one is impaired or the total effect of both is less than that of one chemical when used separately.

**antenna (pl. antennae)**   A pair of appendages located on the head of an insect or mollusk, and two pair on crustacean heads; sensory in function.

**antibiosis**   One organism exerting a toxic or negative effect on another. For example, penicillin is an antibiotic that exerts a negative effect on certain bacteria. In IPM, the term also describes a type of host-plant resistance affecting the physiology of the herbivore.

**antibiotic**   Chemicals produced by microbes that inhibit, or are toxic, to other microbes.

**anticoagulant**   A substance that inhibits blood clotting resulting in internal hemorrhaging; this is the mode of action of a major class of rodenticides.

**antidote**   A practical immediate treatment that alleviates or reverses the physiological damage caused by toxins, including first aid. A remedy or treatment given to reverse the effects of poisoning.

**antifeedant**   Materials that inhibit or stop pest feeding. A control method often used for clothes moth larvae and termites.

**antimetabolite**   Substance that inhibits normal physiology and metabolism, usually with detrimental results. These substances are chemical analogs that block normal absorption and function of nutrients.

**antixenosis**   Plant resistance that affects herbivore behavior.

**apical**   At the tip or end; often applied to the growing points on plants.

**apterous**   Wingless; often used in reference to aphids.

**area-wide IPM**   IPM tactics implemented across large geographic regions to suppress pest populations at the regional rather than field scale; requires cooperation among multiple agencies.

**arthropod (Arthropoda)**   A phylum of invertebrate animals that have paired jointed appendages, an exoskeleton, segmented bodies, and bilateral symmetry. The class Insecta is the largest and most important group of arthropods.

**asexual**   Reproduction without the fusion of gametes, including budding and spore formation; parthenogenic (animals), vegetative (plants).

**attractant**   Chemical substances or devices used to lure insects or other mobile pests to areas where they can be trapped or killed; for example, a light attracts moths at night. Attractants are based on feeding, oviposition, or mating behavior of insects.

**augmentative biocontrol**   The release of beneficial organisms, or biological control agents, to establish or augment a natural population.

**autotrophic**   Organisms that are capable of making their food from carbon dioxide, mineral nutrients, and water.

**auxin**   A plant hormone that stimulates growth.

**avicide**   A pesticide toxic to birds.

**avirulent**   Pathogens that do not cause disease because of the loss of virulence, or of a disease-causing characteristic or change in host susceptibility.

**avoidance**   Growing crops in an area that is largely free of pest species.

***Bacillus thuringiensis (Bt)***   Soil-inhabiting bacterium that produces an insecticide effective against larval stages of many species of Lepidoptera, although some strains are effective against various beetle, mosquito, and blackfly larvae.

**bacterial insecticide**   Bacteria pathogenic to insects (e.g., Bt or *B. popilliae*). Applied using application techniques also used for chemical pesticides.

**bacterium (pl. bacteria)**   Microscopic, prokaryotic, single-celled organisms having cell walls but lacking membrane-bound organelles. Important in decomposition, nitrogen fixation and other symbioses, and as pathogens. A significant component of soil food webs.

**bait (B)**   A pesticide formulated with an attractive food substance and containing a small amount of toxic active ingredient (usually about 5%). Baits are primarily used for control of mollusks, cutworms, and rodents.

**ballooning**   A dispersal method for early instars of lepidopterous larvae, some spiders, and mites. The organism produces a silken thread that is caught by the wind resulting in the organism being transported by air movement.

**band application**   A method of applying pesticides or fertilizer to defined, limited, continuous area, such as in or along the row, rather than over an entire field.

**bed**   (i) A raised ridge of soil for planting crops above the furrows which are located on each side. (ii) An area where seedlings or transplants are grown for later transplanting into the field.

**behavior**   Any response or action of an organism.

**beneficial**   Organism that provides biological control by killing, or competing with, pests.

**beta diversity**   The organisms that occur within a specified region, resulting from the exchange of species among habitats.

**biennial**   A plant with a life cycle that takes two years. In the first year it grows vegetatively, in the second it flowers, produces fruits and seeds, senesces, and dies.

**binomial, Latin**   The two-part formal name of organisms that includes the genus and specific epithet; e.g., *Taraxacum officinale* L. (dandelion).

**bioassay** or **biological assay**   The use of living organisms in experiments to measure the effect of biotic or abiotic stresses or toxins.

**biocide**   A toxin that kills all organisms under a comparable range of exposure.

**biodiversity**   The variety and variability within and among living organisms and the communities in which they occur.

**biofix**   An identifiable phenological event or stage in the life cycle that represents the point for starting degree-day based models.

**biological control** or **biocontrol**   Regulating pest populations by using natural enemies such as herbivores, predators, parasitoids, and parasites.

**biomagnification**   The relative accumulation and concentration of a chemical in the tissues of organisms, that result in increasing tissue concentrations of the chemical in organisms that at successively higher levels in the food chain. An example is DDT.

**biotic**   Living. The biotic elements of an ecosystem are the living organisms therein, including plants, animals, and microorganisms.

**biotic disease**   Disease that is caused by an organism, usually a pathogen, such as a bacterium, fungus, nematode, phytoplasma, or virus.

**biotic insecticide**   Natural or introduced enemies of a pest including predators and parasites that are applied using standard pesticide application techniques.

**biotype**   A designation below the species level for organisms that are distinguished from other members of the same species by morphological, ecological (e.g., temperature or humidity requirements) or physiological characteristics (parasite susceptibility or host preference).

**blast**   The common name for the sudden death of flowers, buds, or young fruit caused by disease.

**blight**   Disease or injury that results in the sudden and conspicuous withering of leaves, fruit, twigs and branches.

**boom**   A horizontal support arm or pipe that allows for mounting applicator nozzles for dispersing liquid mixtures of agricultural chemicals.

**Bordeaux mixture**   A mixture of copper sulfate and calcium hydroxide applied for the control a range of plant pathogens.

**botanicals**   Pesticides derived from plants, such as pyrethrum, rotenone (derris), ryania, and nicotine.

**brand name**   The name that manufacturers use for commercial purposes.

**broad spectrum**   A pesticide that is effective against many pest species, as opposed to a narrow spectrum (see selective) pesticide that is toxic to a limited number of pest species.

**broadcast**   Application of fertilizer, pesticide, or seed over the entire soil surface, rather than in rows or bands.

**brood**   The progeny of any animal. Cohorts of progeny hatched or born at the same time and maturing at about the same rate.

**buffer**   (i) Any chemical that prevents sudden changes in pH. (ii) Area of field where no pesticide is applied to prevent worker exposure or contamination of adjacent areas.

**calibration**   The adjustment of application equipment so that a defined quantity of chemical or other material is applied per unit time or area.

**callus**   (i) Plant tissue that grows over an injury site, such as wound or graft, as a protective response. (ii) Plant tissue grown in tissue culture that remains undifferentiated but retains the potential to differentiate into specific tissues.

**canker**   A localized dry or dead area on a woody part of a plant such as the stem, branch, or trunk, often resulting from pathogen infection.

**canopy**   The foliage of plants.

**carcinogen**   A chemical or biological agent which produces, accelerates or increases frequency of cancers.

**carnivore**   An organism that eats living animals.

**carrier**   (i) A substance added to a chemical compound to improve the uniformity of application. Carriers can be organic (e.g., walnut shells, bark) or mineral (e.g., sulfur, lime, clay). (ii) The water or air used to transport a pesticide from the application machine to the target. (iii) An asymptomatic organism that is harboring an infectious disease agent.

**category I pesticide**   Pesticides that are highly toxic, with $LD_{50}$ between 0 and 50 mg/kg body weight, and are labeled with the signal words DANGER-POISON in large, bold letters with a skull and crossbones logo.

**category II pesticide**   Moderately toxic pesticides with $LD_{50}$ between 50 and 500 mg/kg body weight, that are labeled with the signal word WARNING in large, bold letters.

**category III pesticide**   Pesticides with low order acute oral toxicity and $LD_{50}$ between 500 and 5000 mg/kg body weight, that are labeled with the signal word CAUTION in large, bold letters.

**category IV pesticide**   Pesticides with very low acute oral toxicity and $LD_{50}$ greater than 5000 mg/kg body weight. These present minimal hazard to humans when handled properly, although some are flammable and so must be stored as such.

**caterpillar**   Worms that are the immature stage of lepidoptera (butterflies and moths).

**catfacing**   Fruit deformation caused by insect feeding during fruit development.

**causal agent**   The biotic or abiotic factor(s) responsible for inducing disease.

**caution**   A signal word that appears on pesticide labels to inform users that the pesticide has a low toxicity as defined by FIFRA. See *Category III pesticide*.

**centipede**   Arthropods of the class Chilopoda, have one pair of legs per body segment and a pair of venomous claws on the first segment. Centipedes live in detritus and the soil and are not usually considered pests.

**cephalothorax**   The fused head and thorax of arachnids and crustaceans.

**certified applicator**   Person licensed by the EPA and or state regulators to apply restricted-use pesticides.

**certified seed**   Seed that has its varietal status and freedom from weeds and other pests documented by a government agency.

**certified stock**   Plant propagation material that has been documented by State Departments of Agriculture or other government agencies as free from certain diseases or other pests.

**check**   (i) An earthen barrier that keeps water in a field or orchard. (ii) Units in an experiment that receive either a standard treatment that will result in a predictable response or no treatment.

**chemical control**   The reduction of pest populations by chemical pesticides.

**chemical name**   The name that specifies the chemical structure of a compound such as the active ingredients of a pesticide; e.g., malathion is 0,0-dimethyl-(1,2-dicarb-ethoxyethyl) phosphorodithioate.

**chemigation**   The delivery of pesticides in irrigation water.

**chemosterilant**   A chemical that effectively sterilizes organisms and so prevents reproduction.

**chisel**   A plow with penetrating points used to turn up moist subsoil while leaving dry soil on top, allows deep cultivation, breaks hardpans.

**chitin**   A complex polysaccharide that is a primary structural component of insect cuticle and nematode egg shells, providing mechanical strength and protection.

**chitin inhibitor**   An insecticide that disrupts chitin production.

**chlorinated hydrocarbons**   A category of compounds containing hydrogen, carbon, and chlorine; they have long residual life and are active against a wide range of pests. Examples include endrin, DDT, and toxaphene.

**chlorosis**   A symptom of disease, insect injury, herbicide injury, or physiological disorder in plants, wherein tissues that are normally dark green turn yellow or light green.

**cholinesterase (acetylcholinesterase)**  An enzyme that breaks down acetylcholine as part of normal signal transmission across a nerve synapse; some insecticides and nematicides deactivate cholinesterase.

**chronic toxicity**  A measure that indicates the effects of multiple exposures to a toxin over time.

**chrysalis**  A butterfly pupa.

**circulative (virus)**  Plant and animal viruses that persist, multiply, or accumulate in their vectors.

**clay**  Very fine soil particles that are less than 0.002 mm in diameter. Clay soils have little air space and drain poorly. Clays may be used as carriers in pesticide formulations.

**clone**  Descendants of a single parent or cell that are generally assumed to be genetically uniform because they result from reproduction that does not involve sex and recombination.

**closed-loading system**  Apparatus that allows transfer of pesticide from a mixing tank to an application tank by direct connections such as hoses, pipes, and couplings and so prevent worker exposure to the mixture.

**closed-mixing system**  A procedure that allows a pesticide to be moved from its original container, diluted with water, and transferred to an application tank using equipment that is sealed to prevent worker exposure to the pesticide.

**cocoon**  A silk covering that is produced by an insect larva to protect it during the pupal stage.

**coevolution**  The simultaneous evolution of traits in two or more species because of the selection pressures resulting from the mutual interactions of those species.

**cohort**  Group of individuals of a single species of the same age.

**commercial applicator**  A person licensed by the state, in compliance with an EPA-approved state plan, to apply pesticides for other than private use. Must also be certified to apply restricted use pesticides.

**common name**  The well known name for an organism in colloquial use of the regional language.

**common chemical name**  The formal name assigned to the active ingredient of a pesticide that is usually shorter than the chemical name. Each chemical has a single common name that is used in the scientific literature.

**community**  The interacting assemblage of all organisms in a defined area. Sometimes also used to refer to all members of some particular grouping in a defined area, such as an insect community.

**companion planting**  Growing two or more plant species together so that the different species benefit each other.

**compatible**  Pesticides are compatible if they can be mixed without their properties being altered.

**competition**  The simultaneous demand by two or more organisms or species for the same, limited essential resource.

**compressed air sprayer**  A sprayer with a 1-to 3-gallon capacity, a delivery tube, an air pump for pressure, and a strap so the sprayer can be carried.

**concentrate**  A liquid pesticide formulation that must be diluted prior to application.

**conducive soil**  A soil in which particular plant pathogens are able to cause disease.

**conidium (pl. conidia)**  Generally, any type of that is produced asexually and is nonmotile.

**consumer**   (i) A person who buys agricultural produce. (ii) The ecological term for all nonautotrophic organisms that must consume food.

**contact herbicide**   A herbicide that does not move within the plant and kills only that plant tissue on which it is actually sprayed.

**contact poison**   A pesticide that is lethal when it touches or is absorbed into the body; will act as a stomach poison if eaten.

**control**   (i) In an experiment, an untreated plot used to determine population response in the absence of treatment; also known as a check. (ii) Keeping a pest population below a specified density.

**control action guideline**   Used to define if pest control actions are necessary and when actions should be taken.

**control action threshold**   See *economic threshold.*

**controlled droplet application (CDA)**   Equipment that uses rotary atomizers (spinners) to produce spray droplets of uniform size.

**cotyledon**   A modified leaf that is formed within the seed, and the leaf is present on a seedling at germination; seed leaf. Two are present in dicotyledons; one is present in monocotyledons.

**cover crop**   Non-crop plantings used to protect soil during a fallow period, to add organic matter to soil when plowed under, and in some cases to reduce pest and pathogen numbers. See green manure crop.

**crawler**   Immature mobile stage of scale insects and whiteflies which move about the plants.

**crop rotation**   The practice of sequentially growing different crops in a field to achieve pest management, soil enhancement, and soil conservation.

**cross resistance**   The development of resistance to one pesticide that results in resistance to another pesticide that the organism has not been exposed to.

**crown**   (i) The point where the stem and the root meet in a plant. (ii) The upper branches of a tree.

**cucurbits**   Plants belonging to the family Curcubitaceae, e.g., pumpkin, cucumber, and squash.

**cultivar**   Short for cultivated variety; obtained by crossbreeding or selection. See *variety.*

**cultivator**   Mechanical device attached to a tractor and dragged across the soil to till and aerate soil and kill weeds.

**cultural practices**   The methods and tactics used in crop production.

**cumulative effect**   An effect that increases relative to the summation of a causal stimulus.

**cuticle**   (i) The waxy cutin-containing covering of aerial plant parts. (ii) The external body covering of insects and nematodes that contains chitin in insects and keratin in nematodes.

**cutin**   A waxy substance in the plant cuticle that protects and prevents moisture loss.

**cyst**   The hardened, decay-resistant body of a dead, swollen female nematode that is filled with eggs. The genera *Heterodera* and *Globodera* are cyst nematodes of agricultural importance.

**D-vac**   A portable vacuum device that sucks up insects. It is used to sample insects.

**damage**   Measurable reduction in crop growth or yield.

**damping off**   Infection of newly formed roots and stems of young seedlings by fungi or bacteria, typically at soil level, resulting in decay or seedling death.

**danger**   A signal word used on pesticide labels to inform the user that the pesticide is highly toxic, as defined by FIFRA. The word poison and the skull and crossbones symbol always accompany the signal word danger.

**decline**   Gradual reduction over time in the health and vigor of a plant, eventually ending in death.

**defoliant**   A chemical that causes plant foliage to drop off.

**defoliation**   (i) Reduction in the amount of foliage, possibly due to insects, fungi or other agents as distinguished from natural leaf fall. (ii) Removal of leaves by application of a defoliant.

**degradation**   The process of decomposition to simpler component parts. The breakdown of pesticide residue.

**degrade**   Decompose; break down into less complex constituent parts.

**degree-day** or **day degree (°D)**   A measure used to express time relative to the influence of temperature on the growth rate of cold-blooded organisms. Degree-days are based on cumulative time above and below developmental threshold temperatures.

**dehisce**   The splitting of plant tissues along lines of weakness, such as the opening of a seed pod.

**deposit**   The pesticide remaining on the target immediately following application.

**dermal**   Through, by, of, or about the skin. Dermal pesticide absorption is a measure of potential hazard to workers.

**desiccation**   Dehydration (removal of moisture) by chemical or physical action. Chemicals that promote dehydration are called desiccants.

**detection (detection survey)**   The procedures and protocols used to discover the presence of pests to monitor outbreaks or introductions.

**developmental threshold (lower or upper)**   The coldest temperature below which, or the highest temperature above which, growth, development, or activity of a cold-blooded organism stops.

**dew period**   Length of time dew forms on plants; important in defining opportunities for infection by plant pathogens that require free water to infect.

**diapause**   A dormancy in insects that is initiated in response to environmental conditions. The dormancy permits survival through adverse conditions. It is mediated by hormones and can occur at any growth stage in the life cycle.

**dieback**   Progressive death of twigs, branches, or stems from the tip back toward the main stem of the plant.

**diluent**   Any of several accessory carriers, solvents, emulsifiers, and wetting agents, that are used to formulate pesticides that dilute the active ingredient.

**dimorphism**   Having two distinct morphological forms in the same species; the most common example being sexual dimorphism wherein males and females differ in appearance.

**dioecious**   Plant species in which the male and female flowers are on separate plants.

**directed spray**   Targeting a specific area or part of a plant for pesticide application in order to reduce the amount of pesticide applied to the crop.

**disc**   Soil tillage tool made from multiple groups of concave, circular metal disks. Often used in heavy clay soils because it does not form plowpans.

**disease**   Abnormal physiology and metabolism in an organism. Disease may be caused by pathogens, abiotic conditions, or genetic disorders.

**disease-resistant**   Plant varieties bred for the characteristic that they do not support the development and reproduction of particular parasites.

**disruptive**   Treatment that interferes with the ability of beneficial species to control pest populations.

**dissemination**   The spatial spread of organisms resulting from organism activity, or passive movement by wind, rain, insects, and people.

**diurnal**   Recurring over a 24-hour cycle.

**dormant**   (i) A state in which spores, seeds, or plant organs become inactive and do not germinate or grow. (ii) A state of reduced metabolic activity and movement in some insects, mites, and nematodes.

**dose**   The quantity of a pesticide applied to a target. Also, a constant measure used to define pesticide toxicity.

**drift**   The movement of pesticide to non-target locations by wind or atmospheric circulation of fine droplets or dust.

**drift control**   Modifications of application techniques designed to reduce drift. May include additives to spray mixtures or equipment modifications to reduce fine droplets (those smaller than 100 μm).

**drop-nozzle application**   Lowering nozzles off the spray boom so that they reach into the crop canopy to permit directed spray or basal application.

**droplet**   Spherical body of fluid that is produced by a spray nozzle or spinner.

**duckfeet**   A type of cultivation implement called a sweep used to cultivate fields. Sweeps cut weeds below the soil level to provide mechanical control.

**dust (D)**   A formulation of pesticide that contains an active ingredient combined with dry and finely powdered carrier material such as talc, nut hulls, or clay. The possibility of drift is greatest with dust formulation.

**early postemergence**   The application of herbicide after plant emergence and during the initial growth phase of crop or weed seedlings.

**ecdysis**   The process whereby arthropods shed their exoskeleton, mediated by the hormone ecdysone. Also called molting.

**eclosion**   The emergence of a nematode or insect from its previous life cycle stage, e.g., a larva emerging from an egg, or an adult emerging from a pupa.

**ecology**   The study of the relationships among organisms, their environment, and each other. Study of the distribution and abundance of organisms.

**economic injury level (EIL)**   Pest population density sufficient to cause economic losses that are greater than the economic cost of the control action to reduce pest densities.

**economic poison**   The legal definition of a pesticide; any chemical used to control or mitigate pests.

**economic threshold (ET)**   The pest population density at which a tactic should be initiated in order to stop the pest density from reaching the economic injury level (EIL).

**ecosystem**   The entirety of physical environment and all organisms in a defined region or area.

**ectoparasite**   A parasite that remains on the exterior of the host body while feeding on that host.

**effective dose (ED)**   The $ED_{50}$ is the dose that has an effect with 50% of tested subjects.

**efficacy**   A measure or appraisal of the effectiveness of a control tactic with respect to the extent that the tactic controls the target pest.

**ELISA**   Short for enzyme-linked immunosorbent assay. This is an assay that is used for detecting and identifying pest organisms using the affinity of antibodies for antigens that are specific to particular organisms.

**embryo**   The developing organism before it reaches the stage where it leaves the parent body, emerges from the egg, or germinates from a seed.

**emergence**   The act of moving out of the protective environment of the body of, or location of a previous life history stage, such as when a young plant breaks through the soil surface or when an insect or nematode comes out of an egg.

**emergency exemption**   See *Section 18*.

**emulsifiable concentrate (EC)**   A liquid formulation of pesticide that forms a milky emulsion when added to water.

**emulsifier**   Surface active chemical that produces stable oil-in-water emulsions.

**encapsulated material**   A pesticide formulation where the active ingredient is enclosed inside an inert material to allow for decreased hazard in handling and application, and slow release after application.

**encapsulation**   (i) A type of pesticide formulation in which the active ingredient is enclosed in a support material (often plastic). (ii) A physiological defense mechanism in insects wherein the eggs or larvae of endoparasites are covered by multiple layers of host blood cells and killed.

**endemic**   Organisms that are naturally occurring in a particular area.

**endoparasite**   A parasite that feeds and develops while inside the body of its host.

**endosperm**   The tissue in seeds of angiosperms that contains nutrients and surrounds the embryo.

**entomopathogenic**   Organisms that cause disease in insects.

**entomophagous**   Feeding on insects.

**environment**   The surroundings of an area including the water, air, soil, plants, and animals.

**Environmental Protection Agency (EPA)**   A federal agency that enforces legislation relating to the protection of the environment in the United States.

**enzootic**   A disease in an animal population (cf., epizootic)

**epicuticle**   The outermost layer of the insect body wall (exoskeleton) that protects the insect from drying by a layer of waxes and lipids.

**epidemic**   Severe outbreak of a disease over a large geographic area.

**epinasty**   When more rapid growth on one side of a plant part, particularly leaves, causes it to bend or curl in one direction.

**epiphytotic**   A disease epidemic in a plant population.

**epizootic**   A disease epidemic in an animal population.

**equilibrium position**   The population density at equilibrium.

**eradication**   The complete elimination of all individuals of a species from a particular area.

**evaluation survey**   A systematic survey to evaluate the current and potential significance of a pest outbreak.

**evolution**   The change in the characteristics of organisms from generation to generation. Descent with modification.

**exclusion**   Prevention of pests entering or being moved into an area in which they do not already occur.

**exoskeleton**   The chitinous body wall or external skeleton of insects and other arthropods.

**exotic**   Nonnative organism introduced into an area in which it does not naturally occur.

**experimental use permit (EUP)**   Allow commercial-scale application of experimental pesticides, although the amount that can be applied is restricted.

**exposure**   The amount of contact an organism has with a toxin. Physical external contact is not necessarily the same as exposure.

**exposure period**   That amount of time workers are exposed to pesticides while mixing, loading, applying (including flagging), maintaining, and cleaning equipment. Also, substantial and prolonged contact with pesticides or their residues.

**exudate**   Matter that leaks from, is extruded from, or secreted by an organism.

**exuviae**   The cast-off nymphal or larval skin that results from arthropod molts (ecdysis).

**facultative parasite**   An organism that is usually saprophytic, but under some circumstances lives as a parasite.

**fallow**   Land that is usually used for growing crops that is not planted during the growing season.

**fastidious**   Term applied to parasites that are difficult to isolate or culture on media.

**FIFRA**   Federal Insecticide, Fungicide, and Rodenticide Act.

**filth**   Food contamination by rodents, insects, birds, or parts thereof, or any other objectionable matter.

**flagging**   (i) A branch with yellowing or dead foliage on an otherwise healthy tree as a result of rapid killing by adverse abiotic conditions, disease agents, or insects. (ii) Identification of an area to be treated (or to be left untreated) by aerial application.

**flaming**   Using a tool that emits fire to destroy pests.

**flight peak**   The time of maximum flying during a flight period of adult insects.

**floating row cover**   Layer of lightweight translucent material put over plants as protection from insect attack and environmental stress.

**flora**   The total of plants growing without cultivation in an area. The plants particular to a given geological formation.

**flowable (FL)**   Pesticide formulated as a suspension of finely ground active ingredient mixed with liquids and emulsifiers that can be mixed with water.

**fogger**   Pesticide application equipment that breaks the pesticide into very fine droplets (aerosols or smokes) and blows or drifts the drops onto the target area.

**foliar application**   Spraying a pesticide or fertilizer onto the leaves of target plants.

**FDA (Food and Drug Administration)**   Federal agency that oversees the safety of food, drugs and cosmetics that are available to the public.

**food chain**   A simplified linear depiction of trophic relationships that illustrates the flow of energy from producers to herbivores to carnivores.

**Food Quality Protection Act (FQPA)**   Act instituted in 1996 to improve food safety in the USA.

**food web**   An ordered depiction of the interconnections that occur among the organisms in food chains in a given environment.

**foodspray**   A nutrient-rich liquid food that is sprayed onto foliage to provide food for beneficial insects and so to increase their egg production (e.g., lacewings).

**formulation**   The combination of active pesticide ingredients, inert ingredients, and other additives as sold commercially. Specific formulations are designed to permit accurate and safe pesticide application.

**frass**   The solid feces of insects.

**fruiting bodies**   The reproductive structures fungi that contain spores.

**fumigant**   Pesticides that are a gas under normal temperature and pressure. They are injected into the soil or into a confined area, such as a grain storage structure, to kill pests.

**fungicide**   A substance or pesticide that kills fungi.

**fungus (pl. fungi)**   Single to multi-celled eukaryotic organisms that lack chlorophyll and grow on living or dead organisms, and therefore may cause plant disease. They grow from spores and commonly produce tiny thread-like growths (mycelium) that ramify through a substrate to absorb nutrients.

**gall**   Swellings, deformities, or growths of plant tissue that result from the activities various types of pests, including bacteria, nematodes, and insects.

**gamete**   Reproductive cells that are usually formed by meiosis, such as egg and sperm.

**gamma diversity**   The inclusive organismal diversity of all habitats within an area.

**gene(s)**   The unit of inheritance that consists of a sequence of nucleotides in a DNA molecule, contains the hereditary information for a single characteristic, and occurs at a particular locus on a chromosome.

**generation**   The time required to complete the life cycle of an organism, from birth (hatch, or germination) to sexual maturity and reproduction.

**genetic control**   An approach that emphasizes decreased fecundity instead of increased mortality. See *sterile insect release method (SIRM)*.

**genetic engineering**   Altering the genetic constitution of organisms by removing, modifying, or adding to the DNA in their chromosomes, or modifying the expression of extant DNA in organisms.

**genetically modified organism (GMO)**   Used to inaccurately describe genetically engineered organisms including transgenic crops. Technically, any domesticated organism derived from a wild species by artificial selection is a genetically-modified organism.

**geographic information systems (GIS)**   Computer-based system used for the collection, storage, and analysis of spatial data, including the ability to generate maps.

**germination**   The initiation of growth from a seed or spore.

**gibberellins**   Plant hormones that are involved in regulating cell elongation and other processes in plant tissues.

**girdling**   Damage to, or removal of, a complete ring of outer and inner bark from a shoot, trunk, or root. Girdling prevents the normal downward translocation of photosynthate, resulting in death of tissue either above or below the girdle.

**global positioning system (GPS)**   A system that uses triangulation to multiple satellites to allow a user to determine their geographic location with great accuracy.

**graft union**   Where the rootstock joins the scion or shoot of a grafted tree or vine.

**graft transmissible**   A virus or phytoplasma that can be spread plant-to-plant through grafting.

**granules (G)**   Dry, particulate pesticide formulation wherein active ingredient is incorporated in inert materials such as clay, corn cob, or walnut shell. Granular pesticides are usually applied to the soil surface and mixed into the soil.

**gravid**   An organism that contains fertile eggs.

**green manure crop**   Plants grown and then incorporated into the soil to provide nutrients, organic matter, and to improve soil structure. Often legumes or grasses.

**gregarious**   Living in groups.

**growth regulator**   A chemical that alters the normal growth or reproduction of an organism.

**grub**   A thick-bodied, slow-moving larva that has a well-developed head and thoracic legs, but lacks abdominal prolegs. Usually a beetle.

**guild**   Organisms that have similar resource requirements and trophic habits, consequently filling similar roles in the community.

**gummosis**   The abnormal exudation of gum, latex, or sap from a plant that results from parasite infection.

**gynoecious**   Plants that produce only female flowers.

**habitat**   The environmental conditions and associated organisms that occur in a particular site.

**habitat modification**   Alteration of the environment where an organism occurs, e.g., cultivating land for agriculture.

**half-life**   The time required for 50% of a pesticide to degrade.

**haploid**   Containing only half the chromosome number typical for adult members of a species, as in gametes.

**harrow**   A type of cultivation equipment that is used to level ground, stir soil, and cover seed.

**haustellate**   The sucking mouthparts of some insects composed of stylets and a proboscis.

**haustorium**   The specialized absorbing organ of some pathogens and parasitic higher plants; inserted into host cells to obtain nutrition.

**hazard**   The potential for damage as a function of exposure, for example, to a toxic material.

**head**   In arthropods, the anterior capsule-like body region that has mouthparts and sensory organs.

**headland**   The area at the end of a cultivated field where machinery turns around, or the area that is not treated by aerial application of pesticides to provide a safety zone for flaggers or so aircraft can avoid hazards such as utility lines.

**heat unit accumulation**   The sums of the daily average day degrees above the lower and below the upper developmental threshold temperatures.

**hectare**   Metric unit of land area equaling 10,000 square meters or 2.47 acres.

**hemimetabolous**   Insects that go through three distinct growth stages during their life cycle: egg, nymph, and adult. Nymphs look different than adults and often live in aquatic environments. Also termed incomplete metamorphosis. Examples are mayflies, stoneflies, and dragonflies.

**herbaceous**   Plants that have fleshy tissues rather than persistent woody tissues.

**herbicide**   Pesticide that kills plants. Used to kill weeds.

**hermaphrodite**   Containing male and female reproductive organs and capabilities in the same individual.

**heteroecious**   An organism that must feed on, or parasitize two unrelated host plants in order to complete its life cycle, rust fungus, lettuce root aphid.

**hibernaculum (pl. hibernacula)**   Silken shelter in which larvae of some Lepidoptera hibernate or aestivate.

**hibernation**   A period of dormancy that usually occurs in winter (cf., aestivation).

**holometabolous**   Insect life cycle that includes four growth stages: egg, larva, pupa, and adult. Also known as complete metamorphosis (e.g., beetles, flies, wasps, and moths).

**homeostasis**   The feedback mechanism by which an organism maintains physiological processes within a particular range.

**honeydew**   Exudate produced by sucking insects derived from plant sap, produced by aphids and scale insects.

**horizontal resistance**   Multigenic resistance in plants that confers broad resistance to multiple biotypes of pathogens or phytophagous arthropod (cf., vertical resistance).

**hormesis**   Growth stimulation by exposure to low doses of a toxicant.

**hormone**   A substance produced by an organism that affects the physiology and/or growth of the individual producing it. Generally active at very low concentrations.

**host**   The living plant or animal that acts as a food source for pests, used in reference to parasites and parasitoids.

**host-free period**   The time when the host plant is not grown (e.g., celery and beet-free periods reduce the reservoir of virus available to vectors).

**host-plant resistance**   The ability of a plant to reduce the development and reproduction of a pest that is using the plant as a host. Sometimes used to describe the ability of a plant to tolerate the damage caused by a pest.

**host range**   The organisms that a particular organism can use as food sources.

**hybrid**   The offspring of parents from two different genotypes, cultivars, or species.

**hyperparasite(oid)**   A parasite or parasitoid that uses another parasite as its host.

**hyperplasia**   Proliferation of cell numbers resulting in tissue overgrowth.

**hypertrophy**   Abnormal increase in cell size resulting in increased tissue or organ size.

**hypha (pl. hyphae)**   A single thread or filament of the mycelium that forms the body of a fungus.

**hypocotyl**   Embryo or seedling tissue located between the cotyledons and the root tip.

**hypovirulent**   Reduced virulence in a pathogen strain that can be used to impart resistance to more virulent, destructive strains.

**immigration**   Movement of individuals into an area.

**immunity**   Total and complete resistance of a plant to a pest or pests such that those pests are unable to use the plant for food.

**imperfect stage**   The asexual portion of the life cycle of fungi during which no sexual spores are produced.

**incompatible**   Refers to the decreased effectiveness that results if two or more pesticides are mixed.

**incorporate**   To mix or till fertilizer or pesticide into the soil.

**index**   An assay used to determine if a plant has been infected by a particular pathogen, especially viruses. Suspect infected tissue to inoculated into a susceptible plant that is known to show characteristic symptoms upon infection, thereby confirming the presence of the pathogen.

**indigenous**   Native to a region; endemic.

**induced pest**   See *replacement*.

**inert ingredient**   Biologically inactive substances that are added to pesticide formulations.

**infection**   The entrance and establishment of a pathogen in host tissues.

**infection site (syn. infection court)**   The host location where a pathogen first becomes established.

**infestation**   The physical presence of pest organisms in a location.

**inflorescence**   Collective term for a terminal cluster of flowers.

**injection**   (i) Placement of chemical under tree bark using a device equipped with a reservoir of chemical. (ii) Placement of a chemical below the soil surface with a nozzle-equipped chisel or similar device.

**injury**   The damage caused to plant physiology by pest activity.

**inoculum**   Pest propagules from which new infestations or infections arise.

**inorganic pesticide**   Pesticides that do not contain carbon (e.g., Bordeaux mixture, copper sulfate, sodium arsenite, and sulfur).

**insect growth regulator (IGR)**   Chemicals that mimic the action of insect hormones and can be used to disrupt normal insect growth and development.

**insectary crop**   Plants that attract and provide food for predators and parasites, e.g., untreated alfalfa adjacent to a cotton field.

**insecticide**   Pesticides that kill insects.

**instar**   The period or stage between insect molts.

**integrated pest management (IPM)**   A system that maintains the population of any pest, or pests, at or below the level that causes damage or loss, and which minimizes adverse impacts on society and the environment.

**integument**   An outer covering such as the cuticle of insects and nematodes, or the seed coats of plants.

**internal parasite**   A parasite that feeds from within the host body.

**internode**   Section of stem between two adjacent nodes.

**interplanting or intercropping**   Growing more than one crop simultaneously on a piece of land.

**interspecific**   Interactions between members of different species.

**intraspecific**   Interactions between members of the same species.

**invertebrate**   Animals that do not have an internal skeleton.

**juvenile**   The sexually immature stages of nematodes that hatches from an egg and molts three times before becoming an adult, or the immature stages of ametabolous insects.

**juvenile hormone (JH)**   A hormone, secreted by the insect brain that causes immature characteristics in insects as long as it is produced.

**kairomone**   A chemical produced by an individual of one species that has a beneficial effect on the recipient individual of another species.

**key pest**   Refers to a pest that routinely inflicts economic damage on commercial crops.

**knockdown**   Immediate incapacitation or paralysis of an insect by a quick-acting insecticide such as pyrethrum, allethrin, or tetramethrin.

**label**   Technical information about a registered pesticide in the form of printed material attached to, or printed on, the pesticide container. It is illegal to use the pesticide in ways inconsistent with the label.

**lapse**   A condition when air temperature is highest at ground level and decreases with elevation. Small pesticide particles can be carried aloft when warm air rises.

**larva (pl. larvae)**   The ambulatory, sexually immature, wingless stages between egg and pupa of holometabolous insects; immature nematodes or first instar mites.

**latent**   An infection that produces no visible symptoms.

**latent period**   (i) The time that passes between vector acquisition of a pathogen and when the vector becomes able to transmit the pathogen to a new host. (ii) The time between host infection and production of symptoms or inoculum by the infection.

**lateral movement**   The horizontal movement of an pesticide from the application site.

**layby application**   Applications of an agricultural chemical at the last time in the growing season when it is possible to drive a tractor through the crop.

**LC$_{50}$** (also **median lethal concentration**)   The lethal concentration of a toxicant in the air or water which kills half of the test animals exposed to it over a 24 hour period. LC$_{50}$ is expressed in parts per million for air or water, and in micrograms per liter for a dust or mist. It is often used as the measure of acute inhalation toxicity.

**LD$_{50}$** (also **median lethal dose**)   The dose or amount of an active ingredient which, when taken orally or absorbed through the skin, kills half of the test animals exposed to it. The LD$_{50}$ is generally expressed in milligrams per kilogram. It indicates acute oral or acute dermal toxicity.

**leaching**   The movement of a chemical downward through the soil in water.

**leaf area index (LAI)**   The ratio of total leaf surface area of a plant to the ground surface area covered by the plant.

**leaf miner**   An insect that feeds entirely within leaf mesophyll.

**lepidopterous**   Pertaining to the order Lepidoptera, the moths and butterflies.

**lesion**   A defined necrotic area or localized spot of diseased tissue.

**life cycle**   The sequence of events in the lifetime of an organism.

**life table**   Tabulation of birth and mortality at each life stage for a pest population. Can be used to estimate the rate of population increase.

**lodging**   The failure of grain stems to remain vertical due to wind, rain, disease, or insect damage. Makes harvest difficult; and part of crop is typically lost.

**maggot**   A legless larva without a well-defined head found in Diptera and some Hymenoptera.

**major pest**   Any pest that routinely inflicts economic damage on commercial crops if not controlled, so that management strategies are focused on these pests (see also *key pest*).

**mandibles**   The forward-most mouthparts or jaws of insects.

**material safety data sheet (MSDS)**   Formal description on the chemistry and toxicology of a chemical.

**matrix**   The gelatinous substance into which eggs of some nematodes are deposited, forming an egg mass.

**maximum dosage**   The largest amount of a pesticide that can be applied safely, without excess residues or damage to the host plant. The maximum dosage printed on a pesticide label is the maximum permitted by law.

**maximum residue limit**   The tolerated maximum amount of pesticide allowed to remain in food by the Food and Drug Administration (FDA).

**mechanical agitation**   The stirring, paddling, or swirling action of a device that keeps a pesticide thoroughly mixed in the tank.

**mechanical control**   Pest control tactics based on physical disruption of pests, with methods including cultivation, and plowing (e.g., tillage for weed control).

**median lethal concentration**   See *LC$_{50}$*.

**median lethal dose (MLD)**   See *LD$_{50}$*.

**meiosis**  Cell division that produces cells with half the number of chromosomes, called gametes.

**meristem**  Undifferentiated plant tissue at shoot and root tips and in buds that is capable of division.

**metabolite**  The component parts that result from the breakdown of compound, such as a pesticide, through the physiological processes of an organism.

**metamorphosis**  A marked and abrupt change in morphology through which a holometabolous insect passes in developing from larva to adult (meaning literally "to change form").

**mg/kg**  Milligrams of pesticide (or any toxicant) per kilogram of animal body weight.

**microbial pesticide**  Pesticides that have a living bacterium, virus, fungus, protozoa, or nematode as the active ingredient.

**microclimate**  The climate of a localized area, such as within a plant canopy.

**microhabitat**  The specific habitat of an organism composed of its immediate surroundings and the microclimate.

**microorganism**  An organism of microscopic size, such as a bacterium, virus, fungus, viroid, or mycoplasma.

**microsclerotium (pl. microsclerotia)**  Very small sclerotia, such as those produced by the Verticillium wilt and rice stem rot fungi.

**migrant population**  With reference to insects, a population that has left its place of origin in search of food or shelter, or for other reasons.

**migration**  Long-distance movement of organisms in relation to seasons.

**migratory ectoparasitic nematode**  A nematode that feeds from the exterior of the host and remains capable of movement to other locations on the host or to other hosts.

**migratory endoparasitic nematode**  A nematode that enters host-plant tissue, feeds internally, and continues to move through the tissue as it feeds and reproduces.

**mildew**  A fungus disease in which the mycelium grows over the surface of the plant causing a whitish discoloration.

**millipede**  Arthropod of the class Diplopoda, having two pairs of legs on most body segments. They inhabit debris and are not usually agricultural pests.

**mine**  Tunnel in plant parts produced by the feeding of immature insects.

**minor pest**  Pests that occasionally cause economic damage on commercial crops.

**minor use pesticide**  A pesticide with small potential market due to the small acreages of the commodity it is used on.

**miscible**  Liquids that can be mixed to yield a uniform solution.

**mist blower**  See *air-blast sprayer.*

**mite**  Tiny arthropod of the order Acarina. Closely related to spiders with eight jointed legs, no antennae, and no wings.

**miticide**  A pesticide that kills mites and ticks; same as acaricide.

**mitigation**  The act of reducing pest numbers.

**mode of action**  The physiological mechanism by which a pesticide exerts a toxic effect on the target organism.

**model**  A simplified depiction of reality. May be a mathematical representation of the elements of a population or crop system.

**moderately toxic**  Poisons with $LD_{50}$ equal to 50–500 mg/kg.

**mold**  Fungus with conspicuous mycelial or spore masses growing on damp or decaying matter.

**mollicute**   A class of microorganisms that includes mycoplasmas and spiroplasmas. They lack cell walls and are variable in shape. Spiroplasmas can be cultured but mycoplasmas cannot.

**mollusk**   Snails and slugs (snails without external shells) of the phylum Mollusca, class Gastropoda.

**molluscicide**   A pesticide that kills slugs and snails.

**molt**   With reference to insects, mites, and nematodes, the shedding of the exoskeleton or cuticle before entering the next life stage.

**monitoring system**   A system to track of (i) whether pesticides are escaping into the environment, or (ii) the numbers, life stages, evidences of feeding, damage, and locations of pests in a crop.

**monoculture**   The cultivation of a single crop to the exclusion of all other plants.

**monocyclic**   A disease with one reproductive cycle per year (cf., polycyclic).

**mortality factor**   The cause of death of an organism.

**mosaic**   Virus disease symptoms of variegated yellow and green patterns in leaves.

**motile**   Capable of movement.

**mulch**   A layer of material laid over the soil to prevent weed growth, and reduce water loss.

**multiple feeding bait**   Rodent bait that contains anticoagulant toxins (e.g., warfarin) and requires several feedings to be effective.

**multivoltine**   Any insect that has several generations per year (cf., univoltine).

**mummy**   (i) Unharvested fruit or nut remaining on the tree (also called sticktight). (ii) Remains of an aphid whose body contents have been consumed by a parasite.

**mutagen**   A compound that can cause mutations in the genome.

**mutation**   Heritable change in the genome.

**mycelium (pl. mycelia)**   The vegetative body of a fungus that consists of a mass of slender filaments called hyphae.

**mycoherbicide**   A living fungus applied to kill plants.

**mycoplasma**   Pleomorphic prokaryotic microorganisms that do not have a cell wall; several are plant pathogens.

**Mycoplasma-like organism (MLO)**   See *phytoplasma*.

**mycorrhiza (pl. mycorrhizae)**   The symbiotic association of fungi with plant roots, wherein fungus and plant exchange nutrients in a mutually beneficial association.

**mycotoxin**   Toxins produced by fungi. Such toxins may be present in infected seeds, feeds, or foods, and may cause illness or death in animals that consume the contaminated produce.

**natural enemies**   The predators and parasites that prey on or live in pest organisms.

**necrosis**   Tissue death in a localized region.

**nectary**   A plant gland that secretes nectar.

**negligible residue**   A pesticide residue tolerance set for a food or feed crop.

**nematicide**   A pesticide that kills or incapacitates nematodes.

**nematode**   Unsegmented roundworms that are parasites in plants, animals, and humans. Some species live in soil and water feeding on bacteria and fungi. Sometimes called eel worms.

**nematophagous**   Feeding on nematodes, as in nematophagous fungi.

**no observable effect level (NOEL)**   The highest dose of a chemical that does not produce observable effects when administered to test animals.

**node**   Joint on a plant stem; origin of leaves and branches.

**nodule**   A small swelling on the roots of certain legumes that contain nitrogen-fixing bacteria.

**nonaccumulative**   Refers to pesticides that do accumulated in the environment because they break down rapidly.

**nonpathogenic**   Does not cause disease.

**nonpersistent pesticide**   Pesticides that decompose to nontoxic by-products in less than 1 to 3 days.

**nonpersistent virus**   A virus carried on the mouthparts of its insect vector, but the virus is lost after the vector feeds a few times. Stylet-borne virus.

**non-preference**   Hosts that are unattractive to insects. The opposite of preference.

**nonselective pesticide**   A pesticide that is toxic to a range of organisms.

**nontoxic**   An economic poison with an $LD_{50}$ greater than 5,000 mg/kg. Not poisonous to humans.

**nontarget species**   An organism that is not the intended object of a management tactic.

**no-till**   Growing crops without using tillage.

**noxious weed**   A weed designated by law as particularly undesirable and difficult to control.

**nozzle**   Device that regulates drop size and controls volume flow rate, uniformity or thoroughness of coverage, and safety of liquid pesticide spray applications.

**nuclear polyhedrosis virus (NPV)**   A lethal virus disease of insects, particularly larvae of certain Lepidoptera and Hymenoptera.

**nutrient**   A mineral element necessary for plant or animal growth.

**nymph**   The wingless immature stage of a hemimetabolous insect, e.g., Hemiptera, Orthoptera.

**obligate parasite**   A parasite that requires a living host to survive and reproduce.

**occasional pest**   A species normally present, but infrequently causing damage.

**Occupational Safety and Health Administration (OSHA)**   Federal agency that regulates matters that affect the health and safety of workers.

**oligophagus**   Feeding on a limited range of species.

**omnivorous**   Feeding on a wide range of host species.

**oncogen**   A substance that causes the formation of tumors in animals; may be carcinogenic.

**oospore**   Resting fungus spore; the result of fusion of gametes.

**oral toxicity**   The toxicity of a pesticide when ingested by mouth.

**organic produce**   Commodities produced from plants or animals which are grown without the use of synthetic fertilizers or pesticides.

**organic matter**   The component of soil that is composed of decaying plants and animals.

**organic pesticide**   Pesticides made from carbon-containing chemicals. The two major groups are petroleum oils and synthetic organic pesticides.

**organochlorine**   Synthetic organic insecticide containing carbon, hydrogen, and chlorine. Many are persistent in the environment (e.g., DDT, chlordane, and aldrin).

**organophosphate**   Synthetic organic insecticide containing phosphorus (e.g., parathion, malathion and TEPP (tetraethyl pyrophosphate)).

**outbreak**   High numbers of a particular pest species.

**overwinter** Survival of adverse winter conditions until the next growing season.

**ovicide** A pesticide that kills the eggs of insects and mites.

**oviposit** The act of depositing eggs.

**ovipositor** The morphological structure that insects use to deposit eggs.

**ppb** Parts per billion. One ten millionth of a percent; equal to micrograms per liter.

**ppm** Parts per million. One ten thousandth of a percent; equal to milligrams per liter.

**psi** A measure of pressure, pounds per square inch.

**parasite** An animal or plant that lives for most of its life in or on another organism, the host, from which it obtains nourishment, and to which it causes some damage; usually not lethal to its host.

**parasite guild** A group of parasite species that use a particular host or host stage upon or within which to develop.

**parasitoid** An arthropod parasite that kills its host. Parasitoid adults are free-living, the immature stages are parasitic.

**parthenocarpy** The development of fruit without fertilization and seed.

**parthenogenesis** The development of eggs without fertilization by spermatozoa. Occurs in some nematodes, aphids, beetles, parasitic wasps, and other insects.

**paurometabolous** Insects that go through three stages of growth, egg, nymph, and adult, with nymphs resembling adults except for the absence of wings. Grasshoppers, termites, thrips, true bugs, planthoppers, and aphids are examples.

**pathogen** A parasite that causes disease.

**pathogenic** The ability of a pathogen to cause disease.

**pathovar** In bacteria, a subspecific designation for isolates that infect only certain plants within a genus or species.

**pellets** A dry pesticide formulation with particles over 10 cubic millimeters in size. See *granules*.

**penetrant** A wetting agent that enhances liquid absorption to a surface.

**perennial** A plant that lives for three or more years.

**perfect stage** Part of fungus life cycle during which gametes and sexual spores are produced.

**perithecium** Flask shaped structure of some ascomycete fungi containing asci and ascospores.

**permit** A document required in some states to allow application of restricted-use pesticides.

**persistence** (i) The amount of time after application that a pesticide remains effective and stable in the environment. (ii) A virus that infects its insect vector so that the vector can transmit for the remainder of its life.

**personal protective equipment (PPE)** Clothes, materials, or devices that offer protection from pesticides, especially important when handling or applying highly toxic pesticides. Such equipment can prevent injury or death, and is specified on the label.

**pest** Any organism that interferes with the activities and desires of humans, including pathogens, weeds, nematodes, mollusks, arthropods and vertebrates.

**pest categories** Include pathogens, weeds, nematodes, arthropods, mollusks, and vertebrates.

**pest control**   The actions taken to eradicate, inhibit, or limit pests and to prevent them from interfering with crop profitability or agricultural operations.

**pest control advisor (PCA)**   A person who makes agricultural pest control recommendations, or any person who holds her/himself out as an authority or general advisor on any agricultural use, or solicits services or sales.

**pest control operator (PCO)**   A person or firm that is officially licensed to apply pesticides, or use any method or device for hire, to control pests or prevent, destroy, repel, mitigate, or correct any pest infestation or disorder.

**pest management strategy**   A overall plan for alleviating a pest problem.

**pest management tactic**   A method of controlling a pest or its damage.

**pest resurgence**   The rapid increase in pest numbers after numbers were reduced for a time by a pesticide application. Resurgences may be caused by the destruction of natural enemies.

**pesticide**   Refers to any economic poison intended to control pests. See *FIFRA*.

**pesticide dealer**   Any distributor or retailer that: sells pesticides for agricultural use, methods and devices for the control of agricultural pests, or soliciting pesticide sales.

**pesticide kill**   The death of nontarget organisms due to careless or improper pesticide use.

**pesticide resistance**   See *resistance*.

**pesticide tolerance**   The amount of pesticide residue which may legally remain on or in food at the time of sale. The EPA is responsible for establishing federal residue tolerances.

**petroleum oil**   Refined spray oil classed as paraffin, naphthalene, aromatic, or unsaturated oil based on the hydrocarbons it contains.

**pH**   A measure of relative acidity or alkalinity of aqueous solutions. A pH of 7 is equal to neutrality; above 7 is basic and below 7 is acidic.

**phenology**   The growth stages that an organism passes through over time.

**phenotype**   The observable structural and functional attributes of an organism.

**pheromone**   A chemical released by one organism that influences the behavior of another of the same species.

**phloem**   The tissue that translocates the products of photosynthesis from the foliage to other plant parts.

**phoresy**   One organism being transported on the body of an individual of another species.

**photoperiod**   The duration of light in a daily light-dark cycle.

**photosynthate**   The products of photosynthesis used to support plant growth, respiration, and fruit production.

**photosynthesis**   The process by which green plants use energy in sunlight to convert water and $CO_2$ into carbohydrates.

**physiological disorder**   A disorder caused by factors other than a pathogen; abiotic disorder.

**phytoalexins**   Defense compounds produced by plants as a response to pathogen or arthropod attacks.

**phytopathology**   The study of plant disease.

**phytophagous**   Plant eating.

**phytoplasma**   Mollicutes that infect plants and cause disease.

**phytotoxic**   Toxicant that causes damage or death to plants.

**plant growth regulator (PGR)**   Chemicals that affect plant growth.

**plant pathogen**   A parasite that causes disease in plants.

**plasmid**   A self-replicating, extrachromosomal circular DNA found in certain bacteria and fungi.

**poikilotherm**   Organisms with body temperature determined by ambient temperature; cold-blooded.

**poison**   A chemical or material that can cause injury or death when eaten, absorbed, inhaled, or taken up by plants or animals.

**poisonous bait**   A food or other substance which is mixed with a poison to attract pests so they will consume it.

**poisson distribution**   A function that describes a random distribution, in which the variance is approximately equal to the mean.

**pollute**   To put undesirable chemicals or materials into the environment.

**polycyclic**   A disease with multiple reproductive cycles per season (cf., monocylcic).

**polygenic**   A character determined by the contributions of many genes.

**polymerase chain reaction (PCR)**   A technique to make copies of DNA fragments.

**polymorphism**   Having several forms.

**polyphagous**   Eating many different types of food.

**population**   All the members of a species that occur in a defined area at a specific time.

**population density**   The number of organisms of a species per unit area.

**population dynamics**   Changes in population size and structure over time.

**postemergence application (POST)**   Application of pesticides or fertilizers following the plant emergence.

**PPE**   See *personal protective equipment.*

**precision farming (site-specific farming)**   A farm management technology that uses within-field spatial variability to guide application of management actions. Utilizes GPS and GIS systems to allow application as per site-specific requirements.

**predator**   Animal that attacks and consumes other animals as food.

**preemergence application (PRE)**   The application of pesticides or fertilizers after planting, but before the weeds or crop have germinated.

**preharvest**   The time before the marketable commodity is removed from the field.

**preharvest interval**   The minimum time that must pass after the last pesticide application before the crop is harvested. Specified on the product label.

**preirrigation**   Irrigation before the crop is planted. Used to apply herbicides or moisten ground to allow for fast emergence of seedlings.

**preplant**   Decisions or actions made prior to planting a crop.

**preplant incorporated (PPI)**   Pesticide applied and physically mixed into the soil before planting the crop.

**prepupa**   Nonfeeding, usually inactive larval stage occurring just prior to the transformation to the pupal stage in some insects.

**prey**   An animal eaten for food by a predator.

**primary host**   The main host for a particular disease organism or crop pest.

**primary inoculum**   Pathogen propagules that initiate the first infection during a season.

**proboscis**   Sucking mouth parts of insects through which food is ingested.

**producer**   (i) Farmer or person who grows crops. (ii) Ecological term for green plants that form the base of the food chain.

**prokaryote**   Microorganisms that lack membrane-bound organelles (e.g., bacteria and mollicutes).

**propagative virus**   Viruses that increase in numbers inside their insect vectors.

**propagule**   A part of an organism that can reproduce the organism, such as spores, sclerotia, seeds, tubers, and root fragments.

**protectant**   A chemical applied to the target before pest numbers increase.

**prune**   To selectively remove unwanted branches from plants.

**pupa (pl. pupae)**   Nonfeeding stage of holometabolous insect between the larval and adult stages during which morphological changes occur.

**pupate**   Molt from larval to pupal stage.

**pustule**   A small elevated blister-like area on a leaf where fungal spores are produced.

**pycnidium**   Flask-shaped fungal fruiting structure of some *Fungi Imperfecti* that release spores called conidia when moistened.

**quadrat**   A defined area from which a sample is collected, used in population determination.

**quarantine**   Restrictions to prevent pest spread from infested to non-infested regions.

**race**   A subspecific category that is based on particular attributes, such as behavioral or physiological (other than morphological) differences; similar to a subspecies; often used in describing plant pathogens.

**random sample**   Sample collection wherein every member of the population has an equal probability of being selected.

**rate**   Amount of specific chemical applied per treatment unit, usually area.

**recommendation**   A suggestion from, or advice given by, a farm advisor, extension specialist, pest control advisor, or other agricultural authority that relates to the growing of a crop and the control of pests.

**reduced rate**   A lowered rate of pesticide application to permit survival of beneficials.

**reentry interval**   The time required after a pesticide application so that workers can go back into the area and be safe without wearing personal protective clothing or equipment.

**registered nursery stock**   Plant propagation materials registered by a government agency as free from particular pests. See *certified stock*.

**registered pesticide**   A pesticide approved by the U.S. Environmental Protection Agency (EPA) for use as specified by the label.

**regulated pest**   An organism designated by a state or federal agency as a pest requiring regulatory restriction to protect the host, the public, and the environment.

**release**   Introduction of biocontrol agents, particularly insects, onto a particular site.

**repellent**   Material that deters pests from moving to plants or animals.

**replacement (replacement pest; secondary pest outbreak)**   A minor pest species becomes a key pest after a control tactic targeting a different major pest species allows the minor pest to increase in number and cause damage.

**reservoir**   A site where organisms can survive or increase in number.

**reservoir host**   See *alternative host*.

**resident population**   A pest population in a defined area such as a field or an orchard.

**residual pesticide**   A pesticide that remains in the environment for a relatively long time. The pesticide may continue to be effective for weeks, months, or years.

**residual properties**   The ability of a pesticide to remain effective over time.

**residue**   (i) The amount of toxicant remaining at the target site after application. (ii) Plant debris that remains in the field after harvest.

**respirator**   A face mask that filters poisonous gas and particles from the air, enabling a person to breathe and work safely despite the presence of a toxicant. Used to protect the nose, mouth, and lungs from pesticide poisoning.

**resistance**   (i) Genetically determined capacity of a plant to inhibit pest attack, pest reproduction, or injury due to a pest. (ii) The ability of a significant portion of a pest population to survive a pesticide at rates that once killed most individuals of that population.

**respiratory exposure**   Pesticide inhalation.

**restricted use pesticide (restricted material)**   In the United States, a pesticide that may only be applied by applicators certified by the state in which they work, because the EPA has determined that the pesticide is harmful to the environment or hazardous to workers.

**resurgence**   See *pest resurgence.*

**rhizome**   A horizontal underground stem that forms roots and buds to produce new plants.

**rhizomorph**   A thick strand of fungal hyphae, that can grow between roots and infect adjacent trees.

**rhizosphere**   Describes the zone in the soil located around the roots of plants.

**RNA (ribonucleic acid)**   A polynucleotide that is involved in protein synthesis in cells, it exists in several forms that have specific functions. It serves directly as the primary genetic information in some viruses.

**rodent**   Any animal of the order Rodentia, e.g., mice, rats, squirrels, gophers, and woodchucks.

**rodenticide**   Poison used to kill rodents.

**rogue**   To remove undesirable plants from a crop.

**rootstock**   The root system of a plant to which the scion is grafted or budded.

**rot**   Physical decay of plants and plant products caused by fungi and bacteria.

**rotation**   See *crop rotation.*

**runner**   See *stolon.*

**runoff**   (i) The liquid that does not stay on the target being sprayed. (ii) Water draining off a field.

**rust**   A fungal disease that causes symptoms of reddish brown or black pustules. Caused by fungi in the order Uredinales.

**safener**   A chemical additive that reduces the phytotoxic properties of a pesticide.

**sanitation**   Destruction of crop and weed residues and cleaning of equipment in order to reduce pest inoculum.

**sap**   The term for the liquid in the vascular tissue of plants.

**saprophyte**   An organism that feeds on dead or decaying organic matter.

**scab**   Crust-like disease lesion on the surface of a plant.

**scald**   (i) An injury to the bark of a plant caused by frost or excessive drying by sun or wind. (ii) A fungal disease of cereals that causes leaf browning and plant death.

**scarification**   To cut or abraid the seed coat to stimulate seed germination.

**scion**   The part of the plant above the graft union and rootstock in vegetatively propagated crops; the designated variety.

**sclerotium (pl. sclerotia)**   A vegetative resting body of a fungus composed of a dense mass of thick-walled mycelium, round or irregularly shaped, and able to remain dormant and resistant to unfavorable conditions for long periods.

**scouting**   See *surveillance.*

**secondary host**   A host species attacked less commonly than the principal or primary host.

**secondary infection**   Microorganisms that enter the host through an injury initially caused by another pest.

**secondary pest**   A pest that is normally not a problem until control tactics applied for another pest or pests enable it to increase and cause damage.

**secondary spread**   The spread of a pathogen within a field after the initial or primary infection.

**Section 3 label**   Section within FIFRA that provides full registration of pesticides with a new active ingredient or a subregistration of a material already registered.

**Section 18**   Section in FIFRA that provides for emergency exemption from federal registration; provides the state with the authority to approve new uses of federally labeled pesticides where an emergency is justified and documented.

**Section 24 (c), special local need (SLN) registration**   Issued for five years in an area with specific, verified pest problems either to the chemical formulator (first-party applicant) or to third-party applicants who supply the necessary data.

**sedentary**   Remaining in one location after becoming established; immobile.

**sedentary endoparasitic nematode**   Nematodes that enter the host, moves to a feeding site, and then become immobile.

**seed borne**   Diseases caused by pathogens that survive in, or are carried on, seeds.

**seed treatment**   Substances coated on seeds as slurries, solutions, or as dry mixtures prior to planting. Usually insecticides and/or fungicides are used for control of soil insects, and soil-borne pathogens.

**seedling**   A small plant after it has germinated from a seed.

**selective pesticide**   (i) A pesticide, usually insecticide, that kills a specific target pest but is generally not harmful to most other organisms. (ii) A herbicide that kills weeds but does not harm crop plants.

**semiochemical**   Chemicals that play a role in communication between organisms, such as pheromones, allomones, and kairomones.

**semi-endoparasitic nematode**   A parasitic nematode that becomes only partially embedded in host tissue, with the posterior of the body remaining outside of the plant tissue.

**sensitive**   Low tolerance to pesticide effects.

**sensitive area**   Locations where pesticide applications could be particularly harmful (e.g., stream, residence, park).

**sequential sampling**   Sampling wherein the number of samples taken is variable and depends on whether the results obtained as sampling progresses yield a definite answer about the abundance of a pest.

**sessile**   (i) Refers to immobile stages of some insects, in particular, scales and whiteflies that lack the ability to move. (ii) Without stalk or stem.

**sex pheromone**   A chemical signal from one sex that attracts members of the opposite sex.

**shoot**   The growth of a plant in the form of a stem and its leaves.

**shothole**   (i) A plant disease symptom in which small, round pieces drop out of leaves. (ii) The visible holes left by wood-boring beetles of the family Scolytidae, as they leave their host, shothole borers.

**sign (of disease)**  Characteristic visible indicators of disease visible on a host, for example mildew mycelium on the surface of a leaf (cf., symptoms).

**signal word**  Wording that must appear on pesticide labels describing the relative toxicity of the pesticide. The signal words are "Danger Poison" with a skull and crossbones logo for highly toxic, "Warning" for moderately toxic, and "Caution" for low-order toxicity materials.

**silvicide**  Chemical to kill woody shrubs or trees.

**single-feeding bait**  Rodent bait containing acute toxins such as strychnine or zinc phosphide and requiring only one feeding to be effective.

**SLN**  See *Section 24(c)*.

**slurry**  A thick pesticide suspension made from wettable powder and water.

**smut**  A plant disease caused by fungi in the Basidiomycetes, a sign of which is the presence of sooty spore masses.

**soft rot**  Disease characterized by slimy softening and decay of the affected plant tissues.

**soil application**  Pesticide application to the soil surface rather than to foliage.

**soil borne**  Diseases caused by organisms that live in, or are found in, the soil.

**soil compaction**  Reduced volume of cultivated soils caused by pressure applied to the particles; results in decreased permeability to air and water, thus hindering plant growth.

**soil drainage**  An assessment of the ability of a soil to hold water.

**soil fumigant**  A pesticide that moves through soil as a gas or a vapor to control pests.

**soil incorporation**  Mechanical mixing of a pesticide into the soil.

**soil injection**  Pesticide placement beneath the soil surface without mixing or stirring the soil.

**soil persistence**  The length of time that a pesticide application on or in soil remains effective.

**soil solarization**  Placement of clear plastic tarp over wet, fallow soil to trap the radiant heat of the sun and raise the temperature of soil beneath the tarp to control soil-borne pests.

**soil sterilant**  A pesticide with long residual properties that kills all organisms in the soil; only a few chemicals are sterilants.

**soluble powder (SP)**  A dry pesticide formulation that forms a true solution when added to water; does not need further agitation.

**solution (S)**  Dissolved water-soluble pesticide that can be mixed directly with water; usually contains spreaders and stickers.

**solvent**  Chemical used to dissolve organic pesticides in water.

**sooty mold**  A fungus with dark mycelium that grows on insect honeydew.

**soporific**  A chemical that, once ingested, causes birds to enter a state of stupor and allows capture and removal.

**species**  The basic unit of biological classification. A group of organisms that are similar in morphology and physiology, capable of interbreeding to produce fertile offspring, and reproductively isolated from other such groups.

**speed sprayer**  See *air-blast sprayer*.

**spider**  Any member of the arthropod order Araneae having eight legs and feeding exclusively on animal tissues.

**spiroplasma**  Motile mollicute with helical shape, associated with plant diseases such as citrus stubborn.

**spiracle**  An external opening to the internal system of tubes, or tracheae, that are the respiratory system in insects.

**sporangium (pl. sporangia)**   The structure in which asexual spores are formed.

**spore**   A small reproductive structure produced by certain fungi and other microorganisms, capable of growing into a new individual under proper conditions; analogous to seeds of green plants.

**sporulate**   The production and liberation of spores.

**spot treatment**   Pesticide treatment of small area within a larger locale.

**spray drift**   See *drift.*

**spreader**   Agent added to a pesticide formulation to enhance its ability to increase area of target covered. See *wetting agent.*

**stabilizing agent**   Chemicals that retard pesticide decomposition during storage.

**stand**   A group of plants growing together in one area, usually the density of plants per unit area in a particular field or crop.

**stand decline**   The gradual decrease in the number or vigor of plants in an area.

**stand establishment**   Seed germination and the early phases of seedling growth.

**standard (or label) rate**   The recommended dosage rate of a pesticide.

**sterile insect release method (SIR)**   The introduction of sterile insects into a population to limit reproduction.

**sterilization**   The elimination of living organisms from containers or material by means of heat or chemicals.

**sticker**   Additive that increase the amount of spray adhering to a treated surface.

**sticky trap**   Traps containing a sticky substance that holds insects to be counted.

**stolon**   A modified horizontal stem above ground, creeping and rooting at the nodes, or curved and rooting at the tip.

**stoma (pl. stomata)**   Small aperture in the leaf epidermis through which water vapor is released during transpiration and gases are exchanged between the atmosphere and the plant.

**stomach poison**   Pesticide that acts through the stomach after ingestion.

**strain**   The progeny of a single isolation maintained in pure culture. Sometimes refers to a group of isolates with similar physiological characters or host relationships.

**stress**   Any one or combination of factors that decrease the vigor of an organism.

**stubble**   Plant stems and stalks left in a field after harvest. See *residue.*

**stunting**   A retarding effect on growth and development.

**stylet**   The protrusible oral spear that is the feeding structure of all plant-parasitic nematodes, and also of some other nematodes. Also, the modified mouthparts of certain insects, and mites, which are used to pierce host tissues to obtain food.

**stylet-borne**   Plant pathogens, especially viruses, that are transferred between hosts on the stylets (mouthparts) of their vectors.

**subacute toxicity**   The poisonous or injurious action of a toxin to an organism after repeated sublethal doses.

**succession**   The natural sequence by which one kind of ecological community is gradually replaced by another; the progressive changes in vegetation and animal life that occur over time.

**summer annual**   Plants that germinate in the spring, grow in the summer, produces flowers or seeds, and dies in the fall.

**sunscald**   The destruction of plant tissue caused by exposure to the sun.

**suppressive soil**   Soil in which antagonistic microorganisms prevent plant pathogens from causing disease.

**surfactant**    An additive to liquid mixtures of chemicals to improve emulsifying, dispersing, spreading, or wetting characteristics of the material applied.

**surveillance**    Monitoring to determine pest presence, density, dispersal, and dynamics.

**susceptible**    Unable to inhibit or deter pest feeding and reproduction (cf., resistance).

**suspension**    A liquid pesticide suspension of active ingredient particles. Constant agitation is required to maintain uniform distribution of the suspension.

**swath**    (i) The width of ground covered by one pass of a pesticide application device. (ii) Row of harvested crop placed in the field to dry.

**sweep**    (i) Type of knife used to cultivate. (ii) A net used to collect insects off plants.

**symbiosis**    Two or more organisms living in close association and interacting with one another.

**symptom**    (i) The visible manifestation of disease in an organism. (ii) Plant injury caused by a herbicide.

**symptomless carrier**    An organism infected by a pathogen without showing symptoms.

**synergism**    (i) Response to a pesticide mixture greater than the combined response of each pesticide applied alone. (ii) Damage caused by two pathogens infecting simultaneously that is greater than the sum of the damage caused by each pathogen alone.

**synergist**    A relatively safe chemical that, when added to a pesticide, enhances the toxicity of the pesticide (e.g., piperonyl butoxide).

**synthetic organic pesticide**    Man-made pesticide which contains carbon, hydrogen, and other elements.

**systemic**    (i) A chemical that when applied to an organism spreads throughout the organism through the vascular system. (ii) A pathogen infection that spreads throughout the host.

**tanglefoot**    A sticky substance used in traps to immobilize insects. Sometimes applied in a band around tree trunks to prevent insects from climbing the trunk.

**tank mix**    Two or more formulated pesticides combined in a spray tank for single application.

**taproot**    The large primary root that grows vertically downward, from which lateral roots arise.

**target**    The subject toward which a control tactic, such as pesticide application, is directed.

**taxonomy**    The theory and practice of classifying and naming organisms.

**technical material**    Undiluted, nonformulated pesticide in its pure form as produced by the manufacturer.

**teliospore**    A thick-walled dark spore of rust and smut fungi that is able to survive adverse conditions.

**teratogen**    Something capable of causing birth defects.

**teratogenesis**    The malformation of a fetus, with structural abnormalities present at birth or becoming evident after birth.

**thinning**    The removal of some plants or plant parts from row crops or orchard trees so that the remaining plants have appropriate room to grow and develop.

**thorax**    Middle body region of insects that has the wings and legs.

**till** or **tillage**    To cultivate and prepare the soil for use in agriculture, to kill weeds and make it more favorable for crop growth.

**tiller**   The young vegetative basal shoot in grasses.

**time interval**   The time that is required between the final pesticide application and harvesting; the time interval ensures that the legal residue tolerance will not be exceeded.

**tolerance**   (i) The ability of an organism to withstand unfavorable conditions such as pest attack, extreme weather, or pesticides. (ii) the amount of residue (in ppm) of pesticide that may legally remain in or on any food crops at the time of sale.

**tolerant**   (i) Pests that are not controlled by a particular pesticide treatment. (ii) Plants that can withstand pest attack or pesticide application with no appreciable reduction in yield.

**top sample**   A pest management monitoring sample taken from the top of the tree.

**toxic**   Poisonous. Able to cause injury by contact or systemic action to plants, animals, and people.

**toxicant**   Poison. An agent capable of being toxic.

**toxicity**   The degree to which a chemical is poisonous to an organism; the ability to injure.

**toxin**   A poison.

**trade name**   See *brand name.*

**transgenic**   Plants into which genes have been inserted from another organism using genetic engineering technology. Often inaccurately described by the term GMO.

**translocation**   Movement of water, nutrients, chemicals, pathogens, or photosynthates within a plant.

**transmission**   The transfer or spread of a pathogen from plant to plant.

**transpiration**   The evaporation of water from plant leaves through stomata.

**trap crop**   Plants attractive to certain pests, planted to divert these pests from the main crop.

**treated area**   Location where a pesticide has been applied.

**treatment level**   See *economic threshold.*

**trophic relationships**   Feeding interactions among organisms. A description of who eats whom.

**tuber**   An enlarged, fleshy underground stem, e.g., potato and nutsedges.

**twenty-four (c)**   See *Section 24(c).*

**ultra-low volume (ULV)**   Highly concentrated pesticide applied without dilution at a rate of 0.5 gallons per acre or less.

**uniform coverage**   The even deposition of a pesticide over a target.

**union**   The point at which scion and stock are joined in grafting.

**univoltine**   Having one generation per year (cf., multivoltine).

**USDA**   United States Department of Agriculture.

**vapor drift**   The movement of pesticide vapors to areas where they may cause injury. See *drift.*

**vapor pressure**   The property that determines how easily a chemical changes from the solid or liquid state to the gaseous state.

**variety**   Members of a species that differ from other members of a species in a way that is not sufficient to be recognized as a separate species. Variety is sometimes used synonymously with cultivar.

**vascular wilts**   Plant diseases that cause wilting by the plugging of and damage to the xylem (e.g., *Verticillium* and *Fusarium* wilts).

**vector**   Any carrier that transfers a pathogen from one host to another host.

**vegetative**   Asexual or somatic.

**vegetative growth**   The growth of stems, roots, and leaves rather than of flowers and fruits.

**vermiform**   Worm-shaped, relatively long and slender. Refers to worm-shaped nematode stages and insect larval types.

**vermin**   Pests; usually rats, mice or insects.

**vernalization**   The process by which exposure of seeds or plants to low temperatures stimulates germination or flower development.

**vertebrate**   An animal with a bony spinal column. Examples include mammals, fish, birds, reptiles, and amphibians.

**vertical resistance**   In crop plants, the almost complete resistance to a specific species or strain of pest. Usually determined by a single gene. Some wheat varieties are totally resistant to Hessian fly (cf., horizontal resistance).

**viable**   Capable of germination or growth; alive.

**virion**   The complete individual virus particle that includes the protein coat and the nucleic acid.

**viroid**   An infectious ribonucleic acid that does not have a protein coat.

**virulent**   Vigorously pathogenic; capable of producing disease (cf., avirulent).

**virus**   Microscopic parasites composed of a protein capsule and a nucleic acid core that needs living cells to grow or replicate.

**viviparous**   (i) Bulbs, seeds, or plantlets that germinate before becoming detached from the parent plant. (ii) Giving birth to live young, common in aphids.

**volatility**   The degree to which a substance, usually a liquid, changes to a gas at ordinary temperatures upon exposure to air.

**volume median diameter (VMD)**   A measure of spray droplet diameter at which half the spray volume has larger drops and half has smaller drops.

**volunteer crop**   Growth of undesired plants self-seeded from the previous crop.

**waiting period**   See *time interval.*

**warning**   Signal word used to designate a moderately toxic pesticide; oral $LD_{50}$ = 50 to 500 mg/kg body weight.

**water-dispersable liquid (flowable, F)**   Wettable powder preparations suspended in an oil base when formulated.

**water soaked**   A symptom of certain plant diseases or following herbicide application, where plant tissues appear wet due to loss of cell integrity.

**water-soluble powder (SP)**   Pesticide dust dissolved in water to form a solution. See *soluble powder.*

**weed**   A plant growing where it is not wanted.

**weed control**   The actions taken to eradicate, inhibit, or limit weeds and to prevent them from interfering with crop profitability or agricultural operations.

**wettable powder (WP)**   A dry powder pesticide formulation that is mixed with water for application by sprayer.

**wetting agent**   A chemical that reduces surface tension and allows a pesticide to spread out and more evenly coat a surface. See *surfactant.*

**wick application**   Selective herbicide application by moving a herbicide-soaked wick over the surface of weeds. Rope is often used for wick fabrication.

**wilt**   (i) Loss of turgor or drooping of plant parts because of inadequate water supply or excessive transpiration. (ii) Generic descriptor for vascular plant diseases.

**wind-borne**   Pests that are disseminated by the wind.

**windrow**   A row of hay, grain, alfalfa, beans, etc., cut and left in a field for drying before being baled or processed further.

**winter annual**   A plant that germinates in the fall, lives over the winter, and completes its growth, including seed production, the following spring.

**witch's broom**   An abnormal growth of tufted, small, closely set branches, apparently caused by a fungus or virus, and found on various trees and shrubs.

**xylem**   The water-conducting woody tissue in vascular plants.

**yellows**   Plant diseases in which foliage is yellow and stunted, caused usually by a virus, MLO, or fungus.

**zoospore**   A motile, asexually produced spore; often released in water.

**zero tolerance**   No amount of pesticide residue permitted on commodity when it is offered for shipment.

**zygote**   Diploid cell resulting from the fusion of two gametes; the fertilized egg.

## GLOSSARY REFERENCES

Agrios, G. N. 1997. *Plant pathology.* San Diego, Calif.: Academic Press, xvi, 635.

Anonymous. 2000. UC IPM home page, Calif.: http://www.ipm.ucdavis.edu/

Caveness, F. E. 1964. *A glossary of nematological terms.* Ibadan, Nigeria: International Institute of Tropical Agriculture, 68.

Flint, M. L., and P. Gouveia. 2001. *IPM in practice; Principles and methods of integrated pest management,* Publication #3418. Oakland, Calif.: University of California, Division of Agriculture and Natural Resources, xii, 296.

Herren, R. V., and R. L. Donahue. 1991. *The agriculture dictionary.* Albany, N.Y., Delmar Publishers Inc., vi, 553.

Lincoln, R., G. Boxshall, and P. Clark. 1998. *A dictionary of ecology, evolution and systematics.* Cambridge, U.K.: Cambridge University Press, ix, 361.

Pedigo, L. P. 1999. *Entomology and pest management.* Upper Saddle River, N.J.: Prentice Hall, xxii, 691.

Ricklefs, R. E., and G. L. Miller. 2000. *Ecology.* New York: W. H. Freeman & Co., xxxvii, 822.

Shurtleff, M. C., and C. W. Averre. 1997. *Glossary of plant-pathological terms.* St. Paul, Minn.: APS Press, iv, 361.

Weinberg, H. 1983. *Glossary of integrated pest management.* Sacramento, Calif.: State of California, Department of Food and Agriculture, 84.

# INDEX

Soil mite feeding on *Meloidogyne chitwoodi* eggs. Photo by Renato Inserra.

Abiotic disease, leaf damage caused by frost. Photo by Robert Norris.

Cotton bollworm killed by virus entomopathogen. Photo by David Nance.

Nematode trapping fungus. Photo by Bruce Jaffee.

Predatory mite feeding on mite egg. Photo by Jack Clark.

Base of pheromone trap with trapped codling moths. Photo by Robert Norris.

Onions covered by horse nettle and other weeds. Photo by Robert Norris.

Codling moth larva. Photo by D. Wilson.

Elm trees dying after being infected by Dutch elm disease. Photo by Robert Norris.

Armyworm killed by a polyhedrosis virus. Photo by Robert Norris.

Sclerotia of *Sclerotium rolfsii* in rotting sugar beet. Photo by Robert Norris.

Bark damage caused by rodent gnawing. Photo by Robert Norris.

Soft rot of onion. Photo by Mike Davis.

Cotton showing *Fusarium* wilt. Photo by Edward Caswell-Chen.

Western flower thrips. Photo by Jack Clark.

Peach trees killed by oakroot fungus. Photo by Robert Norris.

Diversity in an agricultural system. Photo by Robert Norris.

Slug. Photo by Robert Norris.

Broadcast flamer in alfalfa. Photo by Robert Norris.

Directed flamer in pears. Photo by Clyde Elmore.

Weeds emerging without preplant herbicide (left) and with preplant herbicide (right). Photo by Robert Norris.

Alfalfa weevil eggs. Photo by Jack Clark.

Apple maggot fly trap. Photo by Ron Prokopy.

Damage from sugar beet cyst nematode. Photo by Edward Caswell-Chen.